New Approaches to the Archaeology of Beekeeping

New Approaches
to the
Archaeology of Beekeeping

Edited by
David Wallace-Hare

ARCHAEOPRESS ARCHAEOLOGY

ARCHAEOPRESS PUBLISHING LTD
Summertown Pavilion
18-24 Middle Way
Summertown
Oxford OX2 7LG

www.archaeopress.com

ISBN 978-1-78969-993-7
ISBN 978-1-78969-994-4 (e-Pdf)

© Archaeopress and the individual authors 2022

This book is available direct from Archaeopress or from our website www.archaeopress.com

Contents

Preface

Current interest in beekeeping, or apiculture, is growing because of the precipitous decline of bees worldwide and the disastrous effect it portends for global agriculture. As a result, all aspects of beekeeping in all historical periods are coming under closer scrutiny. The current volume takes a holistic view of beekeeping archaeology (including honey and associated products, hive construction, and participants in this trade) in one large interconnected geographic region, the Mediterranean, central Europe, and the Atlantic Façade.

Increasingly inventive archaeological work focusing on ancient material remains of apiculture in the Mediterranean, for example, has expanded our knowledge of several areas of seemingly great production intensity, as in the pre-Roman and Roman Iberian Peninsula, and pre-Roman and Roman Crete and mainland Greece. Further work needs to be done now to bridge our growing body of ancient archaeological evidence from the Mediterranean with other areas of ancient Europe in addition to connecting this material to different sorts of evidence from later periods in these same areas, to say nothing of making similar diachronic connections outside Europe. Other research is forging entirely new lines of inquiry hitherto inaccessible in the past through the application of exciting new tools such as organic and pollen residue analysis of surviving hives and apicultural equipment and products from premodern periods (Louveaux, Maurizio, and Vorwohl 1978; Furness 1994; Garnier et al. 2002; Evershed 2008; Garnier 2015; Oliveira et al. 2019).

New Approaches to the Archaeology of Beekeeping focuses on novel approaches to historical beekeeping but also highlights new applications of more established ways of treating apicultural material from the past. The volume is keenly interested in helping readers navigate the challenges inherent in studying beekeeping historically. For example, while numerous ceramic beehives from pre-Roman and Roman Spain and ancient Greece speak to the importance of this industry in the ancient west, such hives tend to disappear in late antiquity, when hives of biodegradable materials like cork, wicker, or logs seem to have more widely replaced them. In fact, these organic hives appear to have been the norm for much of antiquity, with ceramic hives appearing only haphazardly, mostly in cities on the Mediterranean littoral. In temperate and Atlantic Europe, ceramic hives rarely seem to have been used at all. Extant material evidence for apiculture is disproportionately distributed in the ancient world with far fewer material remains, surprisingly, found for the medieval or early modern period. Textually, this situation is the reverse, in that we have an abundance of documentary evidence concerning beekeeping from the medieval and early modern periods and much less for antiquity. One way the volume has attempted to meet this asymmetric array of evidence is through its interdisciplinary and diachronic outlook, allowing current researchers and new voices a chance to see where they can add to the growing conversation on historical beekeeping.

The volume brings together scholars working not only on ancient archaeological evidence of beekeeping but also medieval, early modern, and ethnographic evidence of it. In this sense, *New Approaches to the Archaeology of Beekeeping* is meant to serve as a handbook for current researchers in this field and those who wish to undertake research into the history and archaeology of beekeeping. The array of case studies within are written in such a way as to balance publication of specialist material with inclusive explanation of the methodologies employed in each case.

The arrival of Langstroth's movable frame hive (1852) simultaneously heralded the arrival of modern industrial beekeeping but also the death knell of traditional beekeeping practices over much of the world. Before that time, change in hive technology and associated beekeeping equipment moved, as Suzanne Rotroff has put it, at a 'glacial pace' (Rotroff 2006:126). Studying beekeeping diachronically for much of its history is essential for filling in gaps in any given period. While no standardized beekeeping existed in the premodern world, variation in hive technology and, for instance, smokers, was not extreme in central or southern Europe, North Africa, or the Near East. Those who research the field of historical beekeeping, therefore, in some ways must be versatile and open to looking in other areas and time periods. This specific globalizing approach to historical beekeeping owes itself to the work of a sorely missed scholar to whom the present volume is dedicated, Dr. Eva Crane (1917-2007).

Dr. Crane was a giant in the study of historical beekeeping. Two among her many works on beekeeping have definitively shaped the modern field of apicultural history and archaeology: *The Archaeology of Beekeeping* (1983) and the magisterial *The World History of Beekeeping and Honey Hunting* (1999). With the first volume, Crane established the field of beekeeping archaeology as an international pursuit the holistic study of which could help scholars understand beekeeping developments on a global scale. If that were not ambitious enough, Crane later followed up that volume with a tome representing the capstone of her endeavours in the study of apicultural history. Her

1999 volume surveyed textual and archaeological evidence of beekeeping around the world while also paying close and important attention to a category of evidence long ignored but now quite at home in the historical study of beekeeping, ethnographic evidence of traditional beekeeping.

To date there have been few direct follow-up volumes to Crane's 1983 and 1999 works besides Bortolin's (2008) *Archeologia del miele* and Giuman's (2008) *Melissa: archeologia delle api e del miele nella Grecia antica*. Both focusing, as the titles imply, on archaeological material, and in the one case, on Greece alone. In 2018, in collaboration with the Eva Crane Trust, an edited volume, *Beekeeping in the Mediterranean: From Antiquity to the Present* (eds. Hatjina, Mavrofridis, and Jones, hereafter HMJ) followed in Crane's footsteps more directly. This work was the publication of the proceedings of a conference held at Syros in 2014. That volume continued Crane's work in several ways 1) it brought together contributions of a geographically varied nature within the Mediterranean, 2) considered beekeeping diachronically across time within this macroregion, and 3) considered archaeological, historical, and ethnographic studies of this subject in this zone equally. These were all the hallmarks of Crane's method.

Like Crane's focus, the current volume too takes a combined historical-archaeological approach to the study of the history of beekeeping. The volume's name, *New Approaches to the Archaeology of Beekeeping,* pays homage to Crane's 1983 volume by sharing its disciplinarily holistic perspective. Taken in conjunction with HMJ, the present volume is intended to provide a conspectus view of historical beekeeping in the western Mediterranean and Europe. It is hoped that the conversations engendered by the research within the volume will foster further collaboration and discourse about historical trends in global beekeeping.

Works Cited

Bortolin, R. 2008. *Archeologia del miele*. Mantua: SAP.

Crane, E. 1983. *The Archaeology of Beekeeping*. London: Duckworth.

Crane, E. 1999. *The World History of Beekeeping and Honey Hunting*. London: Duckworth.

Evershed, R. P. 2008. Organic residue analysis in archaeology: The archaeological biomarker revolution. *Archaeometry* 50(6): 895–924.

Furness, C. A. 1994. The extraction and identification of pollen from a beeswax statue. *Grana* 33(1): 49-52

Garnier, N., et al. 2002. Characterization of Archaeological Beeswax by Electron Ionization and Electrospray Ionization Mass Spectrometry. *Analytical Chemistry* 74: 4868–77.

Garnier, N. 2015. Méthodologies d'analyse chimique organique en archéologie, in C. Oliveira, R. Morais, and Á. M. Cerdán (eds) *ArchaeoAnalytics Chromatography and DNA analysis in archaeology*: 13–39. Esposende: Município de Esposende.

Giuman, M. 2008. *Melissa: archeologia delle api e del miele nella Grecia antica*. Rome: Bretschneider.

Hatjina, F., G. Mavrofridis and R. Jones (eds) 2018. *Beekeeping in the Mediterranean from Antiquity to the Present*. Nea Moudania: Division of Apiculture, Hellenic Agricultural Organization 'Demeter'-Greece; Chamber of Cyclades; Eva Crane Trust - UK.

Louveaux, J., A. Maurizio and G. Vorwohl. 1978 Methods of Melissopalynology. *Bee World* 59(4): 139-157.

Oliveira, C., et al. 2019. Chromatographic analysis of honey ceramic artefacts. *Archaeological and Anthropological Sciences* 11: 959-971.

Rotroff, S. I. 2006. *Hellenistic Pottery. The Plain Wares*. (Athenian Agora 33). Princeton: University Press.

Acknowledgements

I would like to express my heartfelt thanks to a colleague, collaborator, mentor, and above all friend, Rui Morais (FLUP, Universidade do Porto) for his help and guidance in bringing this volume to fruition. Rui first befriended me in 2017 when I emailed him about visiting and photographing ceramic beehives in Braga during a short stay in Portugal. Little did I know that so far from giving me some contact information on whom to get in touch with at the Museu de Arqueologia D. Diogo de Sousa and Museu Pio XII, Rui gave me a guided archaeological tour of the city and its museums as only an expert archaeologist and historian of the city could. He has been a dear friend ever since and this book would not have been possible without him in so many ways.

This book would also not have been possible without the generous support of the Department of Classics and Humanities at San Diego State University and the Friends of Classics and Barbara Schuch Endowed Fellowship in Classics and Digital Humanities.

I would also like to thank the many anonymous peer reviewers who strengthened the volume through their close reading and helpful comments.

Chapter 1

A New Approach to the Study of Ancient Greek Beekeeping

Georgios Mavrofridis

University of the Aegean (mavrofridis@geo.aegean.gr)

Summary: In this chapter are examined the ceramic beehives that archaeological investigations in Greece have brought to light, as well as the beekeeping practices that involved them. The beehives belonged to three and possibly four different types. It seems that two of these types were unknown in other Mediterranean regions. The beekeeping practices that involved each of these types are investigated through the scope of: a) the beekeeping practices and methods of the traditional beekeepers that used the respective beehive-types in the Aegean Islands and b) the practice of experimental beekeeping using copies of ancient ceramic beehives.

Key Words: ceramic beehives, horizontal hives, moveable-comb hives, beekeeping practices, Greece

Introduction

Ancient Greek literature offers only scarce evidence regarding the practice of beekeeping and absolutely no hint about the beehive types that were used. On the contrary, ancient Latin texts include enough information about beehives and their use. Archaeological investigations in Greece brought to light types of ceramic beehives that are not mentioned by any of the Roman authors. The use of these beehives demanded different beekeeping methods than those described by the Roman writers. Some of these methods were still applied until several decades ago by traditional Greek beekeepers who used similar beehive types in their beekeeping practices. Other methods disappeared as time went by, but they can be explored and understood through experimental archaeology. In my opinion, one would be able to comprehend ancient Greek beekeeping through a combination of the following: a) the thorough study of the archaeological evidence that is related to beekeeping; b) the study of any respective information appearing in ancient literature; c) the study of traditional beekeeping in Greece, especially that involving beehives reminiscent to their ancient counterparts; d) experimental beekeeping using copies of ancient ceramic beehives. Through this scope, I will try to re-approach and re-discuss the beekeeping practices of ancient Greece.

The Beehives

It has already been mentioned that the surviving works of ancient Greek authors do not offer any evidence regarding the types of beehives used. Roman authors for their part provide detailed information on this matter. According to Varro,

Virgil, Columella, Pliny the Elder and Palladius, beehives were open at both ends and were made of various materials, such as woven sticks, wood, planks, fennel stalks, tree barks, clay, dung, and/or bricks (see Mavrofridis 2018a: 102-5, citing the texts of the Roman authors). Beehives resembling those described by these Roman authors were used in regions of the Mediterranean until the 20th century (Crane 1983: 52-6; 1994: 120-28, 133). The most ambiguous type of all was the beehive made of dung. Several suggestions refer to the process by which these beehives were made, yet none of them seems realistic.[1] Such dung-beehives were made and used until the mid-20th century in Cappadocia (Bodenheimer 1942: 14; Kostakis 1963: 386). I believe that these ethnographic parallels may serve as precise examples of the manufacturing process of their ancient ancestors. The process involved in making these was relatively simple and involved shaping fresh dung into bands that were connected to each other and drying it out (Mavrofridis 2015c: 85-6; 2018a: 103-5).

The use of ceramic horizontal beehives open at both ends is confirmed by archaeological fieldwork in the Iberian Peninsula. Such beehives, dating before and during the Roman period, have been discovered at several sites in the Peninsula (de Almeida and Morín de

---

[1] Frazer (1951: 54-5) maintains that there were no beehives made of dung, but woven beehives that were coated with dung. Crane (1983: 51-2 and 1999: 262), discusses the creation of beehives by a mixture of mud and dung, while Francis (2012: 147, 149) discusses the creation of beehives by dung-bricks or moulds.

Pablos 2012: 732-41; Bonet and Mata 1997: 35-9; Fuentes et al. 2004: 188-194; Morín de Pablos and de Almeida 2014: 283-301; Quixal and Jardón 2016: 52-6). The vast majority of beehives found in the Iberian Peninsula are horizontal, open at both ends. Next to them, a type of beehive open at one end, shaped as a truncated cone is also found (Almeida and Morín de Pablos 2012: 735-6; Morín de Pablos and de Almeida 2014: 292-3). However, its closed edge was pierced allowing the bees to enter and with the beekeepers performing their harvesting and other activities from the open end.

In the Italian Peninsula, to date, no ceramic beehives have been found, or at least, none have been identified aside from a beehive-rim from Naxos in Sicily (Blackman et al. 2010: 153-5). This is probably due to the use of beehives made of perishable materials rather than clay. It is worth mentioning, that the Roman authors did not appreciate ceramic beehives because of their poor insulation properties (Varro, *Res Rusticae* 3.16.17; Columella, *De Re Rustica* 9.6.2; Palladius, *Opus Agriculturae*, 1.38).[2] In Roman Italy the beehives were probably horizontal, open at both ends; the same type of beehive was used in the Middle Ages (Crane 1999: 212-3; Germanidou 2016: 66-76). It was also used later, in the context of traditional beekeeping in south Italy and Sicily (Crane 1999: 186-7, 214; Masetti 2004: 469-71; 2006: 255-8).

Archaeological fieldwork in Greece has brought to light ceramic beehives different than those described by the Roman authors; this difference in shape implies different beekeeping practices, which are not mentioned in the Latin texts. The beehives under discussion could be divided into three main types, according to the beekeeping methods that involved them: a) horizontal, open-at-one-end beehives with quite short extension rings adjusted to their mouth; b) horizontal open-at-one-end beehives with one or more holes pierced at the closed edge; c) vertical open-at-top beehives. There is, however, some evidence for the use of horizontal beehives open at both ends; which, if their use were confirmed, would allow us to add a fourth type to the above division.

Figure 1. 4th century BCE horizontal open-at-one-end beehive and a Hellenistic mould for beehive lids bearing the inscription *EMBIOY*. Attica, Archaeological Collection of the Athens Airport.

---

[2] On this issue see: Francis 2012: 149-50; Mavrofridis 2011: 266-70; 2015d: 352-4; 2018d: 57-8.

Figure 2. Copy of a 4th century BCE horizontal open-at-one end beehive, in use during experimental beekeeping.

The commonest type of beehives in ancient Greece was without any doubt the horizontal open-at-one-end (**Figure 1, 2**). The beehives of this type were often different in terms of length, diameter, and certain morphological features. Beehive fragments that seem to belong to this type were found on Salamis and date between the late 6th century BCE and the mid-5th century BCE (Mavrofridis and Chairetakis 2019: 40). Numerous such beehives have been discovered in Attica and other regions (such as the Northwest Peloponnese, Boeotia, Euboea, Crete, Ephesus, Chalkidiki, many islands of the Cyclades and on Agathonisi); the earliest of them date to the end of the 5th century BCE.[3] They were in use at least until the 6th and 7th century CE, possibly even until the 10th to 12th century CE.[4] Traditional beehives, open at one end that were similar to their ancient counterparts were used in several south Aegean islands until some decades ago (Mavrofridis 2018c: 74) (**Figure 3**). In some cases (e.g. on the island of Paros) there are still beekeepers that use them.

Open-at-one-end beehives were regularly found with short extension rings; two to three and in some cases only one comb could be attached to these rings. Two types of ceramic, disc-shaped lids were also discovered. The first type includes lids with a crescent-shaped flight hole at their edge, two pairs of holes and often one more hole at the centre. In some cases, they would be decorated with floral motifs or relief concentric circles; rarely, they would bear an inscribed name (**Figure 1**) or merely a letter (Bossolino 2016 : 510-2; Lolos 2000: 126). It is suggested that these lids would be adjusted to the beehive with the use of a forked branch, which would be tied with a rope to the lid (through the pair of holes) and all around the beehive's mouth or extension ring (Jones et al. 1973: 410).[5] The second type includes lids pierced by many small holes, through which bees entered into the beehives (Giannas 2018: 82; Mavrofridis 2018b: 849).

---

[3] On the finds before 2006, see: Rotroff 2006: 126-7 (past bibliography is included). On more recent finds, see: Bibliodetis 2017: 159, 169-70; Francis 2012: 150; 2016: 87-99; Giannas 2018: 79-82; Karnava et al. 2015: 120; Kataki 2012: 538, 541; Triantafyllidis 2012: 637-53; Tsigarida et al. 2013: 386.

[4] Germanidou 2016: 87-100. On a ceramic beehive, open at one end, which is possibly dated to the 10th-12th centuries CE, see pp. 92-3 of that work.

[5] However, this suggestion does not explain the central hole that pierces many of these lids.

Figure 3. Traditional ceramic open-at-one-end-beehive from Amorgos Island, Cyclades.

Figure 4. 1st century CE horizontal open-at-one-end beehive with hole at the closed edge, from Georgioupoli, Crete (Photograph: Archive of Ephorate of Antiquities of Chania, courtesy of Michalis Milidakis).

Horizontal open-at-one-end beehives with one or more holes pierced at the closed edge appeared in the Hellenistic period, and seem to have been in use –at least in Crete– into the Roman period as well (**Figures 4-5**).[6] The closed edge of a 3rd century BCE beehive that was discovered in the Athenian Agora has a small central hole and may be classified in this group of beehives (Rotroff 2001: 176-7; 2006: 129). I do not agree with the opinion that this Athenian beehive was vertical (Anderson-Stojanović and Jones 2002: 368; Bosolino 2016: 515). In 2008, conducting a test to determine the verticality or horizontality of this hive, a swarm was put into a facsimile of a vertical ceramic beehive with a central entrance hole pierced at its base. It abandoned the beehive the next day (Mavrofridis 2009a: 120). Furthermore, there is no evidence attesting the use of vertical, open-at-top beehives with a central flight hole at the base. There is one example of a vertical, open-at-top beehive from 19th century Cythera,

---

[6] About this type of beehive used as burial jars in a Hellenistic cemetery in Chania, Crete, see: Kataki 2012: 538, 541 (the cemetery is dated to the second half of the 4th century BCE or the first half of the 3rd century BCE). Concerning similar beehives in Roman Crete, see: Crane 1999: 191; Francis 2016: 88-9; Hayes 1983: 132, 134.

Figure 5. The opening at the closed edge of a 1st century CE horizontal beehive from Georgioupoli, Crete (Photograph: Archive of Ephorate of Antiquities of Chania, courtesy of Michalis Milidakis).

which has a large central hole at the base. However, the flight entrance of this beehive was pierced in the vertical walls of the hive, while the hole at the base by contrast allowed ventilation and prevented humidity (Mavrofridis, 2017: 320).

Vertical, open-at-top beehives seem to have appeared in the Hellenistic period, according to evidence known thus far. Such beehives have been discovered at Isthmia (Anderson-Stojanović and Jones 2002: 255-65) (**Figures 6, 7**) and some fragments found in Attica and on the islands of Agathonisi, Chios, and Delos may be ascribed to this type.[7] These beehives were shaped as reversed truncated cones; this means that their side walls converge towards the base. The entrances of the bees are pierced at the side walls, close to the base or at the circumference of the base itself.

Let us now refer to the horizontal beehives, open at both ends. On Agathonisi, a 40cm long cylinder was found almost complete and it has been interpreted as a horizontal, open-at-both-ends beehive (Giannas 2018: 80). Even though I do not reject this interpretation, I

believe that it is also probable that this cylinder was the extension ring of an open-at-one-end beehive. The length of the object is rather small, compared to that of a horizontal, open-at-both-ends beehive. The small size of the object (its capacity reaches 28.5 liters) does not rule out the possibility that it was used as a beehive; however, it is not so common. Ancient, horizontal, open-at-both-ends beehives that have come to light (in Spain and Portugal) or similar beehives that were used until some decades ago all around the Mediterranean (in Cyprus, the Dodecanese, Lesbos, Crete, Ios, the Near East, and Egypt) were much longer and their capacity was considerably bigger. On Agathonisi, ancient beekeepers used extension rings; such objects are discovered on the island in large quantities, their length varying from 8 to 14cm (Giannas 2018: 81; Triantafyllidis 2012: 643). Similar, a 30cm long extension ring that is dated between the 5th and the 4th century BCE was found in Attica (Jones 1990: 63-5, 70-1); another one, 24cm long that dates to the Roman period was unearthed on Crete (Francis 2016: 88). The length of traditional extension rings, which were used in beehives, open-at-one-end, in the Dodecanese, the Cyclades, as well as on Malta varied from 37 to approximately 50cm (Mavrofridis 2018: 107).

Besides the cylindrical object that is discussed in the previous paragraph and may have served as an open-at-

---

[7] Mavrofridis 2018a: 108 including the relevant bibliography. However, the lack of entrances for the bees, which is observed in every fragment so far discovered, does not allow me to accept without any reservation that these examples indeed belonged to vertical, open-at-top beehives.

Figure 6. Experimental beekeeping using a copy of the Hellenistic vertical beehive found in Isthmia that bears the inscription ΟΡΕΣΤΑΔΑ.

Figure 7. The professional beekeeper Isidoros Tsiminis holding moveable-combs, taken from a copy of a vertical beehive from Isthmia.

both-ends beehive, there are some further indications regarding the use of such beehives in certain regions. P. Triantafyllidis, based on the fact that no closed ended beehives have so far been discovered on Agathonisi, concluded that at least some of the island's horizontal beehives, dating between the late 3rd century BCE and the 1st century CE, were probably open at both ends (Triantafyllidis 2010: 40; 2012: 642; 2014: 649).[8] Likewise, J. Francis, who discusses hive-craft in Chapter 4 of the present volume, expressed a similar thought about the beehives of Roman Crete. She considered that the small number of closed ends from broken beehives was due to the wider use of open-at-both-ends beehives (Francis 2016: 88). This suggestion must remain an assumption as long as no open-at-both-ends beehives are found on Crete. I believe that for the time being there is no solid evidence that would confirm the use of such beehives in ancient Greece.

There is no doubt that in ancient Greece, not only clay but also other, perishable, materials (such as woven sticks, wood, planks, tree barks and hollowed tree trunks) were used to make beehives that are no longer preserved. The Roman authors refer to many such beehives, but some of them were most probably not used in ancient Greece. This is the case of beehives made of cork (*Quercus suber*), a tree that did not grow in Greece and maybe of those made of fennel stalks (*Ferula communis*). Beehives made of fennel stalks were used until recently by traditional beekeepers in Sicily (Crane 1999: 186-7; Masetti 2004: 469-71) and Morocco (Pechhacker et al. 2001: 101). However, they are never reported in Greece and generally in the Eastern Mediterranean. After all, the fennel is not a very common plant in modern Greece and likewise, it was probably not very common in antiquity or the medieval period either.

Finally, the existence in ancient Greece of observation hives, which allowed the observation of the bees' life to a certain extent, is a possibility that cannot be excluded; some kinds of these hives are described by Pliny the Elder (*Naturalis Historia* 11.26; 11.47). At this point, it would be interesting to add an anecdotal story cited by 14th-century Arab writer Al-Damiri al-Din (c. 1344-1405). According to him Aristotle, in his effort to study the honey production, created a beehive made of glass. However, the bees became annoyed by his indiscreetness and disturbed his observations by coating the interior of the glass with mud.[9]

The Practice of Beekeeping

It has already been discussed that three or possibly four types of beehives were used in ancient Greece. Each of these types included several sub-types, differing from each other in terms of shape, size or certain secondary features. Some of these sub-types were merely local varieties and it generally seems that they did not impose different beekeeping practices. In this chapter, I will not deal with the sub-types of the beehives, as I aim to focus more on the main characteristics of beekeeping practices with each of the main types of ceramic beehives in ancient Greece.

1. Beekeeping with horizontal, open-at-one-end beehives

Horizontal open-at-one-end beehives were the most common type of ceramic beehive in ancient Greece (**Figures 1, 2**). Their inner walls were incised, in some cases only at the upper part, for the better and steadier attachment of the honeycomb. However, it is probable that some of the ancient Greek beehives would not have such incisions. There is an example of a horizontal ceramic beehive, dated to the 1st century CE, which was found in Georgioupoli in Crete that remained plain, without any incision.[10] Likewise, 18th century horizontal beehives from Syros, and perhaps from other islands of the Cyclades as well, had similar incisions on the upper part of their interior (Della Rocca 1790: II, 16). However, the traditional ceramic beehives, which were widely used on many Greek islands, until some decades ago, did not have such incisions, save some rare examples (Mavrofridis 2014: 19-20).[11] Hence, it may be concluded that the existence of incisions on the interior of the ceramic horizontal beehives could be helpful, without being necessary for the steadier attachment of the honeycombs.

In these beehives, the honeycombs, which occupied the part extending from the centre to the closed edge of the hive, were never harvested and as a result, they were not replaced by the bees. This was a significant disadvantage, because the non harvested honeycombs became black and hard over time, affecting the health and longevity of the bee colony. Any beekeeping activity demanding the opening of the beehive took place from the side of the bees' entrance. This fact made harvesting difficult for the beekeeper. According to beekeeping expert V. Dermatopoulos, who witnessed harvesting from ceramic horizontal open-at-one-end beehives on Naxos in the 1970s, 'to harvest these beehives is trouble and torture both to the beekeeper and the bees' (Dermatopoulos 1974: 214).

---

[8] Ch. Giannas (2018: 80) believes that all horizontal beehives of Agathonisi were open at both ends.

[9] On this story by Al-Damiri al-Din, see Davies and Kathirithamby 1986: 49.

[10] The beehive was presented by M. Miliadakis and N. Maragkoudakis in the 5th Meeting for the Archaeological Work in Crete, which was carried out in Rethymno (21-24/11/2019).

[11] In some rare cases, instead of incisions, a row of relief bands made of clay was made on the upper part of the interior.

One or more short extension rings were adjusted to the mouth of the horizontal open-at-one-end beehives. The majority of these extension rings were 6 to 9cm long; there are some smaller ones, 3cm long versions (Sackett 1992: 189),[12] and some larger than 10cm long variants (Francis 2006: 382; 2016: 88; Lüdorf 1998/99: 112; Triantafyllidis 2012: 643). Two unique extension rings reached a length of 24cm (Francis 2016: 88), and 30cm respectively (Jones 1990: 63-5, 70-1). At any rate, these extension rings were not too long. Their use involved the creation by the bees of honeycombs running vertical to the axis of the beehive. The use of extension rings would make no sense if honeycombs would be created in parallel with- or diagonal to the axis of the beehive, while additional difficulties would come up during harvesting. In other words, the beekeeper should be aware of the way to make the bees build their honeycombs vertical to the axis of the beehive, i.e. in parallel with its mouth.

Eva Crane realised the deep knowledge and expertise ancient and traditional beekeeping demanded, when she conducted experimental beekeeping, using a copy of an ancient ceramic beehive and extension ring from Vari, Attica. The final outcome was different than what was initially expected because the bees did not build their honeycombs in parallel with the mouth; as a result, 'harvesting combs from the extension was a very messy business'(Crane 1999: 202).

The method of creating honeycombs running vertical to the axis of the horizontal beehive can be described as follows: the beekeeper took a round piece of comb which usually contained brood from a hive and placed it in the centre of an empty one with the help of a cross made from thin sticks. Then a new swarm was placed into the empty beehive. This method was known to beekeepers of Syros in the late 18th century (Della Rocca 1790: II, 488-490 and III: 32-6). It was also applied in the 20th century by traditional beekeepers on Samos (Bikos 2015a: 27), Ikaria (Bikos 2006b: 96-7), Crete (Bikos 2012: 168; Mavrofridis 2009b: 203; 2019a: 8; Ruttner 1979: 213), and perhaps Rhodes (Vrontis 1938/48: 202). On many other Aegean islands, the local beekeepers were unaware of this method.

In ancient beehives, this method alone was not enough for the correct function of the short extension rings. Inside the empty beehive, part of the honeycomb had to be placed in such a position, so that the bees would attach the rest of the honeycombs to the extension rings, not to the junction point of the extension rings and the body of the beehive. This meant that the beekeepers would have take into consideration the so-

called 'bee space,' which is the distance left between honeycombs by bees, during the honeycombs' creation.

It is suggested that the short extension rings of ancient beehives were probably used in the production of the unsmoked honey that is frequently mentioned by ancient authors (Strabo 9.1.23; Pliny, *Naturalis Historia* 11.15; Columella, *De Re Rustica* 6.33.2; cf. Jones et al. 1973: 412). This is the honey which was harvested without the use of smoke. The extensive use of smoke by the beekeeper, during harvest, affects the smell and taste of the harvested honey to such an extent that these two features depend on the fuel that was burnt (Tananaki et al. 2009: 142-4). So the suggestion that these extension rings were used for the production of the unsmoked honey, which would be much more expensive than the usual one, is very interesting; I have offered further supporting arguments in a previous publication (Mavrofridis 2009b: 203). However, I believe that experimental beekeeping, using copies of ancient horizontal beehives, could answer many questions, namely: how were these extension rings used? Was unsmoked honey indeed taken from them? If so, how was it taken?[13]

In some cases, the lids which covered the mouths of horizontal open-at-one-end beehives were pierced by numerous small holes that served as entrances for the bees. The earlier lids of this type are dated to the Hellenistic period and were discovered on Paros (Mavrofridis 2018b: 849) and Agathonisi (Giannas 2018: 82). Two similar lids are found in Attica, dating to the Roman[14] and the Early Byzantine period.[15] One more lid was unearthed on Crete and may be dated to the 6th century (Kalokyris 1960: 324-5). The lids were most probably used to protect the bees from the oriental hornet (*Vespa orientalis*), the greatest enemy of bees in south Greece, especially in regions with a warm, dry climate. Ceramic lids with small holes -bee entrances- were used as covers for horizontal beehives on several islands of the Cyclades and the Dodecanese until some decades ago. The small holes did not allow the oriental hornet to enter the beehive so that the bees were capable of defending themselves more efficiently (Mavrofridis 2018b: 849-852).

---

[12] A single honeycomb could be attached by the bees to a 3cm long extension ring.

[13] In 2019, the author together with the agriculturist of the Institute of Mediterranean Forest Ecosystems in Athens, Dr S. Gounari, started practicing experimental beekeeping using copies of ancient, Byzantine and traditional ceramic horizontal and vertical beehives. The results concerning the horizontal beehives will be forthcoming.

[14] It is exhibited in the Archaeological Collection of the Acharnes Municipality, in Attica.

[15] It is exhibited in the Archaeological Collection of the Athens Airport.

2. Beekeeping with horizontal open-at-one-end beehives with one or more holes pierced at the closed edge

The hole or holes at the closed edge of these beehives could function as bee entrances and all beekeeping activities could be carried out at the back end. This would facilitate harvest and other activities of the beekeeper that involved the opening of the beehive; this way, the beekeeper would not be exposed to the bees that flew in and out of the beehive. In addition, it would be possible to smoke through these holes. Nevertheless, the beekeeper would be unable to harvest the honeycombs that were situated at the first half of the beehive; as a result, these honeycombs would not be renewed, exactly like in the case of horizontal open-at-one-end beehives. The beehives under discussion were used, until some decades ago, by traditional beekeepers in some Near Eastern countries (Crane 1999: 176-8; Komeili 1990: 18-20; Robinson 1981: 94-5), Cyprus (Bikos 2013a: 114-5), Malta (Masetti 2003: 470-1; Walker 2002: 188), and Ethiopia (Kristky 2010: 18-9).

There is another way to practice beekeeping with these beehives. On Ikaria, local beekeepers used –and some continue using– ceramic beehives with a large hole at their back end (**Figure 8**). These beehives were surrounded by stone slabs leaving an opening at the front. So the back hole was not used as a bee entrance, but served to conduct better ventilation and prohibited humidity (Bikos 2006a: 9-10). The bees entered the beehive through the bee entrance at the lid. The harvest and the rest of the beekeeper's activities were undertaken from the front open end. The back hole of the beehives on Ikaria provided better ventilation to the swarm even during the transfer to new flowerings; some of the beekeepers of the island transferred their beehives either inside the island itself or on the islands of Fournoi (Bikos 2006b: 96-7). It is probable that at least some of the ancient beehives with one or more holes at the closed edge would be used in the 'Ikarian way.' It is noteworthy that a 1st century CE beehive found in Georgioupoli in Crete,[16] although a bit smaller (**Figures 4, 5**), shares many similarities with the traditional beehives of Ikaria.

Figure 8. Traditional ceramic beehive with hole at its closed edge from Ikaria Island (Photograph: Th. Bikos /*Melissokomiki Epitheorisi*).

3. Beekeeping with vertical open-at-top beehives

In my opinion, there is no doubt that vertical open-at-top beehives were used as moveable-comb hives. Moveable-comb hives use top-bars at their upper part where the bees attach the honeycombs one by one; these top-bars have a particular width. These hives have sloping walls inclining towards the base; this shape goes against the tendency of the honey bee (*Apis mellifera*) to attach the honeycombs on the side walls when their shape is vertical.[17] As a result, the honeycombs that are created are hanging, merely attached to the top-bars; hence, it is possible to lift and transfer them when necessary, allowing a number of tasks in beekeeping that are impossible or difficult to be applied when working with fixed-comb hives. These design features facilitate the easy and full control of the beehive's interior,[18] the easier multiplication of the bee colonies without hunting swarms,[19] the prevention of swarming[20] and the easy harvest.[21]

Traditional moveable-comb hives are not recorded at all in any of the regions where *Apis mellifera* exists, save in south Greece.[22] The 17th-century travellers Jacob Spon (Spon 1678: II, 224-5) and George Wheler were the first to describe these hives, as they saw them in Attica (Wheler 1682: 412-3). In the same century, Zuanne

---

[16] The beehive was presented by M. Milidakis and N. Maragkoudakis in the 5th Meeting for the Archaeological Work in Crete (Rethymno, 21-24/11/2019) and it will be published in the Proceedings of the Meeting.

[17] The Asian species of bee *Apis cerana* does not attach its combs to the side walls in traditional hives whether they are vertical or sloping.
[18] Examine the population, the brood, the food or possible diseases and so on.
[19] This is achieved by transferring honeycombs with bees and brood (approximately half of them) from a hive to an empty one.
[20] This is achieved by multiplying the bee colonies in spring, during the swarming period.
[21] The beekeeper can simply lift and take the honeycombs that he wants.
[22] Traditional moveable-comb hives are also recorded in northern Vietnam and southern China used in beekeeping with *Apis cerana*. See Crane et al. 1993: 76-84; Crane 1999: 402-4.

Figure 9. Traditional ceramic moveable-comb beehive from central Crete.

Papadopoli referred to these hives in his handwritten memoir (Papadopoli, *L'Occio* 133r-133v). Later, in the late 18th century, Della Rocca and John Hawkins also described the moveable-comb hives (Della Rocca 1790: II, 466; Cotton 1842: 103a-106b. See also Mavrofridis 2012: 401-3; 2017: 311-2). They were made of various materials, such as clay, woven sticks, planks, stone slabs attached to each other and a fragment of porous stone that was carved to obtain the proper shape. They were in use until some decades ago. Traditional, ceramic moveable-comb hives are registered in central Crete (**Figure 9**) and on the islands of Cythera and Kea (**Figure 10**) (Mavrofridis 2017: 300-21).

Ancient vertical beehives discovered in Isthmia – four of them are restored– have the same features as the traditional moveable-comb hives: sloping walls converging towards the base, bee entrance close to the base or the circumference of the base and capacity appropriate for moveable-comb hives (Mavrofridis 2013a: 18-24; 2017: 321). The wooden top-bars that were placed on their mouth are not preserved, given the fact that wood is biodegradable and, save in rare archaeological conditions, is infrequently preserved from the ancient world.

Some of the beehives from Isthmia have incised inner walls, while others do not. The incisions on the inner walls of the vertical beehives were of no actual purpose; it is suggested that they were simply remains of the technique used on the horizontal hives (Anderson-Stojanović Jones 2002: 370). This view seems credible and I would add some further thoughts. Ceramic beehives were manufactured by potters, but they were used by beekeepers. Potters were not necessarily aware of the beekeeping practices; however, they were probably aware that at least most beekeepers would prefer beehives with incisions on the inside. Since these incisions should be created on horizontal beehives, potters could create them on vertical beehives too, without knowing their exact role and significance. This feature did not affect the beekeeper's activity when using vertical beehives; hence, it would be probable that a beekeeper would buy vertical, open-at-top beehives from a potter, without paying any attention to the incisions.

In the early 1980s, it was suggested that vertical ceramic moveable-comb hives evolved from the horizontal ceramic beehives shaped as truncated cones. This evolution took place on Crete during antiquity. Three

Figure 10. Traditional moveable-comb hive from Kea Island, Cyclades.

hypothetical intermediate steps of this evolution were proposed: the first step included the so-called 'moveable-nest hives,' which had a uniform lid, where bees attached their honeycombs; the second step included beehives, which had a lid that was divided into two or three compartments and the third step included the moveable-comb hives (Ifantidis 1983: 81-6). Based on this pattern, some researchers considered the beehives from Isthmia as 'moveable-nest hives' (Crane 1999: 404) or beehives of another intermediate step (Anderson-Stojanović and Jones 2002: 347, 370-2).

In my opinion, it would be possible for a 'moveable-nest hive' to function in the context of an experiment (Ifantidis 1983: 81-3), but in real-life conditions, it would cause the beekeeper many problems that would be difficult to solve. For example, it is difficult to imagine how a beekeeper would take every nest (namely all the honeycombs with the entire bee population) out of the beehives to place them on special supports, harvest, and then put them back into the hive. This would be a very difficult and risky task demanding very precise moves (and possibly luck as well). Furthermore, in such a manipulation there would always be the danger of an accident for the nest with all the associated

consequences to the bee colony. The possibility of using 'moveable-nest hives' should be rejected for two more reasons (cf. Mavrofridis 2013a: 23-5)

First, the landscape of Greece often demands the installation of apiaries on steep and sloping grounds, where it would be difficult to put supports holding the nest horizontally, next to each beehive. Second, in contrast to the moveable-comb hives, the beekeeper should confront the entire population of the hive, dealing with aggressive subspecies of bees (such as those of Crete, the Aegean Islands, and central and south Greece).[23] Lastly, no known examples of the 'moveable-nest hives' are recorded and in my view, their existence in the past is to be doubted.

In an effort to investigate through experimental methods whether the ancient vertical beehives from Isthmia were used as moveable-comb hives, an experiment was conducted using copies of three among the four complete beehives, which were discovered there (**Figures 6, 7**). This effort proved that

---

[23] *Apis mellifera adami* in Crete, *Apis mellifera cecropia* in the other regions.

these beehives could be easily and successfully used as moveable-comb hives.[24]

4. Beekeeping with horizontal, open-at-both-ends beehives

The most important advantage of practicing beekeeping with horizontal, open-at-both-ends beehives is that harvesting could be achieved alternately from the front and the back opening so that the honeycombs would be renewed. Alternately harvesting from both sides of the open-at-both-ends beehives is mentioned by Columella (Columella, De Re Rustica 9.15.11), the same practice was applied by traditional Greek beekeepers who used horizontal, open-at-both-ends beehives in the 20th century on several islands, such as Ios (Bikos 2006c: 222), Kalymnos (Bikos 1999: 76), Rhodes (Vrontis 1938/48: 205), Samos (Bikos 2015b: 134), Herakleia (Bikos 2009a: 16), Amorgos (Bikos 2009b: 415), and eastern Crete (Mavrofridis 2019a: 8).

The user of a ceramic horizontal beehive, which was manufactured by the potter as open-at-one-end, cut the closed end and used it as a back lid; this beehive is found in Attica and it is dated to the 6th-7th century CE (Gini-Tsofopoulou 2002: 135). It is obvious that its user preferred open at both ends beehives. It is not possible to confirm whether this happened even at earlier times. Ethnographic parallels do exist. On Amorgos, many beekeepers transformed open-at-one-end beehives into open-at-both-end types, by carefully breaking their closed end (Bikos 2009b: 415). The same phenomenon occurred on Herakleia, where this practice was followed by all the beekeepers of the island (Bikos 2009a: 16).

Columella (Columella, De Re Rustica 9.14.13) and Pliny (Pliny, Naturalis Historia 21.47) refer to the use of lids, which were moved inside the horizontal open-at-both-ends beehives, reducing their inner space, according to the needs of the bees. However, this practice was unknown to the traditional Greek beekeepers that used this type of beehive.

If it will be proven that the horizontal, open-at-both-ends beehives were indeed used in ancient Greece, the following question comes to mind: were extension rings adjusted to their edges? As far as I know, no extension ring has been found from the Iberian Peninsula, where almost exclusively horizontal, open-at-both-ends beehives were used. Greek traditional beekeepers that

used such beehives did not place extension rings on them. However, there are ethnographic parallels from the eastern Mediterranean and North Africa supporting the use of extension rings on ceramic or other horizontal, open-at-both-ends beehives. In these cases, the extension rings were placed exclusively at the back edge of the hives (Crane 1999: 387-8). In conclusion, if extension rings were used on ancient Greek open-at-both-ends beehives, they were probably placed at the back side of the hives.[25]

5. Characteristics of beekeeping in ancient Greece

It seems that beekeeping in ancient Greece was not merely static, but in some cases at least it was migratory. Columella, citing Celsus, mentions that apiaries from Achaea were transferred to Attica and Euboea and from the Cyclades to Skyros (Columella, De Re Rustica 9.14.19-20). Even though the beehives, which were used, are not defined, it would be possible to carry even ceramic beehives with great caution. Clay, as a material, does not seem to impede migratory beekeeping. There are many examples of traditional migratory beekeeping with ceramic beehives. Traditional beekeepers on Ios and their colleagues on Kea, Amorgos, Ikaria, and eastern Crete used to transfer their ceramic beehives until the mid-20th century, looking for new flowerings.[26] The beehives were transferred on beasts of burden, in boats or even by beekeepers themselves, who carried one or two ceramic hives (Bikos 2006b: 97; 2009b: 417; Mavrofridis 2015b: 178-9).

Focusing on the examples mentioned by Columella, one may assume that the beehives from Achaea would be transferred to Attica towards the end of spring and the beginning of summer due to the famous Attic honey that came from various herbaceous plants, especially thyme (Thymus spp). The apiaries could remain in Attica even after harvesting the thyme honey in the summer, taking advantage of the honeydew of the insect Marchalina hellenica in the Attic pine tree forests and take benefit from the flowering of the autumn heather (Erica manipuliflora) and the strawberry tree (Arbutus unedo). The traditional migratory beekeepers who acted in Attica until the third decade of the 20th century transferred their beehives to these plants (Typaldos-Xydias 1927: 40-4). Concerning the transfer from Achaea to Euboea, it should be noted that at present the beekeeping flora of southern Euboea consists mainly of: ironwort (Sideritis spp.), oregano (Origanum spp), thyme, autumn heather, and strawberry tree; in

---

[24] Mavrofridis 2013a: 26-7; 2013b: 82-4. In 2019, beekeeping was practiced again by the author and the director of the Laboratory of Beekeeping at the Institute of Mediterranean Forest Ecosystems in Athens, Dr Sofia Gounari. Copies of the ceramic vertical beehives from Isthmia were de novo used. The main purpose of this action was not to investigate whether these were moveable-comb hives (this was taken for granted) but to study the local bee in the environment of ceramic hives.

[25] According to Giannas (2018: 80), extension rings were adjusted at both edges of the beehives on Agathonisi; this is a view that I do not find plausible.

[26] Horizontal beehives, open at both ends on Ios and in eastern Crete; horizontal, open at one end with a hole at the closed end on Ikaria; horizontal open at one end on Amorgos and vertical top-bar hives on Kea.

northern Euboea are mainly pine tree forests with their honeydew (Bikos 2010a: 243; Typaldos-Xydias 1927: 43). Probably the same plants were exploited by migratory beekeepers in antiquity. In the pine forests of north Euboea traditional beekeepers from northwest Attica and Boeotia transferred their beehives until at least the third decade of the 20th century (Typaldos-Xydias 1927: 41-4).

Regarding the transfer of beehives from the Cyclades to Skyros, it is more probable that the apiaries were transferred at the end of summer, provided that the flora of these islands has remained almost the same since the antiquity. On most islands of the Cyclades, there are no other significant flowerings after the end of the thyme flowering. On the contrary, in the northern part of Skyros grow nowadays many pine trees, which offer rich honeydew, as well as areas with autumn heather (on the beekeeping flora of Skyros see Bikos 2008b: 224; Liakos 2006: 263-4).

In some cases, it seems that beekeeping was practiced in the confines of the settlements, for instance in Athens, where many ceramic beehives were found. Urban beekeeping in the city probably began in the last quarter of the 5th century BCE, during the Peloponnesian War. This was a period in which Athenian beekeepers could not find access to the hinterland, due to the Spartan invasion. Then, it perhaps became obvious that beekeeping in the city was possible and the beekeepers continued practicing it even after the end of the war (Rottroff 2002: 297; Rotrof 2006: 131). Other settlements of the Greek territory, where urban beekeeping was probably carried out were the following: Salamis on the island of Salamis and Rahi in Isthmia (Mavrofridis 2018d: 58). Similar examples are recorded in the western Mediterranean, in the Iberian settlements, as early as in the 3rd century BCE (Bonet and Mata 1997: 42). The oldest example of urban beekeeping is dated to the period between the mid-10th until the early-9th century BCE and it comes from Tel Rehov, in Israel (Mazar 2018: 46; Mazar and Panitz-Cohen 2007: 210; Mazar et al. 2008: 636). To avoid problems deriving from the large number of bees within the ancient cities, the beehives would be placed in specific locations, such as the roofs of the houses, so that the population would not be irritated (Mavrofridis 2018d: 57-9).

In the countryside, beehives were placed either on the ground, in walls inside bee boles, or in rows on low walls. A fourth-third-century BCE wall equipped with boles, where it is believed that beehives were placed, has been discovered in Attica (Oikonomakou 1995: 60). Walls, inside- or on which beehives were placed, were possibly used on Agathonisi as well (Giannas 2018: 79). The lease of a wall on Salamis is mentioned in an inscription and could be related to the practice of beekeeping on the island (SEG 33.167 / IG II² 1590 and 1591); it could

have been a wall with bee boles or a bee enclosure (Chairetakis 2018: 324). The traditional, ceramic, horizontal, open-at-one-end beehives, which were used on many Aegean islands until the middle at least of the 20th century, were usually placed inside boles (Bikos 2008a: 148-9; 2009b: 413-4; 2010b: 322-4; 2013b, 178-9; 2014, 279-80; Florakis 1971, 129). In some cases, open-at-top hives, were also placed inside boles (Mavrofridis 2015a, 109-10; 2019b, 10). Ceramic, horizontal open-at-both-ends beehives that were traditionally used on Samos (Bikos 2015a, 25-6), on Chios (Bikos 2015c: 251-3), on Kalymnos (Bikos 1999: 74-8) and in eastern Crete (Mavrofridis 2019a: 7-8) were placed on low walls that were built for this purpose.

It is possible that bee enclosures protecting the beehives from thieves and enemies existed in ancient Greece. In the area of Sphakia (Crete) many 'beehive sites' are located, where a significant number of beehives and the associated pottery, which is dated to the Roman period are found. Many of them were surrounded by a low, stone-built enclosure (Francis 2016: 93-4; Price and Nixon 2005: 675).

In the 1st century CE, Columella described a bee enclosure with openings on the wall for the bees that looked like a row of windows (Columella, De Re Rustica 9.5.3); a similar description is cited in the 10th century Geoponica (15.2.9) by Florentinus, who is considered to have lived in the 3rd century CE (Germanidou 2016: 31). Bee enclosures were used until the last century in traditional beekeeping of Greece, both in the mainland and on the islands (Mavrofridis 2016: 196-9).[27] These traditional bee enclosures did not usually have openings on the walls; however, one of them, situated near the village Gouves in the district of Heraklion (Crete), has openings similar to those described by Columella and Florentinus. In certain parts, the height of this particular bee enclosure, which is considered to have been built in the 17th or the 18th century, is over 4m (Anagnostakis 2018: 107; Mavrofridis 2019a: 9; Mavrofridis and Goutzamani 2019: 413).

In ancient Greece, the harvest was probably carried out once a year. However, it seems that it was not a rare phenomenon to harvest twice or even three times a year. Pseudo-Aristotle mentions two seasons when bees offer their honey: spring and autumn (Pseudo-Aristotle, Historia Animalium 626b.29-31). Varro and Pliny refer to three harvests per year, but it is not clear whether this is something that took place in the Greek territory as well (Varro, Res Rusticae 3.16.34; Pliny, Naturalis Historia 11.14-15). In Geoponica, Didymus, a writer who lived between the 3rd and the 5th century CE (Mavrofridis

---

[27] There is also a notion about the use of bee enclosures on Crete in the early 15th century (see Nixon and Moody 2017: 490).

and Goutzamani 2019: 414), also referred to three annual harvests (*Geoponica* 15.5.1).

In my opinion, two or three harvests per year could be carried out in migratory beekeeping. In static beekeeping, however, two harvests would be possible in some areas. I assume that only very few areas, where apiaries would be placed permanently, could offer three annual harvests. In the Greek territory, traditional beekeeping in the migratory way included two or even three harvests, according to the year and region of activity (Mavrofridis 2015b: 176-9; Typaldos-Xydias 1927: 21-3, 35-8, 45). But static beekeeping, in most cases, included only one harvest, usually in the middle of summer. The regions, where local conditions allowed a second harvest to static apiaries, were not many: eastern (Bikos 2012: 168) and central Crete (Papadopoli, L'Occio 132v), Samos (Bikos 2015b: 134), Rhodes (Vrontis 1938/48: 205) and the Athenian territory are some of them (Cotton 1842: 105b).

According to pseudo-Aristotle, after the harvest, a quantity of honey was left for the bees to spend their winter (Pseudo-Aristotle, *Historia Animalium* 626a.2-3); when the quantity of honey was not sufficient, bees were fed with figs and sweet foods (Pseudo-Aristotle, *Historia Animalium* 626b.7-8). Varro mentions that bees were fed with mature figs boiled in water, water sweetened with honey or a mixture of smashed raisins and dry figs that were macerated in boiled wine (Varro, *Res Rusticae* 3.16.28).

For his part, Columella refers to dried figs that were macerated in water, raisins sprinkled with water, must or raisin-wine (Columella, *De Re Rustica* 9.14.15), while Pliny to raisins or crushed dried figs, as well as raisin-wine, boiled must or hydromel.[28] Finally, in the *Geoponica* there is a reference to feeding bees with honey wine and smashed raisins mixed with some savory (*Geoponica* 15.4.4-5). It is interesting that until some decades ago, on Ios and on Rhodes, the bees were fed with similar foodstuffs, to be more precise they were offered a mixture of figs, boiled in sweet wine (Bikos 2006c: 222), or smashed raisins and boiled must (Vrontis 1938/48: 204).

Regarding the bees' illnesses, pseudo-Aristotle knows one, but without naming it; this illness is characterised by the reduced will of the bees to work as well as by the bad smell of the hives (Pseudo-Aristotle, *Historia Animalium* 626b.20-21). It is probably a sort of foulbrood (Liakos 2000b: 330), the European or American foulbrood, that Columella and Pliny also had in mind (Columella, *De Re Rustica* 9.13.1; Pliny, *Naturalis Historia* 11.20). Pliny referred to it as *blapsigonia*, meaning injury of the young. Pseudo-Aristotle (Historia Animalium 626b.17-19) refers to the infestation of the honeycombs by the greater wax moth larvae (*Galleria mellonella*) not as a 'hostile attack,' but as a disease by the name of *cleros* (κλῆρος) (Liakos 2000a: 139). Likewise, Columella refers to this phenomenon as if it were a disease (Columella, *De Re Rustica* 9.13.11), known to ancient Greeks as *phagedaina* (φαγέδαινα), while Pliny (*Naturalis Historia* 11.20) cites the name *claros* (κλᾶρος) (Liakos 2000b: 332).[29]

Ancient beekeepers had many erroneous perceptions. One of them concerned the gender of the queen bee that was thought to be a male.[30] This view was fully accepted until the end of the 16th century (Crane 1999: 569-70; Theodoridés 1968: 23-6). It is interesting to mention that even in the 20th century many traditional beekeepers in Greece, especially on islands, such as Thasos, Syros, Naxos, Folegandros, Samos, Ikaria, and Crete, still believed that the queen bee was male (Mavrofridis, forthcoming). Another wrong idea was that the worker bees carried a small stone to ballast themselves while they were flying through the winds for forage. Respective notions are found in pseudo-Aristotle (Pseudo-Aristotle, *Historia Animalium* 11.40.626b.24-26) and other writers (Pliny, *Naturalis Historia* 11.10; Dio Chrysostom, *Orations* 44.7; Plutarch, *De Sollertia Animalium* 967b; Aelian, *De Natura Animalium* 1.11). Until at least the seventh decade of the 20th century, the beekeepers on Anafi Island maintained the same view, as they thought that the bee carried a small gravel stone, which was left behind when the bee concentrated an equal amount of nectar (Oikonomidis 1966: 634).

To sum up, beekeeping in ancient Greece was considerably different than the beekeeping in other regions in antiquity. Based on the archaeological finds of ceramic beehives, it was practiced with three and possibly even four types of beehives, indicating different methods in various beekeeping activities. Two of these types, namely the horizontal beehives with short extension rings and the vertical, open-at-top, are only found in the Greek territory, meaning that the beekeeping methods related to them were not known in other Mediterranean areas.

Acknowledgements

I would like to express my deepest gratitude to the archaeologists Michalis Milidakis and Nikos Maragoudakis (Ephorate of Antiquities of Chania, Crete) for permitting me to use the photographs of a ceramic beehive from Georgioupoli. In addition, I am thankful to the editor of the journal *Melissokomiki Epitheorisi* for the photograph of the traditional beehives from Ikaria.

---

[28] Pliny, *Naturalis Historia* 21.48. Both, Columella and Pliny, refer also to feeding bees with bird or poultry meat. This is impressive because bees do not eat meat. However this practice that was recorded by the above mention authors is attested in the 20th century in the district of Florina, Greece. See Anagnostopoulos 2000: 308.

---

[29] In the Doric dialect cleros (κλῆρος) was claros (κλᾶρος).
[30] Aristotle first, in several parts of the *Historia Animalium* and *De Generatione Animalium* refers to the king or the governor of the bees.

Sources

Aelian, *De Natura Animalium*: Αἰλιανός, Περὶ ζώων ἰδιότητος, Α-Δ. Αθήνα 1996, Κάκτος [Αρχαία Ελληνική Γραμματεία 'Οι Έλληνες'.

Aristotle, *Historia Animalium*: Ἀριστοτέλης, Τῶν περὶ τὰ ζῷα ἱστοριῶν, Δ-Ι. Αθήνα 1994, Κάκτος [Αρχαία Ελληνική Γραμματεία 'Οι Έλληνες'].

Aristotle, *De Generatione Animalium*: Ἀριστοτέλης, Περὶ ζώων γενέσεως Γ-Ε. Αθήνα 1994, Κάκτος [Αρχαία Ελληνική Γραμματεία 'Οι Έλληνες'.

Dio Chrysostom, *Orations*: Dio Chrysostom, *Discourses* 37-60 (trans. H. L. Crosby). Cambridge, MA 1946, Harvard University Press [The Loeb Classical Library].

Geoponica: *Geoponica sive Casiani Bassi scholastici de re rustica eclogue* (ed. H. Bekh). Lipsiae 1895 [Biblioteca Scriptorum Graecorum et Romanorum Teubneriana].

Columella, *De Re Rustica*: Columelle, *De l'agriculture* (ed. J. C. Dumont), Livre IX. Paris 2001 [Les Belles Lettres].

Palladius, *Opus Agriculturae*: Palladius, *Traité d'agriculture, Livre I et II* (ed. R. Martin). Paris 1976 [Les Belles Lettres].

Papadopoli, *L'Occio*: Zuanne Papadopoli, *L'Occio (Time of Leisure). Memories of seventeenth-century Crete* (ed. A. Vincent). Venice 2007, Hellenic Institute of Byzantine and Post-Byzantine Studies.

Pliny, *Naturalis Historia*: Pline l'Ancien, *Histoire naturelle, Livre XI* (ed. A. Ernout and R. Périn), Paris 1947· Livre XXI (ed. J. André), Paris 1969 [Les Belles Lettres].

Plutarch, *De sollertia Animalium*: Plutarch, *Moralia, Whether land or sea animals are clever* (trans. W. C. Hembold), Vol. 12. Cambridge, MA 1957, Harvard University Press [The Loeb Classical Library].

Pseudo-Aristotle, *Historia Animalium IX*: Ἀριστοτέλης, Τῶν περὶ τὰ ζῷα ἱστοριῶν, Η, Θ, Ι. Αθήνα 1994, Κάκτος [Αρχαία Ελληνική Γραμματεία 'Οι Έλληνες›].

Strabo *Geographica*: Strabo, *Geography* (Trans. H. L. Jones), Vol. IV. Cambridge, MA 1927, Harvard University Press [The Loeb Classical Library].

Varro, *Res rusticae*: Varron, *Économie rurale* (ed. C. Guiraud), Livre III. Paris 1997 [Les Belles Lettres].

Bibliography

Almeida, R. R. and J. Morín de Pablos. 2012. Colmenas cerámicas en el territorio de Segobriga. Nuovos datos para la apicultura en época romana en Hispania, in D. Casasola and A. Ribera I Lacomba (eds) *Cerámicas hispanorromanas II, Productiones regionales*: 725-743. Cádiz: Universidad de Cádiz.

Anagnostakis, I. 2018. Wild and Domestic Honey in Middle Byzantine Hagiography: Some Issues Relating to its Production, Collection, and Consumption, in F. Hatjina, G. Mavrofridis and R. Jones (eds) *Beekeeping in the Mediterranean from Antiquity to the Present*: 105-118. Nea Moudania: Hellenic Agricultural Organization 'Demeter.'

Anderson-Stojanović V. R. and J.E. Jones. 2002. Ancient Beehives from Isthmia. *Hesperia* 71(4): 345-376.

Bibliodetis, B. 2017. Η κεραμική. In Λ. Παλαιοκρασσά-Κόπιτσα (ed) *Παλαιόπολη Άνδρου, τριάντα χρόνια ανασκαφικής έρευνας*: 158-175. Andros, χ.ε.

Bikos, Th. 1999. Μελισσοκομικές καταγραφές. *Μελισσοκομική Επιθεώρηση* 13(2): 72-78.

Bikos, Th. 2006a. Μελισσοκομικές καταγραφές. *Μελισσοκομική Επιθεώρηση* 20(1): 9-13.

Bikos, Th. 2006b. Μελισσοκομικές καταγραφές. *Μελισσοκομική Επιθεώρηση* 20(2): 96-100.

Bikos, Th. 2006c. Μελισσοκομικές καταγραφές. *Μελισσοκομική Επιθεώρηση* 20(4): 220-225.

Bikos, Th. 2008a. Μελισσοκομικές καταγραφές. *Μελισσοκομική Επιθεώρηση* 22(3): 144-149.

Bikos, Th. 2008b. Μελισσοκομικές καταγραφές. *Μελισσοκομική Επιθεώρηση* 22(4): 223-229.

Bikos, Th. 2009a. Μελισσοκομικές καταγραφές. *Μελισσοκομική Επιθεώρηση* 23(1): 14-20.

Bikos, Th. 2009b. Μελισσοκομικές καταγραφές. *Μελισσοκομική Επιθεώρηση* 23(6): 412-417.

Bikos, Th. 2010a. Μελισσοκομικές καταγραφές. *Μελισσοκομική Επιθεώρηση* 24(4): 242-247.

Bikos, Th. 2010b. Μελισσοκομικές καταγραφές. *Μελισσοκομική Επιθεώρηση* 24(5): 322-326.

Bikos, Th. 2012. Μελισσοκομικές καταγραφές. *Μελισσοκομική Επιθεώρηση* 26(3): 166-171.

Bikos, Th. 2013a. Μελισσοκομικές καταγραφές. *Μελισσοκομική Επιθεώρηση* 27(2): 114-117.

Bikos, Th. 2013b. Μελισσοκομικές καταγραφές. *Μελισσοκομική Επιθεώρηση* 27(3): 178-182.

Bikos, Th. 2014. Μελισσοκομικές καταγραφές. *Μελισσοκομική Επιθεώρηση* 28(4): 279-283.

Bikos, Th. 2015a. Μελισσοκομικές καταγραφές. *Μελισσοκομική Επιθεώρηση* 29(239): 24-29.

Bikos, Th. 2015b. Μελισσοκομικές καταγραφές. *Μελισσοκομική Επιθεώρηση* 29(240): 132-137.

Bikos, Th. 2015c. Μελισσοκομικές καταγραφές. *Μελισσοκομική Επιθεώρηση* 29(242): 249-259.

Blackman, D., et al. 2010. Miscellanea Apicula. In N. Sekunda (ed) *Ergastiria: Works presented to John Ellis Jones on his 80th birthday*: 150-158. Gdansk: Gdansk University.

Bodenheimer, F. S. 1942. *Studies on the Honey Bee and Beekeeping in Turkey*. Istanbul: Nümune Matbaasi.

Bonet, H. R. and C.P. Mata. 1997. The Archaeology of Pre-Roman Iberia. *Journal of Mediterranean Archaeology* 10(1): 33-47.

Bossolino, I. 2016. '<Μέλι> πρωτεύει τὸ Ἀττικὸν καὶ τούτου τὸ Ὑμήττιον καλούμενον' Honey Production in Attica, an Antique Excellence, in C. Soares and J. Pinheiro (eds) *Partimonios alimentares de aquem e alem-mar*: 499-519. Coimbra: Coimbra University Press.

Chairetakis, Y. 2018. Οικιστική οργάνωση και χωροταξία στη Σαλαμίνα από τον 6ο έως τον 1ο αι. π.Χ/

Settlement patterns and spatial planning on Salamis from the 6th to the 1st century B.C. Unpublished PhD dissertation, University of Athens.

Cotton, W. Ch. 1842. *My Bee Book*. London: Rivingston.

Crane, E. 1983. *The Archaeology of Beekeeping*. London: Duckworth.

Crane. E., van Luyen, V. and V. Mulder. Traditional Management of *Apis cerana* Using Movable-Comb Hives in Vietnam. *Bee World* 74(2): 75-85.

Crane, E. 1994. Beekeeping in the World of Ancient Rome. *Bee World* 75(3): 118-134.

Crane, E. 1999. *The World History of Beekeeping and Honey Hunting*. London : Duckworth.

Della Rocca, Ab. 1790. *Traité complet sur les abeilles*. Paris: De l'Imprimerie de Monsieur.

Dermatopoulos, V. 1974. Η μελισσοκομία στη Νάξο. *Μελισσοκομική Ελλάς* 24(284): 214-215.

Davis, M. and J. Kathirithamby. 1986. *Greek Insects*. New York and London: Oxford University Press.

Florakis, A. E. 1971. *Τήνος. Λαϊκός πολιτισμός*. Athens: Ελληνικό βιβλίο.

Francis, J. E. 2006. Beehives and Beekeeping in Graeco-Roman Sphakia, in *Proceedings of the 9th International Congress of Cretan Studies, A5*: 379-390. Heraklion: Society of Cretan Historical Studies.

Francis, J. E. 2012. Experiments with an Old Ceramic Beehive. *Oxford Journal of Archaeology* 31(2): 143-159.

Francis, J. E. 2016. Apiculture in Roman Crete, in J. E. Francis and A. Kouremenos (eds) *Roman Crete: New Perspectives*: 83-100. Oxford and Philadelphia: Oxbow Books.

Fraser, H. M. 1951. *Beekeeping in Antiquity*. London, University of London Press.

Fuentes, M., T. Hurtado and A. Moreno. 2004. Nuevas aportaciones al estudio de la apicultura en época ibérica. *Recerques del Museu D'Alcoi* 13: 181-200.

Germanidou, S. 2016. *Βυζαντινός μελίρρυτος πολιτισμός*. Athens: Εθνικό Ίδρυμα Ερευνών.

Giannas, Ch. 2018. Beekeeping Practices in Agathonisi during Antiquity, in F. Hatjina, G. Mavrofridis and R. Jones (eds) *Beekeeping in the Mediterranean from Antiquity to the Present*: 79-84. Nea Moudania: Hellenic Agricultural Organization 'Demeter.'

Gini-Tsofopoulou, E. 2002. Πήλινη κυψέλη, in Δ. Παπανικόλα-Μπακιρτζή (ed) *Ώρες Βυζαντίου - Έργα και Ημέρες στο Βυζάντιο. Η καθημερινή ζωή στο Βυζάντιο*: 135. Athens: Ταμείο Αρχαιολογικών Πόρων και Απαλλοτριώσεων.

Hayes, J. W. 1983. The Villa Dionysos Excavations, Knossos. *The Annual of the British School at Athens* 78: 97-169.

Ifantidis, M. 1983. The Movable-Nest Hive: A Possible Forerunner to the Movable-Comb Hive. *Bee World* 64(2): 79-87.

Jones, J. E. 1990. Ancient Beehives at Thorikos: Combed Pots from the Velatouri. In *Thorikos IX, 1977/1982*: 63-71. Ghent: Comité des Fouilles Belges en Grèce.

Jones, J. E., A.J. Graham and L.H. Sackett. 1973. An Attic Country House Below the Cave of Pan at Vari. *The Annual of the British School at Athens* 68: 355-452.

Kalokyris, K. D. 1960. Συμπληρωματική ανασκαφή της εν Πανόρμω Κρήτης παλαιοχριστιανικής βασιλικής. *Πρακτικά της εν Αθήναις Αρχαιολογικής Εταιρείας* 1955: 321-326.

Karnava, A., E. Kolia and E. Margaritis. 2015. A Classical/Hellenistic Oil Pressing Installation in Foti-Vroskopos, Keos, in A. Diler, K. Şenol and Ü. Aydinoglu (eds), *Olive Oil and Wine Production in Eastern Mediterranean During Antiquity*: 107-123. Izmir: Ege Üniversitesi.

Kataki, E. 2012. Δοκιμαστική ανασκαφική έρευνα στο Εθνικό Στάδιο Χανίων, in Μ. Ανδριανάκης, Π. Βαρθαλίτου and Ι. Τζαχίλη (eds) *Αρχαιολογικό Έργο Κρήτης* 2: 537-547. Rethymno: Πανεπιστήμιο Κρήτης.

Komeili, A. B. 1990. Beekeeping in Iran. *Bee World* 71(1): 12-24.

Kostakis, Th. 1963. Ανακού. Athens: Κέντρο Μικρασιατικών Σπουδών.

Kritsky, G. 2010. *The Quest for the Perfect Hive*. Oxford: University Press.

Liakos, B. 2000a. Αριστοτέλης, ο πρώτος μελισσολόγος ερευνητής. In *Η μέλισσα και τα προϊόντα της*: 134-142. Athens: Πολιτιστικό Τεχνολογικό Ίδρυμα ΕΤΒΑ.

Liakos, B. 2000b. Εχθροί και ασθένειες των μελισσών. Μαρτυρίες από τα παλιά. In *Η μέλισσα και τα προϊόντα της*: 330-336. Athens: Πολιτιστικό Τεχνολογικό Ίδρυμα ΕΤΒΑ.

Liakos, B. 2006. Η μελισσοκομία στη Σκύρο. *Μελισσοκομική Επιθεώρηση* 20(5): 260-264.

Lolos, G. 2000. Σαλαμινιακές έρευνες 1998-2000. *Δωδώνη* 19: 113-165.

Lüdorf, G. 1998/99. Leitformen der attischen Gebrauchskeramik: Der Bienenkorb. *Boreas* 21/22: 41-169.

Masetti, N. 2003. Malta, Isle of the Knights, Island of Honey. *American Bee Journal* 143(6): 468-471.

Masetti, N. 2004. Legends and Reality of Traditional Beekeeping in Sicily. *American Bee Journal* 144(6): 469-472.

Masetti, N. 2006. Stone Hives and Apiaries of Puglia 1 and 2. *American Bee Journal* 146(2): 159-163; 146(3): 255-258.

Mavrofridis, G. 2009a. Μελισσοκομία με αντίγραφα αρχαίων κυψελών. *Μελισσοκομική Επιθεώρηση* 23(2): 120.

Mavrofridis, G. 2009b. Το ακάπνιστο μέλι. *Μελισσοκομική Επιθεώρηση* 23(3): 200-204.

Mavrofridis, G. 2011. Μελισσοκομία στον ρωμαϊκό κόσμο. *Μελισσοκομική Επιθεώρηση* 25(4): 266-271.

Mavrofridis, G. 2012. Η μελισσοκομία στην Αττική στα τέλη του 18ου αιώνα. *Μελισσοκομική Επιθεώρηση* 26(6): 400-404.

Mavrofridis, G. 2013a. Κυψέλες κινητής κηρήθρας στην αρχαία Ελλάδα. *Αρχαιολογική Εφημερίς* 152: 15-27.

Mavrofridis, G. 2013b. Experimental Archaeology: Beekeeping with Copies of Ancient Upright Hives. *Bee World* 90(4): 82-84.

Mavrofridis, G. 2014. Πήλινες κυψέλες και εσωτερικές εγχαράξεις. *Μελισσοκομική Επιθεώρηση* 28(1): 17-21.

Mavrofridis, G. 2015a. Παραδοσιακή μελισσοκομία. *Μελισσοκομική Επιθεώρηση* 29(240): 106-110.

Mavrofridis, G. 2015b. Η νομαδική μελισσοκομία πριν την έλευση της σύγχρονης κυψέλης. *Μελισσοκομική Επιθεώρηση* 29(241): 176-179.

Mavrofridis, G. 2015c. Dung Made Beehives. *Bee World* 92(3): 85-86.

Mavrofridis, G. 2015d. Πειραματικές έρευνες για την κατανόηση του μελισσοκομικού παρελθόντος. *Μελισσοκομική Επιθεώρηση* 29(243): 349-354.

Mavrofridis, G. 2016. Μελισσομαντριά. *Μελισσοκομική Επιθεώρηση* 30(247): 196-200.

Mavrofridis, G. 2017. Οι παραδοσιακές κυψέλες κινητής κηρήθρας. *Πελοποννησιακά Γράμματα* 2: 299-334.

Mavrofridis, G. 2018a. Μελισσοκομία στον ελληνορωμαϊκό κόσμο – οι κυψέλες. *Αρχαιολογία & Τέχνες* 127: 100-111.

Mavrofridis, G. 2018b. Ελληνιστικά πώματα κυψελών για προστασία των μελισσών από τη *Vespa orientalis*. In *Πρακτικά Θ´ Επιστημονικής Συνάντησης για την Ελληνιστική Κεραμική*: 849-856. Athens: Ταμείο Αρχαιολογικών Πόρων και Απαλλοτριώσεων.

Mavrofridis, G. 2018c. Παραδοσιακή μελισσοκομία. *Αρχαιολογία & Τέχνες* 128: 66-79.

Mavrofridis, G. 2018d. Urban Beekeeping in Antiquity. *Ethnoentomology* 2: 52-61.

Mavrofridis, G. 2019a. Traditional Beekeeping in Crete (17th - 20th century). In *Proceedings of the 12th International Congress of Cretan Studies*: 1-15. Heraklion: Society of Cretan Historical Studies.

Mavrofridis, G. 2019b. *Η ελληνική παραδοσιακή μελισσοκομία και η συμβολή της στις διεθνείς εξελίξεις*. Athens: Ινστιτούτο Γεωπονικών Επιστημών.

Mavrofridis, G. forthcoming. Παραδοσιακή μελισσοκομία στις Νότιες Κυκλάδες. Υλικά, μέθοδοι και διαχείριση του χώρου. *Επετηρίς Εταιρείας Κυκλαδικών Μελετών*: forthcoming.

Mavrofridis, G. and Y. Chairetakis. 2019. Η μελισσοκομία της Σαλαμίνας στη διαχρονία. *Μελισσοκομική Επιθεώρηση* 33(263): 40-44.

Mavrofridis, G. and M. Goutzamani. 2019. Μέλισσα και μελισσοκομία στα Γεωπονικά. *Μελισσοκομική Επιθεώρηση* 33(268): 411-415.

Mazar, A. 2018. The Iron Age Apiary at Tel Rehov, Israel, in F. Hatjina, G. Mavrofridis and R. Jones (eds) *Beekeeping in the Mediterranean from Antiquity to the Present*: 40-49. Nea Moudania: Hellenic Agricultural Organization 'Demeter.'

Mazar, A. and N. Panitz-Cohen. 2007. It is the Land of the Honey. *Near Eastern Archaeology* 70(4): 202-219.

Mazar, A. et al. 2008. Iron Age Beehives at Tel Rehov in the Jordan Valley. *Antiquity* 82(317): 629-639.

Morín de Pablos, J. and R. R. de Almeida. 2014. La apicultura en la Hispania Romana: Producción, consumo y circulación. *Anejos de Archivo Español de Arqueología* 65: 279-305.

Nixon, L. and J. Moody. 2017. Cultural Landscapes and Resources in Sphakia, SW Crete: A Diachronic Perspective. *From Maple to Olive. Publication of the Canadian Institute in Greece* 10: 485-504.

Oikonomakou, M. 1995. Λαυρεωτική. Αρχαιολογικόν Δελτίον, Χρονικά 50: 60-61.

Oikonomidis, D. B. 1966. Η μελισσοκομία εν Νάξω και εν Ανάφη. *Επετηρίς Εταιρείας Κυκλαδικών Μελετών* 5: 617-634.

Pechhacker, H., Jochi S. and Chatt, A. 2001. Beekeeping Around the World. *Bee World* 82(2): 99-104.

Price, S. and L. Nixon. 2005. Ancient Greek Agricultural Terraces: Evidence from Texts and Archaeological Survey. *American Journal of Archaeology* 109(4): 665-694.

Quixal, D. S. and P.G. Jardón. 2016. El registro material del colmenar ibérico de la Fonteta Ràquia (Riba-Roja, Valencia). *Lucentum* 25: 43-63.

Robinson, W. J. 1981. Beekeeping in Jordan. *Bee World* 62(3): 91-97.

Rotroff, S. I. 2001. A New Type of Beehive. *Hesperia* 20(2): 176-177.

Rotroff, S. I. 2002. Urban Bees. *American Journal of Archaeology* 106(2): 297.

Rotroff, S. I. 2006. *The Athenian Agora, Vol. 33, Hellenistic Pottery: The Plain Ware*. Princeton: American School of Classical Studies at Athens.

Ruttner, F. 1979. Minoische und altgriechische Imkertechnik auf Kreta. In *Bienenmuseum und Geschichte der Bienenzucht*: 209-229. Bucharest: Apimondia.

Sackett, L. H. 1992. *The Roman Pottery*, in L. H. Sackett (ed.), *Knossos from Greek to Roman Colony: Excavation at the Unexplored Mansion II*: 147-256. London: British School at Athens.

Spon, J. 1678. *Voyage d'Italie, de Dalmatie, de Grèce et du Levant*. Lyon: Antoine Cellier les fils.

Tananaki, C., S. Gounari and A. Thrasyvoulou. 2009. The Effect of Smoke on the Volatile Characteristics of Honey. *Journal of Apicultural Research* 48(2): 142-144.

Theodoridès, J. 1968. *Historique des connaissans scientifiques sur l'abeille. In Traité de biologie de l'abeille*, V. 5: 1-34. Paris: Masson et Cie.

Triantafyllidis, P. 2010. *Το ακριτικό Αγαθονήσι*. Athens, Νομαρχιακή Αυτοδιοίκηση Δωδεκανήσου.

Triantafyllidis, P. 2012. Πήλινες κυψέλες από την αρχαία Τραγαία (Αγαθονήσι). *Δωδεκανησιακά Χρονικά* 25: 535-653.

Triantafyllidis, P. 2014. Πήλινες κυψέλες από την αρχαία Τραγαία, in *Πρακτικά Η´ Επιστημονικής Συνάντησης για την Ελληνιστική Κεραμική*: 635-653. Athens: Ταμείο Αρχαιολογικών Πόρων και Απαλλοτριώσεων.

Tsigarida, E. M., S. Vasileiou and E. Naoum. 2013. Νέα στοιχεία για την οργάνωση και την οικονομία της

Κασσάνδρας κατά την ελληνιστική και ρωμαϊκή περίοδο. *Το Αρχαιολογικό Έργο στη Μακεδονία και στη Θράκη* 23: 377-398.

Typaldos-Xydias, A. 1927. *Η νομαδική μελισσοκομία εν Ελλάδι*. Athens: Παράρτημα Γεωργικού Δελτίου.

Vrontis, A. 1938/1948. Η μελισσοκομία και το μαντατόρεμα στη Ρόδο. *Λαογραφία* 12: 195-230.

Walker, P. 2002. Traditional Stone Apiaries in Malta. *Bee World* 83(4): 185-189.

Wheler, G. 1682. *A Journey into Greece*. London: Cademan.

Chapter 2

Smoke and Bees: From Prehistoric to Traditional Smokers in Greece

Sophia Germanidou

Marie Curie Research Fellow, Newcastle University (sophiagermanidou@yahoo.gr)

Summary: The smoker was a necessary tool in beekeeping; it was used during the early phase of the wild honey collection, as well as during beekeeping activities that involved beehives. Many morphological variations of this ceramic object are found in the Greek territory, dating as early as Prehistoric times. However, there is no published sample dating in the Ancient and Byzantine periods, save only scarce references and representations. This article deals with the methods and the raw materials which were used to smoke the bees; it also investigates the typological evolution of smokers in time, taking into account the main textual and iconographic evidence. Furthermore, the features of smokers used in traditional Greek beekeeping are described, citing relevant ethnographic parallels.

Keywords: smoke, smokers, Greece, un-/non smoked (*akapnisto*) honey

The Beginning: The Rock Paintings/ Engravings Evidence

Already since the first interactions between humans and bees, humans realised the importance of keeping bees away in order to collect their honey without problem. In their effort to repel the bees, smoke proved to be the most efficient and common agent that reduced the aggressiveness of the swarm. To be more precise, smoke repelled the bees, by tranquilising them through inducing them to gorge on honey; smoke was transmitted as a message of fire alert conducting them to gain fuel for escape. Humans were aware of these effects on bees since the very early stages of honey hunting. However, in those early times, no specific pot was used during the smoking of the bees and it seems that 'inventing' such a tool was not a priority. The numerous prehistoric rock paintings or petroglyphs (engravings) show that in the harvest of wild honey the following objects were essential: a leather bag or a basket-container for the harvest of the honeycombs, ladders, sticks and ropes so as to approach the nest when at a height, poles with prongs, and maces and axes for the removal of hanging honeycombs. Only two representations of honey hunting-gathering on rock paintings in India depict the application of smoke using a leafy branch as a flammable material (Crane 2001: 18). A rock painting in southern Zimbabwe (Toghwana Dam/ Matopos National Park, c.8000 BCE) (**Figure 1**) offers a unique image of smoking bees in front of honeycombs (Pager 1973: 61-68; Pager 1976: 9-10, fig. 1; Crane 1999: 58, fig. 8.6a; Crane 2001: 33, fig. 5f.).

Despite the rare iconographic evidence, it is thought that smoke was widely used during that remote era, produced through smouldering vegetation in a bundle or through fibres of a torch. The simple and inexpensive application of smoke through torches or bundles of vegetation continued in traditional beekeeping of modern times along with the survival of wild honey collection. Mainly plants bundle of selected vegetation were used as fuel: fungi, leaves, herbs, (dry) grass or, later on, puffball-fungi, sulphur, tobacco – as well as rotten wood, wet straw, and twigs, even old clothes. The flammable material that prevailed came from animals (oxen, horses) and it was their dung. Otherwise, in later times, sulphur was burned in the flight hole and bees were killed.

It is not known when and under which circumstances the use of a specific pot, the smoker, became necessary and was considered exclusively a practice of beekeeping. What is certain, according to the archaeological evidence from prehistoric Greece, is that its creation did not depend on the discovery and early use of beehives. It is possible that, at a certain moment, incense burners used for various religious, ritual, or other purposes were also used in beekeeping, since their function was similar to that of the smokers. The creation of a specific pot was probably driven by various reasons such as a) the need to burn the flammable material for a longer time b) the more accurate direction of the smoke onto the bees and c) the possibility of placing the pot on the ground so that the beekeeper could work undisturbed using both hands.

The First Representations

The earliest known representation of a smoker is found on a stone bas-relief from the temple of Ne-user-re, in Abu Ghorab, nowadays kept in the Egyptian Museum in Berlin (2400 BCE, **Figure 2**) (Crane 1999: 164, fig. 20.3a; Crane 2001: 87, fig. 11a). An ovoid-shaped object, open at both sides, it depicts a beekeeper blowing smoke

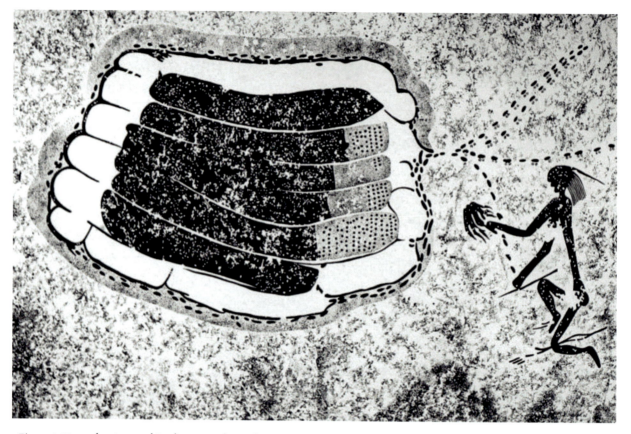

Figure 1. Honey hunter smoking honeycombs, rock painting in southern Zimbabwe, Toghwana Dam/Matopos National Park, 8000 BCE? (Pager 1973: 61-68; Pager 1976: 9-10, fig. 1; Crane 1999: 58, fig. 8.6a; Crane 2001: 33, fig. 5f).

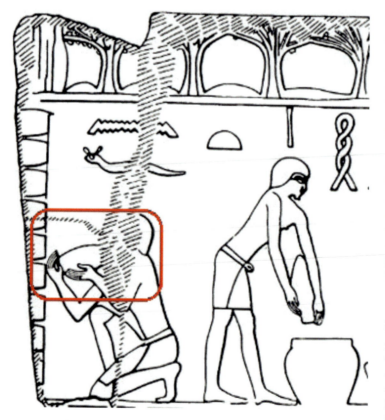

Figure 2. Beekeeper probably smoking beehives through an oval object, stone bas-relief from the temple of Ne-user-re, in Abu Ghorab, today in the Egyptian Museum in Berlin, 2400 BCE (Crane 1999: 164, fig. 20.3a; Crane 2001: 87, fig. 11a).

(?) through a hole towards the hives. However, the features of this object do not correspond to a smoker, despite what was initially argued. It is most probable that the depicted pot was used to induce the queen by 'emitting a breath or a little sound,' a practice known and applied by Egyptian beekeepers for centuries.[1] This practice survived in the traditional beekeeping of the Mediterranean – for instance it was known in Cyprus and in Greece.[2] Support for this reading is provided by a depiction of a beekeeper in a beekeeping scene in the famous tomb of Rekhmire in the West Bank of Luxor (c. 1450 BCE). Here the beekeeper holds a smoker shaped like a small deep bowl (Crane 1999: 165, fig. 20.3b).

---

[1] More details in: Kritsky 2015: 10, 11, 13, figs 2.3, 2.5 with previous bibliography.

[2] Rizopoulou-Igoumenidou 2000: 403. Simple ways of blowing smoke (i.e. using an open tile with lit dung placed inside) towards the bees have been recorded in various places of Greece, for example in Santorini, Skyros, Laconian Mani. Information provided by Giorgos Mavrofridis based on his own research and Thanasis' Bikos explorations.

Figure 3. Graphic depiction of the interior of the prehistoric smoker from Zakros-Crete during experimental beehive smoking (Tyree et al. 2012: 227, fig. 24.5).

Diachronic Research on Greek Smokers

One of the earliest and fundamental contributions to the archaeology of smokers remains the article of K. Davaras (1989) who presented the three most common examples of this form from Minoan Crete; two of them were found at Zakros and one at Knossos. The author cited a list of similar vessels from Faistos, Agia Eirini on Kea and the Ashmolean Museum. He also identified as smokers the so-called 'snake-tubes' from Enkomi in Cyprus and from Crete (Davaras 1989: 1-7). An interesting and significant initiative that Davaras took was the copy of the Minoan smoker from Zakros (**Figure 3**) and its use in the practice of experimental beekeeping (Stamataki et. al. 2009: 165-170; Tyree et al. 2012: 223-230; Mavrofridis 2015: 349-354). Doumas and Aggelopoulou (1997) grouped together the open perforated cups as a certain type of smoker, making this suggestion with reservation citing relevant bibliographic references. They also referred to the 'barrel-shaped' pots, without identifying them as smokers, even though they are (Doumas and Angelopoulou 1997: 543-554). More recently H. Harissis (2009, 2018) gathered the most samples of prehistoric smokers and suggested a typological division, shedding more light on the various aspects of the issue.[3] G.

Mavrofridis (2009) dealt with a specific subject considering the smoke and the ways that it affected the quality of honey. His research revealed the basis for the term *akapnisto* (ακάπνιστο) (un-/ non smoked honey) that appears in various ancient Greek, Roman and Byzantine texts (Mavrofridis 2009: 200-4).

After a vast chronological gap, photographs of smokers appeared in publications dealing with ethnography as well as in articles dealing with modern Greek beekeeping.[4] Ceramic smokers of traditional beekeeping are nowadays part of ethnographic exhibits or private collections.[5] A thorough examination of their types in comparison with their ancient counterparts is still anticipated but remains a desideratum with wide implications.

[3] Harissis and Harissis 2009: 27-30; Harissis 2017: 26-28, figures 8, 9a, 9b; Harissis 2018b: 81-82. These references include respective bibliography and present most of the smokers mentioned below.

[4] See for example Psaropoulou 1986: 43, 71, 73, 250; Vallianos and Padouva 1986: 67, fig. 91; Psaropoulou 1990: 55, 57; Venetoulias 2004: 48. Sparsely and selectively, in the volume: The bee and its products...2000; Germanidou 2016; see also the articles of the late Thanasis Bikos and of Giorgos Mavrofridis in the periodical series *Melissokomiki Epitheorisi*.

[5] For example, in the under preparation Museum of Beekeeping on the Syggrou Estate, Marousi; in the Centre for the Study of Modern Pottery (G. Psaropoulos Foundation), the Museum of Greek Folk Art, the Museum of Modern Greek Culture in Athens, and the Peloponnesian Folklore Foundation in Nafplion.

Archaeological Evidence Concerning Smokers Discovered in Greece

Even though ancient Egypt is considered to be the cradle of beekeeping, since the earliest representations of beekeeping are found in Pharaonic funerary monuments, no object from Egypt is so far identified as a smoker. The same observation applies to the entire Near East, where beekeeping was also developed since very early times. Surprisingly, many ceramic objects presenting a remarkable morphological variety are identified as smokers and are found in Greek territory beyond doubt. They are older or contemporary with those depicted in the Egyptian beekeeping scenes. Some of them were discovered in northern Greek sites, but most of them were found in southern Greece and on the islands. It is striking that many prehistoric smokers are found, despite the fact that beehive fragments or beekeeping representations are not yet confirmed in the Greek territory (Mavrofridis 2006: 268-272; Harissis – Harissis 2009: 27-30; D' Agata and De Agelis 2014: 349-357; Papageorgiou 2016: 1-26; Harissis 2017:18-35). Even more impressive is the almost complete absence of published identified smokers that date to Classical, Late Antique, and Byzantine times.

General Features of Prehistoric Smokers in Greece

According to current published evidence, we can note that most of the smokers used in Prehistoric times were discovered in southern Greece, in the Peloponnese and on Crete, as well as in the islands of the Aegean Sea, on Skyros, Lemnos and Lesvos. The examples found in northern Greece (at Axiochori in Kilkis and Archontiko in Giannitsa) and in central Greece (Sesklo in Volos), are more ancient and rarer. They are mainly found in buildings that have been identified as storerooms, as well as in settlements and remote caves, which suggests the use of smokers during the collection of wild honey.

The morphological diversity of these objects is remarkable, and it may be interpreted as the result of experimentations. Nevertheless, some features remain the same as they are necessary for the main function of each smoker, which is to blow smoke onto the beehives; these features are a) the spherical or cylindrical, even funnel-like, shape, b) a small size, enough to contain the flammable material and making the object easy to carry, c) a large opening for the flammable material, d) a protruding, mouth-shaped, smaller opening for the diffusion of smoke, and e) a handle that allowed the beekeeper to carry the object and work with it. The most elegant samples are equipped with small feet so that the object would stand on the ground. Another object that served as a smoker was the perforated cup, pierced by many small holes around its upper part; these holes allowed for the diffusion of smoke.

Every prehistoric smoker was ceramic and made of rather coarse fabric. Metal is a material that would not have been preferred, due to the high temperatures that would occur within the smoker. It is also worth mentioning that many other objects (incense-burners, sieves, lamp-stands, even pipes) that share some similarities with the smokers are erroneously identified as such.[6]

Brief Review of the Prehistoric Finds[7]

I. The three earliest examples are similar to each other. One is found in Sesklos (Volos in central Greece) and it is dated to the final Neolithic period (4000-3500 BCE, **Figure 4**); the other two are found in northern Greece, at Axiochori (Kilkis) and at Archontiko (Giannitsa) (2300-1900 BCE); the last one is slightly different from the other two, in regard to the form of its body and its elongated nozzle. They have a spherical body, only half of which is perforated, one handle and two openings (one for the flammable material and the other for the diffusion of smoke).

II. The most known and adequately studied smokers were found at Zakros, Crete. The first was found in a cave and the second, which is the only complete object of this sort, was discovered in the storeroom of a house (1525/50-1400 BCE); their use in beekeeping is confirmed by the practice of experimental beekeeping (see **Figure 3**).[8] They have a cylindrical, almost funnel-shaped body, one or two handles, holes at the front side, which appear slightly angled an ovoid-shaped opening at the underside and short feet. Two more samples discovered in houses at Zakros are fragmentary. A particular cylindrical and oblong perforated object from Enkomi in Cyprus represents the so-called 'snake tubes;' however, it could have served as a smoker and therefore, it is included in this group.

III. This group of smokers includes objects dating back to the 3rd millennium BCE; they were discovered in various sites within Greek territory and beyond: Mandalo at Pella (2300-1900 BCE) (Papaefthimiou-Papanthimou and Pilali-Papasteriou 1987: 177, 180, fig. 7), Knossos (nowadays kept in the Ashmolean Museum), Thermi on Lesvos, Poliochni on Lemnos, Troy, Pelopio in ancient Olympia (2200-2000 BCE) and so on (Doumas and Aggelopoulou 1997:

---

6 Example in: Germanidou 2016: 43, note 172.
7 For a detailed presentation of most smokers and relevant bibliography see note 10. Only for the samples that are not presented in the above cited references further references are included. Approximate chronology is suggested for three significant samples.
8 See note 7.

Figure 4. Smoker from Sesklo, Volos-central Greece, 4000-3500 BCE (Photographic Archive of the National Archaeological Museum, Athens, Greece / Department of the Collections of Prehistoric, Egyptian, Cypriot and Near Eastern Antiquities © Hellenic Ministry of Culture and Sports/Archaeological Receipts Fund).

Figure 5. Smoker from Pelopio, Ancient Olympia-Peloponnese, 2200-2000 BCE (Photo: Sophia Germanidou).

547, figs. 3 and 5). Despite their morphological differences, they all have a common feature: the absence of holes perforating their bodies. The smoker from ancient Olympia is impressively plain in shape, yet incredibly effective in terms of function: it is characterised by a circular body with an opening for the flammable material and an oblong nozzle for the diffusion of smoke (**Figure 5**). The smoker from Mandalo at Pella is similar, but more roughly made. This type of smoker prevailed and survived in subsequent ages. Such an object, for example, is described by the Roman author Columella (see below); it is suggested and illustrated by the pioneer Greek Catholic priest and beekeeper Stefano Della Roca in his book *Traité complet sur les Abeilles* (1790) (Crane 1999: 342, fig. 34.2a). It also survived and predominated in Modern Greek traditional beekeeping (**Figure 6**).

IV. Two more exceptional examples of 'pitcher-like' smokers with handles were found at Knossos: one of them has a perforated body, while the other one has openings for the flammable material and the diffusion of smoke.

V. The last and more common type of prehistoric smokers has raised doubts about its possible multiple functions. It includes the perforated cups with or without handles, which were discovered in Palamari on Skyros (2500-1900 BCE), Poliochni on Lemnos (2600-2400 BCE), Ampelofyto (**Figure 7**), the Palace of Nestor in Chora (Messinia), Lerna in Argolis, Kolona on Aegina, Heraion on Samos, Aplomata on Naxos and so on (Doumas and Aggelopoulou 1997: 550, fig. 12). The function of these objects has been the subject of exhaustive discussions: the holes perforating the body of the cups must have served a certain function; based on these holes they are interpreted as braziers, incense burners, or strainers draining cheese or honeycombs. The morphological features of the perforated cups, such as the foot-base, the handle and the holes perforating the upper part of their body, allow one to identify them as smokers. However, their interior is not scorched and no burning traces are left (Rutter 1995: 326-329). Della Roca, in his book that was published in 1790 not only describes the usual type of smokers (Group III); he also provides the valuable information of a second type of smoker 'with many small holes in the top part' and probably with a handle, since it would be too hot to be held. Many ethnographic parallels indeed confirm that this type was used in Mediterranean traditional beekeeping and not only in Greece (Crane 1999: 342, fig. 34.2b).

Figure 6. Smoker, 20th century, provenance unknown (©Peloponnesian Folklore Foundation of Nafplion).

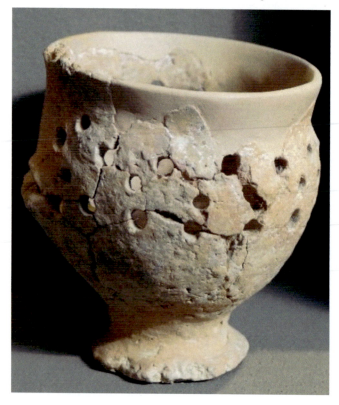

Figure 7. Smoker from Ampelofyto, Messinia- Peloponnese, 1600-1200 BCE (photo: Tina Gerolymou).

Smoking and Smokers in Ancient Greece and Rome

According to the textual evidence, the use of smoke in beekeeping continued in ancient Greece. Aristotle (384-322 BCE), the first beekeeper-scientist, observed that 'bees devoured honey more ravenously' after being smoked (Arist. *History of Animals* 9.40.623b). An epigram of the orator Zonas from Sardis (c. 80 BCE) refers to a beekeeper in the following terms, 'smoking with his skilful hand' (*Greek Anthology* 9. 226). The Greek author, Apollonius Rhodius (3rd century BCE), was aware of the fact that bees were repelled by smoke (Wendel 1953/1974: 135, lines 130-31). Julius Pollux, a Greek scholar (2nd century CE) referred to the smoking of bees, as a method for driving them away (*Onomasticon* I, 15). Despite the confirmed use of smoke in ancient Greece, no Classical beekeeping smoker has yet to be identified, at least as far as we know.

All the Roman authors who referred to beekeeping mentioned the use of smoke. However, only two authors describe the object itself (Crane 1994: 118-134). Varro (1st century BCE) (*de Re Rustica* 3.16.18) gave instructions for smoking the hives gently and lightly not only in harvesting but also when induced to enter a new hive. Columella (1st century CE) (*de Re Rustica* 3.18.31) offered more details about the smoker design: 'earthenware vessel with handles, shaped like a narrow pot. The beekeeper blows the fuel through the one open end and the smoke emerges from the other, more pointed end. The fuel was dried dung or a plant resin...smoke should be applied at the opened back of the hive so that the bees should move to the front end or outside' (*R.R.* 15.5-6).

The Un-/Non Smoked Honey

Strabo's *Geography* (9.1.23) (1st century BCE) offers the first testimony about the well-known Attic honey, which he referred to as 'akapnistos' (un-/non smoked). Pliny the Elder (1st century CE) (*Natural History* 11.15.45) also mentioned that the best honey was the unsmoked one, without going into more details like the other ancient Greek and Roman authors. The term 'unsmoked' is used to designate honey in the Byzantine period as well; it appears even in the Lives of Saints, such as that of John the Almoner (7th century CE), which refers to the high quality of such quality of honey (Festugère and Rydén 1974: 356-7, lines 557-60). It may be inferred by the term itself that this honey was extracted without the use of smoke during harvesting; as a result its taste and smell were not affected. But this was a very difficult task, even for the most competent and experienced beekeeper.

In the past, it was believed that unsmoked honey was extracted from honeycombs that were compressed inside a basket without being heated in a brazier. A more recent and plausible suggestion connects the unsmoked honey to the content of the extension rings, which were applied to the mouth of the horizontal beehives. It was not difficult to collect the honeycombs from the interior of these rings; this was made possible thanks to their small length (max. 0. 075 m.). This technological innovation upgraded the quality of the honey in terms of taste and smell and it is considered to be one of the most important achievements of ancient Attic beekeeping (Mavrofridis 2009: 200-4).

Smoke as a Symbol in Orthodox Patristic Literature

The orthodox patristic literature includes allegories inspired by the use of smoke to repel bees. In most cases, smoke was the main weapon against sin. This allegory is found in texts of the 4th century bishops, Saint Basil the Great, bishop of Caesarea (*Patrologia Greca* 36, col. 620, lines IA 27-30; *Patrologia Greca* 37, col. 1102, lines 1064-65), Saint Gregory the Theologian or Gregory Nazianzen, archbishop of Constantinople (*Patrologia Greca* 32, col. 1197, line 2, col. 1328, 5C). Likewise, in the texts of Saint John Chrysostom, archbishop of Constantinople, smoke is used for the 'purification' of the Christian entity (*Patrologia Greca* 62, col. 105). The representation of Saint Antony on Byzantine wall-paintings offers an interesting detail: an eloquent epigram written on the Saint's scroll mentioning that 'as the smoke drifts away the bees, in such a way psalms drive away the wasps (i.e. the evil spirit)'(Germanidou 2016: 27).

Next to the allegories, certain texts include descriptions of the use of smoke against bees; such references are given by Eustathius, archbishop of Thessalonica in the 12th century (Koukoules 1950: 275. O, 291, 41), or the lexical encyclopaedia *Etymologicum magnum* (11th century), citing that smoke was a basic substance in the beekeeper's activity (Lassere and Livadaras 1992: B 178). An interesting and accurate allegory is provided by the Byzantine government official and historian, Nicetas Choniates (12th century), who mentioned that the Byzantines drove out the Persians, like smoke repels the bees (van Dieten 1972: 136, ll. 25-35).

Testimonies and Representations of Smokers in Byzantine Times

The most important reference concerning the use of beekeeping smokers in the Byzantine period is found in the *Geoponica*. It is a 10th century treatise, a compilation of earlier (6th century) works about agricultural issues. The chapter about beekeeping includes a citation concerning the use of smoke when approaching the beehive, as well as the name of the smoker, which is called there a *chytridion* (small cooking-pot) and

contains dung (*Geoponica* 15, 6. 1-3). No other Byzantine name for this specific object is known; therefore the above descriptive term is helpful in an effort to imagine what this object actually looked like: it was probably a small vessel that looked like a cooking-pot and it had one or two handles so as to be held, carried and used by the beekeeper.

Representations of such objects in Byzantine and generally in medieval art are rare. The beekeeping scenes that decorate the Exultet rolls – the scrolls that include the Latin psalm *Laus Apium* (The Praise of Bees) – show two cases of beekeepers holding a smoker or using smoke while approaching beehives. The first case (Exultet Vat. Lat. Barberini 592: 1070-1100) presents two young assistants of the older and more experienced beekeepers holding open, ovoid-shaped pots with one or two handles, probably serving as smokers (**Figure 8**). In the second case (Exultet Pisa 2: 1000-1100), a beekeeper is presented on his knees in order to remove honeycombs from hives. He holds a lit torch using its smoke, obviously as a way of driving away the bees (**Figure 9**).[9]

The most important representation of a smoker is found in folio 145v of the *Book of Job* in the Par. gr. 135 (1361/2) Byzantine illuminated manuscript. This manuscript is rather special, due to the western influences on its iconography. The miniature in question has been the subject of study, as it includes agricultural scenes such as a woman milking an animal, a man shearing a sheep, as well as beehives made of tree trunks.[10] Next to one of these hives, a small, light blue object is barely visible; it has an impressive protruding nozzle, with wide mouth and narrow neck; the object is depicted standing on the ground and smoke is blown through its top (**Figure 10**). This iconographic detail is extremely important for two

Figure 8. Lit torch used as a smoker, **Exultet** Vat. Lat. Barberini 592, 1070-1100 (Germanidou 2019: fig. 6).

---

9 Germanidou 2019: 4-5, for comments and past bibliography.
10 Germanidou 2016: 65, for comments and past bibliography.

Figure 9. Smoker in the probable shape of a 'chytridion', Exultet Pisa 2, 1000-1100 (Germanidou 2019: fig. 5).

Figure 10. Byzantine smoker, History of Job, Par. Gr. 135, folio 145v, 1361/2 (©Bibliothèque nationale de France, Department of Image and Digital Services).

reasons: it is the only unequivocal Byzantine/ medieval depiction of a smoker and it differs from the image of the 'chytridion' given by *Geoponica*.

The Smokers of Traditional Greek Beekeeping

The smoker became a necessary tool for Greek beekeeping in modern times (19th-20th century), especially before the importation of the so-called American 'hot blast' – made of metal–invented in the United States in 1873. Studies both detailed and general of these objects are lacking. Notwithstanding, one may make two main observations: a) their morphological variety is remarkable, so that it is often difficult to identify them (**Figure 11**); and b) their resemblance to their prehistoric and medieval ancestors (**Figures 12a,b**) is impressive, so that one could

Figure 11. Clay object with open holes in its oval surface probably wrongly identified as smoker (©Peloponnesian Folklore Foundation of Nafplion).

infer that they are the continuation of a very long tradition. The flammable material that was used was cow dung, in certain cases mixed with some pine needles. Occasionally, pulverised asphodel was used to disinfect the pot (Spees 2003: 93-4). The use of plain torches survived as well.

Future Research Trends

New valuable information as to the use of smoke in beekeeping and its influence on the produced honey, missing from other fields, is now being undertaken. In particular, gas chromatographic and organic residue analyses carried out on honey pots and beehives in Portugal, have detected levoglucosan (a biomass burning tracer) which has been connected with the process of fumigation to calm the bees. This finding may result in the secure identification of pots as smokers

Figure 12a. A female beekeeper posing proudly next to beehives holding a smoker, resembling the Byzantine one depicted in figure 10 (©Spata Educational Association).

Figure 12b. Smoker from the island of Sifnos, 20th century, resembling the prehistoric smoker from Pelopio-Ancient Olympia in figure 5 (Photo: Yorgos Kyriakopoulos).

and of the material used to produce smoke (Oliveira and Morais and Araújo 2015: 193-212).

Summing up, clay smokers represent a fascinating vessel shape that is coming to be examined more thoroughly from historical, ethnographic, archaeological, and archaeometric angles. These humble, yet functional objects are rarely exhibited or stored in local museums. Their continued study is sure to add further evidence regarding the sophisticated and influential past of Greek beekeeping.

Acknowledgements

I owe a great deal of the present study to the broad beekeeping knowledge of Giorgios Mavrofridis and his willingness to share his valuable experience and information with me. Roberto Bixio, a vivid, tireless scholar and in-situ researcher of Byzantine apicultural sites, has also been a source of inspiration and guidance. Alexandra Konstantinidou has been a helpful ear for questions of translation but also a helpful collaborator. This chapter could not have been completed without their keen and thorough review, for which I am thankful.

Bibliography

Crane, E. 1994. Beekeeping in the World of Ancient Rome. *Bee World* 75/3: 118-134.

Crane, E. 1999. *The World History of Beekeeping and Honey Hunting.* London: Duckworth.

Crane, E. 2001. *The Rock Art of Honey Hunters.* Monmouth, UK: International Bee Research Association.

D'Agata, A. L. and S. De Angelis. 2014: Minoan Beehives. Reconstructing the Practice of Beekeeping in Bronze Age Crete. in G. Touchais et al. (eds) *Physis. L' environnement naturel et la relation homme-milieu dans le monde égéen protohistorique. Actes de la 14 Rencontre ègéenne international*: 349-357. Leuven-Liege: Peeters.

Davaras, K. 1989. Μινωικά μελισσουργικά σκεύη. In *Φίλια Ἔπη εἰς Γεώργιον Ε. Μυλωνᾶν διά τά 60 ἔτη τοῦ ανασκαφικοῦ τοῦ ἔργου*, III: 1-7. Athens: The Archaeological Society at Athens.

Doumas, C. and A. Aggelopoulou. 1997. The Basic Pottery Types at Poliochni and Their Diffusion in the Aegean, in C. Doumas and V. La Rosa (eds) *Η Πολιόχνη και η πρώιμη εποχή του Χαλκού στο Βόρειο Αιγαίο / Poliochni e l'antica età del Bronzo nell'Egeo settentrionale*: 543-554. Athens: Scuola Archeologica Italiana.

Festugère, A. J. and L. Rydén. 1974. *Léontios de Néapolis Vie de Syméon le Fou et Vie de Jean de Chypre.* Paris: P. Geuthner.

Germanidou, S. 2016. *Byzantine Honey Culture.* Athens: National Hellenic Research Foundation.

Germanidou, S. 2019. Medieval Beekeepers: Style, Clothing, Implements (Mid-11th–Mid-15th Century). *Ethnoentomology* 3: 1-15.

Harissis, H. and A. Harissis. 2009. *Apiculture in the Prehistoric Aegean. Minoan and Mycenaen Symbols Revisited.* Oxford: BAR International Series 1958.

Harissis, H. 2017. Beekeeping in Prehistoric Greece, in F. Hatjina et al. (eds) *Beekeeping in the Mediterranean from Antiquity to the Present. International Symposium*: 14–35. Nea Moudania: Hellenic Agricultural Organization 'Demeter'; Chamber of Cyclades; Eva Crane Trust.

Harissis, H. 2018. Προϊστορικοί μελισσοκόμοι. Archaeology and Arts 126: 78-91.

Koukoules, F. 1950. *Θεσσαλονίκης Εὐσταθίου. Τά Λαογραφικά.* Athens: Society for Macedonian Studies.

Kritsky, G. 2015. *The Tears of Re: Beekeeping in Ancient Egypt.* Oxford: Oxford University Press.

Lassere, F. and N. Livadaras. 1992. *Etymologicum magnum genuinum. Symeonis etymologicum una cum magna grammatica. Etymologicum magnum auctum*, v. 2. Athens: Φιλολογικός Σύνδεσμος Παρνασσός.

Mavrofridis, G. 2015. Πειραματικές έρευνες για την κατανόηση του μελισσοκομικού παρελθόντος. *Melissokomiki Epitheorisi* 29/243: 349-354.

Mavrofridis, G. 2009. Το ακάπνιστο μέλι. *Melissokomiki Epitheorisi* 23/3: 200-204.

Mavrofridis, G. 2006. Η μελισσοκομία στον μινωικό – μυκηναϊκό κόσμο. *Melissokomiki Epitheorisi* 20/5: 268-272.

Oliveira, C., R. Morais and A. Araújo. 2015. Application of Gas Chromatography Coupled with Mass Spectrometry to the Analysis of Ceramic containers of Roman Period: Evidence from the Peninsular Northwest, in C. Oliveira, R. Morais and A. Morillo Cerdán (eds) *ArchaeoAnalytics: Chromatography and DNA analysis in Archaeology*:193-212. Munícipio de Esposende: Esposende.

Pager, H. 1973. Rock Paintings in Southern Africa Showing Bees and Honey Hunting. *Bee World* 54/2:

Pager, H. 1976. Cave Paintings Suggest Honey Hunting Activities in Ice Age Times. *Bee World* 57/1: 9-14.

Papageorgiou, I. 2016. Truth Lies in the Details: Identifying an Apiary in the Miniature Wall Painting from Akrotiri, Thera. *The Annual of the British School at Athens* 111:1-26.

Psaropoulou, B. 1986. *Οι τελευταίοι τσουκαλάδες του ανατολικού Αιγαίου.* Napflio: The Centre for the Study of Modern Pottery – G. Psaropoulos.

Psaropoulou, B. 1990. *Η κεραμική του χθες στα Κύθηρα και στην Κύθνο.* Athens: The Centre for the Study of Modern Pottery – G. Psaropoulos.

Rizopoulou-Igoumenidou, E. 2000. *Η παραδοσιακή μελισσοκομία στην Κύπρο και τα προϊόντα της (μέλι και κερί) κατά τους νεότερους χρόνους. Η μέλισσα και τα προϊόντα της.* Athens: Piraeus Bank Group Cultural Foundation.

Rutter, J. 1995. *The Pottery of Lerna IV, v. III.* Princeton: American School of Classical Studies at Athens.

Speis, G. 2003. *Beekeeping on the Island of Andros. An Ethnographic Approach.* Andros: Kaireios Library-Eva Crane Trust.

Stamataki, P. et. al., 2009. Πειραματική αρχαιολογία χρησιμοποιώντας ένα μινωικό μελισσοκομικό καπνιστήρι. Μελισσοκομική επιθεώρηση 23/3: 165-170.

Tyree, L. et al. 2012. Minoan Bee Smokers: An Experimental Approach. In Mantzourani, E. and Betancourt, P. (eds) *Philistor: Studies in Honor of Costis Davaras:* 223-232. Philadelphia: INSTAP Academic Press.

Vallianou, C. and M. Padouva. 1986. *Τα Κρητικά αγγεία του 19ου – 20ου αιώνα.* Athens: Museum of Cretan Ethnology.

Van Dieten, J. 1972. *Nicetae Choniatae orationes et epistulae.* Berlin: De Gruyter.

Venetoulias, G. 2004. *Τα κεραμικά της Κύθνου.* Athens: En Plo editions.

Wendel, K. 1953/1974. *Scholia in Apollonium,* Berlin: Weidmann.

Chapter 3

Potters and Beekeepers: Industrial Collaboration in Ancient Greece

Jane E. Francis

Concordia University (jane.francis@concordia.ca)

Summary: This chapter surveys the history and physical characteristics of ceramic beekeeping equipment and explores the role of the potter in the invention and manufacture of these containers. As specialized objects used by a small number of consumers, beehives may not have been a frequent product of ceramic workshops and were probably made infrequently and commissioned by beekeepers; these shapes were not part of a potter's usual repertoire. These beehives display significant technological overlap with transport amphorae, and workshops and potters making the latter may also have produced the former. An examination of the relationship between potter and beekeeper may also elucidate the origins of ceramic hives, and explain both their unusually static morphological development and the wide array of variations in small details of rims, sizes, and bases/floors.

Keywords: potters; beekeepers; ancient Greece; industry; ceramic hives; vessel morphology

Introduction

The study of ancient ceramics utilises multiple analytical approaches and methodologies to identify vessel shapes, functions, and chronologies. Fabric analysis, now regularly integrated into pottery research, can help to connect ceramics with specific workshops or clay sources and to separate these from imitations made elsewhere. Distribution studies follow vessels as they travel around the ancient world, often ending up far from their centres of production, and can reveal transport routes and trends in consumption.

Considerations of the role of the potter, about which little is known, are generally lacking in the scholarship on ancient ceramics.[1] Excavations of ceramic workshops do not elucidate the social status of these artisans, the set-up of their establishments, their interactions with traders or local consumers — i.e., the market — or even the economic value of their products (Hasaki and Raptis 2016). The process of making pottery, however, is well known, and was even illustrated on Greek painted plaques and vessels.[2] Many kiln sites have been studied and published, but an area that has not been properly studied is the collaboration between the potter and the consumer of the object. In many cases, this may have been negligible: a potter can easily develop a particular cup shape because, in his own life, he uses cups, knows their requirements innately, and he understands what is popular or will sell at any given moment. At the same time, morphological developments like the trefoil rim on a jug, the vertical handle on a hydria, or a higher or lower pedestal stem on a drinking cup may be initiated by a potter looking to better his products and thus his sales, but they may also have been instigated by collaboration with — or at least suggestions by — the consumer, who finds an existing shape less than satisfactory. Shifts in decorative schemes and themes may have been similarly encouraged. On the other hand, shapes not part of a potter's daily life or ceramic repertoire may have required a different type of input in their manufacture: the production of a transport amphora, for instance, at some point had to entail a discussion between a potter and trader about the most efficient shape for maximizing shipboard space, facilitating transport, and minimizing breakage.

A ceramic shape that must have required such collaboration between potter and consumer is the beehive and its related accessories, all of which were made from terracotta starting in the late 5th century BCE. Research on ancient apiculture with such containers is relatively new, dating back to the early 1970s with the excavation and study of pottery from the Vari House in Attica and the scientific confirmation that a specific type of vessel was used as a beehive (Jones *et al.* 1973). Since this time, various aspects of these beehives and their use have been explored. Examples have been identified in far more areas from both survey projects and excavations (Rotroff 2006: 127 map 2), which are normally published as evidence for apiculture as an agricultural undertaking in reconstructions of the ancient landscape (e.g., Jameson *et al.* 1994: 289–90). Regional studies that

---

[1] An exception may be research on Bronze Age ceramics, where fabric analysis and experimental reconstructions have provided some insights into choices made by the potter, but these do not address collaboration between potter and consumer. See, e.g., Moody *et al.* 2012.

[2] For instance, the Penteskouphia plaques and various black-figure vessels; for a discussion of this evidence, see Hasaki 2002: 31–50.

include beekeeping often rely substantially on later ethnographic parallels and practices (e.g., Mavrofridis 2009c; 2017b; 2017c; 2019). A typology, at least for Attic evidence, has been attempted (Lüdorf 1998–99). Fabric analysis on ceramic has been rare and focused on specific areas (e.g., Karatasios *et al.* 2013; Moody *et al.* 2003: 89–90). Distribution studies continue to provide an ever-widening circle of findspots and thus use and possible manufacture (Rotroff 2006: 127 map 2). Experimental reconstructions have confirmed the viability of these vessels for beekeeping (Mavrofridis 2008; Francis 2012; Kalogirou and Papachristoforou 2018).

Also emerging from this research are several key facts about ceramic beehives that cannot be explained by traditional methodologies employed for the study of other classes of vessels. First, these containers vary in details from region to region, all the while adhering to consistent standards and general morphology. Second, their morphological changes are nearly negligible and can remain the same, repeated in production after production for centuries; they are thus impervious to standard norms for chronological assessment and have also thwarted attempts at clear typologies. Third, there is nothing particularly specific about the clays from which they are made, unlike, for instance, cooking pots; they do not seem to require special mixtures or tempers, at least on available evidence. Fourth, all these prior factors mean that their distribution was very likely to have been local or regional, and it is unfortunate that so few beehives have received fabric analysis of any sort, especially within the context of other, locally manufactured ceramics. None have been recovered from shipwrecks to indicate that they were part of maritime trade.

Beehives also have no natural ceramic antecedents, and their specific, physical characteristics do not appear on other vessels; their function, and thus their appearance, is entirely unique and very much tied to their functionality. Their use does not accommodate a large group of individuals: the beekeeper who installs the hives in the landscape, and the potter who creates the container. The following investigation focuses on the role of this craftsman, and in what ways he may have contributed to the development of the beehive shape, which was repeated without significant change for so many centuries.

Much of the ensuing research is based on the ceramic beehives collected by the Sphakia Survey Project in Crete and studied by the author. As part of this work, I have been able to examine whole and fragmentary hives from many other sites on the island dating from the Hellenistic through early Byzantine periods, and these are introduced as comparanda where applicable.

The Archaeological Evidence

Ceramic Beekeeping Equipment

Scholars have identified two types of ancient ceramic beehives: long tubes set horizontally, and vertically positioned vat-like tubs, which are sometimes referred to as *kalathos* beehives (e.g., Bonet Rosado and Mata Parreño 1997). This chapter is concerned only with the first type, as the second is somewhat controversial and the manner of their use not entirely clear (Rotroff 2006: 128–30).[3]

Beekeeping with ceramic hives can employ three separate parts that can be used together in varying combinations: the hive; short, open-ended sleeves called extension rings; and circular, discoid lids. The hives and rings both display some type of grooves or scoring on at least part of the interior surface, a feature that has been key in their identification during fieldwork. Lids can have ridges and holes but lack the interior treatment.

The beehives are fashioned as long cylinders with a convex wall, in some degree; some are slightly more bulbous than others, but these distinctions are difficult to assess, as the majority of examples are fragments (Figure 1).[4] These tubes taper from a smaller base at one end to a wider open mouth at the other. Complete examples are rare, but lengths can be derived from, for instance, 6th century CE hives from Isthmia, at 0.55 to 0.64 m (Anderson-Stojanović and Jones 2002: 348, 351–54). Attic hives are of a similar length, which offers a general average.[5]

Rims are variously shaped, but normally project in a rolled, everted, or thickened profile.[6] Bases also differ,

---

[3] These vessels were either laid on their sides, thus functioning as horizontal hives, or they are ancient versions of post-antique top-bar hives, the existence of which is attested in Greece only in the 17th century. Experimental reconstructions that converted these vessels into a facsimile of an ancient top-bar hive only prove that a swarm will establish itself inside such a container, not that this occurred in antiquity. The author will address this issue in subsequent research; for existing discussion, see Anderson-Stojanović and Jones 2002: esp. 349–51; also Mavrofridis 2008; 2009b; 2017a.

[4] Compare, for instance, straighter-sided hives from Isthmia (Anderson-Stojanović and Jones 2002: 348 fig. 4) and Athens (Lüdorf 1998–99: B8 fig. 18) with those with slightly more bulbous walls, also from Athens (Lüdorf 1998–99: B16 fig. 21, B23 fig. 26; Sparkes and Talcott 1970: 217–18, 366 no. 1853 pl. 88).

[5] For example, Lüdorf 1998–99: 84 no. B5 at 0.46 m; Jones 1990: 66, at 0.959–0.600 m. The lengths of beehives from pre-Roman Spain are within the same range; see Bonet Rosado and Mata Parreño 1997: 36 table 1.

[6] Beehives from the Athenian Agora excavations furnish good examples of all these rim shapes, which continue through the Roman period: rolled rim (Rotroff 2006: no. 362 fig. 58); everted rim (Rotroff 2006: no. 365 fig. 59); thickened rim (Rotroff 2006: no. 360 fig. 58). The examples from Spain provide a similar range of rim forms: Bonet Rosado and Mata Parreño 1997: 38 fig. 6.

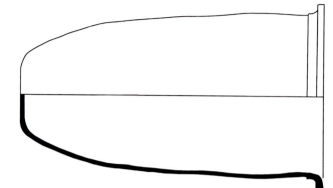

0 ■■■□ 4cm

Figure 1. Horizontal, tubular beehive, from Athenian Agora (after Rotroff 2006: fig. 59 no. 365) (drawing by D. M. Buell).

and their configuration seems to be regionally distinct: in mainland Greece (Attica, Isthmia), they are solid and run from flat to convex and bulbous, but in all cases formed as part of the vessel.[7] In Crete and pre-Roman Spain, tubular hives have both ends open with similar rim shapes on each.[8] In these cases, the hives would have been closed with lids of organic materials, like wood or stone, like schist.

Scoring is present on part of all these containers, in the interior surface normally interpreted as the ceiling of the hive and placed upwards in order to facilitate the bees' attachment of honeycombs (Broneer 1959: 337, citing D. Pallas). This feature has proven problematic in the identification of beehives: only the scored fragments of a hive are recognized as such, and many unscored pieces of these containers have undoubtedly escaped detection. The number of recovered hives may thus be much larger than suspected.

These beehives were laid horizontally, and bolstered on one or both sides by buildings, trees, or rocks to prevent rolling (e.g., Crane 1983: 46 fig. 24, 47 fig. 27). They could be arranged in rows and stacked. There is no evidence in Greece that these hives were set on racks or bases, nor do they seem to have been plastered into walls, as in the early Iron Age apiary at Tel Rehov, in Israel (Mavrofridis 2010; Mazar 2018; Mazar and Panitz-Cohen 2007; Mazar et al. 2008), or in later Mediterranean apiculture, as in Cyprus (Mavrofridis 2009c: fig 2).

Several morphological variations have been identified on Crete and tentatively elsewhere, but in very small numbers. One of these hive types is a shorter, rounded, bell-shaped beehive with a flaring rim and a flattened 'button' in the centre of the closed, domed base. These occur in late-Roman contexts at Eleutherna (Yangaki 2005: 48 no. 61 fig. 61a) and Gortyn (Di Vita 1988–89: 449 fig. 34a, b) but are not common (Figure 2).[9]

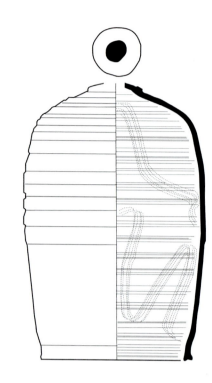

0 ■■■□ 4cm

Figure 2. Bell-shaped beehive, from Eleutherna, Crete (after Yangaki 2005: fig. 61a) (drawing by D. M. Buell).

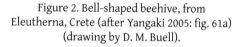
0 ■■■□ 4cm

Figure 3. Beehive with ring base, from Knossos, Crete (after Hayes 1983: fig. 15 no. 177) (drawing by D. M. Buell).

Their relationship to the longer, tubular beehive is unknown and will be explored below. The second variation is a shorter cylinder with a solid, convex floor surrounded by a ring base (Figure 3). These have been identified at Gortyn (Di Vita 1988–89: 449 fig. 33a, b), Phalasarna (Francis 2011), and Knossos (Hayes 1983: 132 no. 177 fig. 14), where the ring is pierced, perhaps for suspension; no examples of this type are complete. Dates for this type of hive range from the late-Classical/Hellenistic through the 2nd or even the early 3rd century CE.

---

[7] The 'Type 1' complete beehives excavated at Isthmia display the range of base shapes, from nearly flat to convex (Anderson-Stojanović and Jones 2002: 348 fig. 4).

[8] At present, complete, open-ended hives are securely known only from pre-Roman Spain (Bonet Rosado and Mata Parreño 1997: 35–36 figs 1, 2). Crete, which preserves many thousands of beehive fragments, has not yet revealed evidence for a solid base on this type of beehive and it is, at present, concluded that these hives were also open-ended.

[9] This shape also occurs in post-antique Andros, where it is termed 'rare' (Bikos and Rammou 2002: 6–7 fig. 3; Mavrofridis 2015b: fig 7).

Extension rings remain a puzzling part of the ceramic beekeeping suite due to the lack of any clear antecedents. These open-ended sleeves seem to have been attached to the mouths of hives in order to expand the size and thus the capacity of the honeycombs and eventual yield of honey and wax; similarities can be drawn with post-antique, wicker ekes (Jones 1990: 69–71). A further benefit of these rings is that they could be removed individually without unduly disturbing the swarm inside, thus allowing more efficient harvesting and without recourse to smoking (Jones 1990: 69–71; Mavrofridis 2009a: 202–203). Diameters of these rings are consistent with those of the hive mouths, although their lengths vary. These smaller, but thick-walled (at least 1 cm wall thickness), densely made objects have a much higher rate of preservation than the hives, and many survive with both rims intact. These can be quite long, as an example from Thorikos at ca 30 cm demonstrates (Jones 1990: 64), although most are shorter; Rotroff cites the rings from Vari and Trachones as between 6.5 and 9 cm in length (Rotroff 2006: 128). Extension rings from Crete are also within this range (Di Vita 1988–1989: 446), although some are considerably shorter (Figure 4).

Jones estimates that a 10.2 cm ring, excavated at Thorikos, could hold three honeycombs (Jones 1990: 70). The rims of these objects are variously shaped. Many slope downwards towards the interior and have a flat top surface (e.g., Martin 1997: pl. 116 nos 4, 7; Vogt 2000: 195 nos 3–5, 197 no. 2), while others are flat and horizontal (Martin 1997: pl. 116 nos 5, 6; Vogt 2000: 197 nos. 1, 3), thickened (Vogt 2000: 197 no. 4) or projecting (Lüdorf 1998–99: fig. 42; Triantafyllidis 2010: 42 fig. 51; Vogt 2000: 197 no. 5). The latter two types may have facilitated the attachment to the beehive — or another ring — with rope, as in the reconstruction by Jones *et al.* (1973: 447 fig. 19). In other cases, these rings could have been attached with mud, clay, or even beeswax or propolis.

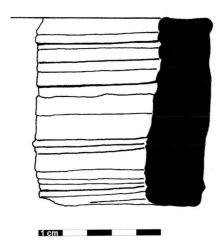

Figure 4. Beehive extension ring, from the Sphakia Survey, Crete (drawing by Anne Bowtell).

Beehives need to be enclosed containers to prevent the entry of predators and disruption to the swarm, yet had to be easily removable by the beekeeper. Hives with solid floors of any configuration only need a closure for their mouths, while open-ended hives require lids at both ends. Round ceramic discs are found alongside beehives at sites in mainland Greece, especially in Attica (Jones *et al.* 1973: pl. 77; Lüdorf 1998–99: figs 43–48). These feature raised bosses or ridges for affixing them to a projecting beehive or extension ring rim and seem to allow for the attachment of a stick as a handle (Crane 1983: 46 figs 24–25). Holes pierced through the surfaces facilitate bee access. Some ceramic lids also feature embossed or stamped decoration on their top surfaces (e.g., Lüdorf 1998–99: fig. 49). The use of these lids seems to be regional: they are very common in Attica, but rare at Isthmia, where ceramic hives are otherwise frequent (Anderson-Stojanović and Jones 2002: 364 no. 34 fig. 5). They do not occur, so far, on the Greek islands where hives and extension rings have been found; the example from Gortyn is a rare instance on Crete (Rendini 1988: 246 no. 236 fig. 208). Where such manmade lids are not employed, beekeepers would have employed discs of stone like schist (Bikos and Rammou 2002: 6).

Regional distinctions can also be observed in the scoring on the interior surfaces of beehives and extension rings. Complete hives show these deliberate grooves made with a sharp stick or comb with multiple teeth on only part of the interior diameter (Lüdorf 1998–99: 84 B4 fig. 14). In some cases, this feature stops well before the rim of the open mouth, but it can also be continued closer to the rim (e.g., Papanikola-Bakirtzi 2002: fig. 157). Attic beehives show the use of multi-toothed combs drawn across the clay in neat, often cross-hatched patterns (e.g., Lüdorf 1998–99: B5 fig. 15; Rotroff 2006: no. 353 fig. 58); but also in individual straight groups (Triantafyllidis *et al.* 2010: fig. 49). On Crete and other islands (e.g., Vogt 2000: 197 nos 1–5 fig. 48), but also in Attica (e.g., Jones 1990: figs 57, 60), scoring was executed in individual lines that are mostly horizontal (i.e., parallel to the rim) but in varying thicknesses, spacing, and depths. In Dalmatia, fragments are scored regularly, with a more corrugated appearance (Kristina Jelinčić, pers. corr.). The scoring on extension rings is usually dense and horizontal (e.g., Lüdorf 1998–99: E9 fig. 38), but the comb can also be observed in Attica rings, occasionally in a cross-hatched pattern (e.g., Lüdorf 1998–99: E3, E6 fig. 38).

Graeco-Roman ceramic beekeeping equipment has a wide distribution across Greece, the islands, and western Asia Minor, but these finds are unevenly spread. At present, this material is attested in mainland Greece throughout Attica (compiled in Lüdorf 1998–1999; also Rotroff 2006: 124–131, 283–285), at Isthmia (Anderson-Stojanović and Jones 2002), the Corinth area (Gregory 1985: 428, 1986: 297), Methana (Mee and Forbes 1997: 121), the Argolid (Jameson *et al.* 1994: 289–289, 445, 447); Megara (Francis 2009), and Boeotia (Vroom 2003: 91, 95–96, 107, 109, 111). On the islands, beekeeping equipment is known from Eretria (Metzger 1993: 112, 116, 1998: 189, 197, 201), Chios (Anderson 1954: 137, 142; Boardman 1956: 51–53), Lemnos (Massa 1992: 227–228), Samos (Tölle-Kastenbein 1974: 120 fig. 198D), Agathonisi (Karatasios *et al.* 2013), Kea (Sutton 1991: 260–263), Paros (Hasaki 2003), Delos (Bruneau 1970: 260–261; Siebert 1988: 763; Peignard-Giros 2012: 245), and Siphnos (Brock and Young 1949: 87 no. 25 pl. 30.3; Ashton 1991: 126). In Ionia, examples come from Ephesos (Gassner 1997: 104 no. 375). Beekeeping equipment on Crete has been identified at excavations at Knossos (Hayes 1983: 110, 132, 134 no. 177; Callaghan 1992: 93 no. 9; Sackett 1992: 176–77), Gortyn (E.g., Martin 1997: 339–41; Albertocchi 2011: 211), Eleutherna (Vogt 2000: 95; Yangaki 2005: 48 no. 61, 82 no. 437a, 345 nos. 344–45), Aghia Galini (Vogt 1993–1995: 70–71), Chania (Kataki 2012: 546 pl. 3), Syvrita (Karamaliki 2011: 296), and in the survey areas of Sphakia (Francis 2001; Francis *et al.* 2000: 442; Moody *et al.* 2003: 89–90), Moni Odigitria (Francis 2010, 42–4), the Akrotiri peninsula (Raab 2001: e.g., 160), Gournia (Hayes and Kossyva 2012:168), Mesara (Watrous and Hadzi-Vallianou 2004: 322), Galatas (Gallimore 2017: 96), and Kommos (Callaghan 1995: 388). The same type of containers was employed in pre-Roman Spain, in the 3rd century BCE (Bonet Rosado and Mata Parreño 1997) and examples have also been attested in Roman Dalmatia.[10] Oddly, few beehives are confirmed from Italy, despite Roman authors producing sometimes-extensive discussions on contemporary apicultural practices. This lacuna can perhaps be explained by a general opposition to ceramic — or clay, as they are termed — hives, which Roman writers derided as lacking appropriate temperature control: too hot in the summer and too cold in the winter. This perception has been debunked (Francis 2012; Kalogirou and Papachristoforou 2018) but may have been sufficient to encourage beekeepers to use hives of other materials such as wicker and wood, which have not survived.

Pre-Classical Ceramic Hives in the Mediterranean

The need for large amounts of honey, the only sweetener of antiquity that did not require processing, and wax, used for numerous artistic enterprises and activities of daily life, was constant throughout Greek and Roman cultures, and these may have been obtained through wild honey hunting, either sporadically or as a supplement to apiculture with organic hives[11].

References to these commodities in ancient literature, as well as observations on bee behaviour, only confirm an interest in bees and their products, not that these were domesticated with manmade hives.[12] However, the date at which Greeks began to manufacture and use ceramic beekeeping equipment is unknown, as is whether all three known pieces were contemporary in their inception. The earliest, securely identified Greek examples come from the late-5th century BCE, but beekeeping could have been practiced with manmade hives in organic materials much earlier and survived alongside the clay examples. The harvesting of honey and wax from wild honeycombs should also be considered.

Early examples of tubular, horizontally placed beehives can be found outside Greece, in ancient Egypt, where paintings in several tombs portray beekeepers at apiaries with multiple stacks of hives (Crane 1999: 163 table 20.3A; Kritsky 2015). The open mouths face the workers, who are shown variously extracting honeycombs, smoking the bees, and decanting honey into containers. The detail with which the artists often rendered both the equipment in use and illustrated the processes of apiculture indicates sufficiently widespread beekeeping knowledge in Egyptian society that even tomb artists could represent the practice accurately.

The earliest of these images is a relief in the Sun Temple of Neuserre at Abu Ghorab of the 5th Dynasty (2465-2323 BCE), which shows a stack of tubular hives (Crane 1999: 164 fig. 20.3a; Kritsky 2015: figs 2.3, 2.4, 2.5). In front, a beekeeper wields an object identified as a smoker (Crane 1999: 164; Kritsky 2007: 64; 2015: 10). Elsewhere in this relief, men pour honey into a large container, while a bee flies in the background. The 18th Dynasty (1550–1307 BCE) Tomb at Rekhmire at Thebes (TT100) depicts a similar scene: cylindrical beehives with rounded bases are stacked vertically on the right, set on top of a rectangular platform. Bees are being smoked, and beekeepers extract honeycombs (Crane 1999: 165, fig. 20.3b; Kritsky 2007: 66; 2015: figs 4.3, 4.4, 4.5, 4.6, 4.10). The poorly preserved Tomb of Amenhotep I at Thebes in the same dynasty (TT73) shows the same type of apiary and smoking process (Crane 1999: 165 fig. 20.3c; Kritsky 2007: 65; 2015: fig. 4.1). Much later, in the 26th Dynasty (664–525 BCE), the Tomb of Pabasa at Luxor (TT279), a painted relief depicts stacked, bullet-shaped hives with rounded ends and a honey-gathering scene (Crane 1999, 167, fig. 20.4a; Kritsky 2007: 66–68; 2015: figs 5.5, 5.6, 5.7). A contemporary, ill-preserved relief in the Tomb of Ankhhor (TT414) is similar, with smaller hives (Kritsky 2007: 68–69; 2015: figs 5.12, 5.13).

---

[10] A perforated vessel from Fishbourne, in Roman Britain, does not belong to this class of vessels (Cunliffe 1971: 210-11).
[11] Other known sweeteners include grape must juice, dates, figs, and barley, but these must be boiled up and processed, as opposed to honey, which can be obtained directly from the honeycomb; see e.g., Dalby 1996: 47.

[12] The ancient sources on bees and bee behaviour are presented and discussed in Giuman 2008.

Absent from the Egyptian archaeological record are the hives themselves. Nineteenth-century excavations at Kahun revealed cylindrical clay objects with one closed end that are tentatively identified as beehives (Flinders Petrie 1890: 25 pl. XIV). Gas liquid chromatography found beeswax inside them, but they have a very small diameter (7.8 cm). Crane (1999: 164) interpreted these as model beehives, while Kritsky (2015: 25) suggests that they may have been used for collecting honey. No other real hives are known, despite the prevalence of apiculture in tomb art.

Two important points may be derived from this Egyptian evidence. First, the artistic renderings depict the same type of beehives positioned in the same manner for nearly 2000 years. This shows a consistency in equipment that must reflect reality. Second, the lack of physical remains for ceramic hives over the very long period represented by the artistic record strongly signifies that the tomb images may refer to — and beekeeping in Egyptian society may have employed — unfired clay hives. This interpretation is supported by the discovery of a 10th–9th century apiary at Tel Rehov in Israel, where tubular hives made out of unfired clay mixed with straw and animal dung were stacked in three tiers; the excavators estimate approximately 100 hives (Mazar and Panitz-Cohen 2007: 206, 207, 211).[13] The accidental preservation of these unfired hives was due to the fire that destroyed the apiary complex.

The evidence for beekeeping in prehistoric Greece is far more elusive. Wax in ceramic containers is attested as early as the Neolithic era, in fragments of perforated pottery from Limenaria on Thasos and Dikili Tash in Macedonia (Decavallas 2007), but these containers cannot be confirmed as beehives and may simply have been containers for wax.

The presumption of beekeeping with ceramic hives in Minoan Crete has been extensively addressed (e.g., Mavrofridis 2006; Harissis and Harissis 2009), but without conclusions satisfactory to all researchers. Honey and wax — and perhaps, even the bee — were of enormous importance in Minoan society and ritual. Many artistic renderings, such as the famous Malia bee pendant (e.g., Bloedow and Björk 1989), and Linear B ideograms for honey and wax are accepted by many scholars as proof of organised apiculture with manmade hives.[14] A structure with rows of dark triangles depicted in the miniature fresco from Room 5, West House at Akrotiri has been interpreted as an apiary with 'upright pointed baskets' — a type of wicker skeps hive coated

with dung or clay (Papageorgiou 2016: 104). Similarly shaped objects represented in Minoan art, such as the Master Impression Seal from Chania (Papageorgiou 2016: 103 n. 10), are also proposed as beehives. Yet none of these images specifically shows beekeeping or bees — unlike the Egyptian paintings —, nor are apiaries of this, or any type, confirmed from other prehistoric evidence. Residue analysis on the interior of Minoan pottery from Mochlos (Evershed et al. 1997) and Pseira (Beck et al. 2008: 62) has revealed beeswax and the use of these vessels as candles, but the source of these products is not known and, like the earlier evidence, may have been obtained and observed through wild honey hunting (Mavrofridis 2016). Physical evidence for beehives is also lacking. Bronze Age Crete (MMII through LM IB) saw the production of ceramic, vat-like containers not unlike the ambiguous, vertical beehives of the Graeco-Roman era. The identification of these vessels as beehives is entirely due to their interior surfaces, which bear wide grooves on at least a portion of the height as well as the floors, where they are often executed in a spiral pattern. This feature has been equated with the scoring on later hives, despite their completely different patterns, thicknesses, and widths. None of these containers have been scientifically tested to confirm that they held beeswax or honey, even when they are claimed as hives by excavations that tested other classes of artifacts.[15] These 'beehives' have been found at Bronze Age and early Iron Age sites, both from excavated contexts and survey projects: Chania (Hallager 2003: 242–43); Nerokourou (Melas 1999: 487); Kommos (Watrous 1992: 25 no. 439 fig. 22); Mochlos (Brogan and Barnard 2003: 56); Chrysokamino (Ferrence and Shank 2006); Gournia survey area (Watrous 2012: 118–19 site 60); Kavousi (Day 2012: 5); Kato Syme (Melas 1999: 487–88); Zakro (Chrysoulaki 2000: 585 fig 3ζ.[16]

The Egyptian evidence discussed above also argues against the use of ceramic hives in prehistoric Greece. The long, horizontal beehive consistently represented for centuries is not found in the Greek world, despite ongoing contact between the Minoans and Egypt, with cultural and artistic influence from the east. A symbol in Minoan hieroglyphics identified as the 'bee sign' and present on five seals is derived from Eyptian hieroglyphics (Evans 1921–36 I: 281). The Egyptian town of Kahun, which has furnished the putative beehive models, preserves Minoan pottery (Fitton et al 1998). The Tomb of Rekhmire, along with its beekeeping

---

[13] Gas chromatography identified beeswax in the walls of these containers as well as remains of burnt honeycombs and many body parts of honeybees (Mazar et al. 2008).

[14] For example, the Linear B ideogram *179 may depict a triangular skeps hive (Papageorgiou 2016: fig. 11); ideogram *168 is interpreted as a cylindrical, ceramic beehive (Davaras 1986).

[15] This was the case at Chrysokamino, where fragments were identified as 'beehives' without scientific study, despite residue analysis performed on other types of pottery from the same excavation; see Ferrence and Shank 2006, but also Day's review (2008: 768).

[16] Melas (1999: 487) cites 18 sites with these types of vessels. D'Agata and De Angelis (2014: 353 n.27) state that they have identified 237 examples, at their date of publication. They cite a further 10 of these 'beehives' dating after the LMIB period (D'Agata and De Angelis 2014: 355).

scene, contains a depiction of Keftiu (Cretans) bearing gifts (Freed 2001: 386c; Wilkinson and Hill 1983: fig. 41). Yet the horizontal type of beehive and their stacked arrangement familiar from Egyptian art find no parallel in prehistoric Greece. The only exception may be the ideogram *168 in a Linear B tablet from Knossos, shaped like a horizontal tube. This symbol has been translated as 'of honey', but this is disputed (Davaras 1986); Linear B tablets from Pylos also refer to honey in ritual contexts (e.g., Chadwick 1976: 124–26; Harissis and Harissis 2009: 13), but these cannot be connected with manmade hives or those made of terracotta.

Another stark difference between Egyptian beekeeping and that claimed for prehistoric Greece can be cited. The tomb images of Egypt depict apiaries with beekeepers engaged in their tasks; their equipment is also depicted — hives, smokers — as are the products, along with honeycombs and bees. These are snapshots of beekeepers at work. Minoan bee imagery, by contrast, depicts only the bee as an independent object and without any context:[17] bees are stand-alone motifs, such as the heraldically positioned bees that form the gold pendant from Malia (e.g., Bloedow and Björk 1989, 29–30; Woudhuizen 1997; Harissis and Harissis 2009, 11–12). The bee is also rendered on seals, which have been understood as indicating the contents of sealed containers as honey and wax and used by magistrates in charge of such commodities, but again, the source of these products is unknown (e.g., Harissis and Harissis 2009, 12). This may be due to differing meanings for the bees, honey and wax — cultural versus ritual, for instance — but it is a striking distinction in the comparison between Egyptian and Minoan beekeeping, where no apiculture scenes were represented.

The available archaeological record strongly argues against the development and use of fired, ceramic beehives in the prehistoric Mediterranean, especially if, as seems logical, the hives employed in ancient Egypt were unfired. There also does not appear to have been any technology transfer from Egypt to Greece. The clear, irrefutable evidence for the use of manmade hives then has a gap until approximately the last quarter of the 5th century BCE, when the fired, horizontal beehives begins to appear in datable contexts in Attica and elsewhere in Greece.[18]

The Case for Ceramic Hives

The distribution of ceramic beekeeping equipment is, at present, limited to the Greek mainland and many islands, a few sites in western Asia Minor, pre-Roman Spain, and Dalmatia; an obvious gap is Italy, where only a few examples have been identified.[19] By the late 5th century BCE, ceramic hives were in use throughout southern Greece and many islands, and within two centuries would be employed in Spain. These hives must have supplemented those made with organic materials already employed throughout Mediterranean landscapes. Roman authors in Italy were well versed in beekeeping practices and were evidently part of a long tradition (Crane 1994; Mavrofridis 2011; 2018a); they cite numerous prior texts, now lost, from earlier Mediterranean writers (Francis 2012). Of particular interest in these writings are the materials used for beehives. These authors recommend beehives out of bark, especially of cork trees (Columella 9.6.1; Palladius 1.38; Pliny HN 21.47.80; Varro 3.16.15; Virgil G. 4.33). Ferula, or fennel stalks, which were plentiful in the Mediterranean, were also used (Columella 9.6.1; Palladius 1.38; Pliny HN 21.47.80; Varro 3.16.15), as were woven plant material like withies and osier (Columella 9.6.1; Palladius 1.38; Pliny HN 21.47.80; pseudo-Quintilian 13.3; Varro 3.16.15; Virgil G. 4.33). Hives were also made from cut boards and hollow logs (Columella 9.6.1; Geoponica 15.2.7–8; Palladius 1.38; Varro 3.16.15). Other cited materials for hives are dung (Columella 9.6.2; also Mavrofridis 2015a), transparent stone, possibly mica (Pliny HN 21.47.80; also Mavrofridis 2018b), and brick (Columella 9.6.2).

Absent from this list of recommended materials is terracotta — or even unfired clay —and authors are instead unanimous in their condemnation of this type of beehive, which are said to lack temperature control for the swarm inside, being too hot in the summer and too cold in the winter (Columella 9.6.2; Palladius 1.38; Varro 3.16.17). While this criticism may explain the lack of these hives anywhere in Greek or Roman Italy, it cannot be reconciled with the enormous number of such hives identified elsewhere in the Mediterranean, especially in Greece; ceramic beehives also continued to be used throughout the region well into the 20th century (e.g., Bikos and Rammou 2002; Mavrofridis 2015b). It may be that ceramic hives supplemented those made of the organic materials described by ancient authors, but at some point, a decision had to be made to do so. And, if this is the case, the scale of apiculture in these areas must have been enormous. Available resources, like appropriate plants and trees that could supply the materials for beehives — or even good-quality clay — may have been a factor in the choice of hives.

---

[17] The identification of objects on several rings as beehives and bees (Harissis 2016: fig. 3) is unsubstantiated.

[18] A vessel and other fragments with interior incisions from Chios (Anderson 1954: 137 no. 28 fig. 5) may be beehives; their context is late-7th century (Archaic) but disturbed and thus not certain. The single rim shape is not attested in later beehives.

[19] These include a possible beehive rim from Sicilian Naxos, dated to the 4th century CE (Blackman et al. 2010: 153–154 figs 4–6), a putative Hellenistic honey strainer from Podere Funghi at Poggia Colla, and a possible beehive used for a child's burial at Locri Epizephyrii (Elia and Meirano 2015: 319). The author thanks Ann Steiner for the reference to the Poggia Colla container.

The benefits of beekeeping with ceramic equipment, however, must also have been realised by ancient workers and may have played a role in their invention and implementation in Greece. Some of these advantages are not provided by hives of other materials.

First, ceramic hives have a body thickness of at least 1 cm and they are generally made from well-refined and hard-fired clay. Harissis and Harissis (2009: 22) cite their weight as a disadvantage, but this need not be the case. They are durable and can be easily moved, as required, perhaps slung over a donkey for transport (Mavrofridis 2015c: 177 fig 1), whether from a potter's workshop to an apiary, around different parts of the landscape for seasonal access to vegetation, for storage when empty, or even for re-use, as in the case of beehives used for childrens' burials.[20] Some ancient authors also cite mobility as a desirable trait in the selection of beehive material (e.g., Columella 9.6; *contra*, Florentinus in the *Geoponica* 15.2; also Dalby 2011: 300). Other materials will not have this benefit: stone hives would be too heavy; unfired clay hives may lose their shape or be prone to breakage; hives of wooden boards, if rectilinear, would not fit easily over a transport animal. Only wicker or woven hives are particularly mobile over long distances.

Second, ceramic hives are sturdy once established in place and are unlikely to be damaged by natural occurrences, like the intrusion of animals; sheep and goats could easily harm or destroy lighter hives of wicker or bark. Ancient authors mention strategies for avoiding this danger, which must have been common (e.g., Florentinus *Geoponica* 15.2). Although the horizontal ceramic hives need to be bolstered to avoid rolling, they are substantial enough to stay in place. They are impervious to rain, and their walls could protect the swarm from wind; ancient authors cite these elements as detrimental to the comfort and productivity of bees (e.g., Columella 9.4.1). Hives of lighter material may be dislodged or blown over in high winds, and a rainy season might result in penetrating damp.

Third, ceramic hives have a longer lifespan than those of organic materials, simply because they would not rot, fall apart, or require structural maintenance. Florentinus, in the *Geoponica* (15.2; Dalby 2011: 300), provides instructions on how to prevent wooden hives from rotting, which must have been an ongoing problem. Ceramic hives will break, but terracotta

containers can be mended, as ancient examples attest (e.g., Lüdorf 1998–1999: 134 B4 Fig. 14).

Fourth, it may be that the longevity of these containers prompted their inheritance as heirloom items for successive generations, much as they are in modern Crete today. The lack of morphological change makes this difficult to recognise, but their constituent elements are few — rim(s), long body, floor/base, rings, scoring, and a lid — and they do not require shape updates or developments. These containers worked appropriately: further shape elaboration was not needed and did not improve their functionality. The bees did not care if a rim was everted or thickened to a triangular profile, and a beehive of a century earlier would have been as efficient as a new one. The only change that might have been desired is capacity, which could have been increased by the addition of one or more extension rings.

Finally, and more hypothetically, ceramic hives could be cleaned. Evidence for diseases within beehives in antiquity is scarce and dependent upon ancient authors who themselves were probably not beekeepers. The types of hives where this occurred are also never mentioned, but references to circumstances and foods that may cause bees to fall ill are rampant in ancient literature (e.g., Columella 9.13). More serious, on the basis of post-antique and modern parallels, are diseases that might invade a beehive. Once established in a hive made of organic material like wicker or bark, these could probably not be eradicated. A ceramic hive, on the other hand, could be cleaned out of its contents, washed, scrubbed, and dried, smoked for fumigation, and the swarm re-installed. Hive fumigation is described by Columella as both a seasonal practice (9.14.7) and a deterrent to disease (9.13.8); Florentinus recommends smoking as a treatment for lice and poor eyesight (*Geoponica* 15.2); he also provides a cure for bee diarrhoea. Crane (2000: 289) equates a 'disorder' in bees described by Aristotle (*HA* 9.40.626b) with a modern 'bacterial infection'. In antiquity, as now, cleanliness of a hive was extremely important (Columella 9.5.2), and ceramic vessels would have provided optimal material for a thorough cleaning job.[21]

This data answers the question about the emergence of ceramic beehives after centuries — or possibly millennia — of apiculture with hives of other materials, traditions that clearly continued in some parts of the Mediterranean. It is unlikely that these terracotta hives completely supplanted those in earlier material, but must have supplemented them. Beekeepers evidently had a range of materials, sizes, and types of hives from

---

[20] The practice of re-employing ceramic beehives as sarcophagi was not widespread but occurred consistently at sites in Attica. Examples come from Athens (Alexandri 1972: 35, undated but amid 5th and 4th century burials), including the Kerameikos cemetery (Knigge 1976: 30–31, Hellenistic); Marathon (Jones 1976: 88–90, 2nd century BCE); Oropos (Pologiorghi 1998: 127–128, undated); Rhamnous (Perakos 1992: 5–6, 4th century BCE). Funerary re-use of beehives has also been documented on Crete, at Chania (Kataki 2012, 4th–3rd centuries BCE) and Phalasarna (unpublished).

[21] Columella (9.14.7) recommending cleaning the hive with a stiff feather from a large bird such as an eagle, although he does not cite the type of hives to receive such treatment.

which to choose, and, clearly, many of them saw the benefits of ceramic beehives.

The Role of the Potter

An understanding of the role of the potter in the invention and manufacture of ceramic beekeeping equipment is based on the objects themselves, not only their morphological characteristics, but also their findspots, number of fragments, and regional variations. As specialised objects required by only a small portion of the population, beehives and extension rings — as well as lids, in the regions that used them — were unlikely to have been a shape spontaneously invented or produced by a potter without some instruction from a beekeeper.

The ceramic hive may have its origins in several possible sources. The unfired, cylindrical hives of ancient Egypt and the Near East may have been transformed through firing into the familiar terracotta shape. Beekeepers in Greece almost certainly knew about the Egyptian and Near Eastern horizontal hives through centuries of sustained contact with these cultures, but chose not to emulate these containers, which had significant drawbacks. These vessels of unfired clay could not sustain an outdoor setting without integration into a shelter, like a porch or roof, as they would be ruined by rain and been vulnerable to breakage; the preserved examples from Tel Rehov, as noted above, are situated within an architectural complex (also Mavrofridis 2018c). Unfired beehives are also not well suited to frequent movement. Beehives are today moved for numerous reasons, and the distribution of beehive fragments in Greece indicates a similar practice in antiquity: to provide the bees with a variety of vegetation; to take advantage of possibly changing water sources; and to move the hives away from seasonal weather disruptions. It is significant in this respect that all other types of hives cited by ancient authors, with the exception of those made of stone, are moveable, and this may have been a prime consideration in the material of an apiary. A solution to this problem may have been to adopt the shape and functionality — including the horizontal positioning — of the unfired hives, which were known to be successful, but to render this shape in a sturdier material. For this, a beekeeper would have required the services of a potter able to fashion the hive on the wheel but also to fire it in a kiln.

Another motivation for the adoption of ceramic hives may have been hollow logs, which are essentially long, tubular containers and frequently the location of wild bee nests (Crane 1999: 53). Ancient authors cite this location as an optimal type of hive (Columella 9.6.1; Varro 3.16.15; Palladius 1.38). Observation of the situations preferred by wild bees may have led to a replication of this natural shape in a sturdier, more permanent material. It is also possible to equate the application of interior scoring with the natural striations of the wood on the interior of the log; this treatment might have been initially provided to encourage bees to nest in these ceramic 'logs' and then repeated in subsequent creations. Once again, however, this transference from a naturally occurring hive to a manmade shape requires the participation of a potter, even if the conception of this hive type came from a beekeeper.

If the potter was not a beekeeper, he would have needed to take direction from one on the details of shape and size during construction: a potter would not intuitively know what morphological details and changes might be either required or desirable for the functionality of the container. Changes to the shape of a cup handle or basin wall may be understood as an improvement, but input from the beekeeper would have been necessary for the potter to make similar alterations to a beehive. These were specialised items, and a significant divide thus separated their manufacture from their use. It is possible that some potters who produced beehives were themselves beekeepers and were thus able to refine improvements in the shape, but no ancient sources — or other type of evidence — indicate such a thing.

Further, changes to the shape of a hive in the rim, diameter, amount of tapering, or wall thickness, were meaningless to the vessel's functionality. The crucial factors — mobility, sufficient wall thickness, length and diameter, a sufficient diameter to contain honeycombs, and the form of the foot/base — were probably established by collaboration between the potter and beekeeper, but then repeated. The potter could have instigated deviations, but within the fairly rigid expectations of the shape. Without close knowledge of apiculture, the pottery may have made changes to suit human needs of the beekeeper, not the preferences of bees, once the essential shape had been employed and seen to work.

It is clear that, from the late-5th century BCE through the late-Roman era, large portions of mainland Greece reproduced the same general concept of the horizontal beehive. The manner in which this information was transmitted to various regions must have involved potters as well as beekeepers: a general description of a hive may have been given to a potter, who then may have produced his best facsimile of it; if this container worked, then it could have been replicated with little deliberate change. Beekeepers in some areas may have felt that narrower hives, as in Crete, were preferable to larger, but several other distinctions can be noted. The open-ended hives of Crete and pre-Roman Spain require two lids rather than one solid base or floor, but neither of these areas preserves evidence for the consistent use of ceramic lids, such as those employed in Attica and some Greek islands (Mavrofridis 2018d). This open-ended arrangement allows for the removal of the honeycombs from the back of the hive as well as the front, which Pliny states is preferable (*HN* 11.10.24; also Rotroff 2006: 129). If this was the case, then the development of this type of hive must be ascribed to the beekeeper, for whom such a configuration was an advantage and whose particular beekeeping practices required it; this is not the sort of alteration likely to have been instigated by a potter.

Some of the regional deviations in beehives can be seen to overlap with 'standard' pottery production. The creation of the shorter, Cretan type with the lower body encircled by a ring base defies understanding, even as to how these vessels were positioned — at present, they are regarded as horizontal — but the type of ring would have been very familiar to a potter from many other types of vessels, such as cups, plates, and bowls. The manufacture of these vessels may have resulted from a misguided notion that they would be set vertically and thus required a base.

A comparison of transport amphorae and ceramic beehives reveals further technological overlap. The Cretan, late-Roman bell-shaped hive type, if positioned upside down, is very similar to some contemporary Cretan amphorae, which were circulated around the Mediterranean from the late Hellenistic through the early Byzantine era (e.g., Gallimore 2016). A complete beehive of this form from Eleutherna (Yangaki 2005: 48 no. 61 fig. 61c) resembles an MRC 2A (*Medio-Romano Cretese*) amphora from the same site, right down to the ribbed exterior surfaces (Yangaki 2004–2005: 508 fig. 2). The wall thickness of both these vessel types is within the same range, and their shapes as large, cylindrical vessel rely on shared potting skills and technological approaches (Karatasios *et al.* 2013: 42). Both these ceramic shapes involve frequent movement and their construction had to provide durability.

Where opportunities for fabric analysis on both shapes are possible, the clays used for beehives and amphorae are also seen to be similar and, in some cases, overlap. An example is the 'Creamy Powder Buff Tan' fabric used for both amphorae and beehives recovered from the Sphakia Survey (Moody *et al.* 2003: 73). The beehives and Cretan amphora from Eleutherna show use of the same clay (Vogt 2000: 186, 194). In addition, many of the Cretan beehives are made from local clays, thus providing an example of another shape generated by a site's ceramic workshops: e.g., Gortyn (Martin 1997: 340 no. 408) and the Akrotiri Peninsula (Raab 2001: 104 no. 135).[22] This situation on Crete may also be true for other

regions, but beehives are not normally subjected to fabric analysis. An exception is Agathonisi (Karatasios *et al.* 2013), where chemical and petrographic analyses of beehives identified high amounts of calcareous grits in the clay, which is proposed as a desirable component for beehive clays. This parallels macroscopic analyses of Cretan amphorae from Sphakia, which also show elevated amounts of calc, perhaps added as to preserve the taste of the wine (Moody *et al.* 2003: 84 and n. 70).[23] A beehive fragment from the Akrotiri peninsula in northwest Crete also contains noteworthy amounts of calc (Raab 2001: 97 no. 85); it may be further significant that this vessel is an example of the ring-base beehive type.

The question of whether the creation of extension rings also required consultation is more problematic. These pieces of beekeeping equipment closely resemble amphora stands, except the diameter does not contract in the center and they include interior scoring (for Crete, see Hayes 1971: pl. 39c; Vogt 2000: 199). The creation of these rings would not necessarily require a detailed model or description, if the potter were already familiar with such stands. On the other hand, their diameters must match those of the hive, which can vary considerably, and these rings may have been manufactured at the same time as the hives to ensure at least that the two pieces would fit together. There is also sufficient variation in the details of the rings — height, rim shape —to suggest that there was not widespread, even regional, consensus on one single, appropriate form or size; like beehives. Some extension rings, identified by their interior scoring, are extremely short; three examples from Sphakia are 5.1 and 5.5 cm from rim to rim.[24] This short length may be ascribed to a potter's misconception about the functionality of what he was making, but enough of these narrow rings are preserved to indicate that they worked.

Despite their sturdy nature and relatively thick walls, ceramic beehives and extension rings are not particularly common in the archaeological record, compared to other coarse ware vessels. The Sphakia Survey on Crete collected 504 fragments, which might seem like a sizeable assemblage, but this number is spread across the region's 470² kms (Francis 2016: 98 Appendix 7.1). While 72 sites preserved fragments, only

---

[22] An intriguing and rare instance of beekeeping equipment from one

area ending up in another is a mould for a beehive lid with the name EMBIOY, dating to the Hellenistic era (3rd–1st centuries BCE) from the excavations of the Athens Airport, and a lid with the same name from the House of Stucco at Delos (Peignard-Giros 2012: 245 fig. 1). No fabric analysis is available for either object, and the place of their manufacture is unknown; it need not be Attica.

[23] This parallel is intriguing, but involves a very small number of sites. More research is needed to determine whether both classes of vessels consistently contain this feature.

[24] In the forthcoming final publication of the Sphakia Survey, these rings are nos 8.72:37, 8.72:18, 6.25:45. Two of these were found at the same site and may represent the same source and may have been made by the same potter, although their fabrics are different.

14 sites had 10 or more fragments, and none of these contained more than 22. It is clear that these were sturdy vessels and, unless a beekeeper was increasing the size of his apiary or had suffered significant breakage, individual beehives were likely to last a long time. These two elements — longevity and the fairly small number required — mean that no potters were likely to have specialised in their manufacture. Instead, these hives may have been manufactured *ad hoc*, upon request, perhaps by potters who were also adept at making amphorae, and with amphora clay on hand. In this scenario, a potter receiving a commission may have never have made a beehive before but have been generally familiar with its salient features, or that a potter may have fashioned one in the past and vaguely recall what was required. In both cases, these potters may rely on knowledge of amphorae, which were more frequently manufactured. The small number recovered in the Greek landscapes suggests that potters did not produce ceramic hives on spec or that they were mass-produced. The large number of small morphological variations that are not chronologically significant, even within one region, also supports this reconstruction.

Conclusion

Evidence for the manufacture of ceramic beehives of the horizontal type indicates that regional potters made at least some of these vessels at individual request rather than as part of a regular repertoire of shapes. Exceptions to this may have been occasional spates of increased manufacture — perhaps in the spring when the bees 'awakened' and resumed production in the hives and when beekeepers may have been looking to replace broken vessels or augment their apiaries for the production season. The existence of new beekeepers setting up an apiary must also be considered, when a potter may have been required to increase his production. Fabric analysis on these vessels, however, reveals that they did not travel very far, at least on evidence from Crete. Some areas with large numbers of extant hives, like Sphakia, do not have any indication of pottery production, although research on possible kiln sites ongoing (Francis *et al.*: forthcoming); the beehive fabrics do not designate a foreign — even as far away as east Crete — source.

On Crete, these vessels can be connected with transport amphorae on the basis of both their shapes and fabrics, and this parallel almost certainly extends to their manufacture as well. Not only would these vessels have been made in the same workshops, from the same batches of clay, but the end results were sturdy, fairly thick-walled, vessels that were created for specific purposes but could be re-used. Unlike amphorae, however, the morphological details of beehives are so far not well understood. But even in this distinction, one can reconstruct the hand of the potter who shapes the vessel on the wheel, presumably at some point under instruction from a beekeeper. Secure knowledge of such collaboration cannot be assumed for most other classes of pottery, and the scant evidence for it is therefore all the more valuable, even if so many other aspects of ceramic beekeeping equipment remain unknown.

Acknowledgements

This research has been supported by an enormous number of individuals. Particular thanks go to the directors of the Sphakia Survey, Jennifer Moody and Lucia Nixon, as well as Simon Price† and Oliver Rackham†. The following institute facilitated field research on Crete: The British School at Knossos, the Canadian Institute in Greece, Institute for Aegean Prehistory (Pacheia Ammos), the Gortyn Excavations and the Scuola Archeologica Italiana di Atene, the Eleutherna Excavations and the University of Crete, the KE' Ephoreia in Chania. Also on Crete, individual assistance was provided by Eleni Hatzaki, Elpida Hadjidaki, Loeta Tyree, Antonino Di Vita†, Kapua Iao, Matt Buell, Jerolyn Morrison, Vance Watrous, Barbara Hayden, Holly Raab, Anaya Sarpaki, Athanasia Kanta, Loeta Tyree, Tanya Yangaki, Stavroula Markoulaki, Vanna Niniou-Kindeli, Maria Vlazaki, and Georgos Farangkakis. I would also like to thank Virginia Anderson-Stojanović, John Ellis Jones, Susan Rotroff, Jutta Stroszeck, Kristina Jelinčić, Anne Bowtell, and, particularly, George W. Harrison. In the UK, I was lucky enough to meet with Eva Crane†, whose insights were particularly welcome. Drawings are by Matt Buell and Anne Bowtell, and I thank them both.

This research was funded by grants from Oxford University, the Eva Crane Trust, Concordia University and the Aylwin Cotton Foundation.

Bibliography

Primary Sources

Aristotle. *History of Animals, Volume III: Books 7-10.* Edited and translated by D. M. Balme. Loeb Classical Library 439. Cambridge MA: Harvard University Press, 1991.

Columella. *On Agriculture, Volume II: Books 5-9.* Translated by E. S. Forster, Edward H. Heffner. Loeb Classical Library 407. Cambridge MA: Harvard University Press, 1954.

Geoponika. Geoponika: Farm Work: A Modern Translation of the Roman and Byzantine Farming Book. Edited and translated by A. Dalby. Totnes: Prospect Books, 2011.

Palladius. *Palladius on Husbandrie. From the Unique MS. of about 1420 A.D. in Colchester Castle.* Edited by B. Lodge. London: Trübner and Co., 1873.

Pliny. *Natural History* (in multiple volumes). Translated by H. Rackham. The Loeb Classical Library, Cambridge MA: Harvard University Press, 1938-1952.

Pseudo-Quintilian. *The Major Declamations Ascribed to Quintilian. A Translation.* Translation by L. A. Sussman (Studien zur klassischen Philologie 27). Berlin: P. Lang, 1987.

Cato, Varro. *On Agriculture.* Translated by W. D. Hooper, Harrison Boyd Ash. Loeb Classical Library 283. Cambridge MA: Harvard University Press, 1934.

Virgil. *Eclogues. Georgics. Aeneid: Books 1-6.* Translated by H. Rushton Fairclough; revised by G. P. Goold. Loeb Classical Library 63. Cambridge MA: Harvard University Press, 1916.

Secondary Sources

Alexandri, O. 1972. Αχιλλέως και Ιάσονος 52 (οικόπεδον Κωνσταντινίδη). *ArchDeltion* 27, B1, *Chronika*: 32–35 pl. 39.

Anderson, J. K. 1954. Excavations on the Ridge of Kofina in Chios. *Annual of the British School at Athens* 49: 128–182.

Anderson-Stojanović, V. R. and J. E. Jones. 2002. Ancient Beehives from Isthmia. *Hesperia* 71: 345–376.

Ashton, N. 1991. *Siphnos. Ancient Towers B.C.* Athens: Eptalophos.

Barnard, K. A., et al.. 2003. The Neopalatial Pottery: A Catalogue, in K. A. Barnard and T. M. Brogan (eds) *Mochlos IB. Period III. Neopalatial Settlement on the Coast: The Artisans' Quarter and the Farmhouse at Chalinomouri. The Neopalatial Pottery* (Prehistory Monographs 8): 33–98. Philadelphia: INSTAP Academic Press.

Beck, C. W., et al. 2008. Absorbed Organic residues in Pottery from the Minoan Settlement of Pseira, Crete, in Y. Tzedakis, H. Martlew and M. K. Jones (eds) *Archaeology Meets Science. Biomolecular Investigations in Bronze Age Greece. The Primary Scientific Evidence 1997-2003*, 48–73. Oxford: Oxbow Books.

Bikos, T. and E. Rammou. 2002. Beehives of the Aegean Islands. *Bee World* 83: 5–13.

Bloedow, E. F. and C. Björk. 1989. The Mallia Pendant: A Study in Iconography and Minoan Religion. *Studi micenei ed egeo-anatolici* 27: 9–67.

Bonet Rosado, H. and C. Mata Parreño. 1997. The Archaeology of Beekeeping in Pre-Roman Iberia. *Journal of Mediterranean Archaeology* 10: 33–47.

Blackman, D., et al. 2010. *Miscellanea Apicula*, in N. Sekunda (ed.) *Ergastiria: Works Presented to John Ellis Jones on his 80th Birthday*, 150–158. Gdansk: Akanthina.

Boardman, J. 1956. Delphinion in Chios. *Annual of the British School at Athens* 51: 41–62.

Brock, J. K. and G. MacWorth Young. 1949. Excavations in Siphnos. *Annual of the British School at Athens* 44: 1–92.

Broneer, O. 1959. Excavations at Isthmia, Fourth Campaign, 1957–1958. *Hesperia* 28: 298–343.

Bruneau, P. 1970. Les objets. Les vaisselle, in P. Bruneau (ed.) *L'îlot de la Maison de Comédiens* (Délos 27): 239–262. Paris: Éditions E. de Boccard.

Chadwick, J. 1976. *The Mycenaean World.* Cambridge: Cambridge University Press.

Chrysoulaki, S. 2000. Εργαστηριακές εγκαταστάσεις στην περιοχή Ζάκρου, in Th. Detorakis and A. Kalokairinos (eds) *Πεπραγμένα του Η´ Διεθνούς Κρητολογικού Συνεδρίου (Ηράκλειο, 9-14 Σεπτεμβρίου 1996)* A3: 581–597. Herakleion: Historical Museum of Crete.

Crane, E. 1983. *The Archaeology of Beekeeping.* London: Duckworth.

Crane, E. 1994. Beekeeping in the World of Ancient Rome. *Bee World* 75: 118–34.

Crane, E. 1999. *The World History of Beekeeping and Honey Hunting.* London: Duckworth.

Crane, E. 2000. Prevention and Treatment of Disease and Pests of Honeybees: The World Picture. *New Zealand Beekeeper* 7: 5–8.

Cunliffe, B. 1971. *Fishbourne. A Roman Palace and its Garden.* Baltimore: The Johns Hopkins Press.

D'Agata, A. L. and S. De Angelis. 2014. Minoan Beehives: Reconstructing the Practice of Beekeeping in Bronze Age Crete, in G. Touchais (ed.) *Physis. 14ème rencontre égéenne internationale, Paris 11-14 Decembre 2012*: 349–358. Louvain: Peeters.

Dalby, A. 1996. *Siren Feasts. A History of Food and Gastronomy in Greece.* London: Routledge.

Dalby, A. 2011. *Geoponika: Farm Work: A Modern Translation of the Roman and Byzantine Farming Book.* Totnes: Prospect Books.

Davaras, C. 1986. A New Interpretation of the Ideogram *168. *Kadmos* 25: 38–43.

Day, L. P. 2008. Review of P. P. Betancourt (ed.) *The Chrysokamino Metallurgy Workshop and its Territory*, in *American Journal of Archaeology* 112: 767–768.

Day, L. P. 2012. Building Complex E, in L. P. Day and K. T. Glowacki (eds) *Kavousi IIB. The Late Minoan IIIC Settlement at Vronda. The Buildings on the Periphery* (Prehistory Monographs 38): 1–45. Philadelphia: INSTAP Academic Press.

Decavallas, O. 2007. Beeswax in Neolithic Perforated Sherds from the Northern Aegean: New Economic and Functional Implications, in C. Mee and J. Renard (eds) *Cooking up the Past. Food and Culinary Practices in the Neolithic and Bronze Age Aegean*: 148–157. Oxford: Oxbow.

Di Vita, A. 1988-1989. Atti della Scuola 1988-1989. *Annuario della Scuola archeologica di Atene e delle Missioni italiane in Oriente* 66-67: 427–483.

Evans, A. 1921-1936. *The Palace of Minos: A Comparative Account of the Successive Stages of the Early Cretan Civilization as Illustrated by the Discoveries at Knossos* (4 vols). London: Macmillan.

Evershed, R. P., et al. 1997. Fuel for Thought? Beeswax in Lamps and Conical Cups from Late Minoan Crete. *Antiquity* 71: 979–985.

Ferrence, S. C. and E. B. Shank. 2006. Evidence for Beekeeping, in P. P. Betancourt (ed.) *The Chrysokamino Metallurgy Workshop and its Territory* (*Hesperia* Suppl. 36): 391–392. Princeton: American School of Classical Studies in Athens.

Fitton, L., H. Granger-Taylor and S. Quirke. 1998. Northerners at Lahun: Neutron Activation Analysis of the Minoan and Related Pottery in the British Museum, in S. Quirke (ed.) *Lahun Studies*: 112–140. Reigate: SIA.

Flinders Petrie, W. M. 1890. *Kahun, Gurob, and Hawara*. London: K. Paul, Trench, Trübner and Co.

Francis, J. 2009. Ancient Ceramic Beekeeping Equipment at the University of Ottawa. *Mouseion* 9, 159–170.

Francis, J. 2011. The Domed Beehives of Roman Phalasarna. Oral Presentation at ΙΑ´ Διεθνές Κρητολογικό Συνέδριο (Ρέθυμνο 21-27 Οκτωβρίου 2011). Rethymnon, Greece.

Francis, J. 2012. Experiments with an old ceramic beehive. *Oxford Journal of Archaeology* 31: 143–159.

Francis, J. E. 2016. Apiculture in Roman Crete, in J. E. Francis and A. Kouremenos (eds) *Roman Crete: New Perspectives*: 83–100. Oxford: Oxbow Books.

Francis, J., E. Nodarou, and J. Moody. Forthcoming. The Fabrics of Roman to Early-Byzantine Cretan Amphorae from the Sphakia Survey, in M. C. Curtis and J. E. Francis (eds) *Change and Transition on Crete: Interpreting the Evidence from the Hellenistic Through to the Early Byzantine Period. Papers Presented in Honour of G. W. M. Harrison*. Oxford: Archaeopress.

Freed, R. E. 2001. The Tomb of Rekhmire, in K. R. Weeks (ed.) *Valley of the Kings: The Tombs and the Funerary Temples of Thebes West*: 384–389. Vercelli: White Star.

Gassner, V. 1997. *Das Südtor der Tetragonos—Agora. Keramik und Kleinfunde* (Forschungen in Ephesos 13/1/1). Vienna: Verlag der Österreichischen Akademie der Wissenschaften.

Giuman, M. 2008. *Melissa: archeologia delle api e del miele nella Grecia antica*. Rome: Bretschneider.

Gregory, T. E. 1985. An Early Byzantine Complex at Akra Sophia near Corinth. *Hesperia* 54: 411–428.

Gregory, T. E. 1986. Diporto: An Early Byzantine Maritime Settlement in the Gulf of Korinth. Δελτίον XAE 12 (1984), Περίοδος Δ´. Στην εκατονταετηρίδα της Χριστιανικής Αρχαιολογικής Εταιρείας (1884-1984): 287–304.

Hallager, B. P. 2003. The Pottery, in E. Hallager and B. P. Hallagher (eds). *The Greek-Swedish Excavations at the Agia Aikaterini Square Kastelli, Khania 1970-1987 and 2001*. Vol. III:1. *The Late Minoan IIIB:2 Settlement* (Skrifter utgivna av Svenska institutet i Athen 40). Jonsered: Åström.

Harissis, V. H. 2016. Beekeeping in Prehistoric Greece, in F. Hatjina, G. Mavrofridis and R. Jones (eds) *Beekeeping in the Mediterranean from Antiquity to the Present. International Symposium, Syros, October 9-11 2014*: 14–35. Nea Moudania: Division of Apiculture, Hellenic Agricultural Organization 'Demeter'-Greece; Chamber of Cyclades; Eva Crane Trust - UK.

Harissis, H. V. and A. V. Harissis. 2009. *Apiculture in the Prehistoric Aegean. Minoan and Mycenaean Symbols Revisited* (British Archaeological Reports—International Series 1958). Oxford: John and Erica Hedges Ltd.

Hasaki, E. 2002. *Ceramic Kilns in Ancient Greece: Technology and Organization of Ceramic Workshops*. Unpublished PhD Thesis, University of Cincinnati.

Hasaki, E. 2003. A Hellenistic Ceramic Workshop on Paros: Hellenistic Ceramic Technology and Production in Context (Abstract). *Archaeological Institute of America, 104th Annual Meeting Abstracts*: 71. Boston: Archaeological Institute of America.

Hasaki, E. and K. T. Raptis. 2016. Roman and Byzantine Ceramic Kilns in Greece (1st – 15th c. CE). Continuities and Changes in Kiln Typology and Spatial Organization of Production, in A. Cycyzza, B. M. Giannattasio and S. Pallecchi (eds) *Archeologia delle produzioni ceramiche nel mondo antico. Spazi, prodotti, strumenti e techniche. Atti del convegna (Genova, 1-2 Dicembre 2014)*: 209–229. Genoa: Aracne.

Hayes, J. W. 1971. Four Early Roman Groups from Knossos. *Annual of the British School at Athens* 66: 249–275.

Hayes, J. W. 1983. The Villa Dionysos Excavations, Knossos: The Pottery. *Annual of the British School at Athens* 78: 97–169.

Jameson, M. H., C. N. Runnels, and T. H. van Andel (eds) 1994. *A Greek Countryside: The Southern Argolid from Prehistory to the Present Day*. Stanford: Stanford University Press.

Jones, J. E. 1976. Hives and Honey of Hymettus. *Archaeology* 29: 80–91.

Jones, J. E. 1990. Ancient Beehives at Thorikos: Combed Pots from the Velatouri, in H. F. Musche (ed.) *Rapport Préliminaire sur les 13e, 14e, 15e et 16e Campagnes de Fouilles* (Thorikos 9, 1977/82): 63–71. Ghent: Comité des fouilles belges en Grèce.

Jones, J. E., A. J. Graham, and L. J. Sackett. 1973. An Attic Country House below the Cave of Pan at Vari. *Annual of the British School at Athens* 68: 355–452.

Kalogirou, K. and A. Papachristoforou. 2018. The Construction of Two Copies of Ancient Greek Clay Beehives and the Control of their Colonies' Homeostasis, in F. Hatjina, G. Mavrofridis and R. Jones (eds) *Beekeeping in the Mediterranean from Antiquity to the Present*: 69–78. Nea Moudania: Division of Apiculture, Hellenic Agricultural Organization 'Demeter'-Greece; Chamber of Cyclades; Eva Crane Trust - UK.

Karatasios, I., et al. 2013. Technological Insights into the Ancient Ceramic Beehive Production of Agathonisi Island, Greece. *Applied Clay Science* 82: 37–43.

Kataki, E. 2012. Δοκιμαστική Ανασκαφική Έρευνα στο Εθνικό Στάδιο Χανιών, in M. Andrianakis, P. Varthalitou and I. Tzachili (eds) *Αρχαιολογικό Έργο*

Κρήτης 2. Πρακτικά της 2ης Συνάντησης. Ρέθυμνο, 26-28 Νοεμβρίου 2010: 537–547. Rethymnon: University of Crete.

Knigge, U. 1976. *Der Südhügel* (Kerameikos 9). Berlin: de Gruyter.

Kritsky, G. 2007. The Pharaoh's Apiaries. *Kmt* 18: 163–169.

Kritsky, G. 2015. *The Tears of Re: Beekeeping in Ancient Egypt*. New York: Oxford University Press.

Lüdorf, G. 1998-1999. Leitformen der attischen Gebrauchskeramik: Der Bienenkorb. *Boreas* 22–22: 41–169.

Martin, A. 1997. Ceramica comune: vasi da mensa e da dispensa, in A. Di Vita and A. Martin (eds) *Gortina II. Pretorio. Il Materiale degli scavi Coloni 1970-1977* (Monografie della scuola archeologica di Atene e delle missioni Italiane in oriente 7): 291–345. Padua: Bottega d'Erasmo.

Massa, M. 1992. *La ceramica ellenistica con decorazione a relievo della bottega di Efestia* (Monografie della scuola archeologica di Atene e delle missioni Italiane in oriente 5). Rome, "L'Erma" di Bretschneider.

Mavrofridis, G. 2006. Η μελισσοκομία στον Μινωικό-Μυκηναϊκό Κοσμό. *Μελισσοκομική Επιθεώρηση* Sept.–Oct. 2006: 268–272.

Mavrofridis, G. 2008. Μελισσοκομία με αντίγραφα αρχαίων κυψελών. *Μελισσοκομική Επιθεώρηση* Jul.-Aug. 2008: 255.

Mavrofridis, G. 2009a. Το ακάπνιστο μέλι. *Μελισσοκομική Επιθεώρηση* May–Jun. 2009: 200–204.

Mavrofridis, G. 2009b. Μελισσοκομία με αντιγραφα αρχαίων κυψελών. Πρώτα συμπεράσματα. *Μελισσοκομική Επιθεώρηση* Mar.–Apr. 2009: 120.

Mavrofridis, G. 2009c. Η Μελισσοκομία στην Κύπρο (από τις αναρχές ως τις μέρες μας). *Μελισσοκομική Επιτεώρηση* Jan.–Feb. 2009: 33–38.

Mavrofridis, G. 2010. An Apiary of the 10th Century BC. *Bee World* 87: 36–37.

Mavrofridis, G. 2011. Η Μελισσοκομία στον Ρωμαϊκό Κόσμο. *Μελισσοκομική Επιθεώρηση* Jul.–Aug. 2011: 266–71.

Mavrofridis, G. 2015a. Dung made Beehives. *Bee World* 92: 85–86.

Mavrofridis, G. 2015b. Παραδοσιακη Μελισσοκομια. *Μελισσοκομική Επιθεώρηση* Mar.–Apr. 2015: 106–10.

Mavrofridis, G. 2015c. Η νομαδική μελισσοκομία πριν την έλευση της σύγχρονης κυψέλης. *Μελισσοκομική Επιθεώρηση* May–Jun. 2015: 176–79

Mavrofridis, G. 2016. Κυνηγοί μελιού κατά την Προϊστορική Εποχή. *Μελισσοκομική Επιθεώρηση* Jul.-Aug. 2016: 279–82.

Mavrofridis, G. 2017a. Μελισσοκομία με αντίγραφα αρχαίων κάθετων κυψελών με βάση τις αρχαίες κυψέλες της Ισθμίας. www.arxaiologia.gr (5/10/2017)

Mavrofridis, G. 2017b. Η παραδοσιακή μελισσοκομία της Θάσου. ΘΑΣΙΑΚΑ. Περιοδικη εκδοση της εταιριας θασιακων μελετων και της θασιακης ενωσης Καβαλας. Τομος δεκατος Ογδοος 2017. Ζ' συμποσιο Θασιακων μελετων. ή Θασος δια μεσου των Αιωνων: Ιστορια – Τεχνη – Πολιτισμος. Πρακτικα: 267-99. Kavala: n.p.

Mavrofridis, G. 2017c. Θασιακή Μελισσοκομία (16ος – 20ος αι.). *Μελισσοκομική Επιθεώρηση* Nov.–Dec. 2017: 432–437.

Mavrofridis, G. 2018a. Μέλισσα και μελισσοκομία στο έργο του Πλίνιου του Πρεσβύτερου. *Μελισσοκομική Επιθεώρηση* Mar.–Apr. 2018: 119–124.

Mavrofridis, G. 2018b. Stone Beehives on the Islands of the Eastern Mediterranean, in F. Hatjina, G. Mavrofridis and R. Jones (eds) *Beekeeping in the Mediterranean from Antiquity to the Present*: 126–135. Nea Moudania: Division of Apiculture, Hellenic Agricultural Organization 'Demeter'-Greece; Chamber of Cyclades; Eva Crane Trust - UK.

Mavrofridis, G. 2018c. Urban Beekeeping in Antiquity. *Ethnoentomology* 2: 52–61.

Mavrofridis, G. 2018d. Ελληνιστικά πώματα κυψελών για προστασία των μελισσών από τη *Vespa orientalis*, in E. Kostou (ed.) *9th Scientific Meeting on Hellenistic Pottery. Thessaloniki December 5-9 2012*: 825–848. Athens: Archaeological Receipts Fund.

Mavrofridis, G. 2019. Η μελισσοκομία της Σαλαμίνας στη διαχρονία. *Μελισσοκομική Επιθεώρηση* Jan.–Feb. 2019: 40–44.

Mazar, A. 2018. The Iron Age Apiary at Tel Rehov, Israel, in F. Hatjina, G. Mavrofridis and R. Jones (eds) *Beekeeping in the Mediterranean from Antiquity to the Present*: 40–49. Nea Moudania: Division of Apiculture, Hellenic Agricultural Organization 'Demeter'-Greece; Chamber of Cyclades; Eva Crane Trust - UK.

Mazar, A. and N. Panitz-Cohen. 2007. It is the Land of Honey: Beekeeping at Tel Rehov. *Near Eastern Archaeology* 70, 202–229.

Mazar, A., D. Namdar, N. Panitz-Cohen, R. Neumann and S. Weiner. 2008. Iron Age Beehives at Tel Rehov in Jordan valley. *Antiquity* 82, 629–639.

Mee, C. and H. Forbes (eds) 1997. *A Rough and Rocky Place: The Landscape and Settlement History of the Methana Peninsula, Greece*. Liverpool: Liverpool University Press.

Melas, M. 1999. The Ethnography of Minoan and Mycenaean Beekeeping, in P. P. Betancourt, V. Karageorghis, R. Laffineur and W.-D. Niemeier (eds) *Meletemata. Studies in Aegean Archaeology presented to Malcolm H. Wiener as he enters his 65th Year* (Aegaeum 20): 485–491. Austin: University of Texas at Austin, Program in Aegean Scripts and Prehistory.

Metzger, I. R. 1993. Keramik und Kleinfunde. In P. Ducrey, I. R. Metzger, and K. Reber (eds). *Le Quartier de la Maison aus mosaïques* (Eretria. Fouilles et recherches 8): 97–143. Lausanne: Payot.

Metzger, I. R. 1998. Keramik und Lampen. In K. Reber (ed.) *Die klassischen und hellenistischen Wohnhäuser im Westquartie* (Eretria. Ausgrabungen und Forschungen 10): 173–227. Lausanne: Payot.

Moody, J., et al. 2003. Ceramic Fabric Analysis and Survey Archaeology: The Sphakia Survey. *Annual of the British School at Athens* 9: 37–105.

Moody, J., J. E. Morrison and H. L. Robinson. 2012. Earth and Fire: Cretan Potting Traditions and Replicating Minoan Cooking Fabrics, in E. Mantzourani and P. P. Betancourt (eds) *Philistor. Studies in Honor of Costis Davaras* (Prehistory Monographs 36): 119–31. Philadelphia: INSTAP Academic Press.

Papanikola-Bakirtzi, D. 2002. *Everyday Life in Byzantium.* Athens: Hellenic Ministry of Culture.

Peignard-Giros, A. 2012. Les céramiques communes à Délos à l'époque hellénistique tardive (IIe-Ier siècles av. J.-C.). In C. Batigne Vallet (ed.) *Les céramiques communes dans leur contexte régional: faciès de consommation et mode d'approvisionnement: actes de la table ronde organisée à Lyon les 2 et 3 février 2009 à la Maison de l'Orient et de la Méditerranée* (Travaux de la Maison d'Orient et de la Méditerranée 60): 243–256. Lyon: Maison de l'Orient et de la Méditerranée.

Perakos, B. X. 1992. Ανασκαφή Ραμνούντος. *Praktika tis Archaiologikis Etaireias 1989*: 5–6

Pologiorghi, M. I. 1998. Ἑμνημεία του δυτικού νεκροταφείου του Ορωπού: Οικόπεδο Οργανισμού Σχολικών Κτιρίων. Athens: Archaeological Receipts Fund.

Raab, H. A. 2001. *Rural Settlement in Hellenistic and Roman Crete: The Akrotiri Peninsula* (British Archaeological Reports—International Series 984). Oxford: Archaeopress.

Rendini, P. 1988. Ceramica acroma, in A. Di Vita (ed.) *Gortina I* (Monografie della Scuola archeologica di Atene e delle missioni italiane in Oriente 3): 229–252. Rome: "L'Erma" di Bretschneider.

Rotroff, S. I. 2006. *Hellenistic Pottery. The Plain Wares.* (Athenian Agora 33). Princeton: University Press.

Siebert, G. 1988. 2. Quartier de Skardhana. A. La maison des Sceaux, in A. Pariente *et al.* Rapport sur les travaux de l'éecole française en Grèce en 1987: 755–767. *Bulletin de correspondance hellénique* 112: 697–791.

Sparkes, B. A. and L. Talcott. 1970. *Black and Plain Pottery of the 6th, 5th, and 4th Centuries B.C.* (Athenian Agora 12). Princeton: University Press.

Sutton, R. F. Jr. 1991. Ceramic Evidence for Settlement and Land Use in the Geometric to Hellenistic Periods, in J. F. Cherry, J. L. Davis and E. Mantzourani (eds) *Landscape Archaeology as Long-Term History. Northern Keos in the Cycladic Islands from Earliest Settlement until Modern Times* (Monumenta Archaeologica 16): 245–263. Los Angeles, UCLA Institute of Archaeology.

Tölle-Kastenbein, R. 1974. *Das Kastro Tigano: Die Bauten und Funde griechischer, römischer, und byzantinischer Zeit* (Samos 14). Bonn, Habelt

Triantafyllidis, P. 2010. *Agathonisi on the Frontier. The Archaeological Investigations as Kastraki (2006-2010).* Athens: Prefecture of the Dodecanese.

Vogt, C. 2000. The Early Byzantine Pottery, in P. G. Themelis (ed.) *Προτοβυζαντινή Ελεύθερνα*, τομέας 1.2: 37–199. Rethymnon: University of Crete.

Vroom, J. 2003. *After Antiquity. Ceramics and Society in the Aegean from the 7th to the 20th Century. A Case Study from Boeotia, Central Greece* (Archaeological Studies Leiden University 10). Leiden, Faculty of Archaeology, Leiden University.

Watrous, L. V. 1992. *The Late Bronze Age Pottery* (Kommos 3). Princeton: University Press.

Watrous, L. V. 2012. Appendix A. Catalogue of Sites, in L. V. Watrous, D. Haggis, K. Nowicki, N. Vogeikoff-Brogan and M. Schultz, *Archaeological Survey of the Gournia Landscape: A Regional History of the Mirabello Bay, Crete, in Antiquity* (Prehistory Monographs 37): 105–133. Philadelphia: INSTAP Academic Press.

Wilkinson, C. K. and M. Hill. 1983. *The Metropolitan Museum of Art's Collection of Facsimiles. Egyptian Wall Paintings.* New York: The Metropolitan Museum of Art.

Woudhuizen, F. C. 1997. The Bee-Sign (Evans no. 86): An Instance of Egyptian Influence on Cretan Hieroglyphic. *Kadmos* 36: 97–110.

Yangaki, A. G. 2004-2005. Amphores crètoises: le cas d'Éleutherna, en Crète. *Bulletin de Correspondance Hellénique* 128–129: 503–523.

Yangaki, A. G. 2005. *Le céramique des IVe-VIIIe siècles ap. J.-C. d'Eleutherna: sa place en Crète et dans le bassin égéen.* Paris: Scripta.

Chapter 4

Etruscan 'Honey Pots': Some Observations on a Specialised Vase Shape

Paolo Persano

paolo.persano@outlook.com

Summary: The present chapter discusses a specific Etruscan vase shape the 'olle stamnoidi a colletto'. This peculiar shape, developed in Etruria during the Hellenistic period, is characterised by a distinctive rib that creates an empty channel around the entire circumference of the vase, close to the mouth. Revisiting two contributions published by the author several years ago, the chapter examines the typological connection of these Etruscan vases with Iberian 'potes meleiros' and Greek 'melopitari' (vases devoted to honey storage according to ethnographic comparisons). It also discusses the contexts of their discovery in Etruria (tombs and, rarely, votive or domestic deposits) and the typological variation of these vases.

Keywords: Honey storage jars; vase typology; Etruria; pre-Roman Italy; modern Greece; Iberian Peninsula; ethnoarchaeology; archaeometry

Towards an Archaeology of Honey in Etruria

In recent years, the archaeology of beekeeping and honey has been the subject of numerous contributions focused on the different aspect of the production cycle of honey, a substance which has been overlooked for too long by archaeologists.[1] The debate on honey production and on the material documentation available to reconstruct it – from hive construction to consumption patterns – has thus been newly enriched by data which can be integrated with literary sources (Morel 1877; Lafaye 1904; Bortolin 2008 appendix I).

This framework stresses the role of honey even outside its main use as an ancient foodstuff, used to sweeten and preserve food (André 1981: 89, 143, 186-189). Indeed we can identify many 'secondary' functions of honey in medicine as an antiseptic, for healing, for the treatment of gastrointestinal afflictions, and as an excipient for other drugs (Nielsen 1993; Cilliers and Retief 2008; Bormetti 2014: 32-33) and in artisan production – for instance, its use as a purple fixative (Vitr. *De Arch.* 7. 13. 1-2; Plut. *Alex.* 36. 2-3, 686b; Kardara 1961: 264; Bortolin 2008: 27). The by–product of wax, fundamental in antiquity, had uses that went beyond mere lighting (Regert et al. 2001; Beretta 2009; Notari 2012: 25).

Archaeological studies on honey have not focused only on technological analysis and on the reconstruction of ancient material culture: the ideological and iconographic aspects of this substance have also been investigated, providing further proofs of its importance for the cultural heritage of the populations of the ancient Mediterranean (Burn 1985; Cherici 1991; Vázquez Hoys 1991; Giuman 2008).

As in the rest of the Mediterranean basin, honey also played a significant role in the Etruscan world. In this case, as with many other perishable products and foodstuffs, we must acknowledge the almost complete lack of any archaeological record: only a few literary sources and general considerations on the ancient world can give us an idea of the importance of honey at the time.

In this regard, it is worth recalling two particularly notable cases, which reflect the high degree of specialisation reached by Etruscan beekeeping already in the archaic period. The excavation of the Etruscan settlement of Forcello–Bagnolo San Vito (Mantua) has brought to light the carbonised remains of a beehive in what has been identified as a workshop (room 3 of building F2). The archaeometric analyses of these remains particularly of the pollen (Castellano et al. 2017) – found in a context that can accurately be dated to between 510 and 495 BCE (De Marinis 2016) – suggest a form of itinerant beekeeping along waterways similar to that documented for nearby Ostiglia by Pliny (*NH* 21.43.73). The hives, installed on river barges, would have collected pollen from different environments: after their voyage, they would have been transferred into a workshop on land for the extraction and storage of the honey.

---

[1] For an overview of archaeological traces of ancient beekeeping: Crane 1983; Balandier 2004; Bortolin 2008 (with a full discussion); Harissis and Harissis 2009; Bortolin 2012; Bormetti 2014; Morín de Pablos and de Almeida 2014; Hatjina et al. 2018.

Shifting from an artisan / domestic context to a funerary one, mention can be made of tomb A in the Casa Nocera cemetery near Casale Marittimo (PI). In this high–status, rich grave from the Orientalising period, a beehive has been discovered inside a specific bronze container with an elongated shape, decorated with a fringe of small pendants (Esposito 1999: 56, 90-91). This discovery bears witness to the ritual importance of honey in connection with burials from as early as the Orientalising period, as also documented by literary sources in the Greek world (Gamurrini 1917; Sacco 1978: 81; Giuman 2014; Baldoni 2013: 144-145).

This contribution aims to examine the stamnoid collared jar (*olla stamnoide a colletto*), a very specific vase shape well attested in Etruria during the Hellenistic period, the use of which is possibly connected with the storage of honey (Figures 1.a-b; Figure 2a-d). Traditionally considered to be cineraria, these vases have been identified as containers for viscous substances such as honey (Shepherd 1992: 161; Jolivet 2010; Bormetti 2014: 35-36). Archaeological and ethnographic comparisons with morphologically related vessels – the so-called Iberian *potes meleiros* and the Greek *melopithari* – reinforce the hypothesis that these Etruscan vases also had a similar function (Jolivet 2010: 16-17 discussing only contemporary Greek vases; Morais 2014a: 96; Morais 2014b; Persano 2014-2015; Persano 2016; Ambrosini 2016: 453-459). I will here discuss the shape of these pots and then examine their (chronological and topographic) diffusion in Etruria, before advancing some general conclusions.

Collared Vases as Honey Pots

Let us start by discussing the shape of these vases. Clearly, according to the current criteria for typological definition, a single formal attribute (in this case, the collar) is not enough to identify a particular vase type. Combined with other elements, however, it can help define an ancient functional form, enabling us to understand the functional aspects of a given group of vases marked by this particular attribute.

By stamnoid collared jar I mean vases that, while typologically different (from a strictly typological point of view), can be traced back to an ideal model, insofar as they present all the following features:

1) they belong to the functional macro–category of storage pots, which is to say common pottery, which is often decorated with red glaze motifs and – in some rare instances – totally covered in black glaze. We must therefore rule out kitchenware, which was meant to be exposed to fire and whose fabric was therefore designed to withstand high temperatures. Two horizontal handles were always added to make it easier to move these vases;

2) they are closed shapes: globular and ovoid pots with a rim that is much smaller in diameter than their point of maximum expansion. In the case of most of the surviving specimens, the upper section of the neck, just below the rim, is ribbed in such a way as to aid the attachment of the lid. In some cases the fastening of the lid was further ensured by the addition of small sockets under

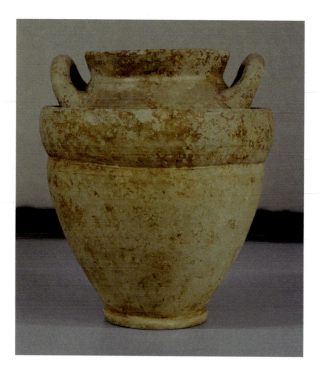

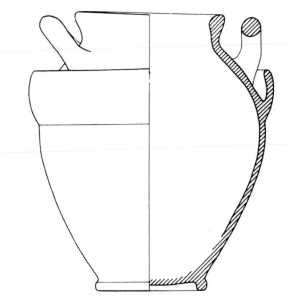

Figure1.a Florence, MAN inv. 11908 (photo courtesy Museo Archeologico Nazionale di Firenze, Ministero della Cultura) – b. from Shepherd 1992, fig. 27.

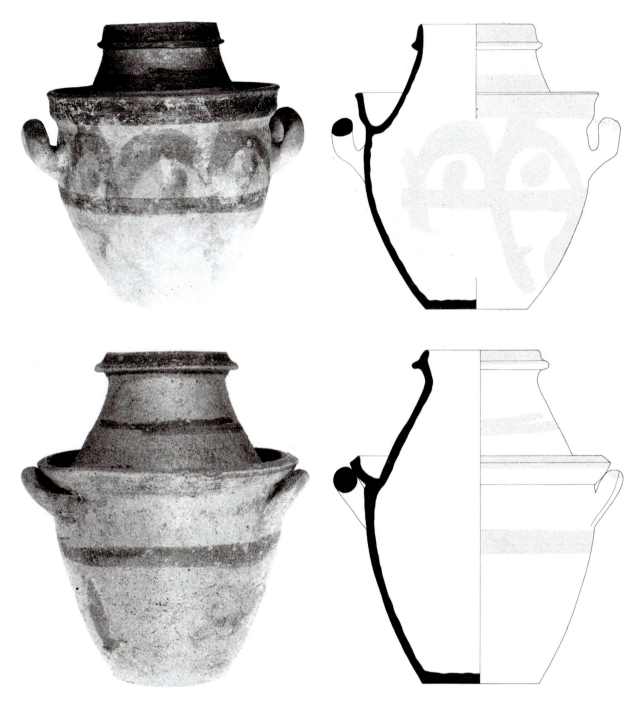

Figure 2.a-b. (from Serra Ridgway 1996: pl. LIII e CL); c-d. (from Serra Ridgway 1996: tav. CLXV).

the ribbing, through which a rope must have been fitted e.g. Florence, MAN, inv. 77512: Persano 2016: 14, no. 52 (Martelli 1976: 47, Fig. 10; Shepherd 1992: 174); Sevres, Ambrosini 2016: 94, no. 10;

3) they are marked by the *fundamental* attribute of a collar. This is an applied clay element, arranged concentrically with respect to the rim and designed to create a seamless channel with a shape and depth suitable for the collection of a liquid substance. The collar is only partly moulded on the body of the vase during its crafting: the vertical walls are created *by adding* more clay.

This collar is a consciously selected formal attribute. Hence, we should not associate these products with the highly common modifications made to the shape of a vase either to aid the attachment of the lid or for aesthetic reasons, to give it a more pleasant shape.

On the basis of good enough evidence, we have thus defined a set of closed containers used for preservation and marked by a collar. This small (and inexpensive) variation during the crafting of an ordinary storage pot i.e. the addition of an external channel – enabled the creation of a new, useful container. But for what purpose?

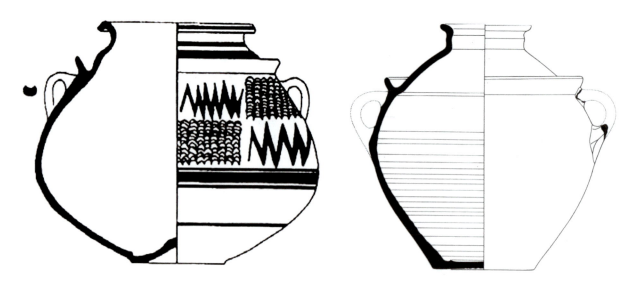

Figure 3.a (from Garcia Cano 1995, fig. 5). b (from Delgado 1998).

One answer may be found by resorting to ethnographic analogy, a crucial tool in archaeological research (Vidale 2004; Gallay 2011; Marciniak and Yalman 2013).[2]

Pots of exactly the same shape are to be found in the traditional (i.e. modern) vase repertoires of Portugal, Spain, and Greece, where their use as honey pots is ethnographically documented – even by the names given to these containers. The collar, filled with water, prevents insects (especially ants) from making their way into the pot through the rim and spoiling its contents.

One puzzling element highlighted by E.J. Shepherd in a personal remark is the evaporation of water, given that we are dealing here with a warm Mediterranean climate. How often would it have been necessary to add water, in order to maintain it at a sufficient level to keep insects out? Probably, the only way to reach valid conclusions on the matter is through an experimental study.

Numerous studies, particularly by Iberian archaeologists, have noted very close formal analogies between these modern vases and some pots documented in the same areas from at least the Iron Age to the Roman Imperial one (the first of these studies was Manuela Delgado's work on specimens from Bracara Augusta proper (Delgado 1996-1997; already on this shape, see also Fletcher Valls 1953). For these vases too, which have been unearthed in archaeological contexts and stand out for the presence

of a collar, a primary use as honey containers has been hypothesised. Archaeometric analyses have further confirmed the presence of honey residues, at least in some of these ancient cases (Oliveira and Morais 2014: 40, 44-45; Oliveira et al. 2014). Collared vases of this sort are found in:

The Iberian Peninsula: continuity in the use of double–handled storage vessels (Figure 3) from the Iron Age to the Roman Age, probably extending through the Middle Ages into the Modern Age (Delgado 1996-1997; Fernandes 2003: 69, 95-99; Morais 2006: 151-156; Gomes 2007; Morais 2011). It is in the Modern Age that we find them defined as honey pots (*potes meleiros, pucheiros miel, cántaros de miel, vasos meleiros, talhas meleiras*). In particular, a collared jar from the museum collection in Ventimiglia has been convincingly interpreted as being an Iberian import and dated to the Roman Imperial Age (Figure 4). The vessel in question bears a notable inscription that confirms its use as a honey pot: *Tes(ta) p(ondo libras) VIII, mel (pondo libras) VII (uncias) II* (Gandolfi and Mennella 2012).

Greece (Crete and Sifnos?): use, only in the modern/ contemporary age, of collared vessels in two different forms: *kouroupa* for household consumption (Figure 5) and *pithari* for consumption on a larger scale. Production is limited to Crete (certainly the centres of Thrapsano and Margarites); specimens are on display at the museums of Neapolis (Aghios Nikolaos) and Boroi (Vallianou and Padouva 1986: 40,45; Persano 2016: 17, figs. 10-12), and at the G. Psaropoulos foundation in Athens – but all these vessels are Cretan (Persano 2014-2015: 117-120). A survey of ethnographic museums in the Cyclades has not yielded any positive result, with the exception of Sifnos, although it is likely that the collared vase here is of Cretan origin (Persano 2016: 17-18; fig. 13).

---

[2] On the concept of analogy, see Stiles 1977; Yalman 2005. The first ethno-archaeological studies have been: Hampe and Winter 1962; Hampe and Winter 1965.

Figure 4. from Gandolfi, Mennella 2012, fig. 1.

Figure 5. Athens, Κέντρο Μελέτης Νεώτερης Κεραμεικής (P. Persano).

Ancient Italy: the stamnoid jars here under discussion, dated to between the 4th and the 2nd cent. BCE. Their shape would not appear to have survived into more recent ages. However, it is important to bear in mind that not all 'collared pots' can be assigned to this vase shape (understood as a theoretical model). In this respect, one might mention some Bronze Age Nuragic pots (Leonelli and Campus 2000: 518, 601, pl. 354) some with a perforated rim, like the jar from the village assembly hut at La Prisgiona (Antona et al. 2010: 1727-1728, fig. 14.); vases unearthed in France, which can be assigned to the *modelée varoise* production (Rivet 1982: 260, shape 21; Bérato 2009: 398-399, shape 1626); and collared vessels from North Africa (Mukai 2016: 178-179, context 115.10, fig. 119.10 (2nd half of the 5th cent. CE) or the Levant in the Middle Ages (Stern 1997: 40-42, nos. 31-33, fig. 5).

What deserve a completely separate treatment are the cylindrical or truncated–conical vases dated to between the late Hellenistic period and the 2nd cent. CE: while having a collar around the rim, they present holes connecting the inside of the vase to the area marked by the collar. The main features of these pots, known as *Milk boiler* or *Milchkocher* (although they do not withstand fire), is the connection between the channel and the material preserved inside the body of the vase (Figure 6) (Santrot and Santrot 1995, 189-192, nos. 511-512; Hayes 2000: 295; Hayes 2008, 255-256, no. 1477; Puppo 2018).

Etruscan collared jars

Considering the undeniable formal similarities between stamnoid collared jars and the Iberian vessels

just described, it has been hypothesised that these Etruscan vases too may have served as honey pots, as E.J. Shepherd had already suggested in the 1980s on the basis of considerations of a different kind (Shepherd 1992). Again without any claim to exhaustiveness, I can now integrate the list I previously presented with these additional items:

1) Sèvres, Cité de la Céramique S.N. 459. Published as being of Cycladic origin. Since the inventory number–and hence provenance–has proven wrong, it is possible to accept E. Di Paolo Colonna and G. Colonna's hypothesis that the vessel is Etruscan (note the marked similarities with the specimen from Orvieto) (Massoul 1934, p. 13, IIBb1, tab. 6.1 (SN 459); Di Paolo Colonna and Colonna 1978: 243; Persano 2014-2015: 119-120, fig. 8).

2-3) Ardea, votive deposit of Casarinaccio (Ten Kortenaar 2005: 262-263). An almost complete specimen (AC65DEP133– h. 11.4 cm; rim diam. 9.1 cm) and a sherd that can be assigned to a different specimen (AC65DEP134).

4) Proceno, Collezione Cecchini (Michetti 2003: 161-162, fig. 31).

5) University of Pennsylvania Museum, inv. MS 2858 (from Tenuta di Montebello ai Poggialti, Tarquinia): h. 26.1 cm; rim diam. 9.6 cm (Becker et al. 2009: 92).

6) University of Pennsylvania Museum, inv. MS 2859 (from Tenuta di Montebello ai Poggialti, Tarquinia): h. 26.4 cm; rim diam. 11.4 cm (Becker et al. 2009: 92-93).

7) University of Pennsylvania Museum, inv. MS 2860A (from Tenuta di Montebello ai Poggialti,

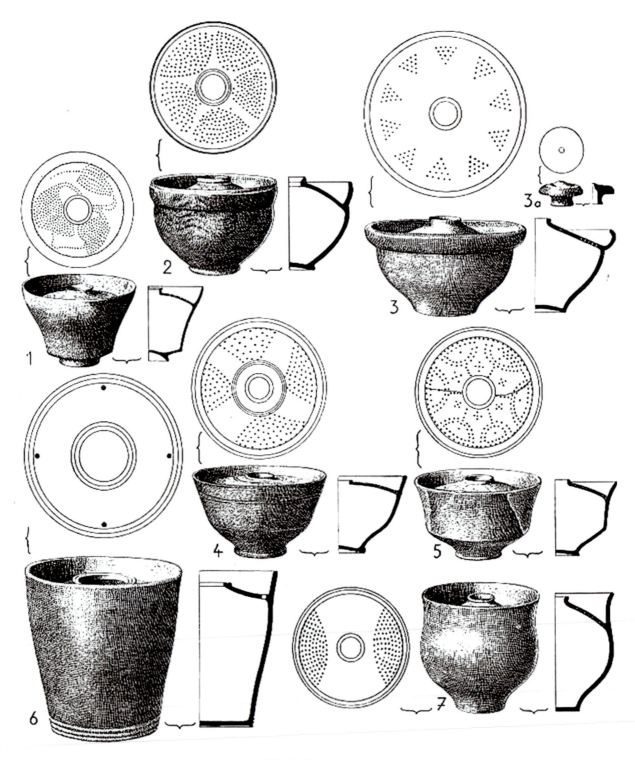

Figure 6. Milkboiler from Puppo 2018, fig. 2.

Tarquinia): h. 29.3 cm; rim diam. 11 cm (Becker et al. 2009: 95).

8)   Rofalco, room 2b, 'honey pot' sherds (Sabbatini 2014: fig. 1.101; Diidel et al. 2016).

9)   Norchia, tomb PA 36, sherds that can be assigned to the same vase from the dromos of the tomb (Colonna Di Paolo and Colonna 1978: 303-304, nos. 47-50).

10-12) Norchia, PB1 (from excavation of the funerary chamber 21, 23, 24/7/1971), sherds that can be traced back to three specimens (Ambrosini 2016: 94, nos. 9, 10 (h. 24.9 cm), 11 (h. 29.2 cm); tabs. 50, 66, 420).

13) Norchia, PB4 (dig opposite the excavation area, 13/9/1972), four sherds from the same sample (Ambrosini 2016: 132, nos. 1-4, tab. 85).

14) Norchia, PB13, wall sherd with a hinge, possibly from a honey pot (Ambrosini 2016: 224, no. 4).

15) Norchia, PB 28 excavation of the funerary chamber, 29/9/1972 (Ambrosini 2016: 314, no. 1, tab. 299).

16) Norchia, PB 35 (excavation of the *dromos*, 29/9/1972) two sherds that can be assigned to the same specimen (Ambrosini 2016: 326, no. 1, tab. 312, 440).

17) Florence, Antiquities Market.[3]

18) Florence MAN, Passerini Collection from Lucignano, Camporsi (Persano 2020).

The stamnoid collared jars are a kind of ware chiefly found in southern Etruria: most specimens come from the city of Tarquinia and its environs, but the overall distribution area extends to interior and northern Etruria. The southernmost find was made in Minturno (Persano 2016: 14, n. 61; Kirsopp Lake 1934–1935: 104–105, type 25, tabs. 15–16; Shepherd 1992: 174, no. 28), the northernmost ones in Lucca (Persano 2016: 14, nos. 58-59; Custer 1928: 29, fig. 3; Shepherd 1992: 174, nos. 35-36) and (not surely) Genoa (Persano 2016: 14, no. 62; Milanese, Mannoni 1984: 143, fig. 9.9; Shepherd 1992, 174, no. 37).

As regards the possible centres of production, as common and household items collared jars were manufactured in several local workshops by different craftsmen, along with other storage pots, the fabric and decorative motifs of which they share.

We find traces of these centres in four major areas: the countryside around Tarquinia, Minturno (a likely kiln discard), the Siena area, and possibly northern Etruria, which is believed to be the place of origin of the specimen found in Genoa. The identification of this sherd as part of a collared jar is far from certain: all that is left of the vase is a rim sherd with some ribbing (the context, moreover, is 5th–century BCE, which would make it the earliest example of this shape in Etruria). While it presents similarities with the rims of collared jars, it could also be assigned to other shapes.

Tarcento is instead unlikely the place of origin of one of the two black-glaze jars (The other black-glaze vase is: Persano 2016: 13, no. 37 (Barbieri 2003: 227, Fig. 2), recently discussed (Figure 7) (Persano 2016: 14, no. 60; van Ingen 1933: 36, tab. XIX.15; Shepherd 1992: 174, no. 38; Morais 2014a: 96–97; Morais 2014b: 257). However, the scholar who published the item was far from certain that it originated from this centre in the Friulian Prealps (van Ingen 1933: 36 'said to be from Tarcento'; Ambrosini 2016, proposing Taranto). We should bear

Figure 7. Courtesy University of Michigan, Kelsey Museum of Archaeology.

its complex ownership history in mind: before its acquisition by the university of Pennsylvania (1923), this vase was part of the archaeological collection of the University of Marburg.

On the basis of the chronology of the archaeological context, these vases have been dated to the Hellenistic period, between the late 4th and the 2nd cent. BCE. A more precise dating is usually impossible: frequently each tomb contains several depositions, and it is difficult to assign each artefact to an individual context.

The excavation data available for some specimens chiefly document their provenance from graves, where the conservation circumstances have enabled the survival of almost complete vases. The many almost complete vases that have found their way into museum collections without any clear information about their archaeological context must have the same origin. In the case of sherds, collar fragments may be mistaken for rims from larger vessels, as has occurred with the specimens from Cosa (Persano 2016: 14, nos. 55-56; Dyson 1976: 88, fig. 28; Shepherd 1992: 174, nos. 32-33), and as is frequently the case with unusual vase shapes (Di Gennaro and Depalmas 2011: 59).

Because of the presence of charred human bones inside some of these pots, it has been suggested that the vase shape in question was assigned the function of a cinerary urn (Martelli 1976). In the Fondo Scataglini necropolis, excavated in relatively recent times and documented in detail, the best-preserved specimens

³ Collection declared by the Ministry 'di eccezionale importanza' (D.M. 29.08.1997), Itineris sale catalogue 28.06.2018: 27-34 (lot. 38), especially fig. pag. 30.

were filled with bones and ashes (as well as personal ornaments in many cases), and closed with lids that were not specifically suited to or properly matching the vessels: generally, lids of the common conical type (Serra Ridgway 1996: 266). I shall mention only a few cases, by way of example.

Tomb 4857 (83), combining inhumation and cremation in a jar (Linington and Serra Ridgway 1997: 57; Persano 2016: 12, no. 15); tomb 5070 (112), in the intact lower chamber of which–with 13 inhumations and 4 cremations–two cremations in a jar occur alongside one in a single–handled jar, another in a jar with an enlarged rim and only the last in a collared jar (Linington and Serra Ridgway 1997: 72-73; Persano 2016: 12, nos. 27-28). The same phenomenon also occurs in tomb 4823 (163) of the same necropolis, where one of the three cremations found is in a collared jar, while the other two are in ordinary jars (Linington and Serra Ridgway 1997: 108; Persano 2016: 12, no. 12).

Again within the Tarquinia area, the cremations from Tenuta di Montebello ai Poggialti, dated to around 300 BCE, have been closely investigated (Becker et al. 2009: 90-96): three were in collared jars (MS 2858- man ? around the age of 65 / MS2859 – around the age of 28 / MS2860A – woman 45 +/- 10 years), one was in an *olpe* (MS2860 – mature adult male), and possibly another was in a bronze container. It is likely that these urns are associated with two conical lids (similar to ones found elsewhere): according to archival documents, these urns were sealed (Becker et al. 2009: 91).

Collared jars were also used for depositions in the burials at Campo della Madonna near San Geminiano

di Moriano: they contained bones and ashes, and one of them still had a conical lid with a button handle, similar to those found on the jars from Castel d'Asso (Persano 2016: 13, no. 38; Hayes 1985: 163; Shepherd 1992: 174, no. 18.Two cases are particularly notable on account of the presence of inscriptions: the collared jar (Figure 8) from the Tommasi Collection (found in the Cvelne tomb at Montaperti) (Persano 2016: 14, no. 51; Cristofani 1979: 180, no. 15, Fig. 143/15; CIE: 42, no. 237 on the context Cristofani 1979: 179-183; Belfiore 2015: 25, n. 16, pl. 5) and the soon–to–be published one from the Passerini collection (*supra*). The latter was retrieved in the late 19th century on the Farnetella farm at Camporsi, near Lucignano, with other pots. It had been used as a cinerary urn in the Sethrni tomb. Described as 'biancastro, che presenta un'altra sporgenza, come orlo, nella curva media e maggiore del corpo,' the vase can only be identified through its inscription (Gamurrini 1899: 218).

The small collared jar from tomb XX at Musarna was part of the grave goods (Figure 9); the archival documents record the discovery of burned human remains in other pots, generic ovoid jars of a larger size (Persano 2016:13, no. 40; Emiliozzi 1974: 56-57, tab. XIX; Shepherd 1992: 174, no. 19).[4] The same intended use can be assumed for the specimen from Populonia, which was found in a pit grave (Shepherd 1992: 174, endnote 66). Also in the case of tomb 1/1984 at Monteriggioni (Casone, Poggio Malabarba), the collared jar is only part of the grave goods, as the cremated ashes were stored in normal ovoid jars (Del Segato 2018: 141, no 143, pl. XXI.1).

The funerary purpose of many of these jars has been safely established, and it is shared by other household

Figure 8.a from Cristofani 1979, fig. 143/15 – b from Belfiore 2015, pl. 5.

---

[4] On the context, see Emiliozzi 1974:55: 'tali olle di forma ovoidale a larga bocca (alt. 0,25-0,30) e di argilla rossigna mal depurata, racchiudevano ancora i residui di ossa umane bruciate'.

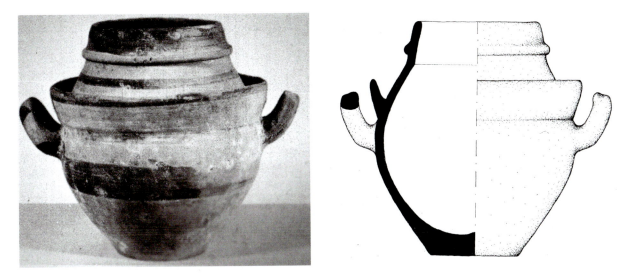

Figure 9. Stamnoid collared jar from Musarna t. XX (from Emiliozzi 1974, pl. XIX).

vessels, such as jugs and the more common globular or ovoid jars, which clearly had a secondary use as cineraria (Serra Ridgway 1996: 266). The funerary purpose of the vessels, therefore, appears to depend more on the archaeological context than on any specificity of their shape compared to similar vessels (jars) of ordinary ceramic used as urns (Martelli 1976: 47; Serra Ridgway 1986: 327; Serra Ridgway 1996: 266; Cavagnaro Vanoni 1996).

At the end of their life–cycles, some of these jars came to be used as containers for burned human remains: but it is likely that in this context the importance of honey in relation to funerary rituals must be assigned considerable weight (Gamurrini 1917; Sacco 1978: 81; Giuman 2014; Baldoni 2013: 144-145). The deposition of human remains in pots preferentially reserved for the storing of this substance used in rites of passage and connected to the afterlife might be explained by its deeper symbolic significance.

Mention can be made of further finds from domestic, or at any rate non–funerary, contexts. One unusual case is the votive deposit of Casarinaccio (Ardea), where the only fully preserved specimen is the smallest stamnoid jar ever discovered (possibly a miniature artefact serving a votive function?). One specimen has been found at Musarna, in the fill of cistern 642 (Persano 2016: 13, no. 39; Jolivet 1999: 482, fig. 2 MU642044; Jolivet 2010: 16-17). Other sherds have instead been found in context at Cosa (Persano 2016: 14, nos. 55-56; Dyson 1976: 88, fig. 28; Shepherd 1992: 174, nos. 32-33) and Pyrgi (Persano 2016: 13, nos. 43-44; Serra 1970: 525, no. 97; Shepherd 1992: 174, no. 29 / Serra 1970: 646, no. 7; Shepherd 1992: 174, no. 30).

The most interesting case, however, is the findspot in Rofalco: inside a room in the housing district in an area SW of the fortress (room 2B), some honey pot

sherds have been found *in situ*. Preserved in the same context were both *dolia* for the storing of foodstuffs and *gliriaria*, pots used for the raising of dormice. It is possible, therefore, to reconstruct what must have been the usual context for stamnoid collared jars, namely interior domestic spaces set aside for activities related to the production, storing or processing of foodstuffs.

When Multifunctionality is the Rule: Honey Pots in Antiquity

As we have seen, it may be assumed that the Etruscan collared vases were preferentially designed for the storage of honey. However, it must be stressed, once again, that there is little point in regarding this as their *one and only* purpose. It is necessary to reflect on the fluidity between vase shapes and vase functions in the pre–industrial world, particularly in the case of storage vessels, which were produced locally and in a non–standardised way. The clear–cut correspondence between given vessel shapes and specific functions derives from contemporary standardised and industrial mass–production and does not really apply to Antiquity.

The circulation and use of domestic objects was based on practicality: products frequently changed function according to people's needs. In the life–cycle of pottery (production–consumption contexts–obliteration), change of function would appear to have been the norm, and it is sometimes documented by alterations resulting from the use of the vessels themselves (Sullivan 1989; Mannoni and Giannichedda 1996: 148-154). In light of the available evidence, therefore, the hypothesis of the existence of a standard shape for honey pots, *exclusively* reserved for the storage of this substance, seems unlikely, given the following reasons:

1) the almost complete loss of vessels made of perishable materials, such as wood or organic animal materials,

and the systematic under-representation of vessels made of recyclable materials, such as glass and metal. Containers made of organic materials probably played an important role in the storing and transportation of foodstuffs. In the case of beekeeping, we should bear in mind that the sources unanimously recommend the use of hives made of perishable materials such as bark, ferula, and wicker (Plin., *HN* 21. 80; Columella, *R.R.* 9. 6. 1-2; Varro, *R.R.* 3. 16. 15), whereas the only objects of this kind to have reached us are terracotta or stone ones (Bormetti 2014: 16-17);[5]

2) the actual use of a much wider and more generic range of pots for the transportation, storage, and selling of honey. The literary sources only focus on unusual elements in the production cycle. Columella, for instance, describes a ceramic utensil for smoking hives, whereas he only mentions *vasa fictilia* in relation to the storage of honey (Columella, *R.R.* 9. 15. 13). Although the written sources often refer to pots designed for the storage of honey, at the present state of the research no references to specific types of vases have been found (Bortolin and Bruno 2006). Even the *vasa mellaria* mentioned by Pliny (*HN* 21. 82) are in all likelihood hives, judging from a Varro passage.[6] Macrobius by contrast uses the same term (*mellarium*) to refer to a container for storing honey (Macrob., *Sat.* 1. 12. 25).

Inscriptions sometimes indicate the contents of a vessel (and sometimes even its weight), but do not allow any generalisations. In the Greek world, amphoras with the engraved inscription M / ME have been interpreted as containers reused for honey or honeyed wine, equivalent to the Romans' *mulsum* (Lawall 2000: 18-19). Not even the small amphora from Syracuse with a painted inscription (MELIKRATA), and reserved for the storage of a beverage made from a mixture of milk and honey, can be considered a honey pot in and of itself (Biondi and Manganaro 2010: 58-61, fig. 7).[7] The engraved inscription on a large pot from Halos, in Thessaly, presents a clearer indication of its contents but, given the size of the vessel, could not have been designed for the storage or transportation of limited amounts of the substance (Reinders and Prummel 2003: 264, P257, appendix 1, fig. 6.3).

In the Roman world too, we find few precious inscriptions referring to honey: they only occur on roughly a dozen specimens (Bortolin and Bruno 2006; Bortolin 2008: 175-180; Bassi 2008). These pots vary considerably in size, ranging from the small amphora from the House of Menander in Pompeii (*CIL* IV, 9421;

Allison 2006: 109, no. 588, tab. 43.1) to the double-handled jar from Arcole (Bortolin and Bruno 2006), from the kantharoid shape from Somma Lombardo (Olcese 1998: 202-203, tab. CXXXII) to the aforementioned jar from Ventimiglia. In the case of those specimens which, on the basis of surviving texts, can be regarded as reserved for the storage of honey, the inscriptions would appear to have been engraved *post cocturam*, meaning that they were added at the moment of using the vessels, to clarify their contents or their weight during commercial transactions.

While these engraved inscriptions are not always easy to read or to interpret, they *unambiguously* attest to the fact that a widely diverse range of vessels, lacking any obvious shared features, were used as honey pots at some point in their life cycles (Bortolin and Bruno 2006; Peña 2007, 103-105; Bassi 2008; Gandolfi and Mennella 2012). This sample offers us a glimpse of the *actual* circulation of a commonly used product in a world where reuse was the norm.

In order to get an overall picture of what pots were used for the storage of honey, it is important to bear in mind that other vessels in the ancient world are believed to have served as honey pots, as in the case of the Iberian *kalathoi-sombreros de copa* (Del Chiaro 1973: 65). However, the intended use of these vessels, which are also documented in Italy, is far from clear: their rim would appear unsuited to ensuring a tight seal and the hypothesis that these pots contained salted products and dried fruit seems equally plausible.[8] The archaeometric analyses conducted on some of these pots from the site of Torrelló del Boverot de Almazora (Castellón) have identified residues of a concoction made from honey and fleshy fruits or must (Juan-Tresserras 2000).

Wax traces have also been found in *kalathoi* either imported from Iberia or of Iberian imitation in Milan (Casini et al. 2015). Other storage vessels labelled as 'honey pots' in the past (Mercedes Vegas Vegas 1973: 115-117, fig. 41, 'tipo 48'; *CIL* XIIII 10008.44, and later described as *Honigkrüge* in the bibliography, are actually ordinary double-handled multifunctional artefacts (Schindler Kaudelka 1989: 43-44).

These caveats do not change the fact that, within the range of possible functions assigned to collared vases (Cf. Szilágyi 2009: 867), the *preferential* use of these

---

[5] On beehives: Morais 2006: 156-157. Grecia: Jones *et al.* 1973: 443-452; Lawall *et al.* 2001: 176-177; Anderson-Stojanović and Jones 2002; Rotroff 2006: 124-131 (with a full discussion); Francis 2012; Mavrofridis 2013.

[6] Varro, *Rust.* 3.16.12: '*melittona ita facere oportet, quos alii melitrophia appellant, eandem rem quidam mellaria*' (ed. C. Guiraud, 1997).

[7] On this specific beverage: Bortolin 2008: 24.

[8] On Iberian *kalathoi* in Italy: Bruni and Conde 1991; Bencivenga Trillmich 1985 (Velia); Conde Berdós 1998 (Liguria, plus general remarks); Bruni 1993 (Etruria); Bertini 2004 (Populonia); Muscolino 2006 (Sicily); Casini, Tizzoni 2012 (northern Italy). On the contents of *kalathoi*: Tarradell and Sanmartí 1980: 315; Montanya Maluquer and Puig Ochoa 1979. For the 'honey hypothesis', see Bonet Rosado and Mata Parreño 1997: 43-44; Bonet Rosado and Mata Parreño 2008: 156-158.

vessels was to store and transport valuable liquids. Honey, the main sweetener in Antiquity, no doubt played a primary role among such products; however we should not underestimate the versatility of ancient vessels.

Conclusions

What model explains the presence of pots of this sort in different areas, in different periods? Are we dealing with a case of polygenesis or are we to imagine that these models (both physical models–i.e. artefacts–and mental ones, i.e. the idea of adding particular attributes to the standard shape) spread out from one specific area?

It is difficult to define a historical framework for Etruscan collared vases: for we risk either downplaying the possibility that contacts of this kind took place, leading to the transition of the pot shape in question from one area to another, or assuming that it is impossible for the idea of a trivial alteration of this sort to have come up in different areas and at different times.

In the specific case of Etruria, collared vases crop up at a specific moment and are chiefly found in the area around Tarquinia, a major centre for Mediterranean trade in the 4th and 3rd centuries BCE (Bruni 1992; Conde 2005; Jolivet 2010). To the best of my knowledge, there is no evidence of the existence of this vase shape in previous periods, nor have any traces have been found of the endurance of the *olla a colletto* shape in Etruria in the Roman age. Given the well–documented importance and stability of the trade relations between Etruria and the Iberian world (e.g. Gori 2006; Della Fina 2007), it seems difficult not to attribute the spread of this shape to cultural interaction between the two areas, although further research is required to shed light on this hypothesis.

Acknowledgements

I would like to express my gratitude to the editor of this book for his kind invitation to participate. In my research on collared jars, on which this present chapter builds, I have become deeply indebted to many people whom I would now like to thank, such as: A. Ciacci, D. Gandolfi, L. Gambaro, M. Iozzo, E. Giannichedda, N. Liaros, M.R. Luberto, C. Nervi, and E.J. Shepherd.

Bibliography

Allison, P. M. (ed.) 2006. *The Insula of the Menander at Pompeii III: the Finds a Contextual Study*. Oxford: Oxford University Press.

Ambrosini, L. 2016. *Le necropoli dell'Etruria meridionale, 3. Norchia II*. Rome: CNR - Consiglio Nazionale Ricerche.

Anderson–Stojanović, V. R. and J.E. Jones. 2002. Ancient Beehives from Isthmia. *Hesperia* 71.4: 345–376.

André, J. 1981. *L'alimentation et la cuisine à Rome*[2]. Paris: Belles Lettres.

Antona, A., M. D. M. Corro and S. Puggioni. 2010. Spazi di lavoro e attività produttive nel villaggio nuragico La Prisgiona in località Capichera (Arzachena), in C. Vismara M. Milanese and P. Ruggeri (eds) *L'Africa romana. I luoghi e le forme dei mestieri e della produzione nelle province africane. Atti del XVIII convegno di studio, Olbia, 11-14 dicembre 2008*: 1713-1733. Rome: Carocci.

Balandier, C. 2004. L'importance de la production du miel dans l'économie gréco–romaine. *Pallas* 64: 183–196.

Baldoni, D. 2013. Riti, usi e corredi funerari a Iasos in epoca ellenistica, in D. Baldoni, F. Berti and M. Giuman (eds) *Iasos e il suo territorio: Atti del Convegno internazionale per i cinquanta anni della Missione Archeologica Italiana (Istanbul, 26-28 febbraio 2011)*: 135-160. Rome: Bretschneider Giorgio.

Barbieri, G. 2003. Considerazioni sulla ceramica in uso a Norchia nel III secolo a.C. attraverso un corredo inedito da una tomba del fosso Pile. *Rivista di Studi Liguri* LXIX: 225–255.

Bassi, C. 2008. Un contenitore per miele da Tridentum, in P. Basso, A. Buonopane, A. Cavarzere and S. Pesavento Mattioli (eds) *Est enim ille flos Italiae... Vita economica e sociale nella Cisalpina romana*: 287-294. Verona: QuiEdit.

Becker, M. J., J. Machintosh Turfa and B. Algee-Hewitt. 2009. *Human Remains from the Etruscan and Italic Tomb Groups in the University of Pennsylvania Museum*. Pisa-Rome: F. Serra.

Belfiore, V. 2015. Le iscrizioni della tomba dei Cvenle a Montaperti (SI) nella letteratura del XVIII e XIX sec.: nuove acquisizioni. *Studia Oliveriana* 2013-2015: 11–35.

Bencivenga Trillmich, C. 1985. La ceramica iberica da Velia. Contributo allo studio della diffusione della ceramica iberica in Italia. *Madrider Mitteilungen* 25: 20–33.

Beretta, M. 2009. Usi scientifici della cera nell'antichità. *Quaderni Storici* 130: 15–33.

Bertini, E. 2004. Sombreros de Copa, in M. L. Gualandi and C. Mascione (eds) *Materiali per Populonia* 3: 143-146. Florence: Ed. all'Insegna del Giglio.

Biondi, G. and G. Manganaro. 2010. Relitti epigrafici per la storia del vino, di droghe e del miele nella Sicilia tardoellenistica. *Epigraphica LXXII* 1/2: 51–67.

Bonet Rosado, H. and C. Mata Parreño. 1997. The Archaeology of Beekeeping in Pre-Roman Iberia. *Journal of Mediterranean Archaeology* 10.1: 33–47.

Bonet Rosado, H. and C. Mata Parreño. 2008. Las cerámicas ibéricas. Estado de la cuestión, in D. Bernal Casasola and A. Ribera i Lacomba (eds) *Cerámicas hispanorromanas. Un estado de la cuestión*: 147-169. Cádiz: Universidad de Cádiz.

Bormetti, M. 2014. Api e miele nel Mediterraneo antico. *Acme. Annali della Facoltà di Lettere e Filosofia dell'Università degli studi di Milano* 67.1: 7–50.

Bortolin, R. 2008. *Archeologia del miele.* Mantua: SAP.

Bortolin, R. 2012. Arnie, miele e api nella Grecia antica. *Rivista di Archeologia* XXXV: 149–166.

Bortolin, R. and B. Bruno. 2006. Il graffito 'melis' su un vaso di Arcole (VR). Considerazioni sui contenitori da miele nell'antichità, in E. Bianchin Citton and M. Tirelli (eds), *... ut... rosae ponerentur. Scritti di archeologia in ricordo di Giovanna Luisa Ravagnan*: 113–124. Rome: Quasar-Canova.

Bruni, S. 1993. Presenze di ceramica iberica in Etruria. *Rivista di Studi Liguri* LVIII: 37–65.

Bruni, S. and M. J. Conde. 1991. Presencia ibérica en Etruria y el mundo itálico a través de los hallazgos cerámicos de los siglos III–I a.C. In J. Remesal and O. Musso (eds) *La presencia de material etrusco en la Península Ibérica*: 543–576. Barcelona: Universitat de Barcelona/Sezione di Studi Storici "Alberto Boscolo."

Burn, L. 1985. Honey pots. Three white-ground cups by the Sotades Painter. *Antike Kunst* 28: 93–105.

Casini, S. and M. Tizzoni. 2012. Kalathoi iberici e loro imitazioni nella Mediolanum celtica. *Notizie Archeologiche Bergomensi* 18: 165–178.

Castellano, L., et al. 2017. Charred honeycombs discovered in Iron Age Northern Italy. A new light on boat beekeeping and bee pollination in pre-modern world. *Journal of Archaeological Science* 83: 26–40.

Cavagnaro Vanoni, L. 1996. Tombe tarquiniesi di età ellenistica. Catalogo di ventisei tombe a camera scoperte dalla Fondazione Lerici in località Calvario. Rome: "L'Erma" di Bretschneider.

Cherici, A. 1991. Granaî o arnie? Considerazioni su una classe fittile attica tra IX e VIII secolo a.C. *Atti dell'Accademia Nazionale dei Lincei. Classe di scienze morali, storiche e filologiche. Rendiconti* 44 (1989): 215–230.

Chiesa, F. 2005. *Tarquinia. Archeologia e prosopografia tra ellenismo e romanizzazione.* Rome: L'Erma" di Bretschneider.

Cilliers, L. and F. P. Retief. 2008. Bees, Honey and Health in Antiquity. *Akroterion* 53: 7–19.

Conde Berdós, M. J. 1998. La cerámica ibérica de Albintimilium y el tráfico mediterráneo en los siglos II–I a.C. *Rivista di Studi Liguri* LXII (1996): 115–168.

Crane, E. 1983. *The Archaeology of Beekeeping.* London: Duckworth.

Cristofani, M. (ed.) 1979. *Siena: le origini. Testimonianze e miti archeologici.* Florence: Olschki.

Custer, A. 1928. Lucca–Rinvenimenti archeologici a Ponte a Moriano. *Notizie degli scavi di antichità. Accademia Nazionale dei Lincei* 1928: 28–31.

Del Chiaro, M. A. 1973. An Iberian Sherd in Yougoslavia. *American Journal of Archaeology* 77: 65–66.

Delgado, M. 1996–1997. Potes meleiros de Bracara Augusta. *Portugália* n.s. XVII–XVIII: 149–165.

Della Fina, G. M. (ed.) 2007. *Etruschi, Greci, Fenici e Cartaginesi nel Mediterraneo centrale.* Atti del XIV Convegno Internazionale di Studi sulla Storia e l'Archeologia dell'Etruria, Rome: Quasar.

Del Segato, V. 2018. IV.2.1 Casone, podere Malabarba, tomba 1/1984, in G. Baldini et al. (eds) *Monteriggioni prima del castello. Una comunità etrusca in Valdelsa*: 139–156. Pisa: Piacini Editore.

De Marinis, R. C. 2016. La datazione della fase F del Forcello di Bagnolo San Vito, in S. Lusuardi Siena, C. Perassi, F. Sacchi, M. Sannazaro (eds) *Archeologia classica e post-classica tra Italia e Mediterraneo. Studi in onore di Maria Pia Rossignani*: 159–172. Milan: Vita e Pensiero.

Di Gennaro, F and A. Depalmas. 2011. Forni, teglie e piastre fittili per la cottura: aspetti formali e funzionali in contesti archeologici ed etnografici, in F. Lugli, A. S. Stoppiello, S. Biagetti (eds) *Atti del 4° Convegno Nazionale di Etnoarcheologia (Roma 17–19 maggio 2006), Oxford*: 56–61. Oxford: Archaeopress.

Di Paolo Colonna, E. and G. Colonna. 1978. *Norchia I.* Rome: Consiglio Nazionale delle Ricerche.

Djiel S, C. Phelps and M. Sabbatini. 2016, Uno spazio abitativo nella fortezza di Rofalco: primi risultati dall'ambiente 2B, poster in *Incontro internazionale di studi-Società e innovazione nell'Etruria di IV-III secolo a.C. Bolsena 21-22 Ottobre 2016.*

Dyson, S. L. 1976. Cosa: the Utilitarian Pottery. *Memoirs of the American Academy in Rome* 33: 3–175.

Emiliozzi, A. 1974. *La collezione Rossi Danielli nel Museo civico di Viterbo*, Rome: Consiglio nazionale delle ricerche, Centro di studio per l'archeologia etrusco-italica.

Esposito, A. M. (ed.) 1999. *Principi Guerrieri. La necropoli etrusca di Casale Marittimo.* Milan: Electa.

Fletcher Valls, D. 1953. Una nueva forma en la cerámica ibérica de San Miguel de Liria (València). *Zephyrus* IV: 187–191.

Francis, J. 2012. Experiments with an Old Ceramic Beehive. *Oxford Journal of Archaeology* 31: 143–159.

Gamurrini, G. F. 1899. III. Sinalunga–Tombe etrusche con oggetti della suppellettile funebre, scoperte nella fattoria della Farnetella. *Notizie degli Scavi* 1899: 217–219.

Gamurrini, G. F. 1917. Di una iscrizione del territorio di Venosa. *Atti dell'Accademia Nazionale dei Lincei. Classe di scienze morali, storiche e filologiche. Rendiconti* s.5 XXVI: 98–102.

Gandolfi, D. and Mennella, G. 2012. Un vaso 'meleiro' con iscrizione graffita da Ventimiglia. In G. Baratta e S. M. Marengo (eds) *Instrumenta Inscripta III. Manufatti iscritti e vita dei santuari in età romana*: 327–342. Macerata: EUM Edizioni Università di Macerata.

Giuman, M. 2008. *Melissa. Archeologia delle api e del miele nella Grecia antica.* Rome: G. Bretschneider.

Giuman, M. 2014. Γλαυκος πιων μελι ανεστη. Ritualità e simbologia del miele nel mito di Glauco, in I. Baglioni (ed.) *Sulle rive dell'Acheronte. Costruzione e percezione della sfera del post mortem nel Mediterraneo antico, 2. L'antichità classica e cristiana*: 75-87. Rome: Quasar.

Gomes, M. V. 2007. Vaso meleiro, de idade sidérica, dos Arrifes do Poço (Aljezur, Algarve). *Conimbriga* 46: 73–88.

Gori, S. (ed.) 2006. *Gli Etruschi da Genova ad Ampurias, atti del XXIV convegno di Studi Etruschi ed Italici, (Marseille-Lattes, 26 settembre - 1 ottobre 2002).* Pisa: Istituti editoriali e poligrafici internazionali.

Hampe, R. and A. Winter. 1962. *Bei Töpfern und Töpferinnen in Kreta, Messenien und Zypern.* Mainz: Mainz Verlag des Römisch-Germanischen Zentralmuseums.

Hampe, R. and A. Winter. 1965. *Bei Töpfern und Zieglern in Süditalien, Sizilien und Griechenland.* Mainz: Mainz Verlag des Römisch-Germanischen Zentralmuseums.

Harissis, H. V. and A. V. Harissis. 2009. *Apiculture in the prehistoric Aegean. Minoan and Mycenaean symbols revisited.* Oxford: BAR IS 1958.

Hatjina, F., G. Mavrofridis and R. Jones (eds) 2018. *Beekeeping in the Mediterranean from Antiquity to the Present.* Nea Moudania: Division of Apiculture, Hellenic Agricultural Organization 'Demeter'- Greece; Chamber of Cyclades; Eva Crane Trust - UK.

Hayes, J. W. 1985. *Etruscan and Italic Pottery in the Royal Ontario Museum. A catalogue.* Toronto: Royal Ontario Museum.

Hayes J. W. 2000. From Rome to Beirut and beyond. Asia Minor and eastern Mediterranean trade connections. *ReiCretActa* 36: 285–297.

Hayes J. W. 2008. *The Athenian Agora, 32. Roman pottery, fine-ware imports.* Princeton, NJ: The American School of Classical Studies in Athens.

Jolivet, V. 1999. De la prospection géophysique à la publication : tombes hellénistiques de Tarquinia. *Journal of Roman Archaeology* 12.2: 476–482.

Jolivet, V. 2010. Commerce, échanges, objets erratiques come marqueurs de rapports culturels? *Bollettino di Archeologia on line*: 12–19.

Jones, J. E., A. J. Graham and H. Sackett. 1973. An Attic Country House below the Cave of Pan at Vari. *The Annual of the British School at Athens* 68: 355–443.

Juan-Tresserras, J. 2000. Estudio de contenidos en cerámicas ibéricas del Torrelló de Almazora (Castellón). *Archivo Español de Arqueología* 73: 181–182, 103–104.

Kardara, C. 1961: Dyeing and Weaving Works at Isthmia. *American Journal of Archaeology* 65: 261–266.

Kirsopp Lake, A. 1934–1935. Campana Suppellex (The Pottery Deposit at Minturnae). *Bollettino dell'Associazione internazionale degli Studi Mediterranei* V, 4–5, 1934–1935: 97–114.

Lafaye, G. 1904. Mel. Daremberg–Saglio III.2: 1701–1706. Paris: Paris, Librairie Hachette.

Lawall, M. L. 2000. Graffiti, Wine Selling, and the Reuse of Amphoras in the Athenian Agora, ca. 430 to 400 B.C. *Hesperia* 69.1: 3–90.

Lawall, M. L., et al.. 2001. Notes from the Tins: Research in the Stoa of Attalos, Summer 1999. *Hesperia* 70.2: 163–182.

Leonelli, V. and F. Campus. 2000. *La tipologia della ceramica nuragica. Il materiale edito.* Viterbo: BetaGamma.

Linington R.E. and F. R. Serra Ridgway. 1997. *Lo scavo nel fondo Scataglini a Tarquinia. Scavi della Fondazione Ing. Carlo M. Lerici del Politecnico di Milano per la Soprintendenza archeologica dell'Etruria meridionale,* Milan: Fondazione Ing. Carlo M. Lerici del politecnico di Milano.

Mannoni, T. and E. Giannichedda. 1996. *Archeologia della produzione.* Turin: Einaudi.

Martelli, M. 1976. Review of Emiliozzi 1974. *Prospettiva* 4: 42–49.

Massoul, M. 1934. *Corpus Vasorum Antiquorum, France. Musée National de Sèvres.* Paris: H. Champion.

Mata Parreño, C. and H. Bonet Rosado. 1992. La cerámica ibérica. Ensayo de tipología. Estudios de Arqueología Ibérica y Romana, in *Homenaje a Enrique Pla Ballestrer*: 117-174. Valencia: Servicio de Investigación Prehistórica, Diputación Provincial de Valencia.

Mavrofridis G. 2013. Κυψέλες κινητής κηρήθρας στην αρχαία Ελλάδα. *Αρχαιολογική Εφημερίς* 152: 15–27.

Michetti L. M. 2003. Proceno. Un insediamento di confine tra i territori di Vulci, Orvieto e Chiusi. *AnnFaina* 10: 153–189.

Milanese, M. and T. Mannoni. 1984. Gli Etruschi a Genova e il commercio mediterraneo. *Studi Etruschi* 52: 117–146.

Montanya Maluquer, R. and M. R. Puig Ochoa. 1979. La cerámica ibérica pintada tardía y sus perduraciones. *Rivista di Studi Liguri* 45: 221–230.

Morais, R. 2006. Potes meleiros e colmeias em cerâmica: uma tradição milenar. *Saguntum* 38: 149–161.

Morais, R. 2014a. Notícia sobre vaso grego destinado ao transporte e conservação de mel. *Portugália* 35: 95–100.

Morais, R. 2014b. News about a Greek vase used to transport and conserve honey. P. Bádenas de la Peña et al. (eds) *Per speculum in aenigmate. Miradas sobre la antigüedad. Homenaje a Ricardo Olmos*: 256–258. Madrid: Asociación cultural hispano-helénica.

Morel, Ch. 1877. Apes. in C. Daremberg and E. Saglio (eds) *Dictionnaire des antiquities grecques et romains* I.1: 304–305. Paris: Hachette.

Morín de Pablos, J. and R. R. de Almeida. 2014. La apicultura en la Hispania romana: producción, consumo y circulación, in M. Bustamante Álvarez and D. Bernal Casasola (eds) *Artífices idóneos : artesanos, talleres y manufacturas en "Hispania": reunión científica, Mérida (Badajoz, España), 25-26 de octubre, 2012*: 269-294. Mérida: Consejo Superior de Investigaciones Científicas, Instituto de Arqueología.

Muscolino, F. 2006. Kalathoi iberici da Taormina. Aggiornamento sulla diffusione della ceramica iberica dipinta in Sicilia. *Archivo Español de Arqueología* 79: 217–224.

Nielsen, E. R. 1993. Honey in Medicine, in J. Leclant (ed.) *Sesto congresso internazionale di Egittologia, (Torino,*

1-8 settembre 1991): 415–419. Turin: International Association of Egyptologists.

Notari, S. F. 2012. *Ceramica e alimentazione. L'analisi chimica dei residui organici nelle ceramiche applicata ai contesti archeologici*. Bari: Edipuglia.

Olcese, G. (ed.) 1998. *Ceramiche in Lombardia tra II a.C. e VII d.C. Raccolta dei dati editi*. Mantua: SAP.

Oliveira, C. and R. Morais. 2014. Estudos de cromatografia aplicados à arqueologia romana: apresentação de resultados preliminares, Ciências e técnicas do Património. *Revista da Faculdade de Letras* XIII: 37–60.

Oliveira, C., et al. 2014. Análise de fragmentos cerâmicos de potes meleiros e colmeias por cromatografia gasosa acoplada à espectroscopia de massa, in R. Morais, A. Fernández and M. J. Sousa (eds) *As produções cerâmicas de imitação na Hispania, II Congresso Internacional da SECAH-EX OFFICINA HISPANA*: 599–610. Porto: FLUP.

Peña, J. T. 2007. *Roman Pottery in the Archaeological Record*. Cambridge: Cambridge University Press.

Peroni, R. 1998. Classificazione tipologica, seriazione cronologica, distribuzione geografica. *Aquileia Nostra* LXIX: 10–27.

Persano, P. 2014–2015. Vasi 'a colletto' come contenitori per miele. Alcune considerazioni. *Rivista di Studi Liguri* 80–81: 113–124.

Persano, P. 2016. Vasi da miele in Etruria. Confronti archeologici ed etnografici per le olle stamnoidi 'a colletto'. *Archivo español de arqueología* 89: 9–24.

Persano, P. 2020. 100. Olla a colletto acroma (vaso da miele), in M. Iozzo and M.R. Luberto (eds) *Tesori dalle terre d'Etruria. La collezione dei conti Passerini, Patrizi di Firenze e Cortona*: 197-199. Livorno: Sillabe.

Puppo, P. 2018. Milk cookers o semplicemente vasi-filtro? Una problematica ancora irrisolta, in C. Boschetti and M. Cavalieri (eds) *Multa per aequora. Il polisemico significato della moderna ricerca archeologica. Omaggio a Sara Santoro*: 309-320. Louvain: Louvain-la-Neuve / Presses Universitaires de Louvain.

Regert, M., et al. 2001. Chemical Alteration and Use of Beeswax through Time. Accelerated Ageing Tests and Analysis of Archaeological Samples from Various Environmental Contexts. *Archaeometry* 43.4: 549–569.

Reinders, H. R. and W. Prummel (eds) 2003. *Housing in New Halos. A Hellenistic Town in Thessaly, Greece*. Exton, Pa.: Swets & Zeitlinger Publishers.

Rotroff, S. 2006. *Hellenistic Pottery. The Plain Wares, The Athenian Agora XXXIII*. Princeton: American School of Classical Studies at Athens.

Sabbatini, M. 2014. I granai di Vulci. I magazzini della fortezza di Rofalco, in S. Neri (ed.), *Tecniche di conservazione e forme di stoccaggio in area tirrenica e Sardegna*: 111-126. Rome: Officina Edizioni.

Sacco, G. 1978. Due note epigrafiche. 2. Il miele e la cera nelle iscrizioni funerarie. *Rivista di Filologia e d'Istruzione Classica* 106.1: 77–81.

Schindler Kaudelka, E. 1989. Die gewöhnliche Gebrauchskeramik vom Magdalensberg. Klagenfurt: Klagenfurt Verlag des Landesmuseums für Kärnten.

Serra, F. R. 1970. La ceramica fine con tracce di decorazione dipinta etc. *Notizie degli scavi di antichità. Accademia Nazionale dei Lincei* 1970 (II suppl.) II: 504–552.

Serra Ridgway, F. R. 1986. B) La necropoli del Fondo Scataglini (Villa Tarantola), in M. Bonghi Jovino (ed) *Gli Etruschi di Tarquinia*: 324–340, 350–353. Modena: Panini.

Serra Ridgway, F. R. 1996. *I corredi del fondo Scataglini a Tarquinia. Scavi della Fondazione Ing. Carlo M. Lerici del Politecnico di Milano per la Soprintendenza archeologica dell'Etruria meridionale*. Milan: Ed Et.

Shepherd, E. J. 1992. Ceramica acroma, verniciata ed argentata, in A. Romualdi (ed) *Populonia in età ellenistica. I materiali dalle necropoli, atti del seminario (Firenze 30 giugno 1986)*: 152-178. Florence: Il Torchio.

Sullivan, A. P. 1989. The Technology of Ceramic Reuse: Formation Processes and Archaeological Evidence. *World Archaeology* 21.1: 101–114.

Tarradell, M. and E. Sanmartí. 1980. L'état actuel des études sur la céramique ibérique, in *Céramiques Hellénistiques et Romaines*, Tome 1: 303-330. Besançon: Université de Franche-Comt.

Ten Korteenar, S. 2005. Ceramica depurata acroma, in F. Di Mario (ed) *Ardea. Il deposito votivo di Casarinaccio*: 261-270. Rome: Soprintendenza per i beni archeologici del Lazio.

van Ingen, W. 1933. *Corpus Vasorum Antiquorum*. United States of America, University of Michigan, Fascicule 1. Cambridge (Massachusetts): Harvard University Press.

Vallianou C. and M. Padouva. 1986. Τα κρητικά αγγεία του 19ου και 20ου αιώνα Μορφολογική-κατασκευαστική μελέτη, Βώροι 3. Athens: Μουσείο Κρητικής Εθνολογίας.

Vázquez Hoys, A. M. 1991. La miel, alimento de eternidad. Alimenta. Estudios en homenaje al Dr. Michel Ponsich. *Gerión Anejos* III: 61–93.

Vegas M. 1973. *Cerámica común romana del Mediterráneo occidental*. Barcelona: Universidad de Barcelona, Instituto de Arqueología y Prehistoria.

Chapter 5

Palynological Insights into the Ecology and Economy of Ancient Bee-Products: A Contribution to the History of Beekeeping

Lorenzo Castellano

Institute for the Study of the Ancient World, New York University (lc2995@nyu.edu)

Roberta Pini, Cesare Ravazzi, Giulia Furlanetto, Franco Valoti

Research Group Vegetation, Climate and Human Stratigraphy, Laboratory of Palynology and Paleoecology, Institute of Environmental Geology and Geoengineering, National Research Council (Italy)
(roberta.pini@cnr.it; cesare.ravazzi@cnr.it; giulia.furlanetto@unimib.it; francovaloti@gmail.com)

Summary: The aim of this chapter is to provide an up-to-date review of the contribution of palynology to the reconstruction of the exploitation of ancient beekeeping and bee-products. After having introduced the reader to some of the main concepts in the palynology of bee-products (§2), the available evidence from Europe and the Mediterranean basin will be presented and discussed (§3). In the concluding section (§4) the palynological approach to ancient bee-products will be conceptualized, and its limits and potentials further discussed.

Key words: Palynology; bee-products; honey; mead; archaeometry

Introduction

From prehistory to the modern world, organic materials and products are at the very center of human activities, encompassing most aspects of subsistence and economy (e.g. Van der Veen 2018). Notwithstanding this centrality, the presence of macroscopic organic materials in the archaeological record is strongly biased by their high susceptibility to decay (Huisman 2009), mitigated only by pre- or post-depositional transformation processes hampering decomposition (e.g., carbonization, mineralization, coalification), or by favorable depositional environments inhibiting oxidation (e.g., hyper-arid, anoxic) (Gallagher 2014). These preservation issues are well known to scholars studying the history of beekeeping and bee-products. Honeycombs and the main products extracted from them—honey and beeswax—have, in fact, an extremely low probability of being identified in the macroscopic archaeological record, due to their perishable nature and a very low melting point (62–64 °C for beeswax). It is, thus, not surprising that the material evidence of bee-products in antiquity is extremely rare (Crane 1999). Despite the paucity of direct archaeological evidence, it is clear that beekeeping and honey hunting represented crucial activities in past societies, as unequivocally indicated by a rich textual and iconographic record (Crane 1999).

The relatively recent systematic incorporation in archaeological research of methods and techniques borrowed from the natural sciences has redefined our understanding of the archaeological record itself, which now includes for all intents also biological microremains (Nesbitt 2006) and biomolecular evidence (Evershed 2008; Weiner 2010). In the study of ancient bee-products, the far-reaching implications of these new research directions are well-exemplified by the widespread application of ceramic residue analysis (Evershed 2008), which allows us to detect in a relatively systematic way biomolecular evidence of beeswax, supporting the importance of bee-products in ancient economies (Heron et al. 1994; Evershed et al. 1997 and 2003; Mayyas et al. 2012; Roffet-Salque et a. 2015; Rageot et al. 2016, and 2019; Drieu et al. 2018; Oliveira et al. 2019).

In this chapter we will focus on palynological evidence (Moore, Webb and Collinson 1991), a type of microremain crucial to the study of ancient beekeeping activities and bee-products, which represents a contribution still waiting to be fully explored and systematically addressed. Although, as previously noted, most media preserving bee-related pollen deposits are doomed to be degraded, pollen is nonetheless resistant to oxidation and withstands comparatively higher charring temperature (Pini et al. 2018). The presence of bee-products can be corroborated or reconstructed on the basis of distinctive palynological associations, characterized by concentrations of pollen from taxa known to be foraged by honeybees (e.g., Kvavadze et al. 2007, Castellano et al. 2017). Furthermore, the relative abundances of the different pollen types can provide direct quantitative insights on honeybees' pollination ecology and associated aspects of beekeeping practice.

Our interest for the application of a palynological approach in tracing ancient beekeeping arose from the exceptional discovery of charred honeycombs at the Etruscan Po Plain site of Forcello near Bagnolo San Vito (Mantua, Italy). At the Forcello site, sometime between 510 and 495 BCE, a violent conflagration destroyed a large housing complex. Scattered on the floor of a workshop housed within the complex, we brought to light remains of charred honeycombs, bee-bread, honeybees, and melted honey and beeswax. This exceptional discovery prompted a multidisciplinary study, the results of which have been already published elsewhere (Castellano et al. 2017, Pini et al. 2018). The honeycombs from Forcello are a rather unique case, comparable in the published literature only to the palynological study of the Iron Age apiaries from Tel Rehov, in the Jordan Valley (Weinstein-Evron and Chaim 2016, Mazar and Panitz-Cohen 2020). If the archaeological preservation of honeycombs and pure bee-products is a rare instance, significantly more common is the palynological evidence of honey and honey-based products from crusts (Dickson 1978, Rösch 1999, Rösch 2005, Kvavadze 2006, Kvavadze et al. 2007, Chichinadze et al. 2019), latrine deposits (Deforce 2010, Deforce 2017), and human coprolites (Moe and Oeggl 2014). In joining the call of this volume towards a multidisciplinary approach to the study of ancient beekeeping, the aim of this contribution is to review the available palynological evidence. The survey of the available literature, geographically limited to Europe and the Mediterranean, is followed by a conceptualization of the palynological approach to ancient beekeeping and by a discussion of the potential contribution of integrative palynology to some of the key questions on beekeeping in the ancient world.

Notions of bee palynology

In seed plants fertilization results from successful pollination, i.e. the transfer of pollen (the male gametophyte) from its site of production to the female, ovule-bearing, organ. Through evolution, seed plants developed different effective means to increase the likelihood of successful pollination, by favoring the entire process to occur within a single perfect flower (self-pollination), optimizing pollen dispersal through wind (anemophily) or animals (zoophily), or by the benefiting of more than one vector (facultative pollination type) (Evert and Eichhorn 2013). Zoogamy

represents one of the best-known examples of co-evolutionary processes originated by mutualism, a cooperative interaction between species (Bronstein et al. 2016). Reciprocal adaptive developments chiefly involve anthophilous insects and entomophilous flowers (Willemstein 1987). Among insect pollinators, bees have a crucial role both in ecological and economic terms (Klein et al. 2007). The complex topic of the relationship between bees and flowering plants lies beyond the scope of this contribution. On the other hand, our interpretation of the palynological evidence in relation to ancient beekeeping needs to be based on an understanding of the factors affecting the quantities and sources of the pollen grains found in the different parts of a honeycomb and in the products extracted from it.

Pollen foraging and storage

Honeybees forage on pollen and nectar, providing the basis for the two main food sources stored within a comb: bee-bread and honey. These foraging activities are carried out separately, as part of a highly organized subdivision of the labor (Crane 1990). Pollen is collected by forager bees by brushing their hairy body on mature flower anthers or by scraping them with their jaws and front legs. The pollen collected is mixed with small amounts of nectar or honey (Harano and Sasaki 2015), to facilitate the creation of pollen pellets, which are packed into the 'pollen basket' (corbicula, Figure 1b) thanks to a structure known as pollen press (Figure 1d), and thus transferred from the floral source to the nest. In the comb, pollen is stored in proximity to the brood, where the individual pollen loads are stacked one on top of the other in the hexagonal cell (Figure 1c) (Hepburn et al. 2014). The resulting product is known as bee-bread: a hexagonal structure composed by a stratification of pollen loads mixed with lower amounts of nectar, honey, and bee secretions (Figure 1e). Fresh pollen and bee-bread are consumed by nurse bees, providing them the nutrients (proteins, lipids, and minerals) necessary to the secretion of royal jelly—in turn used to feed the brood and the queen. Recent research (Anderson et al. 2014) supports an interpretation of bee-bread as an effective strategy to preserve the pollen loads: by adding nectar, honey, and bee glandular secretions, the microbial activity is strongly reduced, allowing the preservation of pollen properties over time. Because of the preservative environment, significant morphological differences are not found between freshly collected and hive-stored pollen grains (Anderson et al. 2014).

Several palynological studies have investigated the composition of modern bee-bread and pollen loads, aiming to determine bees' foraging behaviors (e.g., Steffan-Dewenter and Kuhn 2003; de-Sá-Otero et al. 2005 and 2007; Beil et al. 2008; Mondro et al. 2009;

Figure 1. Honeybees' pollen foraging and pollen deposition in the comb. (a) Honeybee landing on an inflorescence of *Sanguisorba dodecandra* (an endemic barnet from the Italian Alps), Rosaceae family, with mature anthers on display; (b) honeybee collecting pollen from an inflorescence of *Rubus* (brambles), on the leg of the insect it is possible to notice the loaded 'pollen basket'; (c) nursing honeybees feeding the brood, to be noted the presence of bee-breads, the yellow/brown substance filling the cells surrounding the brood; (d) right hind leg of a honeybee while foraging on a *Centaurea* (knapweeds) inflorescence, showing the tibia (t), bearing empty pollen basket on its outer side (pb), the metatarsus (mt), and the pollen press structure (pps); (e) bee-bread extracted from a comb, to be noted the different layers composing each single bee-bread, representing the different pollen pellets deposited in the cell.

Da Luz and Barth 2012; Sajwani et al. 2014; De Freitas et al. 2015; Marchand et al. 2015; Layek et al. 2020). Overall, these studies point to a tendency towards flower constancy—i.e., pollinators restrict their visits to a specific flower type, even when other rewarding sources are available (Chittka et al. 1999). This behavior results in the high frequency of unifloral pollen loads, accompanied by fewer mixed pellets. The latter might be expected to result from a reduction in the number of available flowers at the end of a pollination season, or by the emergence of new flowering species in proximity to the main source under exploitation (Sajwani et al. 2014). Insect pollinated (entomophilous) flower types are quantitatively dominant in pollen loads, as expected on the basis of nutrient properties, pollen physiology, and flower anatomy. Likewise, taxa pollinated by more than one vector (amphiphilous; e.g. *Castanea*–chesnut) may also form significant proportions in pollen loads. In addition, pollen of wind pollinated taxa is relatively

frequently documented, although generally in small quantities (de-Sá-Otero et al. 2005 and 2007; Da Luz and Barth 2012; Ponnuchamy et al. 2014; Sajwani et al. 2014; Layek et al. 2020). The presence of pollen from wind-pollinated taxa is explained either by an exploitation of these species in the off-season period—when more rewarding floral sources are not available (Da Luz and Barth 2012)—or by a tendency of the colony to favor a diversified diet (Sajwani et al. 2014). The mutualism between flowers and honeybees is, in fact, facultative (Ketcham 2020), other food sources may be thus explored in addition to entomophilous flowers.

Nectar and honey

Nectar is a sugar rich (sucrose, fructose, and glucose) solution released by specialized flower glands (nectaries) to attract pollinators (De la Barrera and Nobel 2004). Floral sources can accordingly be divided into melliferous (i.e., nectariferous—producing nectar) and non-melliferous (i.e., nectarless—nonproducing nectar). Forager bees extract droplets of nectar with their proboscis, storing the sugar secretion in a modified part of their gut—the 'honey sac'. Once returned to the nest, the nectar is regurgitated and further processed by worker bees. The repeated process of swallowing and regurgitation alter the composition of the fluid, reducing the moisture content, and by means of an enzyme (invertase) converting sucrose into fructose and glucose (White 1978). The processed nectar is stored in cells, which are sealed with a wax cap once the water content is further lowered. At this stage, the ripe product can be properly considered honey: a high-sugar food resource no longer subjected to fermentation and thus storable for future uses.

Although in variable quantities, both nectar and honey contain amounts of pollen, the presence of which results from different episodes of contamination (e.g., Louveaux et al. 1978; Jones and Bryant 1992; Ricciardelli D'Albore 1998; Molan 1998; Bryant 2014). These processes—not to be confused with post-depositional abiotic contamination discussed later—take place at different stages, namely: (a) in the floral source (primary source of contamination); (b) in the hive during nectar processing and storage (secondary source); and (c) while the beekeeper is extracting the honey from the combs (tertiary source). The primary source of contamination is largely dependent on the specific characteristics of the floral species, such as the sexuality and shape of the flower, the size of pollen grains, and the synchronous/asynchronous timing of pollen shedding and nectar secretion (Ricciardelli D'Albore 1998). Beside floral characteristics, honeybees' behavior also impacts quantities and types of pollen grains found in the resulting honey. The transportation and processing of nectar implies a sequence of ingurgitations, during which a significant portion of the pollen grains is filtered-out by the proventricular valve (Molan 1998). This filtration process is particularly efficient with large pollen grains, which consequently tend to be underrepresented in honeys. Altogether, we can thus reconstruct the expected quantities of pollen present in the nectar from specific floral sources, and accordingly classify melliferous species into hyper-represented (large quantities of pollen expected in the nectar), hypo-represented (low quantities of pollen expected), and normally-represented (Bryant and Jones 2001).

Beekeeping practice further affects pollen deposition in honey, thus leading to potentially important discrepancies when comparing the palynological content of ancient and modern honeys. For example, in current beekeeping practice—based on the so-called

Langstroth Hive (Crane 1990)—a 'honey super' is added to the hive column before the harvest season. A flat perforated rack excludes the queen from this section of the beehive, which consequently is usually free of brood. Considering that pollen is generally stored in close proximity to the brood, this apicultural technique might impact the quantity of pollen circulating in the combs from which honey is extracted. The way in which honeycombs are processed is a further important apicultural variable to be considered. In Western modern apiculture, honey is extracted using a centrifuge device (the honey extractor), an innovation of the mid-19th century CE (Crane 1990). Different techniques were otherwise used (Crane 1999), including pressing the entire combs, which might result in higher quantities of pollen in the extracted honey. Furthermore, we do not know if bee-bread was removed before the pressing, crushing honeycomb containing bee-bread would significantly increase the pollen content in the resulting honey (Rösch 1999).

Beeswax

Honeybees secrete liquid wax from specialized glands present on their abdomen (Bogdanov 2004). The resulting wax scales are processed through mandibulation, ultimately producing the material used for comb-building (Hepburn et al. 2014). A vast literature exists on the physiological relationship between pollen and wax production (Hepburn et al. 2014, pp. 149-153), considering pollen as an important dietary requirement (protein source) for the function of the honeybee wax-gland-complex (Hepburn et al. 2014). On the other hand, little attention has been given to the presence and preservation of pollen grains in comb-wax.

Throughout its lifespan, a comb changes coloration: new white cells progressively turn toward different shades of yellows, browns, for finally reaching a very dark coloration (Bogdanov 2004). This process is considered resulting from the progressive incorporation in the comb structure of several exogenous materials, such as pollen, propolis, larvae silk, and various excretions altogether improving the structural properties of the comb (Hepburn et al. 2014). Pending a systematic study of the pollen content of comb-wax of different ages and origins, it is thus reasonable to attribute the pollen grains found therein to processes of contamination occurring inside the beehive. Following the extraction of beeswax from the comb, other taphonomic pathways leading to pollen deposition must be considered—e.g., contamination during the process of extraction, deposition of atmospheric pollen, and the incorporation of grains connected to specific uses of the product (Martinelli et al. 2019).

A review of the palynological evidence of beekeeping and bee-products

In the following subsections the different types of archaeological honeycomb-related materials that have been subjected to palynological analysis are reviewed (Table 1). The evidence is organized into four different groups, reflecting the macroscopic nature and the archaeological origin of bee-products: (a) archaeological honeycombs, beehives, and associated deposits; (b) archaeological beeswax; (c) crusts associated with archaeological materials and interpreted as residues of honey and honey-based products; and (d) evidence of honey and honey-based products consumption from cesspit deposits and human coprolites. This review is geographically limited to Europe and the Mediterranean (Figure 2). Data published only on a preliminary basis (e.g., Bueno Ramirez et al. 2005, Prieto et al. 2005) and more tentative identifications of mixed bee-products residues (e.g., Bosi et al. 2011, Kozáková et al. 2017) are not considered.

Archaeological honeycombs

To our knowledge, palynological studies of archaeological honeycombs are to date limited to the finds from Forcello near Bagnolo San Vito (Italy; Castellano et al. 2017) and Tel Rehov (Israel; Weinstein-Evron and Chaim 2016) (Figure 2). In both instances, the honeycombs have been found charred as part of destruction levels. Remarkably the charred conditions did not hamper the preservation of pollen grains, as discussed by Pini et al. (2018).

Forcello near Bagnolo San Vito (Italy)

Forcello near Bagnolo San Vito (Mantua, N-Italy) was the most important Etruscan settlement north of the Po River in the 6th and 5th century BCE, functioning as a major trade hub strategically located along long-distance trade routes connecting the Mediterranean region to Central Europe (de Marinis and Rapi 2007) (Figure 2). The honeycomb evidence originates from

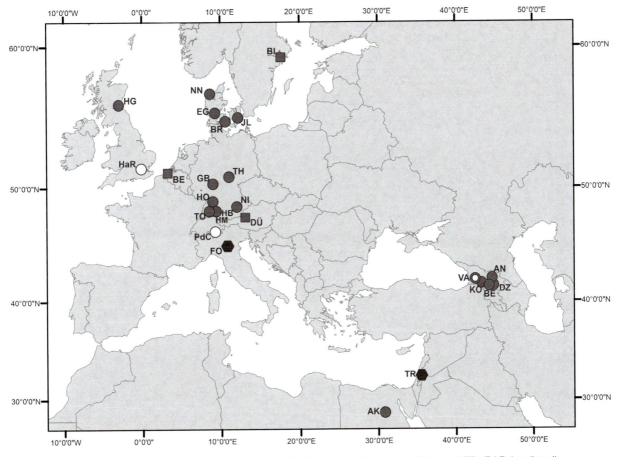

● Pollen from honeycombs, bee -bread, charred honey clots: FO= Forcello near Bagnolo San Vito (Italy); TR= Tel Rehov (Israel).

○ Pollen from beeswax: PcD= Pace di Chiavenna (mobile object); HaR= Horse and Rider Statue (mobile object); VA= Vani (Georgia).

● Pollen from honeybee products found in vessel crusts, body remains, and other burial goods: HG= Ashgrove (UK); JL= Juellinge (Danmark); NN= Nandrup (Denmark); EG= Egtvedt (Danmark); BR= Bregninge (Danmark); TH= Thuringia (Germany); GB= Glauberg (Germany); HO= Hochdorf (Germany); TO= Trossingen (Germany); HB= Heuneburg-Speakhau (Germany); HM= Heuneburg-Hohmichele (Germany); NI= Niedererlbach (Germany); VA= Vani (Georgia); KO= Kodiani (Georgia); BE= Bedeni (Georgia); DZ= Dzedzveb (Georgia); AN= Ananauri (Georgia); AK= Al-Kom al-Ahmar (Egypt).

■ Pollen of honeybee products detected in human coprolites and cesspit deposits: DÜ= Dürrnberg (Austria); BI= Birka (Sweden); BE = Bruges (Belgium).

Figure 2. Location of the sites discussed in the text (see Table 1 for further information).

Table 1. List of palynological studies discussed in the text. For each site we report the acronyms used in Figure 7-8-9, chronology, context type, the interpretation of the analyzed material provided by the original authors and the literature.

siteID	Site	Chronology	Context type	Analyzed materials	Interpretation	Literature
FO	Forcello near Bagnolo San Vito (Italy)	Iron Age	Settlement (workshop)	Honeycomb, bee-bread, honey/wax	Honeycomb burnt during a destruction event	Castellano et al. 2017; Pini et al. 2018
TR	Tel Rehov (Israel)	Iron Age	Settlement (apiary)	Honeycomb, sediment inside beehives	Apiary	Weinstein-Evron and Chaim 2016
VA	Vani (Georgia)	Hellenistic	Settlement (hoard)	Beeswax	Block of molding beeswax	Chichinadze and Kvavadze 2013
PdC	Unknown - *Pace di Chiavenna*	Medieval	Object	Beeswax	Beeswax used as adhesive	Martinelli et al. 2019
HaR	Unknown - *Horse and Rider*	Modern	object	Beeswax	Beeswax model for lostwax casting.	Furness 1994
AG	Ashgrove (UK)	Bronze Age	Burial	Organic debris	Mead flavored with meadowsweet	Dickson 1978
JL	Juellinge (Denmark)	Roman	Burial	Unknown	Alcoholic beverage containing honey and flavored with meadowsweet	Moe and Oeggl 2014
NN	Nandrup (Denmark)	Bronze Age	Burial	Crust from beaker	Alcoholic beverage containing honey and flavored with meadowsweet	Dickson 1978; Moe and Oeggl 2014
EG	Egtvedt (Denmark)	Bronze Age	Burial	Crust from Birch-bark bucket	Alcoholic beverage containing honey and flavored with meadowsweet	Dickson 1978; McGovern et al. 2013; Moe and Oeggl 2014
BR	Bregninge (Denmark)	Bronze Age	Burial	Unknown	Alcoholic beverage containing honey and flavored with meadowsweet	Dickson 1978; Moe and Oeggl 2014
NI	Niedererlbach (Germany)	Iron Age	Burial	Crust from bronze ladle	Unfiltered fresh mead flavored with meadowsweet	Rösch 2005
GB	Glauberg (Germany)	Iron Age	Burial	Crusts from two pitchers found in two separated graves	Unfiltered fresh mead flavored with meadowsweet (pitcher 1); old mead or drink sweetened by honey (pitcher 2).	Rösch 1999
HB	Heuneburg-Speckhau (Germany)	Iron Age	Burial	Crusts from bronze cauldron	Mead	Rösch and Rieckhoff 2019
HM	Heuneburg-Hohmichele (Germany)	Iron Age	Burial	Crusts from bronze cauldron	Mead	Goppelsröder and Rösch 2002
HD	Hochdorf (Germany)	Iron Age	Burial	Crusts from bronze cauldron	Unfiltered fresh mead	Körber-Grohne 1985
TH	Thuringia (Germany)	Medieval	Settlement	Crust from large pot	Mead flavored with meadowsweet	Jacob 1979
TO	Trossingen (Germany)	Medieval	Settlement	Crust from wooden bottle	Beer sweetened with honey and grape mash	Rösch 2008
KO	Kodiani (Georgia)	Bronze Age	Burial	Crust from pots	Honey or mead flavored with meadowsweet	Kvavadze 2006; Kvavadze et al. 2007.
VA	Vani (Georgia)	Iron Age	Burial	Crust from pitcher; sediment embedding bracelet	Honey or honey-based product; honey used for embalming or other aspects of the ritual	Chichinadze et al. 2019
DZ	Dzedzvebi (Georgia)	Bronze Age	Burial	Crust from pots	Honey	Kvavadze and Martkoplishvili 2018
AN	Ananauri (Georgia)	Bronze Age	Burial	Crust from wooden vessel; sediment associated to other burial goods and to the skeleton	Honey; honey used for embalming or other aspects of the ritual	Kvavadze 2016
BE	Bedeni (Georgia)	Bronze Age	Burial	Residues associated to various items in burials 2, 5, and 10	Honey used for embalming or other aspects of the ritual	Kvavadze et al. 2015
AK	Al-Kom al-Ahmar (Egypt)	Coptic	Settlement	Residue from bottom of amphorae	Wine sweetened with honey	Rösch 2005
BI	Birka (Sweden)	Viking Age	Settlement	Human coprolites	Mead flavored with meadowsweet	Moe and Oeggl 2014
DÜ	Dürrnberg (Austria)	Iron Age	Settlement	Human coprolites	Mead flavored with meadowsweet	Moe and Oeggl 2014
BE	Bruges (Belgium)	15th century CE	Settlement	Cesspit deposit	Honey	Deforce 2010
BEL	various sites (Belgium)	12th to the 17th century CE	Settlement	Cesspit deposit	Honey	Deforce 2017

the Phase F building complex: two houses that were part of the same building unit, destroyed by a violent conflagration sometime between 510 and 495 BCE (de Marinis 2016; Quirino 2011). Soon after the fire, the destruction deposit was covered by a thick reclamation layer, on top of which a new occupation phase took place (de Marinis and Rapi 2007). These archaeologically fortunate circumstances allowed the *in-situ* preservation of the original content of the houses—albeit affected to various extents by the temperature of the fire—including honeycomb found scattered on the floor of House F2-Room 3, a space interpreted as an artisan workshop (Castellano et al. 2017) (Figure 3a). Full-

sampling of the fire debris allowed for the identification of bee-bread (Figure 3b), charred honeycombs (Figure 3c), various anatomic parts of the honeybees (heads, abdomens, thoraxes, legs, and various soft tissues) (Figure 3d-e), and large amounts of clots of an amorphous substance—interpreted as resulting from the melting and subsequent solidification of wax and honey (Castellano et al. 2017) (Figure 3d). The honeycombs are associated with a wooden (fir, *Abies alba*) worked object, possibly representing a beehive or a container for the combs. One possibility is that the honeycombs were introduced into the workshop in order to extract honey and beeswax from them.

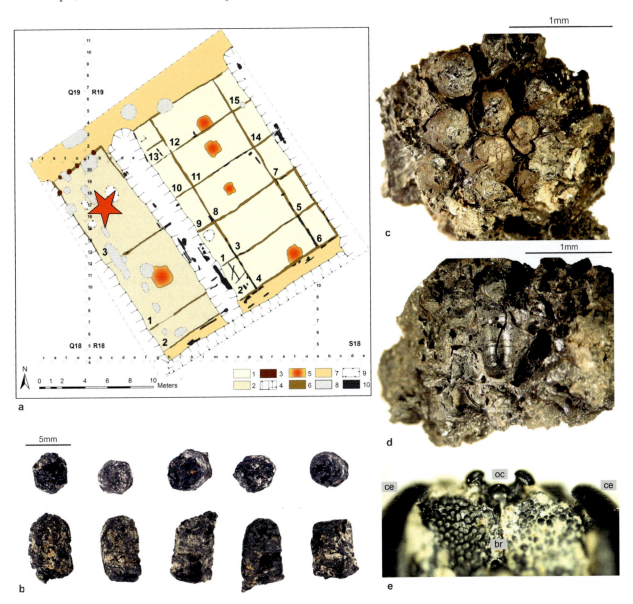

Figure 3. Forcello near Bagnolo San Vito (Italy), honeycomb remains from the House F complex (510-495 BCE). (a), Plan of the house complex, the red start indicates the honeycombs find spot (house F2, room 3). Keys: 1. F1 house; 2. F2 house; 3. Pole pit; 4. negative structure; 5. hearth; 6. foundation wall duct; 7. house outside; 8. not in phase pit; 9. drainage ditch; 10. wooden structural element; (b), charred bee-bread in section and frontal view; (c), charred honeycomb fragment with the hexagonal, thin-walled cell structure exceptionally clearly observable; (d), adult of *Apis mellifera* embedded in a clot of charred vitrified porous honeycomb components; (e), fragments of *Apis mellifera* head, showing the compound eyes (co), ocelli (oc), and brain tissue (br).

Palynological analysis has been carried out on honeycomb fragments (3 samples), bee-bread (24 samples), charred clots (2 samples), and the sediment embedding the charred clots (1 sample) (Castellano et al. 2017). Furthermore, the pollen spectra from honeycomb-related materials have been compared to a sample collected from the infill of a channel surrounding House F2 (1 sample) and to the off-site palynological record from a nearby undisturbed limnic sequence (Ravazzi et al. 2013). Based on the charring degree of pollen grains (Castellano et al. 2017; Pini et al. 2018), bee-bread samples are divided into three categories: (a) 6 bee-bread cylinders are entirely composed by highly-combusted organic matter, in which pollen grains cannot be recognized; (b) 11 bee-bread cylinders contain only opaque pollen grains, in which case the sole identification of fringed water-lily type (*Nymphoides peltata* type; Figure 4a,b) pollen was possible, thanks to preservation of its distinctive shape and size; (c) 7 bee-bread cylinders contain pollen grains characterized by different charring degrees, enabling high pollen counts (Figure 4a). In this latter case, the very high pollen concentrations (5780 to more than 31 million grains/g) are consistent with items composed of pressed, almost pure, pollen (Castellano et al. 2017, Pini et al. 2018).

The presence of bee-bread containing large amounts of preserved pollen allows quantitative considerations. Two out of the seven bee-bread cylinders containing well-preserved pollen are unifloral, composed almost exclusively of pollen of the fringed water-lily type (*Nymphoides peltata* type; Figure 4a,b). The remaining five are of mixed floral sources, containing pollen of brown knapweed (*Centaurea jacea* group), mint type (*Mentha* type), and a variable combination of secondary taxa: knotted hedge-parsley type (*Torilis nodosa* type), coltsfoot type (*Tussilago farfara* type), cheeses type (*Malva sylvestris* type), aster type (*Aster* type), vervain (*Verbena* sp.), and thistles (*Cirsium* sp.) (Castellano et al. 2017). The taxa identified are all known to be insect pollinated and considered to be highly palatable for honeybees (Proctor et al. 1996). With the sole exception of coltsfoot type (*Tussilago farfara* type), which is produced by only one early spring bloomer species (*T. farfara* L.), the flowering times indicate that these foraging activities took place in the summer. The presence of small quantities of pollen from the previous season might suggest that to some extent pollen loads of different ages were stored in the same cell of the comb, resulting in one single bee-bread. As discussed by Castellano et al. (2017), this peculiar palynological association—dominated by fringed water-lily type (*Nymphoides peltata* type) pollen, belonging to only one aquatic plant species, *N. peltata* (Gmelin) Knutze—supports the hypothesis of a form of itinerant boat beekeeping or floating apiaries (Figure 4d), a practice reported for this specific region of northern Italy by the Roman author Pliny the Elder (*Natural History* 21.43.73).

The interpretation of the pollen spectra from the charred clots is complicated by the mixed nature of these materials, originating from the solidification of melted honey and wax. In the charred clots, fringed water-lily type (*Nymphoides peltata* type) pollen is only sporadically attested, while pollen of brown knapweed (*Centaurea jacea* group), mint type (*Mentha* type), and aster type (*Aster* type) is more abundant. The latter taxa are all known to be good nectariferous species, the pollen of which is well-attested in modern honeys (Ricciardelli D'Albore 1998). It is thus possible that the Forcello's bees were visiting these floral sources for both pollen and nectar. Puzzling are the large amounts of grapevine pollen (*Vitis* type) found in the charred clots, in their embedding sediment, and in the analyzed honeycomb fragment. By contrast, *Vitis* pollen is missing from the analyzed bee-bread (Castellano et al. 2017). On a preliminary basis, this atypical presence of large amounts of *Vitis* pollen can be potentially explained by: (a) pollen from grapevine nectar collected by the Forcello honeybees. The function of *Vitis vinifera* 'nectaries', in terms of actual secretion of nectar and its potential exploitation by honeybees is, however, unclear (Free 1993: 529-530); (b) *Vitis* pollen originating from sources other than the honeycombs, for example from wine stored in the same room which might have fallen over the combs during the fire event (see Rösch 1999 for the presence of *Vitis* pollen in wine). It would remain, however, unclear why *Vitis* pollen has not been found in the other specimens from the same area of the room; or (c) the presence of an undocumented beekeeping practice involving the feeding of honeybees with grape-based products. Modern apicultural practices often supplement honeybees' diet with external sources of sugar, in order to compensate during the off-season for the honey harvested from the colony. The artificial feeding of honeybees' colonies is attested also in Graeco-Roman textual sources, including a passage in Varro (*De Re Rustica* 3.16.28) reporting the use of a mixture obtained from pounded raisins and figs soaked in boiled wine. Pending experimental evidence of the actual presence of grapevine pollen in these sorts of mixtures and in the resulting honey, this hypothesis remains to date speculative. It should be noted that although possible, modern apiculture discourages the practice of feeding honeybees with pure grape syrup, which is reported as cause of dysentery in bee colonies (Barker and Lehner 1978).

Tel Rehov (Israel)

The second instance of archaeological honeycombs subjected to palynological analysis comes from the archaeological site of Tel Rehov, in the Jordan Valley (Israel), a large, mounded site flourishing during the Early Monarchy Period (Figure 2). The site gained international attention following the exceptional discovery of an industrial-scale apiary, located in

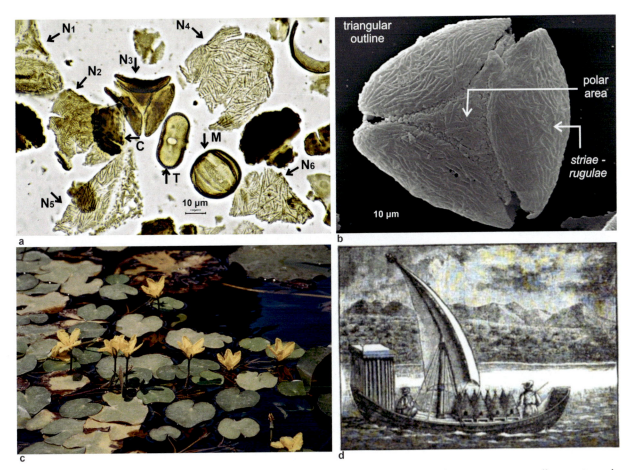

Figure 4. (a) Palynofacies from fossil bee-bread R18_b18_b recovered in the archaeological excavation at Forcello near Bagnolo San Vito (Italy). Pollen grains were not acetolyzed (i.e. chemically stained), the colors observed in this picture are the result of charring induced by the fire that destroyed the building where honeycombs were stored. Arrows indicate pollen grains of: *Centaurea jacea* group (C), *Nymphoides peltata* type (N), *Mentha* type (M), and *Torilis nodosa* type (T). Scale bar: 10 microns. While some grains are preserved as a whole (*Mentha* type, *Torilis nodosa* type), others are broken (N1, N2, N4, N5 and N6). This does not prevent their identification to *Nymphoides peltata* type due to the preservation of diagnostic features, such as the pervasive striato-rugulate ornamentation on the pollen wall. The overall triangular outline and the concave-sided triangular polar areas are well visible on entire grains (N3). See Figure 4b as a reference. (b) SEM picture of *Nymphoides peltata* type, 2200x. The grain shows the features diagnostic for the identification of broken grains in Figure 4a; (c) fringed water lily (*Nymphoides peltata*) flowering in its lotic aquatic environment along a big river in N-Italy. Several bee-bread cylinders from the site of Forcello were almost completely composed by pollen of this aquatic plant; (d) drawing representing floating apiaries observed on the Nile (Egypt) in 19th century CE, from Kellen 1889. Floating apiaries and itinerant boat beekeeping are comparatively well-attested in ethnographic and historical sources.

the heart of the citadel (Mazar et al. 2008; Mazar and Panitz Cohen 2007; Bloch et al. 2010). The structure was destroyed by a fire, allowing the preservation of charred honeycombs and honeybees-related remains (Bloch et al. 2010). The palynological study of these remains was carried out by Weinstein-Evron and Chaim (2016) on the basis of: a) deposits admixed with bee products sampled inside the beehives (8 samples); b) clay from the beehives and 'near-hive soil' (4 samples); and c) honeycomb fragments (2 samples). Bee-bread is not mentioned as being found or analyzed.

Regrettably, detailed information on the composition of the sediment sampled from the beehives is not available (Weinstein-Evron and Chaim 2016), thus depositional and taphonomic processes cannot be reconstructed. The ubiquitous attestation of pollen from nectarless

and anemophilous taxa—such as olive (*Olea europea*), oaks (*Quercus calliprinos* and *Q. ithaburensis*), and pines (*Pinus* sp.)—might suggest that part of the grains resulted from atmospheric deposition. Alternatively, we could hypothesize that the bees from Tel Rehov were intensively foraging on wind pollinated floral sources. More direct is the interpretation of the significant quantities of jujube (*Ziziphus* sp.) pollen, representing the dominant arboreal type in the samples. Honeybees, in fact, forage intensively on jujube, both for pollen and nectar (Weinstein-Evron and Chaim 2016). The identification at the family level of most of the non-arboreal taxa does not allow us to pinpoint precisely the exploited floral sources. Nevertheless, most of these families host a number of well-known melliferous and entomophilous species, supporting their deposition in connection with honeybees' foraging activities.

Archaeological beeswax

Beeswax contains pollen originating from contamination occurring within the beehive. Furthermore, pollen unrelated to bees' activities can be deposited in the wax while in the beehive, during its extraction and use, or in the post-depositional environment, as discussed in section 2.3. It should be stressed that to date we lack published detailed experimental studies on pollen deposition and preservation in modern beeswax, thus hampering our ability to fully evaluate the evidence from archaeological materials.

Chichinadze and Kvavadze (2013) published the palynological study of a block of wax (420 g) found in a hoard of elaborated bronze and iron objects at the site of Vani (Georgia) (Figure 2), dated to the Late Hellenistic Period. According to Chichinadze and Kvavadze (2013), the wax melted during the excavation and in some places it mixed with brownish organic remains and soil (Figure 5). The identification of this material as beeswax is based on morphoscopic considerations. Pollen was extracted from samples collected from the beeswax block and the embedding sediment; results were compared to materials sampled from different loci of the hoard (Chichinadze and Kvavadze 2013), without, however, a specific test of their palynological significance. Beeswax and the associated sediment are distinctively characterized by high quantities of chestnut (*Castanea sativa*) pollen, a type otherwise very rare in the other samples from the hoard—the latter dominated by nectarless and obligatory anemophilous plants, such as pines (*Pinus* sp.), spruce (*Picea* sp.), and alder (*Alnus* sp.). Chestnut is a well-known nectariferous species, abundantly attested in honey residues (Ricciardelli D'Albore 1998). A second common pollen type from beeswax samples is attributed to the rose family (Rosaceae), which includes several melliferous

and entomophilous sources. The presence of reworked pollen of Tertiary age (e.g., *Podocarpus*, *Dacrydium*, *Cedrus*, Taxodiaceae) is explained by the authors as originating from the use of the beeswax block in lost-wax casting: the wax was used to realize the model of the final object and covered by clay in order to obtain the mold. According to the authors, this process eventually caused the incorporation in the wax of pollen of geological age originated from the clay itself (Chichinadze and Kvavadze 2013)—thus, representing a type of contamination resulting from the specific use of the product (see section 4.1). This interpretation is consistent with the archaeological context, if we were to interpret the hoard as belonging to a metal-working artisan. It remains, however, necessary to corroborate this hypothesis with a better understanding of the taphonomy of pollen depositions in beeswax. Furthermore, the preservation of pollen grains in wax used for lost-wax casting would also imply that the burnout of wax from the mold took place at relatively low temperatures, allowing the preservation of pollen grains—the shape of which is preserved up to 350 °C according to Pini et al. (2018)—and of the beeswax itself—the flash point of which is ca. 204.4 °C (Coggshall and Morse 1984). The temperature to which the mold is exposed at this stage of the process (burnout) are highly variable, especially in relation to the specific materials used (Davey 2009).

Large blocks of beeswax preserved in the archaeological record are rather a unicum. On the other hand, beeswax can relatively speaking be more commonly preserved in smaller, yet macroscopically observable, quantities. Martinelli et al. (2019) analyzed pollen extracted from beeswax used as adhesive in a medieval Evangelistary cover known as *Pace di Chiavenna*. The palynological study attempted to circumscribe the geographic origin of the beeswax—and consequently very likely the origin of the object itself—on the basis of the palynological

a
b

Figure 5. (a) 1st century BCE hoard from Vani (Georgia), to be noted the large block of beeswax lying on a metal stand; (b) detail of the beeswax block (from Chichinadze and Kvavadze 2013).

flora detected in it. A similar approach was employed by Furness (1994) for the study of the provenience of a chronologically later beeswax cast model of unknown origin, allegedly attributed to Leonardo da Vinci.

Honey and honey-based products in vessels residues

The attestation of honey and honey-based products in organic crusts is far more common than the aforementioned instances of the preservation of honeycomb and macroscopic beeswax. The palynological analysis of organic residues associated with archaeological objects is of particular importance in the cases of undisturbed contexts, such as burials. Promising is the analysis of residues associated with bronze or copper objects, in which the preservation of fossil pollen is enhanced by corrosion products released by the cuprous artifact (Greig 1989). The identification of honey and honey-based substances relies on the identification of high proportions and concentrations of pollen originated from melliferous and/or entomophilous species (e.g., Rösch 2005). The presence of flower depositions and offerings is an alternative possibility, which needs to be evaluated through an examination of the archaeological context and the palynological record (Tipping 1994). Pollen contamination by fossorial insects may also produce entomophilous-rich pollen assemblages, a process well-known in cave sediments (Bottema, 1975). It might be useful to check for the presence of pedological features suggesting activities of ground-nesting insects (Ponomarenko et al. 2019), and to verify the occurrence of pollen clumps and eventually to consider their taxonomic composition (e.g., Perego et al 2011). Furthermore, in interpreting the pollen content of discrete samples associated with archaeological features, such as organic crusts, we should stress the importance of including in the experimental design a null hypothesis test (Strong 1982)—i.e., to compare the pollen composition of the unknown substance with the composition of the background sediment or media. Thus, allowing, in the cases here discussed, for the discrimination of the honey-derived components from the other pollen components (e.g., atmospheric, floated, reworked).

Besides the identification of honey-based products, pinpointing the specific product producing the organic residue is also problematic: pure honey, honey-based fermented beverages (mead), cocktails of ferment products (e.g., 'Nordic grog'; McGovern at al. 2013), or other liquids (e.g., ale or wine) sweetened with honey. Each of these possibilities needs to be considered in light of the available contextual and bio-archaeological evidence. It should be noted that the process of fermentation of honey to produce mead does not appear to cause significant alteration in pollen quantities and morphology (Dozier 2016).

Honey-based products from prehistoric Scotland and Denmark

A cornerstone in the analysis of honey-related residues from burial contexts is Dickson's (1978) palynological study of a Bronze Age burial from Ashgrove (Scotland). On the basis of an atypical presence of large quantities of lime (*Tilia* sp.) and meadowsweet (*Filipendula ulmaria*) pollen in a residue from the burial, Dickson proposed the presence of mead prepared with lime honey and flavored with meadowsweet flowers. Lime is a well-known important melliferous source, intensively exploited by bees for both pollen and nectar (Ricciardelli D'Albore 1998). Lime trees are excluded from the natural vegetation of the region at the time of the burial, on the basis of current phytogeography and of the paleoenvironmental record. It is, thus, suggested that the mead was produced using an exotic honey, possibly harvested more to the south in the British Isles (Dickson 1978). Meadowsweet is an insect pollinated plant, frequently visited by bees for pollen but not producing any nectar. Consequently, meadowsweet pollen is present in honeys only in small quantities (Zander 1935). As discussed by Dickson (1978) and Moe and Oeggl (2014), ethnographic and historical sources indicate that flowers and leaves of meadowsweet were traditionally used as flavoring ingredient in mead preparation. This former use of the plant is evident in its vernacular names attested across much of central and northern Europe, which often translate as 'mead herb' (Dickson 1978, Moe and Oeggl 2014).

Dickson (1978) brought attention also to earlier unpublished pollen analysis of organic crusts from prehistoric burials sites in Denmark (Juellinge, Nandrup, Egtvedt, Bregninge). These pollen spectra are characterized by high percentages of lime, meadowsweet, and other anemophilous and melliferous floral sources (Moe and Oeggl 2014). More recently, McGovern et al. (2013) on the basis of biomolecular evidence interpreted these residues as originating from 'Nordic grog'—a mixed alcoholic beverage produced by the fermentation of honey, various fruits (e.g., bog cranberry and lingonberry), and cereals.

Mead from Iron Age central European burial contexts

Palynological evidence of honey-based products, including mead, is comparatively abundant from Iron Age central Europe. (Rösch 1999 and 2014; Stika 2010; Rösch and Rieckhoff 2019). The Late Hallstatt/Early La Tène culture, characteristic of this region during the Iron Age, is well-known for the extremely rich funerary mounds in which their elites were buried. A recurrent feature in these funerary contexts is the presence among the burial goods of ceramic and bronze drinking sets, composed of precious imports from the Mediterranean and high-quality local productions. Mead and honey-based products are attested by palynological studies of crusts from bronze vessels retrieved from several of these burial sites, such as Hochdorf (Körber-Grohne 1985), Heuneburg-Hohmichele (Goppelsröder and Rösch 2002), Heuneburg-Speckhau (Rösch and Rieckhoff 2019), Glauberg (Rösch 1999), and Niedererlbach (Rösch 2005). All these sites are located in Germany and dated between the 6th and 5th century BCE.

A first identification of mead in a Late Hallstatt burial site was carried out by Körber-Grohne (1985), on crust residues extracted from the bronze cauldron found in the Hochdorf tumulus (Baden-Württemberg) (Figure 6). This bronze vessel, a precious import from a Greek center in southern Italy, has the impressive capacity of 500 liters (Biel 1981). Palynological analysis from these residues indicated dominance of melliferous and entomophilous taxa: thyme type (*Thymus* type), a very good nectariferous source, is the dominant taxon (c. 15%), followed by other entomophilous taxa, such as the lesser knapweed type (*Centaurea nigra* type), sheep's-bit (*Jasione* sp.), kidney vetch (*Anthyllis vulneraria*), common heather (*Calluna vulgaris*), and cinquefoils (*Potentilla* sp.). Several other floral sources are found in lower quantities, including the meadowsweet genus (*Filipendula* sp.) (Körber-Grohne 1985). The interpretation of these residues as mead was further confirmed by the high concentration of beeswax detected by chemical analysis (Körber-Grohne 1985; Stika 2010). The amounts of pollen and beeswax present in the residues suggest the presence of fresh unfiltered mead, obtained by mixing honeys of different geographic provenience (Rösch 1999).

Palynological evidence of mead comes also from two tumuli associated with the hilltop site of the Heuneburg (Baden-Württemberg). In the largest of these tumuli, the Hohmichele, mead was identified in the organic residues from a large (20-liter) bronze cauldron (Rösch and Goppelsröder 2002). The palynological spectrum is characterized by a high floristic diversity (137 pollen types), dominated by non-arboreal entomophilous taxa, such as the rose family (Rosaceae), legumes (Fabaceae), mint family (Lamiaceae), figwort family (Scrophulariaceae), aster family (Asteraceae), and the St. John's wort genus (*Hypericum* sp.). Mead was recently identified also in residues from a bronze cauldron (14-liter volume) brought to light in a burial from the nearby Speckhau Tumulus (Rösch and Rieckhoff 2019). The organic residue extracted from this latter vessel is dominated by pollen of mint type (*Mentha* type; c. 10%), sheep's-bit (*Jasione montana*; c. 8%), and the meadowsweet genus (*Filipendula* sp; c. 8%).

Further instances of honey-based liquids come from two graves in the Glauberg tumulus (Hesse) (Rösch 1999). In the first grave, crusts were extracted from a bronze beak-spouted pitcher (*Schnabelkanne*). Pollen analysis documented a high floristic diversity (180 pollen types), with a preponderance of bee-pollinated and melliferous taxa, including lesser knapweed type (*Centaurea nigra* type), restharrows type (*Ononis* type), clover type (*Trifolium* type), and mint type (*Mentha* type). On the basis of quantities and types of floral sources, distributed across the entire pollination year, Rösch (1999) interpreted these residues as originating from fresh unfiltered mead. In Glauberg-Grave 2 residues were samples from a bronze tube-spouted pitcher (*Röhrenkanne*). In this second case the pollen spectrum is characterized by an atypical concentration of nectarless taxa, such as hazelnut (*Corylus* sp.; c. 10%) and grasses (Poaceae; c. 6%) (Rösch 1999). The presence of these taxa in a palynological spectrum otherwise dominated by entomophilous and melliferous species (Rösch 1999) might suggests that honeybees were foraging on wind pollinated species—hazelnut in particular is known as an important anemophilous source of pollen exploited by bees in the early spring (Piotrowska 2008). If we accept this hypothesis, a quantity of pollen from the comb was voluntary or accidentally incorporated during the preparation of the mead. The presence of mead, either pure or mixed, is supported also by the abundant attestation of meadowsweet (*Filipendula* sp.).

Finally, mead was identified also in the residues from a bronze ladle found in a burial from Niedererlbach (Lower Bavaria) (Rösch 2005). Meadowsweet (*Filipendula* sp.) is the

Figure 6. The Lion's cauldron from Hochdorf (Germany, Late Hallstatt D).

most abundant taxon documented in these residues (c. 16%), followed by brown knapweed type (*Centaurea jacea* type; c. 9%), mint type (*Mentha* type; c. 7%), St. John's wort genus (*Hypericum* sp.; c. 6%), thyme (*Thymus* sp.; 3%), and maple (*Acer* sp; c. 3%).

Honey products from funeral context in Georgia

Data on honey and honey-based residues are available from several burial contexts in the southern Caucasus (Kvavadze 2006 and 2016, Kvavaze et al. 2007 and 2015, Kvavadze and Martkoplishvili 2018, Chichinadze et al. 2019). Among the earliest attestations, honey-product residues have been identified in 3 vessels found in the burial mound of Kodiani (Georgia), dated to the mid-3rd millennium BCE (Kvavadze 2006, Kvavaze et al. 2007). Entomophilous taxa dominate this record: meadowsweet genus (*Filipendula* sp.) is the most abundant pollen in all the three containers, along with pollen of strawberry type (*Fragaria* type), dog-rose type (*Rosa canina* type) (*Fragaria* type and *Rosa canina* type are summed in the Rosaceae group, Figs. 8 and 9), and umbellifers family (Apiaceae). Kvavadze (2006) and Kvavaze et al. (2007) interpreted these residues as originating from honey, Moe and Oeggl (2014) on the basis of the abundance of meadowsweet pollen considered it possible that these pots contained mead. If confirmed, this later hypothesis would extend the practice of flavoring mead with meadowsweet flowers further back in time and in a geographic context other than central and northern Europe. Honey or honey-based products are palynologically attested in other Bronze Age burial mounds in the region (Table 1), such as Dzedzvebi burial 1 (Kvavadze and Martkoplishvili, 2018), Ananauri (Kvavadze 2016), and Bedeni burial 5 (Kvavadaze et al. 2015).

A later attestation of honey from burial contexts comes from Vani (Georgia) grave 24 (Chichinadze et al. 2019), a rich burial dated to the 4th century BCE. Organic crusts extracted from a bronze beak-spouted oinochoe support the presence of honey or honey-products, as suggested by the atypical quantities of chestnut (*Castanea sativa*) and lime (*Tilia* sp.) pollen. Wind-pollinated species are attested as well, possibly indicating different pollen sources contributing to the formation of the deposit. Entomophilous types were found also in association with a bracelet worn by the deceased, interpreted by the authors as resulting from the application of a honey-based balming product to the deceased body (Chichinadze et al. 2019).

Cesspits and coprolites

Cesspits deposits and coprolites (Shillito et al. 2020) are a further important source of information on honey consumption in the ancient world. In this case honey-based products can be hypothesized on the basis of the presence in sub-fossilized human feces of large quantities of entomophilous and melliferous pollen types, which can be potentially interpreted as incorporated via honey consumption. However, other sources might be considered as well, such as plant-based food consumption (e.g. Bosi et al. 2011).

A strong argument for the identification of honey in cesspit materials is made by Deforce's (2010) study of latrine deposits from the palace of the Dukes of Burgundy in Bruges (Belgium), dated to late 14th or 15th century CE. The author pointed to the presence of significant quantities of pollen from entomophilous and melliferous taxa non-native in the local flora: strawberry tree (*Arbutus unedo*), tamujo (*Securinega tinctoria*), white-leaved rock rose type (*Cistus albidus* type), crimson spot rock rose (*Cistus ladanifer*), rock rose family (Cistaceae), dwarf Spanish heath (*Erica umbellata*), French lavender type (*Lavandula stoechas* type), mock privet (*Phillyrea*), and sun rose (*Helianthemum*). This peculiar association was interpreted as resulting from consumption of exotic honeys, on the basis of diagnostic pollen types likely produced in the Iberian Peninsula (Deforce 2010). Similar pollen associations, characterized by Mediterranean melliferous and entomophilous pollen types, were detected in several other cesspit deposits in central and northern Belgium, dating between the 12th and 17th century CE (Deforce 2017).

Evidence from human coprolites in relation to honey-based product consumption was presented by Moe and Oeggl (2014), on the basis of materials from the Medieval site of Birka (Sweden) and the Iron Age salt-mining center of Dürrnberg (Austria). At both sites, determinant to the reconstruction of mead consumption was the presence of abundant pollen of entomophilous taxa, including large quantities of meadowsweet (*Filipendula* sp.) pollen (Moe and Oeggl 2014). The contribution of this type of evidence for assessing the level of consumption and circulation of honey in ancient times still awaits to be fully explored.

As a final more general hint about coprolites, an argument still to be addressed in the paleoecological literature is the importance of hive exploitation by bears, which might lead to deposition of pollen derived from bee-products in natural sequences with fossil bear's feces-rich sediments, such as cave deposits.

Pollen analysis of ancient bee-products: synthesis, problems, and perspectives

A critical assessment of the palynological evidence

The lack of an accepted categorization of bee-products in archaeological and natural media prompted us to propose the schematic classification presented in Figure 7. Bee-products are ordered according to the underlying genetic processes and their microbotanical (palynological) evidence. Each category of bee-product listed in Figure 7 is referred to the specific site (Table 1) where it was found, along with pollen concentration values (if reported in the literature). Following the uppermost distinction between materials in primary and secondary deposition, the right side of the scheme displays unprocessed bee-products in primary deposition—i.e., resulting from direct deposition by bees, an instance to date documented only at the sites of Forcello near Bagnolo San Vito (Castellano et al. 2017) and Tel Rehov (Weinstein-Evron and Chaim 2016). Pure bee-products might experience physical and chemical changes due to post-depositional natural or anthropic processes (e.g., fire, decantation). On the left, these materials are followed by bee-products still in primary deposition but revealing contamination and/or admixing with a supporting medium: examples are fossil beeswax samples from the Pace di Chiavenna (Martinelli et al. 2019) and Vani (Chichinadze and Kvavadze 2013), and residues of honey used as balming media detected on textiles and yarn (Bedeni; Kvavadze et al. 2015) and on human skeletons (Ananauri; Kvavadze 2016). In the left side of the scheme are located bee-products in secondary deposition—i.e., dispersed in a transporting medium and preserved in an embedding matrix, often rich in organic matter. The original pollen content of these bee-products is still

recognizable, along with a major pollen component related to diet (e.g., abundance of cereal pollen in human coprolites from Dürrnberg, Moe and Oeggl 2014; and cesspits from Bruges, Deforce 2010 and 2017) and to local anemophilous plants. The classification proposed in this scheme is used to differentiate the samples shown in Figs. 8 and 9.

Figure 8 presents a synoptic frame of palynological evidence of bee-products from the sites discussed in the previous section, following the categorization proposed in Figure 7. Samples are ordered according to the study site, in order to account for the similarity of the pollen sourcing vegetation. Pollen proportions within and among sites account for the type of bee-product and for the expected high variability in the taphonomical processes involved. Only samples clearly recognized as natural sediments can be used to test the 'null hypothesis' (Strong 1982), corroborating the occurrence of bee-products among the analyzed materials.

Multivariate analysis provides hints about the possibility of discriminating different categories of bee-products on a statistical base. At the Forcello site (Figure 9a; Castellano et al. 2017) natural aquatic sediments marked by windborne tree pollen are clearly separated (CA axis 1) from honeycomb components

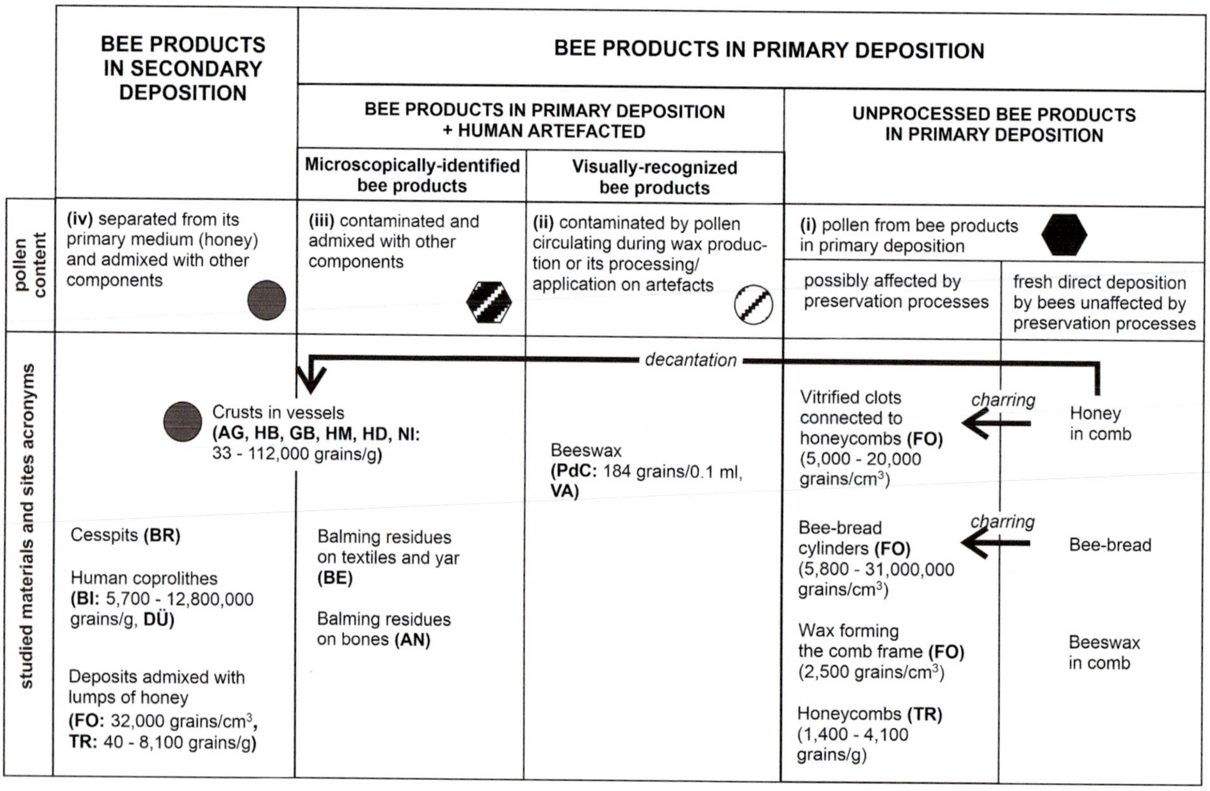

Figure 7. Classification of the reviewed bee-product remains preserved in archaeological and natural media, based both on genetic processes and on microbotanical (palynological) evidence. The classification proposed in this scheme, as well as patterns in symbols, is used to differentiate samples shown in Figs. 8 and 9. For sites acronyms and references, please see Table 1. Sites location is shown in Figure 2.

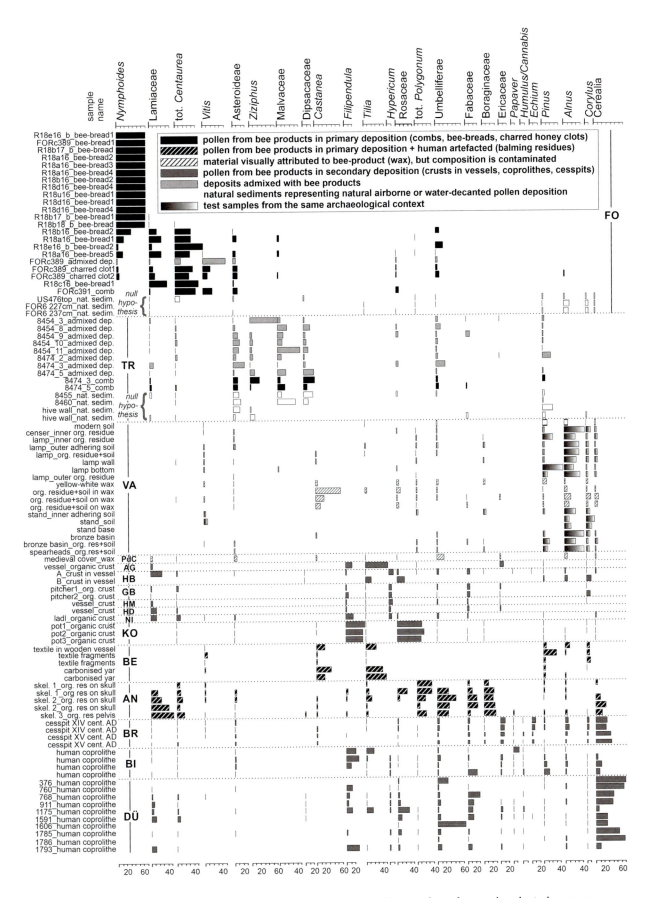

Figure 8. Synoptic frame of published pollen spectra yielding evidence of bee-products from archaeological contexts across Europe and the Levant and referred to the period from the Bronze Age to the 15th century CE. Gray tones and patterns are used to differentiate bee product depositions and subsequent taphonomical processes (see conceptualization in Figure 7).

dominated by pollen from insect-pollinated flowers. Bee-bread composition reflects pollen preferences (at the extreme high *Nymphoides*), while honeycomb composition is believed to originate mostly from honey and thus from nectariferous flowers (at the extreme low *Vitis*) (CA axis 2). A principal component analysis (Figure 9b) was performed on the pollen spectra presented in Figure 8. Archaeological bee-products in primary deposition are dominated by a few pollen taxa of entomophilous plants (e.g. Lamiaceae, *Centaurea*, Dipsacaceae, Malvaceae, Fabaceae) and do not contain a dominance of pollen from obligatory anemophilous plants (e.g. *Alnus*, *Corylus*). This feature is discriminated along the first PCA axis. The occurrence of bee-products in samples from the sites of Vani (Chichinadze and Kvavadze 2013) and Bedeni (Kvavadze 2016) needs to be critically assessed and corroborated by evidence other than palynology: the abundance of pollen from species with facultative pollination (*Castanea*) does not allow

us, in fact, to indicate unequivocally the occurrence of bee-products. The biodiversity range of entomophilous taxa is very diverse in archaeological bee-products, a fact depicted along the second PCA axis, likely reflecting floristic diversity (Rösch 2005) and the number of different biomes encompassed in this review. Secondary-deposited bee-products may contain abundant pollen from anemophilous plants (e.g. Cerealia in samples of human coprolites); yet they are widely distributed along the second axis because of the biodiverse entomophilous flora identifying their bee-derived component. In addition to these quantitative considerations, a qualitative assessment of the flora detected provides further insights into the nature of the evidence here discussed, well-exemplified by the attestation of pollen types absent in the local flora and palaeoflora (e.g., Medieval Belgian latrine deposits, Deforce 2010 and 2017; and Bronze Age burial of Ashgrove, Dickson 1978), which suggests the presence of exotic honey.

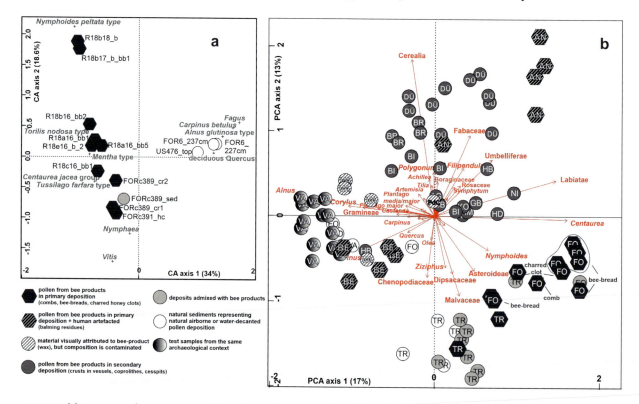

Figure 9. (a) Correspondence Analysis (CA) of the pollen assemblage from Forcello of Bagnolo San Vito (Italy; Castellano et al. 2017); (b), Principal Component Analysis (PCA) of the pollen spectra presented in Figure 8 (see Table 1 for acronyms and literature). Samples are grouped on the basis of criteria detailed in Figure 7, distinguishing between: **(i)** Fossil bee products in primary deposition (black hexagons), either unaffected or affected by preservation processes; **(ii)** material visually recognized as a primary bee-product, but its composition is contaminated (white circles with black transversal lines); **(iii)** Fossil bee-products in primary deposition identified after the analysis of their microscopic botanical content (black circles with white transversal lines) - honey identified on textiles and bones, and interpreted as residues of balming; **(iv)** Secondary-deposited bee-products (dark grey circles)—these bee-products were dispersed in a transporting medium and preserved in an embedding matrix (i.e. decantation crusts in vessels, human coprolites, cesspits, etc.) and were identified by their pollen composition. **(v)** deposits admixed with bee-products (cf type (i), above) (light gray circles); **(vi)** Natural sediments testifying to atmospheric or water-decanted pollen deposition (white circles) - *a priori* classified as devoid of bee-products. Samples included in the analysis originated from the same areas of (i) and (v) and served to test the suitability of the pollen criteria to challenge the *a priori* classification of samples (i) and (v). The use of archaeological deposits, on-site and near-site depositions for this test is to be avoided. **(vii)** test samples from the same archaeological context as (ii), but with no relation with bee products (black to white fading). See also the classification scheme in Figure 7.

Open questions and future research

Palynology offers concrete means to detect the presence of bee-products in archaeological and natural deposits. In addition to corroborating the importance of bee-products in antiquity, this evidence allows us to obtain insights into ancient apicultural practice (e.g., Castellano et al. 2017), circulation and trade of bee-products (e.g., Deforce 2010 and 2017; Dickson 1978), their use in human diet (e.g. Deforce 2010 and 2017; Moe and Oeggl 2014), burial rituals (e.g., Kvavadze et al. 2015; Chichinadze et al. 2019; Rösch 1999), and artisan activities (e.g., Chichinadze and Kvavadze 2013). The full contribution of palynology to the reconstruction of ancient beekeeping and bee-product exploitation still awaits, however, to be fully explored. Emblematic in this regard is the geographic distribution of the studies published to date (Figure 2), which is most likely to be considered in large part an artifactual result of an uneven research coverage, driven by different research priorities and agendas.

In addition to highlighting the efforts undertaken by several authors in detecting and characterizing bee—products in archaeological deposits and materials, the available literature also suggests the need to deploy rigorous scientific protocols for sampling, analytical procedures, and data elaboration. In our opinion, the main efforts to be undertaken for a proper development of experimental palynology for the archaeology of beekeeping are: (i) collection of test samples, both onsite and offsite, to account for the visibility of bee products and verify the null hypothesis; (ii) increase our understanding on taphonomy, preservation and degradation through time of microscopic organic particles in bee products in primary and secondary deposition; and (iii) to develop open access palynological databases on modern honey and wax compositions, allowing for a proper comparison with fossil evidence. It is aim of future palynological research to address these issues. We are confident that the development of an appropriate experimental approach and, especially, a more systematic incorporation of palynological analysis in archaeological research will significantly increase our understanding of beekeeping practices in ancient times, shedding much needed light on this crucial yet somehow archaeologically elusive component of past subsistence and the ancient economy.

Bibliography

Anderson, K. E. et al. 2014. Hive-stored pollen of honeybees: Many lines of evidence are consistent with pollen preservation, not nutrient conversion. *Molecular Ecology* 23 (23): 5904-5917.

Barker, R. J. and Y. Lehner. 1978. Laboratory comparison of high fructose corn syrup, grape syrup, honey, and sucrose syrup as maintenance food for caged honeybees. *Apidologie* 9 (2): 111-116.

Beil, M., H. Horn and A. Schwabe. 2008. Analysis of pollen loads in a wild bee community (*Hymenoptera: Apidae*) – a method for elucidating habitat use and foraging distances. *Apidologie* 39: 456-467.

Biel 1981, J. 1981. The late Hallstatt chieftain's grave at Hochdorf. *Antiquity* 55: 16-18.

Bloch, G. et al. 2010. Industrial apiculture in the Jordan valley during Biblical times with Anatolian honeybees. *PNAS* 107(25): 11240-11244.

Bogdanov, S. 2004. Beeswax: quality issues today. *Bee World* 85(3): 46-50.

Bosi, G. et al. 2011. Seeds/fruits, pollen and parasite remains as evidence of site function: piazza Garibaldi e Parma (N Italy) in Roman and Mediaeval times. *Journal of Archaeological Science* 38: 1621-1633.

Bottema, S. 1975. The interpretation of pollen spectra from prehistoric settlements (with special attention to *Liguliflorae*). *Palaeohistoria* 17: 17-35.

Bronstein, J. L., R. Alarcón and M. Geber. 2016. The evolution of plant–insect mutualisms. *New Phytologist* 172: 412-428.

Bryant, V. M. 2014. The basics of honey identification. *Bee Culture*: 59-63.

Bryant, V. M. and G. D. Jones. 2001. The R-Values of Honey: Pollen Coefficients. *Palynology* 25: 11-28.

Bueno Ramirez, P., R. Barroso Bermejo and R. De Balbin Behrmann. 2005. Ritual campaniforme, ritual colectivo: la necropolis de Cuevas Artificiales del Valle de Las Higueras, Huecas, Toledo. *Trabajos de Prehistoria* 62(2): 67-90.

Castellano, L. et al. 2017. Charred honeycombs discovered in Iron Age Northern Italy. A new light on boat beekeeping and bee pollination in pre-modern world. *Journal of Archaeological Science* 83: 26-40.

Chichinadze, M. and E. Kvavadze. 2013. Pollen and non-pollen palynomorphs in organic residue from the hoard of ancient Vani (western Georgia). *Journal of Archaeological Science* 40: 2237-2253.

Chichinadze, M. et al. 2019. Palynological evidence for the use of honey in funerary rites during the Classical Period at the Vani. *Quaternary International* 507: 24-33

Chittka, L., J. D. Thomson and N.M. Waser. 1999. Flower constancy, insect psychology, and plant evolution. *Naturwissenschaften* 86: 361–177.

Coggshall, W.L. and R. A. Morse. 1984. *Beeswax: Production, Harvesting, Processing, and Products*. Cheshire, Conn.: Wicwas Press.

Crane, E. 1990. *Bees and Beekeeping: Science, Practice and World Resources*. Oxford: Heinemann Newnes.

Crane, E. 1999. *The World History of Beekeeping and Honey Hunting*. New York: Routledge

Da Luz, C. F. P. and O.M. Barth. 2012. Pollen analysis of honey and beebread derived from Brazilian mangroves. *Brazilian Journal of Botany* 35(1): 79-85

Davey, C. J. 2009. The early history of lost-wax casting, in J. Mei and T. Rehren (eds) *Metallurgy and Civilisation: Eurasia and Beyond*: 147-154. London: Archetype.

De Freitas, A. et al. 2015. A melissopalynological analysis of Apis mellifera L. loads of dried bee pollen in the southern Brazilian macro-region. *Grana* 54(4): 305-312.

De la Barrera, E. and P. Nobel. 2004. Nectar: Properties, floral aspects, and speculations on origin. *Trends in Plant Science* 9(2): 65-69.

De Marinis, R. C. 2016. La datazione della fase F del Forcello di Bagnolo San Vito (MN), in S. Lusuardi Siena et al. (eds) *Archeologia classica e post-classica tra Italia e Mediterraneo. Scritti in ricordo di Maria Pia Rossignani*: 159-172. Milan: Vita e Pensiero.

De Marinis, R.C., and M. Rapi. 2007. *L'abitato etrusco del Forcello di Bagnolo San Vito. Le fasi di eta' arcaica.* Florence: Edizioni Latini.

Deforce, K. 2010. Pollen analysis of 15th century cesspits from the palace of the dukes of Burgundy in Bruges (Belgium): evidence for the use of honey from the western Mediterranean. *Journal of Archaeological Science* 37: 337-342.

Deforce, K. 2017. The interpretation of pollen assemblages from medieval and post-medieval cesspits: new results from northern Belgium. *Quaternary International* 460: 124-134.

de-Sá-Otero, P.M., S. Armesto-Baztán, and E. Díaz-Losada. 2005. Initial data on the specific heterogeneity found in the bee pollen loads produced in the "Baixa Limia-Serra do Xurés" nature reserve. *Acta Botanica Gallica* 152 (3): 361-375.

de-Sá-Otero, P.M., S. Armesto-Baztán, and E. Díaz-Losada. 2007. Initial data on the specific heterogeneity found in the bee pollen loads produced in the Pontevedra region (north-west Spain). *Grana* 46(4): 300-310.

Dickson, J.H. 1978. Bronze age mead. *Antiquity* 52: 108-113.

Dozier, C. A. 2016. Saccharomyces cerevisiae Fermentation Effects on Pollen: Archaeological Implications. *Ethnobiology Letters* 7(1): 32–37.

Drieu, L. et al. 2018. Domestic activities and pottery use in the Iron Age Corsican settlement of Cuciurpula revealed by organic residue. *Journal of Archaeological Science: Reports* 19: 213-223.

Evershed, R. P. 2008. Organic residue analysis in archaeology: the archaeological biomarker revolution. *Archaeometry* 50(6): 895-924.

Evershed, R. P. et al. 2003. New Chemical Evidence for the Use of Combed Ware Pottery Vessels as Beehives in Ancient Greece. *Journal of Archaeological Science* 30: 1-2.

Evershed, R. P. et al. 1997. Fuel for thought? Beeswax in lamps and conical cups from Late Minoan Crete. *Antiquity* 71: 979-85,

Evert, R. F. and S.E. Eichhorn. 2013. *Raven Biology of Plants.* Eighth Edition. New York: W.H. Freeman and Company Publishers.

Free, J.B. 1993. *Insect Pollination of Crops.* 2nd Edition. London: Academic Press.

Furness, C. A. 1994. The extraction and identification of pollen from a beeswax statue. *Grana* 33(1): 49-52.

Gallagher, D. E. 2014. Formation Processes of the Macrobotanical Record, in J. Marston, J. d'Alpoim Guedes and C. Warinner (eds). *Method and Theory in Paleoethnobotany*: 19-34. Boulder: University Press of Colorado.

Goppelsröder, A. and M. Rösch. 2002. Pflanzliche Funde aus dem keltishen Grabhügel Hochmichele, Gemeinde Altheim (Kreis Biberack), in S. Kurz and S. Schiek (eds) *Bestattungsplätze im Umfeld der Heuneburg. Forschungen und Berichte zur Vor- und Frühgeschichte in Baden-Württemberg*, 87: 163-203. Stuttgart: Theiss.

Greig, J. R. A. 1989. Pollen preserved by copper salts. In Körber-Grohne, U. and Küster, H.-J. (eds) *Archäobotanik*. Dissertationes Botanicae 133, 11–24. Berlin: J Cramer.

Harano, K. and M. Sasaki. 2015. Adjustment of honey load by honeybee pollen foragers departing from the hive: the effect of pollen load size. *Insectes Sociaux* 62: 497-505.

Hepburn, H.R., C. W. W. Pirk, and O. Duangphakdee. 2014. *Honeybee Nests: Composition, Structure, Function.* Berlin-Heidelberg: Springer-Verlag.

Heron, C., N. Nemcek and K. M. Bonfield. 1994. The Chemistry of Neolithic Beeswax. *Naturwissenschaften* 81: 266-269.

Huisman, D.J. 2009. *Degradation of Archaeological Remains.* Den Haag: SDU Uitg.

Jacob, H. 1979. Pollenanalytische Untersuchung von merowingerzeitlichen Honigresten. *Alt Thür* 16: 112–119.

Jones, G. D. and V. M. Bryant. 1992. Melissopalynology in the United States: A review and critique. *Palynology* 16(1): 63-71.

Kellen, T. 1889. Floating Apiaries in Egypt. *Gleanings in Bee Culture* 17(3): 83.

Ketcham, C. 2020. *Flowers and Honeybees: A Study of Morality in Nature.* Rodopi: Brill.

Klein, A. M. et al. 2007. Importance of pollinators in changing landscapes for world crops. *Proceedings of the Royal Society* B 274: 303-313.

Körber-Grohne, U. 1985. Die biologischen Reste aus Hochdorf, Gemeinde Eberdingen (Kreis Ludwigsburg), in H. Küster et al. (eds), *Hochdorf. Die biologischen Reste aus dem hallstattzeitlichen Fürstengrab von Hochdorf, Gemeinde Eberdingen / Udelgard Körber-Grohne / 1 Neolithische Pflanzenreste aus Hochdorf, Gemeinde Eberdingen (Kreis Ludwigsburg)* (Baden-Württemberg 19): 85-265. Stuttgart: Theiss.

Kozáková, R. et al. 2017. Food offerings, flowers, a bronze bucket and a waggon: a multidisciplinary approach regarding the Hallstatt princely grave from Prague-Letňany, Czech Republic. *Archaeological and Anthropological Sciences* 11: 209-221.

Kvavadze, E. 2016. Palynological study of organic remains from the Ananauri kurgan, in *Ananauri Big Kurgan*: 156-192. Tbilisi: Shota Rustaveli National Science Foundation.

Kvavadze, E. and I. Martkoplishvili. 2018. The significance of pollen and non-pollen palynomorphs in archaeological material for human paleodiet reconstruction, in A. Batmaz et al. (eds) *Context and Connection: Essays on the Archaeology of the Ancient Near East in Honour of Antonio Sagona*: 749–767. Leuven ; Bristol, CT: Peeters.

Kvavadze, E. et al. 2015. The hidden side of ritual: New palynological data from Early Bronze Age Georgia, the Southern Caucasus. *Journal of Archaeological Science*: Reports 2: 235-245.

Kvavadze, E.V. 2006. The use of fossilized honey for paleoecological reconstruction: a palynological study of archeological material from Georgia. *Paleontological Journal* 40(5): 595-603.

Kvavadze, E. et al. 2007. The first find in southern Georgia of fossil honey from the Bronze Age, based on palynological data. *Vegetation History and Archaeobotany* 16: 399-404.

Layek, U., S. S. Manna and P. Karmakar. 2020. Pollen foraging behavior of honeybee (*Apis mellifera L.*) in southern West Bengal, India. *Palynology* 44(1): 114-126.

Louveaux, J., A. Maurizio and G. Vorwohl. 1978. Methods of Melissopalynology. *Bee World* 59(4): 139-157.

Marchand, P., A. N. Harmon-Threatt and I. Chapela. 2015. Testing models of bee foraging behavior through the analysis of pollen loads and floral density data. *Ecological Modelling* (313): 41-49.

Martinelli, E. et al. 2019. Pollen from beeswax as a geographical origin indicator of the medieval Evangelistary cover "Pace di Chiavenna", Northern Italy. *Palynology* 43(3): 507-516.

Mayyas, A. et al. 2012. Beeswax preserved in a Late Chalcolithic beveled-rim bowl from the Tehran Plain, Iran. *Iran* (50): 13-25.

Mazar A. and N. Panitz-Cohen. 2007. It is the land of honey: Beekeeping in Iron Age IIA Tel Rehov—culture, cult and economy. *Near Eastern Archaeology* 70:1094-2075.

Mazar A. and N. Panitz-Cohen. 2020. *A Bronze and Iron Age City in the Beth-Shean Valley. Volume II, The Lower Mound: Area C and the Apiary*. QEDEM 60. Jerusalem: The Hebrew University of Jerusalem.

Mazar, A. et al. 2008. The Iron Age beehives at Tel Rehov in the Jordan Valley: Archaeological and analytical aspects. *Antiquity* 82: 629-639.

McGovern, P. E., G. R. Hall and A. Mirzoian. 2013. A biomolecular archaeological approach to 'Nordic grog'. *Danish Journal of Archaeology* 2(2): 112-131.

Moe, D. and K. Oeggl. 2014. Palynological evidence of mead: a prehistoric drink dating back to the 3rd millennium B.C. *Vegetation History and Archaeobotany* 23: 515-526.

Molan, P. C. 1998. The limitations of the methods of identifying the floral source of honeys. *Bee World* 79(2): 59-68.

Mondro, A. F. H. et al. 2009. Analysis of pollen load based on color, physicochemical composition and botanical source. *Anais da Academia Brasileira de Ciências* 81(2): 281-285.

Moore P.D., J.A. Webb and M.E. Collinson. 1991. *Pollen analysis*. Oxford: Blackwell Scientific Publications.

Nesbitt, M. 2006. Archaeobotany, in M. Black, J.D. Bewley and P. Halmer (eds) *The Encyclopedia of Seeds. Science, Technology and Uses*: 20-22. Wallingford, UK: CABI.

Oliveira, C., et al. 2019. Chromatographic analysis of honey ceramic artefacts. *Archaeological and Anthropological Sciences* 11: 959-971.

Perego, R. et al. 2011. L'origine del paesaggio agricolo pastorale in nord Italia: espansione di *Orlaya grandiflora (L.) Hoffm.* nella civiltà palafitticola dell'età del Bronzo nella regione del Garda. *Notizie Archeologiche Bergomensi* 19: 161-173.

Pini, R. et al. 2018. Effects of stepped-combustion on fresh pollen grains: morphoscopic, thermogravimetric, and chemical proxies for the interpretation of archaeological charred assemblages. *Review of Palaeobotany and Palynology* 259: 142-158.

Piotrowska, K. 2008. Ecological features of flowers and the amount of pollen released in *Corylus avellana (L.)* and *Alnus glutinosa (L.) Gaertn. Acta Agrobotanica* 61(1): 33–39.

Ponnuchamy, R. et al. 2014. Honey Pollen: Using Melissopalynology to Understand Foraging Preferences of Bees in Tropical South India. *PLoS ONE* 9(7): e101618.

Ponomarenko, E. et al. 2019. A multi-proxy analysis of sandy soils in historical slash-and-burn sites: A case study from southern Estonia. *Quaternary International* 516: 190-206.

Prieto Martinez, M. P., J. Tresserras and J. C. Matamala. 2005. Ceramic production in the northwestern Iberian Peninsula: studying the functional features of pottery by analyzing organic materials, in M. I. Prudencio, M. I. Dias and J. C. Waerenborgh (eds) *Understanding people through their pottery. Trabalhos de Arqueologia* 42: 193-199.

Proctor, M., P. Yeo and A. Lack. 1996. *The Natural History of Pollination*. Oregon: Timber Press.

Quirino, T. 2011. Le case F I e F II del Forcello di Bagnolo San Vito (MN): analisi preliminare di due abitazioni etrusche di fine VI secolo a.C. Il filo del tempo. Studi di preistoria e protostoria in onore di Raffaele Carlo de Marinis. *Notizie Archeologiche Bergomensi* 19: 379-390.

Rageot, M. et al. 2019. The dynamics of Early Celtic consumption practices: A case study of the pottery from the Heuneburg. *PLoS ONE* 14(10): e0222991.

Rageot, M. et al. 2016. Exploitation of beehive products, plant exudates and tars in Corsica during the Iron Age. *Archaeometry* 58 (2): 315-332.

Ravazzi, C. et al. 2013. Lake evolution and landscape history in the lower Mincio River valley, unravelling drainage changes in the central Po Plain (N-Italy) since the Bronze Age. *Quaternary International* 288: 195-205

Ricciardelli D'Albore, G. 1998. *Mediterranean Melissopalynology*. Perugia: Università degli studi di Perugia, Facoltà di agraria, Istituto di entomologia agraria.

Roffet-Salque, M. et al. 2015. Widespread exploitation of the honeybee by early Neolithic farmers. *Nature* 527: 226-231.

Rösch, M. 1999. Evaluation of honey residues from Iron Age hill-top sites in south-western Germany: implications for local and regional land use and vegetation dynamics. *Vegetation History and Archaeobotany* 8: 105-112.

Rösch, M. 2005. Pollen analysis of the contents of excavated vessels - direct archaeobotanical evidence of beverages. *Vegetation History and Archaeobotany* 14: 179-188.

Rösch, M. 2008. New aspects of agriculture and diet of the early medieval period in central Europe: waterlogged plant material from sites in south-western Germany. *Vegetation History and Archaeobotany* 17(Suppl 1): 225-238.

Rösch, M. 2014. Direkte archäologische Belege für alkoholische Getränke von der vorrömischen Eisenzeit bis ins Mittelalter, in J. Drauschke, R. Prien and A. Reis (eds) *Tagungsbeiträge der Arbeitsgemeinschaft Spätantike und Frühmittelalter 7. Produktion, Vorratshaltung und Konsum in Antike und Frühmittelalter. Studien zu Spätantike und Frühmittelalter*: 305-326. Hamburg: Verlag Dr. Kovac.

Rösch, M. and S. Rieckhoff. 2019. Alkohol in der Eisenzeit. Anmerkungen aus botanischer und archäologischer Sicht, in P.W. Stockhammer and J. Fries-Knoblach (eds.) *Was tranken die frühen Kelten? Bedeutungen und Funktionen mediterraner Importe im früheisenzeitlichen Mitteleuropa: BEFIM 1*: 101-112. Leiden: Sidestone Press.

Sajwani, A., S. A. Farooq and V. M. Bryant. 2014. Studies of bee foraging plants and analysis of pollen pellets from hives in Oman. *Palynology* 38(2): 207-223.

Shillito, L.-M. et al. 2020. The what, how and why of archaeological coprolite analysis. *Earth-Science Reviews*, 103196.

Steffan-Dewenter, I. and A. Kuhn. 2003. Honeybee Foraging in Differentially Structured Landscapes. *Proceedings: Biological Sciences* 270: 569-575.

Stika, H.-P. 2010. Früheisenzeitliche Met- und Biernachweise aus Süddeutschland. *Archäologische Informationen* 33(1): 113-121.

Strong, D.R. 1982. Null Hypotheses in Ecology, in E. Saarinen (ed) *Conceptual Issues in Ecology*: 245-259. Dordrecht: Springer.

Tipping, R. 1994. "Ritual" floral tributes in the Scottish Bronze Age — Palynological evidence. *Journal of Archaeological Science* 21: 133-139.

Van der Veen, M. 2018. Archaeobotany: the archaeology of human-plant interactions, in W. Scheidel (ed) *The Science of Roman History. Biology, Climate, and the Future of the Past*: 53-94. Princeton: Princeton University Press.

Weiner, S. 2010. *Microarchaeology: Beyond the Visible Archaeological Record*. Cambridge: Cambridge University Press

Weinstein-Evron, M. and S. Chaim. 2016. Palynological investigations of tenth- to early ninth-century BCE beehives from Tel Rehov, Jordan Valley, northern Israel. *Palynology* 40(3): 289-301.

White, J. W. 1978. Honey. *Advance in Food Research* 24: 287-374

Willemstein, S.C. 1987. *An evolutionary basis for pollination ecology*. Leiden Botanical Series 10. Leiden: Brill.

Zander, E. 1935. Beiträge zur Herkunftsbestimmung bei Honig. I. Pollengestaltung und Herkunftsbestimmung bei Blütenhonig. Verlag der Reichsfachgruppe Imker E.V, Berlin.

Chapter 6

La apicultura en el *ager* de Segóbriga (Cuenca, Spain): Un caso de estudio en el centro peninsular (ss. I d.C. al XI d.C.)

Jorge Morín de Pablos

Departamento de Arqueología. Audema (jmorin@audema.com)

Rui Roberto de Almeida

Fundaçao para a Ciencia e Tecnología, Portugal (rui.dealmeida@gmail.com)

Isabel Sánchez Ramos

Seminario de Arqueología, Departamento de Geografía, Historia y Filosofía, Universidad Pablo de Olavide de Sevilla (imsanram@upo.es)

Resumen: El presente estudio aborda la producción, el consumo y la circulación de la miel en la *Hispania* romana. En cierta medida, se puede decir, de una forma genérica, que la actividad apícola, es decir, la que se entiende y se refiere no tanto a la recolección, sino a la producción controlada y orientada a la extracción de la miel como una actividad más del calendario agrícola, se viene interpretando en la actualidad como una actividad complementaria, con cuyo desarrollo se consiguen rentabilizar mayoritariamente las zonas de escasa productividad agrícola. Sin embargo, el reciente descubrimiento de numerosos ejemplares de colmenas cerámicas en el reborde oriental meseteño, concretamente en el entorno rural de la antigua ciudad de *Segobriga*, en la provincia de Cuenca, complementado con otros hallazgos en el *ager* de la ciudad, así como en el entorno de la ciudad de Braga en Portugal, permite aportar nuevos datos al conocimiento de la producción de miel y una nueva dimensión a la actividad apícola, particularmente en lo que se refiere al periodo romano. En este sentido, no se puede asegurar que se trate de una actividad propia de zonas de *Hispania* poco productivas desde el punto de vista agropecuario, sino que hay que ver la producción apícola como una actividad más, que se concibe con criterios intensivos en la época y el territorio que nos ocupa.

Palabras clave: Apicultura, colmenas, *Segobriga*, *Hispania*.

Summary: This study focuses on the production, consumption and circulation of honey in Roman Hispania. To a certain extent, it has been assumed that beekeeping activity, which was a controlled production and oriented to the extraction of honey, was one of the agricultural calendar activities. At present it has been interpreted as a complementary agricultural activity too, one which made possible the development of those areas of scarce agricultural productivity. However, recent discoveries of ceramic hives on the border of the Iberian Eastern central plateau, specifically in the rural area belonging to the ancient Roman city of *Segobriga* (province of Cuenca), alongside with other findings in the rural areas of the city of Braga in Portugal, provide new data related to honey production and a new dimension of beekeeping during the Roman period. In this sense, it cannot be confirmed that it is a specific activity developed in areas that are not very productive from an agricultural point of view. On the contrary, beekeeping production must be understood as just another agricultural activity, in light of intensive criteria at the time and in the territory here under investigation.

Keywords: Beekeeping, ceramic beehives, *Segobriga*, Hispania, Roman Spain

Metodología

Es bien conocida y globalmente aceptada la importancia de la apicultura a lo largo de toda la antigüedad preclásica y clásica, así como durante el período medieval, siendo la miel el principal producto que se obtenía de dicha actividad, uno de los productos de la alimentación más importante y el único edulcorante conocido hasta finales del s. XV, fecha en que empezó a introducirse la caña de azúcar en Europa (Carmona 1999: 131). La miel era un bien relativamente escaso y poco frecuente, apenas asequible para unos pocos, aunque era conocida desde etapas prehistóricas.

Pese al valor económico de esta actividad, reconocido por los estudiosos del mundo clásico, son pocos los trabajos que se han detenido en analizar la miel, la apicultura o los elementos propios de su actividad en épocas antiguas para el área peninsular (Martín Tordesillas 1968; Fernández Uriel 1990; Vázquez Hoys 1991) y fuera del ámbito ibérico (Bortolin 2008).

La propia naturaleza «poco visible» de la actividad apícola, con tendencia a desarrollarse en zonas serranas o marginales, presumiblemente de menor potencial agrícola y difícilmente utilizables para la realización de otras actividades, conlleva que se entienda como una actividad secundaria. Por otra parte, la escasa representatividad arqueológica de sus productos y aperos, quizás haya contribuido igualmente de forma indirecta a su desvalorización, colocándola en un segundo plano, en detrimento de otras actividades productivas/transformadoras, como la olivícola, la vitivinícola o la salazonera, de cuyas instalaciones productivas/de procesado, de los productos manufacturados envasados y respectivos contenedores, así como de sus ejes de distribución y circuitos de comercialización, existen abundantes testimonios y evidencias directas.

En cierta medida, se puede concluir, de una forma genérica, que la actividad apícola, es decir, la que se refiere, no tanto a la recolección, sino a la producción controlada y orientada a la extracción de la miel como una actividad más del calendario agrícola (Bonet y Mata 1995: 277), se viene interpretando en la actualidad como una actividad complementaria, con cuyo desarrollo se consiguen rentabilizar mayoritariamente las zonas de escasa productividad agrícola. Sin embargo, el reciente descubrimiento de numerosos ejemplares de colmenas cerámicas en el reborde oriental meseteño, concretamente en el entorno rural de la antigua ciudad de *Segobriga*, en la provincia de Cuenca, permite aportar nuevos datos al conocimiento de la producción de miel y una nueva dimensión a la actividad apícola, particularmente en lo que se refiere al periodo romano. En este sentido, no se puede asegurar que se trate de una actividad propia de zonas de *Hispania* poco productivas desde el punto de vista agropecuario, sino que hay que ver la producción apícola como una actividad más, que se concibe con criterios intensivos en la época y el territorio que nos ocupa (de Almeida y Morín 2012: 63-81). De hecho, en el mismo espacio, el *ager* segobricense hemos trabajado para otros períodos como el tardorromano, hispanovisigodo y andalusí, siendo incapaces de documentar en el registro arqueológico dicha actividad, que si está documentada etnográficamente en la zona. Finalmente, trabajos monográficos sobre la Antigüedad Tardía en la región tampoco han sido capaces de detectar la actividad apícola (Barroso 2019), en el territorio de las ciudades de Segóbriga, Ercávica, Valeria y la ciudad palatina de Recópolis.

En el presente trabajo se dejarán sin abordar muchos de los aspectos relacionados con la importancia de la miel en la Antigüedad y otros relacionados con la apicultura e historia de la misma, no solo porque en los últimos años varios autores se han ocupado de forma objetiva y detallada, sino también porque excede en gran medida el objeto del presente estudio: la producción, el consumo y la distribución de la miel en esta zona central de *Hispania*. Sin embargo, para el correcto entendimiento de la apicultura en época romana en *Hispania* se hace necesario abordar, aunque de forma preliminar, aspectos de esa temática, así como incluir de manera sintética y retrospectiva algunas de las líneas de investigación y principales resultados alcanzados en las dos últimas décadas en este campo del mundo productivo rural.

La miel y la apicultura en el Mediterráneo Antiguo

La miel constituía un producto bastante apreciado en la Antigüedad por sus innumerables cualidades y atributos. Prueba de ello es el acervo que se dispone actualmente de representaciones iconográficas, siendo además numeroso el conjunto de referencias por parte de los autores clásicos que resaltan y halagan esta sustancia natural, destacando sus excelsas y beneficiosas propiedades. Plinio, entre otros, se refiere a la miel como «*un jugo dulcísimo, ligerísimo y salubérrimo...*» (Plin. *NH* 11.1. 5), «*...que aporta el gran placer de su naturaleza celestial...*» (Plin. *NH* 11. 1. 12).

Esas y otras características confirieron a la miel un *status* particular y motivaron su utilización de forma recurrente en la cocina, al constituir el alimento edulcorante y energético por excelencia, y debido a que sus propiedades antisépticas la hacían de igual modo idónea para utilizarla en la conservación de frutos y en el mundo de la medicina (Fernández Uriel 1990: 136; Vázquez Hoys 1991: 67). En el *Medicamina* de Ovidio, por ejemplo, se menciona el uso de la miel y la cera para fines médicos y cosméticos. Por otra parte, tampoco debe desestimarse el uso de la cera para alumbrado de calidad, para encerar maderas o papiros, y menos aún como soporte de escritura, siendo la materia prima indispensable para las tablillas de escritura.

Tal y como sucede hoy en día, se conocían distintos tipos y calidades de miel, según la flora de cada región, y según la temporada en la cual se cosechaba. Del Ática procedía la miel más apreciada de la antigüedad, seguida de la de Creta, Rodas, Chipre, África, Córcega, Italia y España (Saglio 1900: 1704).

En lo que se refiere al proceso de extracción/cosecha de la miel, se sabe, a través de los relatos que compilaron y que nos legaron los autores clásicos, que se realizaba en distintas etapas, y que de cada una de esas fases resultaban calidades distintas: en la primera se obtenía la miel denominada «*mel optimum*», simplemente dejándose escurrir los panales: en la segunda, se prensaban los mismos para extraer la miel que restase en su interior, que se designaba «*mel secundum*», y que era considerada de segunda calidad (Dosi y Schnell 1992: 19; Blanc y Nercessian 1994: 29). Sin embargo, la extracción de productos melíferos o de derivados de la miel no se agotaba con ese proceso. Una vez prensados los panales se podía aún realizar un lavado de éstos con agua templada, del cual resultaba un líquido dulce que se utilizaba posteriormente en la preparación de hidromiel (que se consumía ampliamente en la Antigüedad), de vinos melados o de refrescos avinagrados (vd. Cruz 1997; *apud* Morais 2006). De ese modo, la miel era habitualmente utilizada en el vasto mundo de los productos y subproductos vitivinícolas, haciéndose indispensable como complemento en la elaboración de vinos dulces (extremadamente apreciados en momentos iniciales del Imperio romano, particularmente en época de Augusto) o de otras bebidas como el «*mulsum*»[1] y el mosto (Morais 2006: 149-150).

Entre los autores latinos clásicos que más versaron sobre la miel y sus usos destacan Plinio, Varron o Columela. Plinio dedicó muchas páginas de su «*Historia Natural*» a la miel y la apicultura, coincidiendo con Aristóteles en cuales eran las mejores mieles de la época. Varron y Columela procedieron a la observación directa para elaborar sus teorías que, aún en la actualidad, son una fuente útil de información. Es sobre todo este último, quien, en el noveno de sus «*Doce Libros de Agricultura*», trata de forma detallada y minuciosa todos los aspectos entonces valorados acerca de la cría de abejas y de la miel, de su correcta producción y recolección, calendario de cosechas, cuidados de las abejas, así como acerca de los principios, técnicas y útiles a utilizar en su práctica.

La evidencia iconográfica de mayor antigüedad que ilustra objetivamente la actividad apícola se remonta al antiguo Egipto, con distintos dibujos representativos de las abejas en las diferentes dinastías, así como numerosas escenas de trabajo apícola, recolección de miel o contabilización de las cosechas (Crane 1983: 35-39); aunque en la Península Ibérica contamos con escenas de recolección más antiguas, como las de la Cueva de la Araña -Bicorp, Valencia-, pertenecientes a las representaciones esquemáticas del Epipaleolítico levantino. A pesar de éstas y de otras innumerables referencias, son raros los hallazgos arqueológicos conocidos en el área mediterránea donde se documente de forma inequívoca la actividad apícola, es decir colmenares o colmenas, puesto que, tal como bien señalaron Bonet y Mata (1995: 279): «*la colmena es precisamente la pieza que permite distinguir la apicultura de la simple recolección de la miel...*». Así mismo, la propia naturaleza de las colmenas se convierte en un factor importante y en un enorme condicionante a la hora de estudiar los recipientes de la producción melífera. Conviene recordar que, tal como señala Columela, las colmenas podían estar realizadas en distintos materiales (*R.R. 9. 6. 76*):

«(...) *de corcho, porque no estarían muy frías en el invierno ni muy calientes en el verano..., si no hubiere... se harán con mimbres entretejidos; y si estas no se encuentran se fabrican con troncos de árboles excavados... Las peores de todas son las de barro cocido, ya que se encienden con los calores del estío y se hielan con los fríos del invierno. Las demás especies que hay de colmenas son dos, unas que hacen con boñiga y otras se construyen con ladrillos (...)*».

De ese modo, es comprensible y justificable la escasez e «invisibilidad» de dichos vestigios, y tal hecho se debe mayoritariamente al tipo de materiales que se pueden encontrar en el registro arqueológico. Consecuentemente, la arqueología solo ha podido constatar, hasta la fecha, aquellas piezas elaboradas con material no perecedero, es decir, las colmenas cerámicas.

En ese sentido, son excepcionales los hallazgos del apiario de Tal Rehov (Israel). En este colmenar, hasta la fecha el más antiguo conocido (Figura 1), se practicaba ya de forma aparentemente industrial, una apicultura intensiva en el s. IX a.C.[2] (Mazar y Panitz-Cohen 2007: 202-203). Otros hallazgos de colmenares y colmenas conocidos son los procedentes de yacimientos que se han convertido en una referencia obligatoria, como la villa griega de Cave of Pan –Vari- (Jones et al. 1973), del Ágora de Atenas, Marathon, Corintos (Crane 1983), así como en las islas de Chios o Creta (Bortolin 2008: 73-78, apud Morais 2008: 81).

---

[1] Se trata del vino mezclado con miel, que fue acompañamiento obligado de la *gustatio*, *promulsis*, y al que aludió Celso. Su forma de preparación fue variada y se escogían los más valiosos vinos y la miel de mejor calidad. El mejor *mulsum vetus*, se hacía con vinos añejos, en los que se mezclaba una parte de miel por dos de vino -según Dioscórides y Macrobio-. Otras recetas de *mulsum* mencionan 10 libras de miel por 13 litros de vino, o una de miel por cuatro de mosto fermentado. *Apicius* usó el *mulsum* en determinadas salsas, empleándose en dicho sentido para hacerlas más dulces o espesarlas. El *mulsum* se usó en múltiples recetas y combinados -ensalada de malvas cocida en salsa al vino dulce, salsas diluidas con *mulsum* y aceite, *patinae* bañadas con *mulsum*, etc.- (Beltrán Lloris 2002: 199-200).

[2] Las colmenas cerámicas eran de forma cilíndrica, con 80 cm de longitud por 40 cm de anchura, y se encontraban superpuestas en varias hiladas sobre un muro de base hecho con adobes, en tres calles paralelas. En las colmenas mejor preservadas se identificó, por un lado, el cierre de arcilla con el agujero utilizado para el paso de las abejas, y por otro la tapadera cerámica, que permitía la apertura y extracción de los panales. Los análisis realizados sobre las colmenas comprobaron su uso y revelaron la presencia de moléculas de cera en su interior (Mazar y Panitz-Cohen 2007: 205).

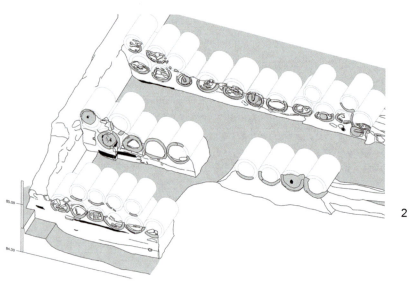

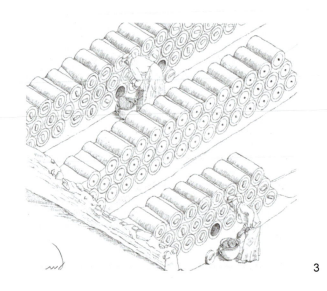

Figure 1. Colmenar de Tal Rehov –Israel-, s. IX a.C. 1: Hilera oriental; 2: Vista axonométrica;
3: Reconstrucción –según Mazar y Panitz-Cohen 2007; 2005-2007.

No obstante, a pesar de las reservas y desventajas que apuntaban los tratadistas latinos a las colmenas hechas con cerámica, al considerar que eran las menos recomendables[3], su uso es un hecho incuestionable desde la Antigüedad, tal como ha quedado reflejado hasta momentos bien cercanos en el tiempo, según se puede observar por paralelos conocidos desde el lejano oriente -Nahaf, Galilea- (Mazar y Panitz-Cohen 2007: 214-215), Jordania, Egipto, Chipre, Grecia (Jones et al. 1973; Crane 1983), Mallorca (Rosselló 1966: 34 y 74) o en Andalucía, ya en territorio peninsular (Martín Morales 1981).

Es bastante probable que el inconveniente térmico se pudiera solventar con una protección adecuada (Bonet y Mata 1995: 280), como ramas, broza, o arcilla, materiales fáciles de trabajar y transportables (Bortolin 2008: 66-69, *apud* Morais 2008: 81) o inclusive que se viera compensado por el aislamiento proporcionado por la arcilla/adobe utilizado en el emparedamiento de las colmenas en el colmenar. Según Bortolin la gran ventaja de las colmenas cerámicas asentaba en su carácter móvil, que evitaba tener que recorrer vastas áreas en busca de fuentes directas para captar/recolectar la miel, a la vez que permitía colocarlas en locales estratégicos inmediatos para la producción según cada momento (Bortolin 2008, *apud* Morais 2008: 82).

En lo que se refiere a la ubicación de los colmenares, parece ser que el lugar idóneo debía ser cercano a las zonas de vivienda, resguardado del frío y del calor, debiendo existir agua en las zonas más cercanas, bien como plantas aromáticas y árboles frutales (Bonet y Mata 1995: 280). Virgilio refleja en su «*Geórgicas*» la importancia de una localización adecuada: (...) «*lo primero es buscar un lugar acomodado para las abejas, en que ni penetren los vientos, ni vayan las ovejas y los cabritillos a pisotear las flores*» (...).

Resulta interesante verificar que en los pocos casos conocidos de colmenas *in situ*, éstas se localizaban en zonas internas de los poblados, como por ejemplo en sitios tan lejanos y cronológicamente distintos como son Tal Rehov –Israel-[4] (Mazar y Panitz-Cohen 2007: 214-215) o Puntal dels Llops (Bonet y Mata 2002).

De una forma genérica, la iconografía antigua, los relatos de los autores latinos, los vestigios arqueológicos y los mismos paralelos etnográficos

muestran que existen notables similitudes entre los recipientes y las posibles prácticas apícolas entre ámbitos geográficos tan lejanos, como son el extremo oriente y el occidente mediterráneo, o cronologías tan distintas, como la Primera Edad del Hierro y la época moderna/contemporánea. Así mismo, parece hoy evidente que se trata de una tradición con varios milenios de existencia que se puede haber desarrollado desde Oriente a Occidente y que todavía continúa con la práctica tradicional de estas explotaciones (Figura 2).

No sólo en el ámbito mediterráneo, como parecía deducirse de los ejemplares conservados en la cultura ibérica, sino también en el interior peninsular, gracias al hallazgo de las colmenas cerámicas en el territorio de la ciudad de Segóbriga. En este sentido, la supervivencia de la práctica tradicional de la apicultura en regiones del interior es un argumento a favor de lo que hasta ahora hemos expuesto. También en la apicultura tradicional del interior vamos a encontrarnos colmenares horizontales, que parecen propios de tradiciones mediterráneas intensivas, junto a colmenares verticales, más propios de una práctica trashumante individual (Figura 3).

Los espacios de producción: colmenas y colmenares en la Península Ibérica. De la época ibérica a la romana.

Las referencias literarias clásicas certifican no solo la existencia de miel producida en *Hispania*, sino que ésta era además exportada y gozaba de gran prestigio. De este hecho dan testimonio varias citas de Plinio (XXI, 74) o Estrabón, particularmente de los productos béticos. Es sobre todo el segundo autor quien proporciona la lista más completa de productos que *Hispania*, y, más concretamente, Turdetania, exportaba a Roma: (...) «*De Turdetania se exporta trigo, mucho vino y aceite; éste, además, no sólo en cantidad, sino de calidad insuperable. Exportase también cera, miel* (...)*»* (Est. 3. 2. 6). A pesar de estas referencias, no se sabe exactamente cual era el papel real de la miel en el escenario de las exportaciones hispanas hacia las demás provincias del Imperio, ni tampoco cuales eran los contenedores adoptados para su transporte (Morais 2006: 149).

No obstante, el primer aspecto que resulta complejo a la hora de referirnos a los envases romanos destinados a la práctica apícola en la Península Ibérica, es el evidente desconocimiento que tenemos de los mismos. El rastreo de muchos de los conjuntos arqueológicos de referencia y la realización de un inventario de esos recipientes refleja un total vacío en la mayor parte de las distintas provincias hispanas.

Paradójicamente, el panorama de la apicultura, de las colmenas en cerámica, y en buena parte las bases para una mejor comprensión de la actividad en época

---

[3] Varrón (*R.R.* 3. 16.16-17) y Columela (*R.R.* 9. 6. 1-4) aluden a la calidad inferior de la miel que se producía en estos recipientes, alegando que se debía al hecho de que no mantenían una temperatura constante.

[4] El colmenar se ubicaba en el espacio interior no amurallado del poblado, en el centro de un área abierta rodeada por una gran densidad de construcciones, tanto de carácter doméstico como público (Mazar y Panitz-Cohen 2007: 209-210).

Figure 2. Ejemplos de colmenares tradicionales modernos y contemporáneos. 1: Colmenar familiar en Nahaf –Israel-, según Mazar y Panitz-Cohen 2007: 214; 2. Colmenas tradicionales del Museum of Settlement de Kibbutz Yifat –Israel-, según Mazar y Panitz-Cohen 2007: 215); 3: Colmenar del Egipto Medio, según Mazar y Panitz-Cohen 2007: 216; 4: Chipre, colmenas horizontales, según Jones et alii 1973; 5: Colmenares de Caeres de Sa Porrasa, Son Ferrer, Mallorca, según PGOU de Mallorca, nº 346.

Figure 3. Ejemplos de colmenares tradicionales en la provincia de Soria. 1: Colmenas horizontales, Arnedillo –Soria-; 2: Colmenas verticales, Santa Cruz de Yanguas –Soria-; 3: Colmena tradicional, Diustes –Soria-; 4. Detalle colmenas Santa Cruz de Yanguas –Soria-, fotografías R. Barroso Cabrera.

romana, proviene de la investigación desarrollada en los últimos años para los momentos anteriores a la presencia romana en la Península, es decir, para el mundo prerromano, más concretamente el ibérico. De ese modo, aunque exceda en parte el ámbito cronológico de este estudio, entendemos que se hace obligatorio una lectura diacrónica y la síntesis de sus principales líneas de conocimiento y aportaciones. En este sentido, tenemos que señalar que es muy probable que la utilización de recipientes cerámicos en los colmenares desde época ibérica nos está probando la existencia de una apicultura intensiva con la utilización de muros colmeneros, que difiere enormemente de la

práctica tradicional trashumante que utiliza colmenas realizadas con materiales perecederos –corcho, mimbre, madera, etc.-. En la práctica trashumante las colmenas se transportaban a hombros de los apicultores o en caballerías –burros y mulas- y su objetivo era garantizar y ampliar la producción de miel. Esta trashumancia comenzaba en el mes de mayo y finalizaba con la llegada de las primeras lluvias y fríos. Las colmenas de disponían de manera vertical y se protegían de los vientos con la construcción de muretes, que todavía hoy en día se pueden ver en algunas regiones de Portugal (Henriques et al. 1999-2000: 329-363). Era habitual que un miembro de la comunidad viviera junto a las colmenas para protegerles de posibles hurtos. Las colmenas se «castraban» una vez al año, aunque en los años de bonanza climática se podían castrar hasta dos y tres veces.

A pesar de las alusiones puntuales a la apicultura en época ibérica (Blázquez 1968), la identificación de ésta arqueológicamente se produce en fechas recientes, cuando la investigación arqueológica aliada a la comparación etnográfica logró identificar los recipientes que se destinaban a albergar en su interior los respectivos panales de miel, las colmenas cerámicas (Bernardes et al. 2014: 507-519). Éstas fueron debidamente caracterizadas y posteriormente reconocidas con particular profusión a partir de mediados de la década de los noventa, destacándose los trabajos de Mata y Bonet (1992), Bonet y Mata (1995; 1997; 2002), de Fuentes Albero, Hurtado Mullor y Moreno Martín (2004), con especial incidencia en el levante peninsular, en el área edetana.

Se dieron a conocer varios ejemplares completos de colmenas, provenientes de las excavaciones del fortín edetano de *Puntal dels Llops*, Monravana y del Tossal de Sant Miquel (Bonet y Mata, 1995; 1997; 2002), y un conjunto muy numeroso de fragmentos con orígenes y procedencias diversas (cerca de 80 yacimientos referenciados, conocidos por excavaciones, prospecciones o recogidas superficiales), en los cuales se apreciaba una considerable variedad de morfologías de bordes (Bonet y Mata 1995: 278-279).

Las colmenas cerámicas ibéricas consisten en unos recipientes cilíndricos abiertos en ambas extremidades, con diámetros comprendidos entre los 24 y 29 cm, una altura entre 53 y 58 cm, y una capacidad de cerca de 48 litros (Bonet y Mata 1995: 280), medidas y capacidades coincidentes con las que tradicionalmente se suelen estimar para las romanas y tradicionales (Crane 1983: table 2, p. 17). Respecto a las diferentes morfologías de los bordes, establecen que los moldurados y los salientes pertenecen al Ibérico Pleno, mientras que en la época iberorromana se asiste a su mayor variedad (Bonet y Mata 1995: 280-281).

Presentan como principal elemento diagnosticable unas típicas estrías incisas y paralelas realizadas en fresco, anteriores a la cocción, o surcos/acanaladuras igualmente paralelos y profundos, en la práctica totalidad de su superficie interior, destinadas a garantizar una mejor fijación de los panales. Según los paralelos arqueológicos y etnográficos conocidos, se trata de un tipo de colmenas fijas que se disponen en una hilera aislada o apiladas sobre el suelo o sobre un murete construido al efecto. Con base en la evidencia de los departamentos del Puntal dels Llops, y, una vez más, también por comparación con los modelos etnográficos de la Kabilia argelina, las Baleares y la

villa griega de Cave of Pan, Bonet plantea la hipótesis de que las colmenas podrían guardarse en el interior de los espacios domésticos mientras no estuvieran en funcionamiento, o bien estar funcionando en las terrazas de las casas (Bonet 1995: 415).

La gran densidad de colmenas del área edetana, y su ausencia o presencia puntual en otras zonas, ha llevado a proponerlas como un tipo propio de dicha área (Bonet y Mata 1995: 282-283). A la par, se consideró que el conjunto de datos documentado permitía considerar que el uso de la miel y la práctica de la apicultura estarían ampliamente difundidos en época ibérica.

Esa imagen de una actividad apícola circunscrita en el espacio se vio redefinida con el descubrimiento de cerámicas análogas en el área albacetense (Soria Combadiera 2000), en la cuenca del Júcar. Dichos descubrimientos ampliaron el contexto geográfico de la difusión y uso de dichos recipientes a los territorios interiores, aportaron nuevos datos para el conocimiento de la cultura apícola, al tiempo que permitieron intuir una actividad más generalizada y con mayor potencia e importancia económica que la que entonces se suponía para época ibérica.

Cronológicamente no se pueden adscribir los distintos tipos de bordes[5] a las diferentes fases dentro de la etapa ibérica o del período romano, puesto que dichos materiales proceden de prospecciones en superficie. Así mismo, la comparación de estos materiales con los conocidos para el área edetana permitió concluir que la mayoría de los tipos de borde identificados eran coincidentes, sobre todo de los que se podían atribuir al horizonte Ibérico pleno, los de bordes triangulares y algunos de los engrosados, y que era a partir del horizonte Ibérico tardío cuando se registra una mayor diversidad formal. Por otra parte, según la misma autora, la existencia de colmenas en esa región se debería bien a contactos mantenidos con el territorio de *Edeta*, o bien se vinculaban a la existencia de una tradición local (Soria Combadiera 2000: 176).

Los últimos hallazgos, publicados por Fuentes Albero, Hurtado Mullor y Moreno Martín (2004), del territorio de *Keilin*, procedentes de prospecciones que han documentado una evolución del patrón de asentamiento que, a grandes rasgos, sigue una tendencia similar a *Edeta*, permitieron añadir más puntos al mapa de distribución, particularmente de los que tenían colmenas en el período altoimperial (Fuentes, Hurtado y Moreno, 2004: 183).

---

[5] Los ejemplares del área albacetense -Casilla del Mixto, Los Charcos, Las Hoyas, Cabezo de los Silos y La Cueva- presentan diámetros de boca que oscilan entre los 26 y 30 cm, con bordes diferenciados y paredes rectas. La morfología de los bordes es sobre todo engrosada, aunque también están presentes los bordes triangulares, salientes y redondeados. Sus superficies de cocción oxidante no presentan ningún tipo de tratamiento (Soria Combadiera 2000: 175).

Los autores recopilaron todos los fragmentos de colmenas cerámicas documentados en el área edetana y albaceteña y realizaron un análisis morfológico y estadístico que reveló aspectos bastante interesantes, que de algún modo permiten arrojar alguna luz al entendimiento de la pervivencia/continuidad de estos tipos cerámicos entre la Edad del Hierro y el período romano (Fuentes, Hurtado y Moreno 2004: 188-194).

Con dicho análisis se pretendió determinar si las diferentes variantes de borde tenían algún significado cronológico, regional o de taller y, por consiguiente, conocer el momento en el que se inicia la fabricación de ciertas formas, su perduración y declive, intentando establecer así qué borde era propio de cada período y área. Asimismo, se definieron cuatro grandes grupos según sus atributos formales: el de los bordes engrosados, externa o internamente; el de los bordes de tipo moldurado; el de los bordes pendientes y, por último, el de los bordes salientes simples, un tipo de borde que apenas sobresale, excediendo el diámetro máximo de la pieza (Fuentes, Hurtado y Moreno 2004: 191-192).

Se concluyó que la producción de las colmenas de tipo «borde engrosado» se inicia en el Ibérico Pleno, incrementándose a lo largo del Ibérico Final, hasta que en época altoimperial se observa su retroceso (Fuentes, Hurtado y Moreno 2004: 190). En lo referente a las colmenas con el borde «moldurado», se trata de una forma con altos índices de producción, tanto en el Ibérico Pleno como en el Final, aumentando en este último tanto los yacimientos como el número de fragmentos hallados, y persistiendo en época Imperial, aunque en menor número. En cuanto a las colmenas con el «borde pendiente», son escasas en los yacimientos del área edetana, y nulas en la provincia de Albacete, siendo por lo tanto una forma típica del Ibérico Final. Por último, que las de «borde saliente simples» aunque podrían tener un origen en el Ibérico Pleno, se deberían adscribir al Ibérico Final, principalmente debido a su mayor presencia en esta fase en cuanto a número de yacimientos y ejemplares (Fuentes, Hurtado y Moreno 2004: 190-194) (Figura 4A).

Recientemente, Quixal y Jardón han publicado la Fonteta Ràquia -Riba-Roja del Túria, Valencia- (Quixal y Jardón 2016). Se trata de un pequeño asentamiento rural en el territorio de Edeta, ocupado entre finales del siglo V al III/II a.C. Destaca por su especialización en la apicultura, siendo uno de los mejores registros del Mediterráneo Occidental, localizándose hasta las tapaderas de las colmenas (Figura 4B).

En lo que concierne a la época romana, la evidencia conocida es más parca. Son relativamente pocos los fragmentos claramente atribuibles a colmenas, motivo por el cual es obligatorio recurrir a los escasos casos conocidos. Están documentados fragmentos de colmenas y de extensiones de colmenas por ejemplo en

Sphakia (Creta) o en Gortyna, donde están presentes en cantidad tanto en momentos tardo-helenísticos, como romanos (como en Cap. 3 de este volumen), en Creta (Hayes 1983: 132, No.177), o aún en Knossos, también de época romana (Forster 2009: 146).

Este panorama es aún más reducido si nos circunscribimos al ámbito geográfico que nos ocupa en el presente trabajo, Hispania. De hecho, fue precisamente la ausencia generalizada de este tipo de recipientes cerámicos en los inventarios peninsulares lo que motivó a Rui Morais a realizar la reciente y aún hoy actualizada síntesis sobre la problemática de la miel y de las cerámicas relacionadas con su producción/almacenamiento en territorio peninsular (Morais 2006), y posteriormente, análisis selectivos de residuos de los contenedores cerámicos de Braga y Martinhal (Sagres) con la aplicación de cromatografía de gases acoplada a espectrometría de masas (Oliveira et al. 2015: 193-212).

En ese trabajo, una referencia obligada para el tema que ahora nos ocupa, se presenta un ejemplar completo procedente de un estrato de amortización/destrucción de la insula de las Carvalheiras en Bracara Augusta (Braga). Se trata de una pieza cilíndrica, con un diámetro máximo de 17,4 cm y mínimo de 13 cm, y una altura de 42 cm, de menores dimensiones que las ibéricas y que las romanas del área de Segobriga (Morais 2006: 157) (Figura 5).

El mismo autor menciona que ese ejemplar que se encontraba depositado en el museo local fue inicial y erróneamente clasificado como un elemento de tubería. Paralelamente, destaca que se diferencia bastante de éstos, no solo por su forma, sino por las estrías presentes en su superficie y por su manufactura particular con arcillas distintas a las de las canalizaciones y de los demás materiales latericios, y afirma que, con gran probabilidad, muchas colmenas de esta ciudad y de otras ciudades hispánicas se encontrarán «camufladas» en el registro cerámico bajo la designación de tuberías (Morais 2006: 157). En este sentido, resulta paradójico que en el territorio segobricense en época tardía se reutilizan como tuberías las cubiertas de material latericio del acueducto a la ciudad en el vicus de Las Madrigueras, pero no las colmenas (Urbina y Morín 2011).

A raíz de ese descubrimiento en el noroeste de la Hispania Tarraconensis, Morais advierte para el desconocimiento existente sobre este tipo de cerámicas en época romana, alude a la necesidad de rever muchos de los fragmentos tradicionalmente clasificados como canalizaciones cerámicas, e incita a su identificación y recopilación en las demás regiones y provincias hispánicas, de tal modo que se pueda cuantificar y valorar debidamente, en base a los vestigios arqueológicos, el papel que la producción y comercialización de la miel desempeñaron en las provincias occidentales (Morais 2006: 157-158). Recientemente, hemos podido identificar la producción

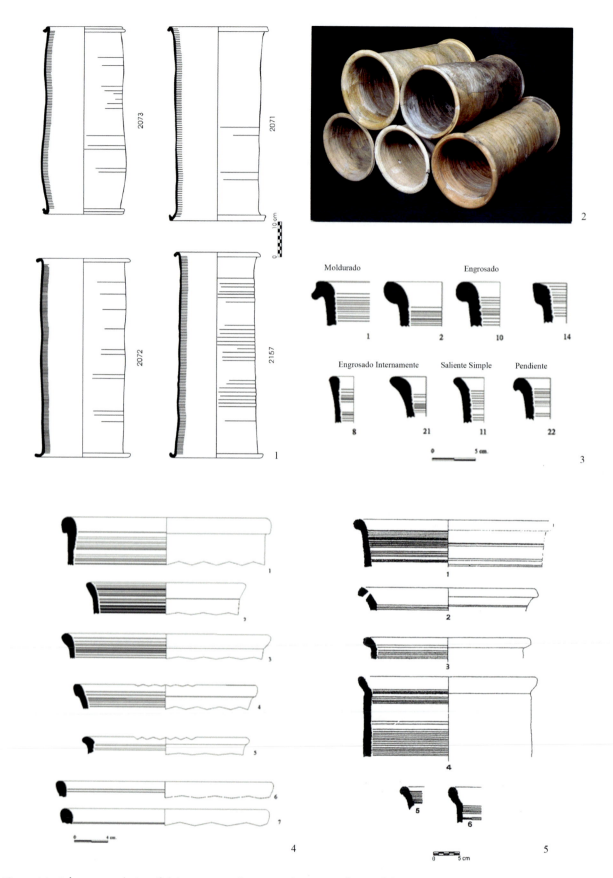

Figure 4.A. Colmenas cerámicas ibéricas. 1: Ejemplares completos procedentes del Puntal del Llops, según Bonet y Mata 2002; 2: Fotografía de colmenas ibéricas del área edetana –Museo de Prehistoria de Valencia-; 3: Principales tipos de bordes de colmenas ibéricas del área edetana, según Bonet y Mata 1995 y recopiladas por Fuentes, Hurtado y Moreno 2004: 190; 4: Tipos de bordes de colmenas del territorio de Kevin, según Fuentes, Hurtado y Moreno 2004: 190; 5. Tipos de bordes de colmenas de la comarca del Júcar –Albacete-, según Soria 2000: 176.

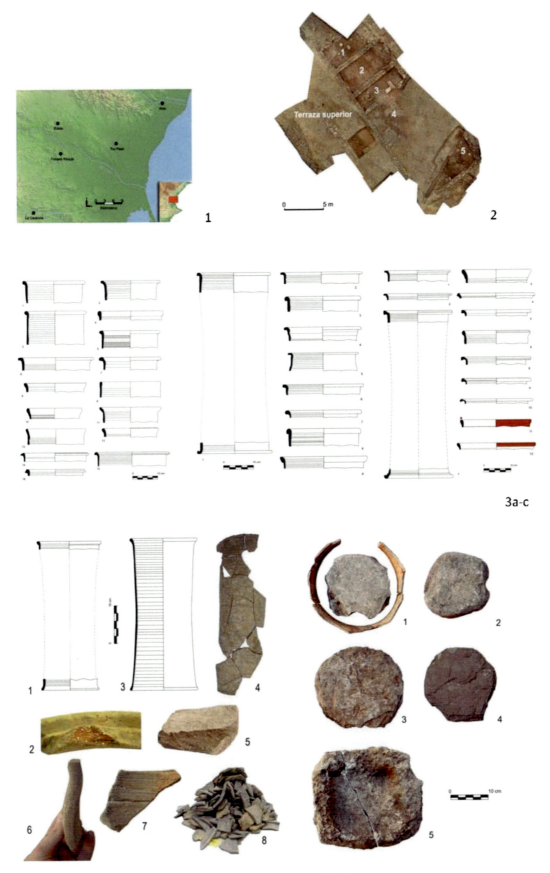

Figure 4.B. 1: Mapa de localización del asentamiento; 2: Ortofoto del asentamiento.3: Colmenas cerámicas I, II y III.
4: Colmenas cerámicas con pastas extrañas, rubefactadas o defectos de cocción. 5: Tapaderas de colmena y posible pileta
(según Quixal y Jardón 2016).

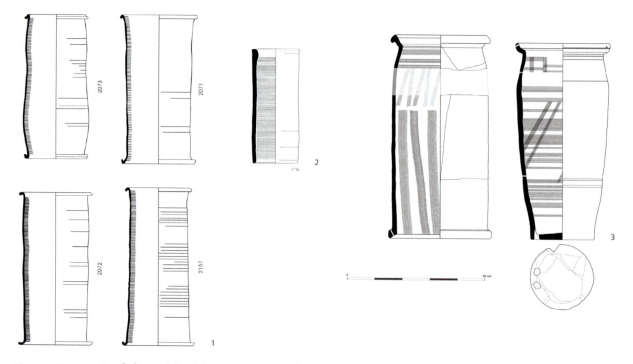

Figure 5. Comparativa de los modelos ibéricos y romanos en la Península Ibérica. 1. Ejemplares completos procedentes del Puntal del Llops, según Bonet y Mata 2002: 2. Colmena cerámica de la insula de las Carvalheiras –Bracara Augusta-, según Morais 2006: 157; 3. Colmena de «tipo tubular» procedente de La Laguna –Segóbriga, Cuenca-; 4: Colmena de «tipo troncocónico» procedente de Los Vallejos –Segóbriga, Cuenca-.

y uso de colmenas cerámicas en el territorio de la ciudad de Segóbriga –Saelices, Cuenca-, lo que nos permite trazar un panorama más preciso sobre la práctica de la apicultura en la *Hispania* romana. En este sentido, la producción y el uso de colmenas cerámicas sólo se pudieron atestiguar en el *ager* segobricense para la época altoimperial. La práctica totalidad de los enclaves excavados no sobrepasan cronológicamente el s. III d.C. Constituye una excepción el *vicus* de Las Madrigueras, en la vega del Valdejudíos, a medio camino entre *Segóbriga* y *Opta*. Este espacio cuenta con una fase bajoimperial de los ss. III y IV d.C., pero no se han atestiguado colmenas cerámicas para este período. Esta ausencia en este caso parece real, no nos encontramos con una incapacidad para reconocer el recipiente productor, como parece ser una tónica en el panorama hispano. La ausencia nos está indicando su sustitución por recipientes fabricados en materiales perecederos y, seguramente, el abandono de la práctica intensiva y el paso de nuevo a una apicultura de grupos familiares especializados con una práctica trashumante.

Para finalizar, no queremos dejar de abordar la problemática de las colmenas cerámicas de cronología andalusí. En este período nos volvemos a encontrar con la fabricación de colmenas cerámicas, seguramente por los talleres locales que son muy activos en la elaboración de todo tipo de recipientes. Creemos que el problema para este período no es la «invisibilidad» del tipo cerámico, sino su confusión con otro tipo

bastante habitual en los yacimientos de la época. Nos referimos a los recipientes destinados a funciones culinarias, como son los hornos o *tannur* destinados a la elaboración del pan. Se trata de un horno portátil destinado a la cocción de pan, aunque también puede servir para la cocción de otros alimentos panificables, como los dulces. Normalmente el *tannur* cuenta con una apertura superior para la evacuación de humos y otra inferior –boca-, que sirve para cargar el horno. Este tipo de forma puede ser confundida con colmenas cerámicas verticales. El aspecto técnico que marca la diferencias entre una y otra serie es la presencia de las estrías interiores, que sólo aparecen en las colmenas cerámicas y no tienen ninguna función en los *tannur*. Estas estrías se realizan con el objetivo de ayudar a que «amarren» los paneles, también se colocan en el interior de los mismos «cruces» o «viros» realizados con maderas locales flexibles en un número de cinco a seis. En las colmenas andalusíes la tapa o «témpano» se realizaría en un material perecedero, mientras que la puerta de entrada de las abejas –«piquera» o «biquera»-, se realizaría por unos pequeños orificios que se han interpretado erróneamente como pequeños tiros de los hornos cerámicos (Figura 6).

Esta posibilidad fue recogida por S. Gutiérrez Lloret a través de una sugerencia que le realizó personalmente el profesor Paul Arthur, quien le planteó la posibilidad de que algunos de estos portátiles fueran colmenas debido al parecido con las colmenas áticas. La autora

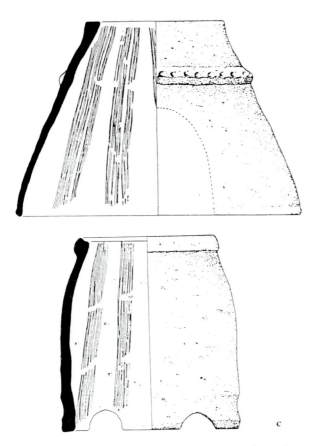

Figure 6. Colmenas andalusíes. 1: Colmena procedente de la Rábita de Guardamar del Segura –Alicante-; 2: Colmena de Sompo –Concentaina, Alicante-, según Gutiérrez Lloret 1993: 164.

no desechó totalmente esta posibilidad y las relaciona con el hallazgo de numerosos fragmentos de *tananir* de la forma M9.1 en la *Llometa del Sifó* –Elche, Alicante-, recogidos por P. Ibarra en 1926 y más recientemente por el Grupo Ilicitano de Estudios Arqueológicos (Gutiérrez Lloret 1996: 145).

La apicultura en el territorio segobricense: producción, consumo y circulación. Una aproximación a su práctica en la Hispania romana.

Los trabajos arqueológicos realizados en diferentes yacimientos en el territorio de Segóbriga entre los años 2009 al 2011, así como otras intervenciones en el *hinterland* de la ciudad, año 2012, nos permiten trazar un panorama relativamente preciso de la práctica de la apicultura en época romana, aunque tengamos sobre todo datos de la producción de miel, mientras que el consumo y la circulación lógicamente cuentan con un mayor número de interrogantes. Sin embargo, creemos que los datos obtenidos en el estudio del territorio que se ha realizado en estos últimos años sería perfectamente extrapolable al de otras ciudades hispanas. La invisibilidad peninsular es una consecuencia clara de una correcta metodología, así como de la ausencia de trabajos sistemáticos sobre los contextos cerámicos romanos.

La producción en época romana altoimperial

En las líneas precedentes ya se ha abordado la producción de la miel en el mundo antiguo mediterráneo y los antecedentes de la Edad del Hierro en el territorio peninsular, en especial en el área levantina. Parece claro que existe una producción intensiva que no tiene relación con la práctica tradicional trashumante, aunque no cabe duda que ésta sería coetánea con la que a continuación vamos a detallar. La localización de colmenas cerámicas sería una prueba de esta práctica intensiva estabilizada en el terreno, ya que el peso de las colmenas imposibilitaría su itinerancia fuera de los espacios de fácil accesibilidad con carruajes o caballerías. Por otro lado, no cabe duda que la ubicación de los colmenares en las zonas de producción agrícola estaba buscando optimizar el rendimiento de las producciones agrícolas de la ciudad –cereales, vid y olivo-.

Las piezas segobricenses proceden de varios yacimientos excavados en el ámbito de la reciente intervención preventiva de la «Conducción Principal del Abastecimiento de Agua Potable a la Llanura Manchega», y localizados en el *ager* de la ciudad de *Segobriga*. En dichos yacimientos (La Peña II, Llanos de Pinilla, Los Vallejos, La Laguna, Casas de Luján II y Rasero de Luján[6]) se pudo recuperar no solo una gran cantidad de fragmentos de bordes, cuerpos y fondos, que testimonian el uso de las colmenas cerámicas en los yacimientos rurales del entorno de la ciudad, sino también la posible producción en, al menos uno de ellos, Rasero de Luján.

Todos los yacimientos intervenidos se inscriben en los típicos yacimientos rurales hispanorromanos de tipo agrícola/transformador, con una *pars rustica* y *fructuaria*, algunos con posible *pars urbana*, lo que permite sin mayores problemas clasificarlos como de tipo *villae*. Se ubican bien en las proximidades de un tramo de acueducto recientemente identificado, que abastece la ciudad desde los manantiales localizados al norte, concretamente desde unos 10 km hacia el norte, o bien a lo largo de las orillas del Cigüela, que discurre hacia el sur, en un radio aproximado de unos 4 kilómetros en torno a la ciudad.

Las colmenas son uno de los elementos constantes en los repertorios cerámicos de los yacimientos segobricenses, constituyendo cerca del 10% de los grandes recipientes. En lo relativo a sus atributos morfológicos, la totalidad de fragmentos de bordes y cuerpos permite caracterizarlas como piezas de tendencia cilíndrica, entre 28 y 32 cm de diámetro máximo en el cuerpo/boca.

---

[6] De todos los espacios se ha publicado la correspondiente Memoria Final, así como un trabajo de síntesis de todos los trabajos: Morín de Pablos 2014.

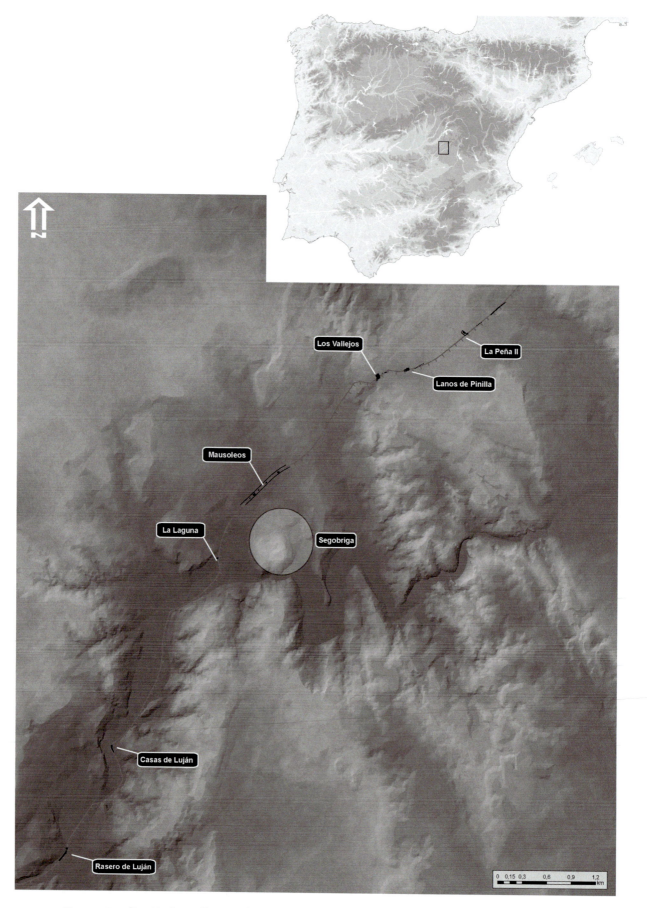

Figure 7. Localización de Segóbriga en la Península Ibérica y ubicación de los yacimientos excavados sobre modelo digital del terreno (J. Morín y R. de Almeida).

Entre las piezas recuperadas no se ha recuperado ningún ejemplar completo, pero, sin embargo, un ejemplar proveniente de La Laguna y otro de Los Vallejos permiten una reconstrucción de la forma que, según creemos, puede ser bastante verosímil. Por lo tanto, se estima una altura próxima a los 70 cm. De este modo, calculando estas dimensiones medias estimadas, estas colmenas tendrían una capacidad aproximada entre 56 y 41 litros respectivamente. La segunda con una capacidad cercana a las capacidades típicas, y la primera con algo más de capacidad que las conocidas para el mundo ibérico, para algunas romanas e, inclusive, otras de carácter tradicional (Bonet y Mata 1995: 280; Crane 1983: 17 y Tabla 2).

Las dos colmenas de mayores dimensiones, recuperadas en La Laguna y en Los Vallejos, permiten también verificar que existen dos morfologías distintas al nivel del cuerpo. Del mismo modo, existe un tipo que denominamos «tipo tubular», que presenta un cuerpo cilíndrico y abierto en ambas extremidades, semejante a un *tubuli* y otro tipo que designamos «tipo troncocónico», apenas abierto en uno de sus lados, siendo el otro lado directamente cerrado en el proceso de moldeado del recipiente, como si de un fondo se tratara, presentando dos perforaciones (Figura 8).

Común a estas piezas es la presencia obligatoria del estriado interior, realizado con la cerámica en fresco *ante*

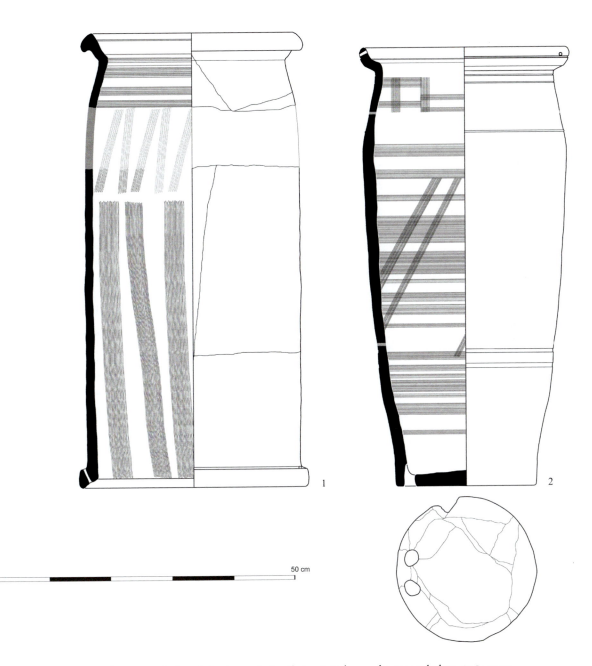

Figure 8. Colmenas cerámicas del entorno de Segóbriga. 1: Colmena de «tipo tubular», La Laguna; 2: Colmena de «tipo troncocónico», Los Vallejos (J. Morín y R. de Almeida).

cocturam. Ese detalle técnico, que permite diferenciar este tipo de objetos de otros como los soportes o las tuberías, con los cuales son frecuentemente confundidos (Bonet y Mata 1995: 280-281; Morais 2006), sirve, tal como ya hemos indicado, para la correcta y fácil adhesión de los panales a las paredes de las colmenas. El estriado de las piezas segobricenses presenta diferentes técnicas de manufactura y diferentes acabados finales, llegando algunas veces a conjugar distintas técnicas en su realización.

El más frecuente es un estriado fino de líneas perfectamente paralelas, que se consigue a través de la aplicación de un objeto rígido, probablemente hecho de madera o de metal, que entendemos sería semejante a un peine, dejando de esa forma la superficie interior con un aspecto «peinado». Esas bandas incisas «peinadas» se alternan con áreas reservadas lisas.

Este tipo de estriado también se conjuga con otro de idénticas características, de mayor o menor dimensión, según los ejemplos, pero realizado en sentido vertical o en diagonal, cortando el primero. Como regla general, el «peinado» vertical se realiza siempre en un segundo momento. Ocasionalmente, en algunos ejemplares con los dos tipos, se documentó que las distintas bandas de «peinado» con dirección cambiante se encontraban separadas por un surco ancho y profundo, que marcaba claramente dichos cambios de dirección de las estrías.

Otro tipo de estriado, menos frecuente, es el que se obtiene de la aplicación en la superficie interior de un elemento puntiagudo o con una extremidad plana bien definida, probablemente un punzón o una espátula fina, que genera unas incisiones profundas y marcadas con aristas muy pronunciadas que llegan a ser cortantes.

Estos estriados lo que buscaban era facilitar el agarre de los panales al recipiente cerámico (Figura 9) Además, hay que señalar que en el interior de las colmenas se disponían las «cruces» o «viros» realizadas con varas de maderas flexibles. Estas «cruces» se colocaban en un número máximo de cinco en el interior de la colmena.

En algunos ejemplares, independientemente de la morfología del borde, se identificaron perforaciones con sección circular y entre 2 a 4 mm de diámetro, que interpretamos como orificios para pasantes de sujeción/cierre de las tapaderas (de madera, cerámica o corcho), que serían amovibles y que permitirían acceder al interior de la colmena y proceder a la extracción de los panales. En este sentido, hay que señalar las diferencias existentes entre la tapa o «témpano» y la puerta de entrada de las abejas, conocida como «piquera» o «biquera». Por último, indicar que en las colmenas griegas la «biquera» estaba realizada en material cerámico (Jones et al.1973) (Figura 10).

Común a todas las colmenas es la existencia de unos bordes bien diferenciados de las paredes. No obstante, su morfología está lejos de estar estandarizada. Es evidente que existen y que coexisten distintos tipos de borde, pero que, a pesar de todo, mantienen un rasgo recurrente, que consiste en una línea de ruptura de la boca de la colmena con relación a la pared, y que se manifiesta como una marcada inflexión interior en la transición hacia el cuerpo, presentando un perfil algo sinuoso y con tendencia cóncava, que puede ser más o menos marcado, en función de la propia forma del borde. El análisis de las distintas morfologías de bordes permite constatar una gran variabilidad, que en nuestro entender es apenas aparente, porque se compone de pequeños matices formales, pudiendo minimizarse estadísticamente y agruparse en cuatro grandes tipos.

El primero de los cuatro tipos se caracteriza por bordes exvasados, engrosados y acentuadamente cóncavos en su interior; el segundo, por bordes apenas engrosados al exterior, con perfiles de sección triangular, redondeada o algo engrosada internamente, pero que se diferencian menos de la pared, con la parte interna menos cóncava; el tercer grupo presenta bordes que son bastante engrosados al exterior y casi pendientes, con la parte interna prácticamente recta, pero bastante inclinados hacia el exterior, haciendo un ángulo cercano a los 45º respecto al cuerpo; el cuarto, y último grupo, se caracteriza por bordes con sección de tendencia rectangular, con un ligero espesamiento en su parte terminal que confiere a la boca un perfil triangular, y con un grosor aproximado al de las paredes, pero destacándose de éstas por una apertura al exterior y por una marcada inflexión interna en la parte final de la pared interior.

En el caso de las colmenas de tipo tubular, el borde de la mitad inferior no se presenta exactamente igual. Según hemos podido observar, principalmente a partir de la colmena reconstituida de La Laguna, el borde inferior es ligeramente más exvasado, más corto y cóncavo. A estas características se suma la presencia de «peines» incisos en la vertical hasta la base, algo que no sucede nunca en las mitades superiores, pues cuando están presentes, se inician siempre después de algunas bandas incisas horizontales.

Independientemente de la morfología de las colmenas, en los dos tipos apenas se procedería a abrir uno de los laterales para la extracción de los panales en época de cosecha y para la limpieza de su interior. Las bocas de las colmenas de tipo tubular se cerrarían en ambos lados con tapones de corcho, madera, arcilla, o tapaderas cerámicas, a la semejanza de las colmenas ibéricas y de otros paralelos conocidos, en los cuales se abriría un orificio que permitiese la circulación de las abejas –«piquera» o «biquera»-, mientras que en las colmenas de «tipo troncocónico» únicamente se cerraría

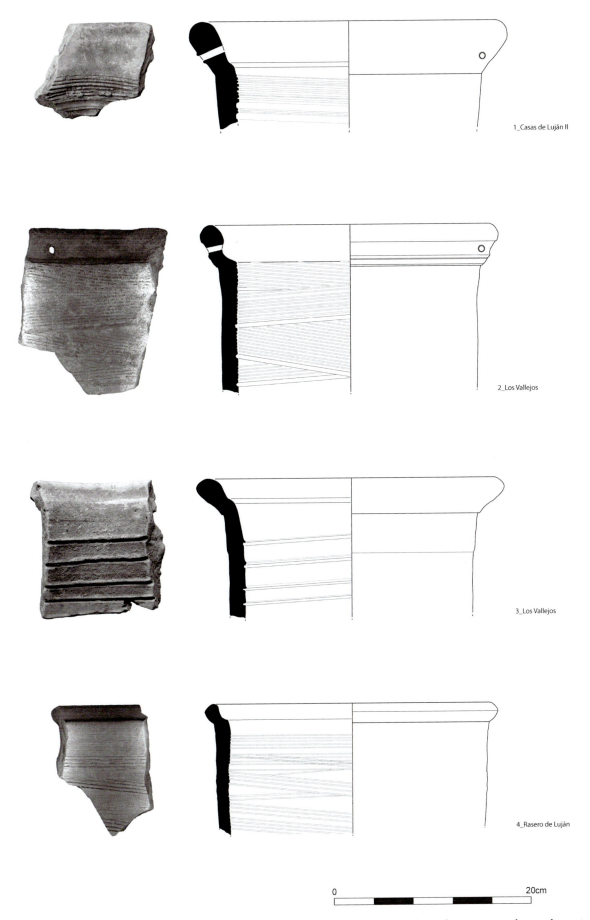

Figure 9. Colmenas cerámicas del entorno de Segóbriga. Fragmentos de borde y detalle de los distintos estriados en el interior (J. Morín y R. de Almeida).

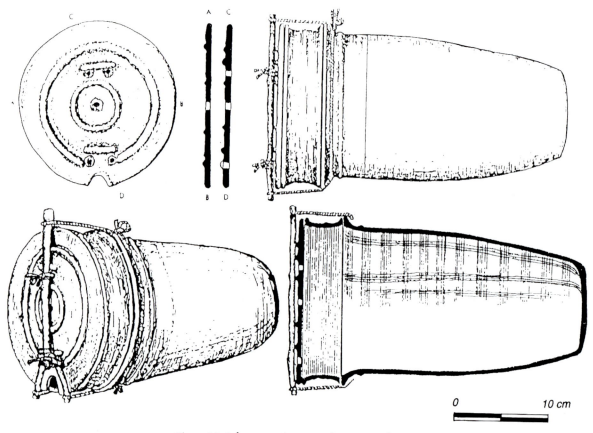

Figure 10. Colmenas griegas, según Jones *et al* 1973.

uno de ellos con la «piquera», puesto que el fondo estaba ya cerrado y perforado.

En el interior de la colmena habitan tres clases de individuos: la reina o maestra (única hembra fecunda), las abejas u obreras con cometidos especializados y los machos o zánganos quienes, además de fecundar a la reina, tienen la misión de dar calor a la colmena y a las crías. Éstos son aniquilados en su mayoría cuando se acaba la flor por las propias abejas. El origen de la colmena lo establece el enjambre o parte del «ganado» que abandona la colmena madre, conducido por la reina vieja, en los meses de marzo o abril, cuando se encuentra más poblada y hay abundante flor en los campos. A veces salen hasta dos y tres enjambres, más pequeños que el principal de cabeza, que se conocen como «jabardos» o «jabardillos».

En lo que concierne a su cronología, la totalidad de las morfologías de cuerpos de colmenas identificadas en los ejemplares del entorno de *Segobriga* están en uso desde momentos iniciales del s. I d.C. hasta finales del s. II/inicios del III d.C., por lo que no es posible afirmar que alguna de las variantes de borde sea exclusiva o más típica de determinado momento. La colmena de «tipo troncocónico» identificada en Los Vallejos procede de un estrato de amortización provisional fechado en torno a finales del s. II/inicios del III d.C. Por ello,

en función de la aparente ausencia de ejemplares de este tipo en contextos más antiguos, puede que este tipo concreto de colmena esté apenas presente en momentos más avanzados.

Con respecto al área de ubicación de las colmenas y de las probables zonas de trabajo/explotación, la información que puede ser extraída de los yacimientos rurales de *Segobriga* es bastante limitada. En primer lugar, la naturaleza de sus procedencias estratigráficas. Prácticamente todos los fragmentos fueron recuperados en contextos de amortización, bien en espacios intramuros, bien en espacios exteriores, mayoritariamente en ambientes de circulación o en formaciones con origen detrítico. A favor de eso habla el generalizado estado de fragmentación de los ejemplares recuperados. Apenas en Los Vallejos y La Peña II este panorama es algo diferente, aunque esté lejos de ser concluyente.

Así, el ejemplar completo de «tipo troncocónico» recuperado en Los Vallejos apareció en el interior de una habitación de la parte urbana de la *villa*, al lado de un ánfora vinaria del tipo Dr. 2-4, ambos bajo el derrumbe de la techumbre de dicha habitación. Parece tratarse de una zona inapropiada para su uso, por lo que entendemos que debería encontrarse almacenada o guardada para limpieza. Por otra parte,

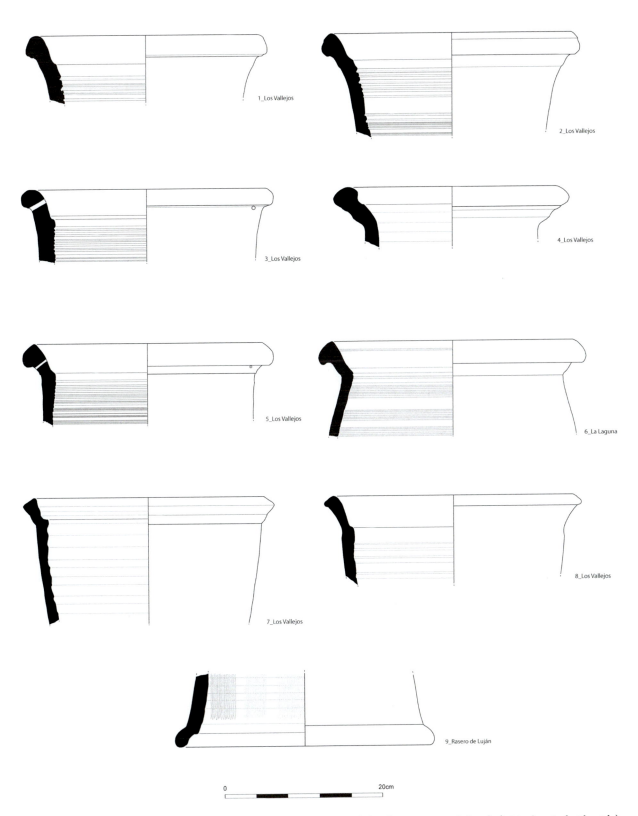

Figure 11. Colmenas cerámicas del entorno de Segóbriga. Fragmentos de los distintos tipos de borde (J. Morín y R. de Almeida).

en el yacimiento de la Peña II, del cual se excavó únicamente su parte *rustica/fructuaria*, gran parte de los fragmentos recuperados preceden de los Ámbitos 4, 5 y 13, los dos primeros pequeños espacios abiertos tipo patio, anexos al lagar y a los almacenes, y el tercero una gran habitación abierta, probablemente

destinada a labores agrícolas y ganaderas. A pesar de que éstos sean quizás los lugares más apropiados para la colocación de colmenas y para el desarrollo de la actividad apícola, tal como en los referidos paralelos conocidos, el hecho de que no se haya recuperado algún ejemplar entero nos lleva a tener algunas dudas y reservas. Finalmente, tenemos que señalar que en el yacimiento de La Laguna sólo se pudo excavar una parte mínima de la explotación. Un espacio que se disponía de forma aterrazada, muy habitual en las construcciones rurales de la ciudad, como en Casas de Luján, donde los diferentes espacios productivos estaban ubicados en diferentes niveles, como el alfar y la almazara. Así parece que sucede en el enclave de La Laguna, en ese espacio de tránsito es probable que se pudiera emplazar el colmenar. Sin embargo, en la práctica tradicional de la apicultura la ubicación de los colmenares se realiza en los campos, y aunque

los cerámicos no tuvieran una trashumancia anual, no cabe duda de que éstos sufrirían traslados cada cierto período de tiempo. De este modo, su masiva presencia en los establecimientos agrarios habría que relacionarlos con labores de limpieza/reparación y acondicionamiento antes del traslado (Figura 12).

Por otro lado, en las cercanías del territorio de Segóbriga hemos podido identificar producciones segobricenses de colmenas. Nos referimos a los yacimientos de Fuente de la Calzada (Santa Cruz de la Zarza, Toledo) y San Blas (Las Pedroñeras, Cuenca).

Figure 12. Localización de las colmenas cerámicas en los yacimientos del territorio segobricense. 1: Los Vallejos; 2: La Peña II; 3: La Laguna; 4: Casas de Luján; 5-6: Propuesta de restitución de la disposición de las colmenas examinadas (J. Morín y R. de Almeida).

En el primero de los enclaves citados, el de Fuente de la Calzada, las colmenas cerámicas proceden de una prospección[7]. Se trata de diferentes bordes con perforaciones que parecen corresponderse al tipo de colmena de cuerpo «troncocónico», como los localizados en Los Vallejos y Casas de Luján, aunque ya hemos señalado que éstos no se encuentran estandarizados, pero si mantienen unos rasgos recurrentes. Además, se localizan perforaciones con sección circular, que hemos interpretado como orificios para pasantes de

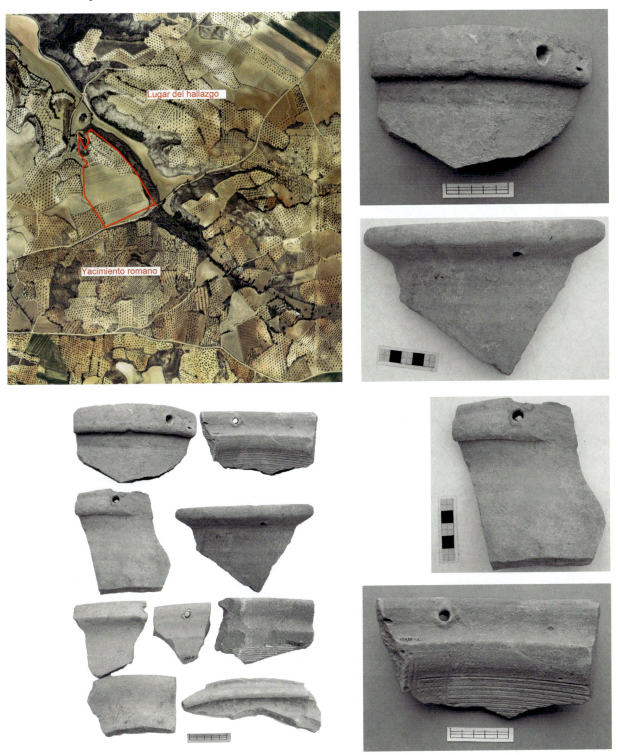

Figure 13. Colmenas de Fuente de la Calzada –Santa Cruz de la Zarza, Toledo-. 1. Plano de localización de las colmenas en relación con el yacimiento romano 2-6. Colmenas cerámicas de tipo «troncocónico» (J. Morín y R. de Almeida).

[7] Queremos agradecer a D. Urbina y C. Urquijo la información que nos han facilitado procedente de la elaboración de la Carta Arqueológica de Santa Cruz de la Zarza.

sujeción/cierre de las tapaderas. La localización de estos ejemplares distantes más de 25 kms en línea recta de la ciudad de Segóbriga, resulta especialmente interesante el lugar de hallazgo, ya que las colmenas no se han localizado en un yacimiento, sino en los campos cercanos al enclave romano. Tendríamos así el primer testimonio de colmenas situadas en los campos, como suele ser habitual en la práctica de la apicultura tradicional (Figura 13).

El otro asentamiento en el que se han localizado colmenas cerámicas procedentes de las alfarerías segobricenses es San Blas –Las Pedroñeras, Cuenca-, distante a más de 45 kms. de la ciudad. El yacimiento es un enclave dedicado a la extracción y el procesado de la arcilla, aprovechando las vetas existentes en la llanura de inundación del arroyo del Cerrojo de San Blas, aunque posteriormente se desarrolla un *vicus* con cronología altoimperial (Morín y Sánchez Ramos 2017).[8] Los fragmentos de colmenas se localizaron como relleno de un «barrero» o fosa de extracción de arcilla que se encontraba colmatada como basurero. No ha sido el único material cerámico procedente de las alfarerías de la ciudad que se ha localizado, ya que se encuentran fragmentos de *dolia*, que hemos definido localmente como subtipo CAT 2. Los fragmentos de colmena localizados pertenecen a las colmenas de morfología «tubular», como las localizadas en el enclave de La Laguna. El hallazgo de estas piezas como material amortizado no aporta datos sobre su emplazamiento en la zona, aunque parece claro que era una costumbre llevar las colmenas a los espacios de

Figure 14. Colmenas de San Blas –Las Pedroñeras, Cuenca-. 1: Planta de la excavación con la localización de la fosa de extracción U.E. 100; 2: Fotografía aérea de la zona excavada; 3. Colmenas cerámicas de tipo «tubular». (J. Morín y R. de Almeida).

---

[8] La intervención se desarrolló en los meses de junio a agosto de 2012 y la Memoria Final se publicó en 2017.

habitación, seguramente para ser preparadas antes de ser reparadas de nuevo (Figura 14).

De las colmenas los romanos obtenían, como sucede en la actualidad en la apicultura tradicional, tres productos: miel virgen, miel y cera. Todo comenzaba con la «castración» de las colmenas. Ésta se realizaba una vez al año, aunque en época de bonanza se podían **«castrar»** las colmenas hasta dos y tres veces hasta la **«cruz»** del medio, que estaba dispuesta a la mitad de la colmena. Los colmeneros cubrirían sus cuerpos y como herramientas utilizarían el **«afumador»** o **«ahumador»**, que producía humo con el que se controlaba a las abejas. Éste podía ser de metal, pero también se conocen recipientes cerámicos con esta función (Bortolin 2008), aunque no se conocen en *Hispania*. Los panales se extraían con las manos y con la ayuda de cuchillos y una rasqueta metálica, que ayudaba al apicultor a extraer los panales. Desde el punto de vista material de todo este proceso sólo dejan huella en el registro arqueológico el afumador, los cuchillos y la rasqueta. El primero es fácilmente identificable, pero no se han localizado en *Hispania*, mientras que los otros dos son objetos comunes que tienen diferentes usos, no son herramientas especializadas.

En cuanto al procesado de la miel, éste es relativamente sencillo. La miel virgen se extrae con el simple prensado de los panales con las manos, que se vierte en un embudo que precipita la miel a un recipiente. La miel y la cera se obtienen mediante el prensado y la utilización del calor. Este proceso tampoco deja huellas claras en el registro arqueológico, ya que la utilización de embudos cerámicos o metálicos es habitual en el trasiego de cualquier tipo de líquido y frecuente en el registro arqueológico hispano. En cuanto al recipiente utilizado para recoger la miel, no es específico, y en la apicultura tradicional se utilizan recipientes cerámicos abiertos y cerrados.

La producción de la Antigüedad Tardía a la Alta Edad Media

En el territorio de Segóbriga se realizaron excavaciones de otros espacios con una cronología posterior. Es el caso del *vicus* de Las Madrigueras II (Morín y Urbina 2011), en la cuenca del Valdejudíos, que tiene una fase de ocupación bajoimperial que resulta muy significativa para entender la desaparición de la apicultura intensiva desde el siglo II d.C. Se localizaron dos cloacas que reutilizaban los materiales latericios que cubrían el acueducto de La Peña que abastecía de agua a la ciudad de Segóbriga. Para ello se recortaron las alas de la *tegula* utilizada como cubierta. Esta mecánica de expolio y reaprovechamiento en este período es una evidencia clara e inequívoca que la práctica apícola intensiva con recipientes cerámicos ha desaparecido

en época bajoimperial, ya que si no se utilizarían *tubuli* cerámicos o las propias colmenas (Figura 15). En este período la práctica apícola estaría reducida a la actividad tradicional, con colmenas construidas en materiales perecederos e itinerantes. En el entorno inmediato de la ciudad de Segóbriga, en plena llanura de inundación del Cigüela, se excavó otro espacio Camino del Escalón/Mausoleos con cronología bajoimperial en el que tampoco se localizaron evidencia de colmenas cerámicas (Barroso, Carrobles y Morín 2013).

De época hispanovisigoda y andalusí también se han excavado asentamientos, como el de la Quebrada II en los que no se han localizado evidencias de colmenas cerámicas (Malalana, Barroso y Morín 2011). Lo mismo sucede con el enclave Medieval de Ermita Magaceda (Malalana y Morín 2011a) (Figura 16). A este respecto señalar, que en el curso bajo del Cigüela se excavaron otros asentamientos andalusíes como Villajos (Malalana y Morín 2011b) y Arroyo Valdespino (Urbina, Malalana y Morín 2014), en los que tampoco se hallaron evidencias de colmenas cerámicas, a pesar de los estudios minuciosos de material. En este sentido, el enclave republicano de Pozo Sevilla (Morín 2013), también en el curso bajo del Cigüela, tampoco cuenta con colmenas cerámicas, lo que indica claramente que su uso se limitó a la época altoimperial, ss. I-II d.C.

En el caso que nos ocupa, el territorio de la ciudad de Segóbriga, en un sentido amplio, la intensidad de los trabajos arqueológicos en el territorio, así como la minuciosidad en el estudio de los materiales, nos permiten descartar por completo el uso intensivo de colmenas cerámicas a partir de la segunda mitad del siglo II d.C. En Oriente, conocemos bien la perduración de este uso intensivo, así como la fabricación de colmenas cerámicas (Figura 17), bien estudiadas por Sophia Germanodiu (2012, 2013 y 2018). Hasta el siglo XI va a convivir la práctica intensiva con colmenas cerámicas y la práctica tradicional con colmenas itinerantes construidas con materiales perecederos (Germanidou 2018), como puede verse en la iconografía de mosaicos y miniaturas orientales (Figura 18).

Otra problemática es la práctica de una apicultura intensiva, pero con colmenares construidos con materiales perecederos, como era habitual en muchas de las comarcas apícolas hispanas, entre ellas la Alcarria y la Manchuela, espacios que ocupa el *ager* de *Segobriga*. Estos espacios son de muy difícil interpretación arqueológica (Figura 19), ya que al desaparecer los muros donde iban insertas las colmenas, resulta prácticamente imposible diferenciar la estructura apícola de un hábitat oportunista o un eremitorio. En este sentido, ya habíamos trabajado previamente en la zona de Ercávica en relación con el

1a-c

2a-b

CLOACA MADRIGUERAS

ACUEDUCTO

3a-b

Figure15. 1: a. Madrigueras II; b y c: Cloacas vista desde el este. 2: a-b. Cloaca, detalle. 3: a. Comparativa entre el material latericio reaprovechado de Madrigueras II y el del acueducto de La Peña. b. Reconstrucción del acueducto de la Peña con su cubierta latericia (J. Morín y R. de Almeida).

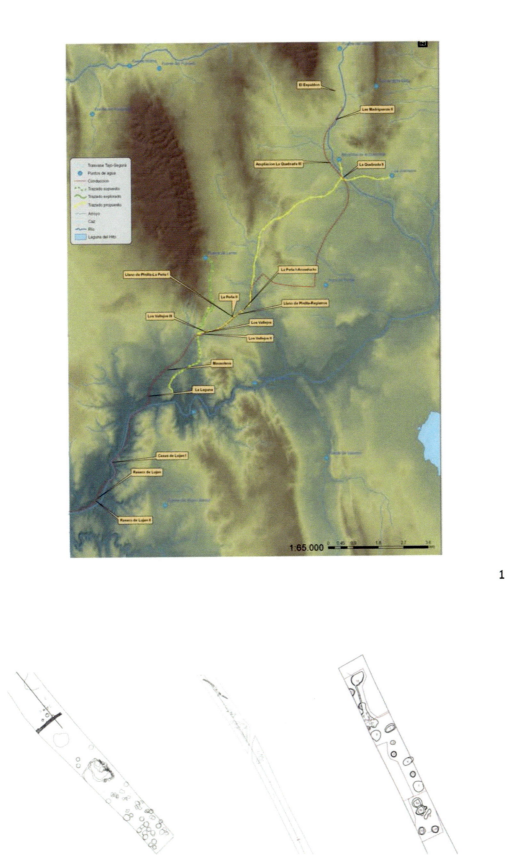

1

2a-c

Figure 16. 1: Mapa con todos los asentamientos excavados en el territorio Segobricense, de la Prehistoria a época Medieval.
2: a. Madrigueras II. b. Camino del Escalón/Mausoleos. c. Ermita Magaceda (J. Morín y R. de Almeida).

Figure 17. 1: Colmena cerámica. Isthmia, Corinto, s. VI d.C. -fot. S. Germanidou, 2018-; 2: Colmena cerámica con pintura de bandas al exterior. Delfos, s. VI d.C. -fot. S. Germanidou, 2018-.

Figure 18. 1: Mosaico de la Basilica Theorokos, Madaba, Jordan -s. VI d.C.-; 2: Sacra Parallela, cod. Par. gr. 923 -s. IX d.C.-; 3: Cod. Taphou 14, 1075-1085; 4: Miniatura de Exultet roll Troia 3, 1150-1200; 5: Miniatura de Exultet roll Vat. Lat. Barberini 592, 1070-1100 -Fot. S Germanidou, 2018-.

Figure 19. Colmenar. Valle del Sedano -Burgos- (J. Morín).

monasterio servitano, donde se identifican claramente las estructuras eremíticas y la tumba del fundador del monasterio, el monje norteafricano Donato (Barroso 2019: 82 ss.).

El consumo y la circulación

Para el adecuado entendimiento de la producción colmenera y de la actividad apícola en época romana en *Hispania*, y en particular en el área del territorio de *Segobriga* y en la zona sur oriental de la *Hispania Tarraconensis*, es forzoso e inevitable, una vez más, remontarnos a la época ibérica. En función de lo que se ha mencionado con anterioridad, la lectura transversal de la evidencia arqueológica actualmente disponible para el territorio que discurre desde el área edetana a la actual región albacetense nos habla en favor de una actividad apícola, que estaría ya ampliamente difundida entre los ss. III y I a.C., como actividad complementaria a una economía agropecuaria y del uso de la miel en la alimentación y como bien de comercialización/exportación.

Dicha actividad productiva es patente en la gran cantidad de yacimientos que presentan este tipo de material, y su potencial excedente se materializa en la esfera de la comercialización extra regional en la exportación los *kalathoi*, típicos recipientes a los que actualmente varios autores atribuyen, entre otras,

una función de contenedores de miel (Becinvenga 1985; Bonet y Mata 1995: 384; Conde 1996; Fernández Mateu 2000: 91; Fuentes *et alii* 2004: 196; Guerin 2003: 313; Muscolino 2006), situación ya avanzada con anterioridad por Emeterio Cuadrado (1968: 128).

Lo que parece igualmente claro es la existencia de dos etapas en la producción apícola ibérica: una inicial, en el Ibérico Pleno, con un sistema de producción doméstico y unas relaciones comerciales en el ámbito local o regional; y otra, en el Ibérico Final, donde el incremento de colmenas sugiere que parte de los productos pudieron ser comercializados a larga distancia (Bonet y Mata 1997: 45; Bonet y Mata 2002: 186). Así pues, parece ser el Ibérico Pleno la fase inicial de fabricación de colmenas cerámicas, mientras que el Ibérico Final y la Época Imperial serían fases de consolidación y perduración, respectivamente. Por tanto, la materialización de un comercio exterior únicamente se produciría en la segunda fase (Ibérico Final) cuando la producción ya se ha estandarizado, y cuando la llegada y conquista romana pudo condicionar e incrementar los contactos comerciales hacia el exterior (Mata, Moreno y Quixal 2008: 42).

Dicha posibilidad comercial y la estrecha relación entre colmenas cerámicas/producción apícola y *kalathoi*, se ve corroborada por la presencia conjunta de colmenas y *kalathoi* en contextos portuarios, como es el caso del puerto de Arse-Saguntum[9] (Aranegui 2004b: 80), bien como por los análisis de contenido de estas vasijas, que revelaron la existencia de frutos carnosos, como los higos, con miel (Juan-Tresserras 2000: 103-104). Ambos aspectos son determinantes en la confirmación de su funcionalidad, permitiendo valorar la exportación de los productos protagonistas de ese comercio, la miel -el contenido- y de otros productos derivados de la práctica apícola, más que de la colmena en sí misma -el contenedor- (Fuentes, Hurtado y Moreno 2004: 196), más allá del ámbito ibérico y explicar el hallazgo de *kalathoi* en pecios en la costa valenciana (Fernández Izquierdo 1995) y fuera de la Península Ibérica, con especial concentración en el Mediterráneo Occidental (Conde 1996; Cuadrado 1968: 129).

Ya se ha mencionado que en las primeras investigaciones sobre apicultura ibérica, las colmenas cerámicas eran entendidas como territorialmente circunscritas, y a raíz de los trabajos de Soria (2000), se amplió la extensión y la utilización de éstas a otros territorios. Las morfologías de las colmenas permiten observar que dos de las variantes presentan una tipología claramente de influencia edetana, mientras que las restantes formas se alejan sustancialmente de éstas. Esta tipología diferenciada parecía corroborar para las segundas una de las hipótesis planteadas por Soria (2000: 177), la de una posible producción local.

No obstante, si es cierto que los bordes de los tipos 4 y 6 de la región de Albacete (Soria 2000: 176) representan una morfología claramente novedosa, y que no se ve representada en ninguno de los estudios dedicados al área edetana, a nuestro entender más bien presentan evidentes similitudes con los fragmentos recuperados en el área segobricense, por lo que, a nuestro juicio, se deberán atribuir a producciones ya de ámbito romano.

La observación comparada de las colmenas de época ibérica tardía del área edetana, y de las de época ibérica tardía/romanas de Albacete con las de época romana de *Segobriga* revela claras afinidades morfológicas y técnicas entre ellas, no solo a nivel de los bordes, sino también al nivel de los cuerpos y acabados de los mismos. Además, las colmenas identificadas en los yacimientos rurales del entorno de la ciudad de *Segobriga*, particularmente las de «tipo tubular», parecen plasmar una continuidad de las técnicas y modelos alfareros de las cercanas regiones ibéricas, a falta de datos para el mismo período en la región segobricense, pero hay que considerar también la posibilidad de su producción local en época prerromana.

Para las colmenas de «tipo troncocónico» es indudable su carácter innovador, pero tampoco aquí creemos que

se deba valorizar excesivamente un modelo autóctono. De hecho, son ya conocidos ejemplos de colmenas cerámicas en Ática, del s. IV a.C., con una de las extremidades cerradas (Jones *et alii* 1973: fig. 79d-80a). También, cabría la posibilidad de un desarrollo local, posiblemente en un momento posterior, ya atribuible a un pleno s. II d.C., de un modelo que podríamos designar como mediterráneo.

A tenor de los datos anteriores, y de los que ahora damos a conocer, la producción colmenera y apícola en época romana en la región inferior de la Meseta, concretamente en Albacete y ahora en el área de *Segobriga*, no parece responder a nuevas prácticas, sino más bien a la aplicación de nuevos estímulos y si acaso nuevos parámetros, así como a la intensificación y a la ampliación de la extensión territorial de una tradición alfarera y de un *know-how* de explotación de recursos del medio ya existentes en época precedente.

Parece ser incuestionable en el área segobricense el fuerte incremento que sufre la actividad apícola y su materialización en todos los yacimientos rurales identificados, en mayor o menor escala, desde el momento de su instalación generalizada en el *ager* inmediato a la ciudad, desde la primera mitad del s. I d.C.

Los yacimientos aquí referidos, por su ubicación cercana a la ciudad y en zonas con evidentes recursos naturales de calidad, son asentamientos rurales dedicados a actividades relativas a la explotación agrícola y ganadera, pero en los cuales las condiciones naturales (flora, recursos hídricos, climatología, etc.) deberían ser también las apropiadas para la actividad apícola.

Su generalizada práctica en los yacimientos segobricenses parece indicar que la explotación/producción de la miel se encontraba vinculada al dominio privado, bajo la responsabilidad de colmeneros o *mellarius* que podrían ser en este caso igualmente los propietarios de cada una de las instalaciones rurales, pudiendo tener también directamente a su cargo la cosecha y procesado/tratamiento de la miel[10], o bien tener a su servicio de manera estacional y itinerante *apiarius* especializados en dichas actividades, en el caso de que lo justificara en volumen de la producción.

Es sabido que la apicultura constituía un papel importante en la economía agrícola romana. Varron por ejemplo, indica, de una forma quizás algo exagerada, que en su época existían apicultores capaces de producir 5.000 libras de miel por año (Morais 2006, *apud* Manacorda 1999: 97). A su

---

[9] En las últimas campañas de excavación se documentó la existencia de colmenas cerámicas ibéricas en niveles republicanos (Aranegui 2004b: 80).

[10] La producción apícola requeriría un trabajo consciente, continuado y una organización previa, pero no siendo necesario ningún tipo de alta especialización ni técnica ni instrumental (Bonet y Mata 1995), tal y como constatan los paralelos etnográficos y las fuentes clásicas (Col. *R. R.* 11.5 y 15).

Figure 20. Mapa con la localización de las colmenas cerámicas en la Península Ibérica: Bracara Augusta, Martinhal, Segóbriga, Santa Cruz de la Zarza y Las Pedroñeras (J. Morín y R. de Almeida).

vez, Virgilio en sus *Geórgicas*, indica también que un colmenar puede llegar a aportar tanto como un viñedo. Lo que en cierta medida nos indican indirectamente estos autores, es que el potencial de la explotación y de producción de la miel en esos lugares conlleva que no se descarte y que podría aportar beneficios económicos considerables.

Así, la producción de miel en el entorno segobricense sería una actividad más, probablemente destinada al autoconsumo, complementaria de las principales actividades agrícolas y de otras transformadoras, como la vitivinícola, o industriales, como la alfarera, ambas igualmente bien representadas en al área en estudio. No obstante, en función de la aparente densidad de hallazgos de colmenas, el volumen de producción podrá, de alguna forma, haber superado el del autoconsumo local, y haberse convertido en una plusvalía económica, y pudiendo su producción excedentaria haber sido comercializada a nivel regional o supra-regional.

Además de la repercusión local y de las implicaciones económicas que pudiera tener la actividad colmenera en *Segobriga* en época romana, los ejemplares de colmenas cerámicas ahora identificados abren a la vez nuevos límites espaciales, posibilidades interpretativas y vías de trabajo de esta particular actividad económica en el interior peninsular, y en la propia Provincia.

Estando ya significativamente documentada en el sureste peninsular para los momentos prerromanos la producción de envases cerámicos destinados a la producción de la miel, y caracterizada en líneas generales la actividad apícola, se impone ahora rastrear para la época romana en las demás áreas peninsulares e identificar esos mismos objetos, que perduraron en territorio peninsular hasta tiempos bien recientes, llegando la miel y la cera a convertirse en época moderna en el segundo producto de exportación castellana, haciendo que los espacios utilizados para la apicultura se preservaran y surgieran normativas que procuraban su correcto desarrollo (Carmona 1999: 132).

De ese modo se podrá valorar debidamente la importancia de una actividad que se hizo tan afamada en época romana, con la miel como sustancia *ex libris* y de la cual la hispana dió pruebas de calidad y cantidad (Figura 20).

Bibliografía

Aranegui, C. 2004. El Grau Vell, puerto de Arse-Saguntum, en P.P. Ripolles Alegre y M. M. Llorens Forcada (eds) *Opulentissima Sagvntum*: 87-97. Sagunto: Fundación Bancaja.

Aranegui, C. 2004b.*Sagunto: Oppidum, Emporio y Municipio Romano*. Barcelona: Bellaterra.

Barroso Cabrera, E. 2019. *De la provincia de Celtiberia a la Qura de Santabarriyya. Arqueología tardía en la provincia de Cuenca (siglos V-VIII d.C.)*. Oxford: Archaeopress.

Barroso Cabrera, R., J. Carrobles Santos y J. Morín De Pablos. 2013. *Camino del Escalón-Mausoleos (T.M. Saelices, Cuenca). Una propuesta de interpretación del 'suburbium' segobricense: la "basílica" de Cabeza del Griego*, en MArqAUDEMA/Serie época romana-Antigüedad Tardía. Madrid: AUDEMA.

Ballester, I. et al. 1954. *Corpus Vasorum Hispanorum: Cerámica del Cerro de San Miguel, Liria*. València: Consejo Superior de Investigaciones Cientificas.

Bencivenga Trillmich, C., 1985. Observaciones sobre la difusión de la cerámica ibérica en Italia. *XVII Congreso Nacional de Arqueología*: 551-556. Zaragoza: Universidad de Zaragoza.

Beltrán Lloris, M. 2002. *Ab ovo ad mala: cocina y alimentación en el Aragón romano*, Institución «Fernando el Católico» (C.S.I.C.), Zaragoza.

Bernardes, J. P. et al. 2014. Colmeias e outras produções de cerâmica do Martinhal (Sagres), en R. Morais, A. Fernández y M. J. Sousa (eds) *As produções cerâmicas de imitação na Hispania*. Tomo I: 507-519. Porto: FLUP.

Blanc, N. y A. Nercessian. 1994 *La cuisine Romaine Antique*. Grenoble: Éditions Glénat, Grenoble.

Blázquez, J.M. 1968. Economía de los pueblos prerromanos del área no ibérica hasta época de Augusto, *Estudios de economía antigua de la Península Ibérica*: 191-269. Barcelona: Vives-Vicens.

Bonet, H., 1995. *El Tossal de Sant Miquel de Llíria: La antigua Edeta y su territorio*, València.

Bonet, H. y C. Mata. 1995. Testimonios de apicultura en época ibérica. *Verdolay* 7: 277-285.

Bonet, H. y C. Mata. 1997. The Archaeology of Beekeeping in Pre-Roman Iberia. *Journal of Mediterranean Archaeology* 10.1: 33-47.

Bonet, H. y C. Mata. 2002. *El Puntal dels Llops. Un fortín edetan*. Serie de Trabajos Varios del SIP, 99. Valencia: Diputación de Valencia.

Bortolin, R. 2008. *Archeologia del Miele*, Documenti di Archeologia, 45. Mantua: SAP.

Carmona Ruiz, Mª. A. 1999. La apicultura sevillana a fines de la Edad Media. *Estudios Agrosociales y Pesqueros* 185: 131-154.

Columela, L. J. 1959. *Los Doce Libros de Agricultura*, vol. II, (Trad. Carlos J. Castro), Colección Obras Maestras. Barcelona: Editorial Iberia.

Conde Berdos, J. M. 199. La cerámica ibérica de Albintimilium y el trafico mediterráneo en los siglos II-I a.C. *Rivista di studi liguri* 62: 115-168.

Crane, E. 1983. *The Archaeology of beekeeping*. London: Duckworth.

Crane, E. 1999. *The World History of Beekeeping and Honey Hunting*. London: Duckworth.

Cuadrado, E. 1968. Corrientes comerciales de los pueblos ibéricos, en *Estudios de Economía Antigua de la Península Ibérica*: 117-142. Barcelona: Vives-Vicens.

Dosi, A. y F. Schnell. 1992. *Vita e costumi dei romani antichi. 3. I Romani in cucina*, Museo della Civiltà Romana 19. Rome: Edizioni Quasar.

Estrabón 1992. *Geografía*, Libros II-IV (Trads. M.J. Meana y F. Piñero. Biblioteca Clásica de Gredos, 169. Madrid: Gredos.

Fernández Izquierdo, A. 1995. Presencia de *kalathoi* en yacimientos submarinos valencianos. *Saguntum* 29: 123-191.

Fernández Mateu, G. 2000. *El kalathos "sombrero de copa" ibérico en el País Valenciano. El kalathos "de cuello estrangulado" del Museo Arqueológico de Villena. Dos bases para un sistema métrico ibérico*. Villena: Fundación Municipal José María Soler.

Fernández Uriel, P. 1993. La evolución mitológica de un mito: la abeja, en C. Blánquez Pérez, J. Alvar y C. G. Wagner (eds) *Formas de difusión de las religiones antiguas*: 133-159. Madrid: Ediciones Clásicas.

Forster, G. 2009. Roman Knossos: the pottery in context. A presentation of ceramic evidence provided by the Knossos 2000 Project (1993-95), Unpublished PhD dissertation, University of Birmingham.

Fuentes, Mª de la, T. Hurtado, y A. Moreno. 2004. Nuevas aportaciones al estudio de la apicultura en época ibérica. *Recerques del Museu d'Alcoi* 13: 181-200.

Germanidou, S. 2012. Ένας δογματικός συμβολισμός σε παραστάσεις Γέννησης του Χριστού στα ιταλοβυζαντινά χειρόγραφα exultet (10ος-13ος αι.), *Deltion of the Christian Archaeological Society* 33: 257-264.

Germanidou, S. 2013. Μαρτυρίες ιστορικών πηγών και αρχαιολογικών ευρημάτων για μια μορφή «βιολογικού» πολέμου στο Βυζάντιο με τη χρήση μελισσών. *Byzantina Symmeikta* 23: 91-104.

Germanidou, S. 2018. Honey culture in Byzantium: an outline of textual, iconographic and archaeological evidence, en F. Hatjina, G. Mavrofridis, y R. Jones (eds) *Beekeeping In the Mediterranean - From Antiquity to the Present*: 93-104. Nea Moudania: Division of Apiculture, Hellenic Agricultural Organization 'Demeter'-Greece; Chamber of Cyclades; Eva Crane Trust - UK.

Guérin, P. 2003. *El Castellet de Bernabé y el Horizonte Ibérico Pleno Edetano*. Valencia: Diputación de Valencia.

Gutiérrez Lloret, S. 1996. *La cora de Tudmir. De la Antigüedad Tardía al Mundo islámico. Poblamiento y cultura material*. Madrid – Alicante: Casa de Velázquez.

Hayes, J.W. 1983. The Villa Dionysos Excavations, Knossos: the Pottery. *The Annual of the British School at Athens* 78: 97-169.

Hernández-Pacheco, E. 1924. *Las pinturas prehistóricas de las cuevas de la Araña (Valencia)(Evolución del arte rupestre en España)*, Comisión de Investigaciones Paleontológicas y Prehistóricas, memoria nº 34. Madrid: Museo Nacional de Ciencias Naturales.

Henriques, F. et al. 1999-2000. Muros-apiários da vacia do medio Tejo (Regiões de Castelo Branco e Cáceres). *Revista Ibn Maruán* 9/10: 329-363.

Jones, J. F., A. Graham y L. H. Sackett. 1973. An Attic Country House Bellow the Cave of Pan at Vari. *The Annual of the British School at Athens* 68: 355-452.

Juan-Tresserras, J. 2000. Estudio de contenidos en cerámicas ibéricas del Torrelló del. Boverot (Almazora, Castellón). *AEspA*, 73: 103-104.

Manacorda, D. 1999. Il mestiere dell'archeolologo. Il ronzo delle api. *Archeo. Attualità del passato* XV. 3.169: 97-99.

Martí Bonafé, M.A., 1998. *El área territorial de Arse-Saguntum en época ibérica*. Alfons el Magnànim, 72. València: Institució Alfons el Magnànim Diputació de València.

Martín Morales, C. (coord.) 1981. *La cerámica popular de Andalucía: Catálogo de la Exposición*. Madrid: Dirección General de Bellas Artes, Ministerio de Cultura.

Martín Tordesillas, A. 1968. *Las abejas y la miel en la Antigüedad clásica* Madrid: Gráficas Condor.

Mata, C. 1998. Las actividades productivas en el mundo ibérico, en C. Aranegui (coord.) *Los iberos. Príncipes de Occidente*: 95-101. Barcelona: Fundación La Caixa.

Mata, C. y H. Bonet. 1992. La cerámica ibérica: Ensayo de tipología, en *Homenaje a D. Enrique Pla Ballester. Estudios de Arqueología ibérica y romana*. Serie de Trabajos Varios del SIP 89: 117-173. Valencia: Diputación Provincial de Valencia.

Mata, C., A. Moreno, y D. Quixal. 2008. Hábitat rural y paisaje agrario durante la segunda Edad del Hierro en el este de la Península Ibérica (Internacional Congreso of Classical Archaeology Meetings Between Cultures in the Ancient Mediterranean). *Bollettino di Archeologia Online*, Volume Speciale, Roma: Ministero Per I Beni e Le Attività Culturali.

Malalana Ureña, A., Barroso Cabrera, R. y J. Morín De Pablos, J. 2011: *La Quebrada II. Un hábitat de la tardoantigüedad al siglo XI. La problemática de los "silos" en la Alta Edad Media hispana*, en MArqAUDEMA. Serie Arqueología Medieval. Madrid: AUDEMA.

Malalana Ureña, A. y J. Morín De Pablos. 2011a. *Ermita de Magaceda II (T.M. Villamayor de Santiago). Un asentamiento de la primera repoblación en el territorio de Uclés: las actividades de manufacturas suburbiales*, en MArqAUDEMA. Serie Arqueología Medieval. Madrid: AUDEMA.

Malalana Ureña, A. y J. Morín De Pablos. 2011b. *Villajos (T.M. Campo de Criptana). Un asentamiento andalusí en la periferia de la Küra de Santaver. MArqAUDEMA. Serie Arqueología Medieval*. Madrid: AUDEMA.

Mazar, A. y N. Panitz-Cohen. 2007. It Is the Land of Honey: Beekeeping at Tel Rehov. *Near Eastern Archaeology* 70.4: 202-219.

Morais, R. 2006. Potes meleiros e colmeias em cerâmica: uma tradição milenar. *Sagvntvm Papeles del Laboratorio de Arqueología de Valencia* 38: 149-162.

Morais, R. 2008. A rota atlântica do mel bético e os contextos de autarcia: vasa mellaria e colmeias em cerâmica. *XIV Congreso de Ceramología*: 73-86.

Morín de Pablos, J. 2014. *Los Paisajes Culturales en el valle del Cigüela*. Madrid: AUDEMA.

Morín de Pablos, J. y D. Urbina Martínez. 2011. *Madrigueras II. Un vicus en el territorio segobricense*. Memoria de Arqueología AUDEMA. Madrid: AUDEMA.

Morín de Pablos, J. 2013. *Pozo Sevilla (Campañas 2008-2010). ¿Una casa-torre en La Mancha?*, en MArqAUDEMA. Serie época romana/Antigüedad Tardía. Madrid: AUDEMA.

Morín De Pablos, J. y I. M. Sánchez Ramos. 2017. *San Blas. Las Predroñeras, Cuenca. Longhouses de época romana en el territorio segobricense. MArqAudema2017. Serie época romana*. Madrid: AUDEMA.

Muscolino, F. 2006. Kalathoi iberici da Taormina. Aggiornamento sulla diffusione della ceramica iberica dipinta in Sicilia. *AEspA* 79: 217-224.

Oliveira, C., R. Morais, y A. Araújo, A. 2015. Application of Gas Cromatography coupled with Mass Spectrometry to the Analysis of Ceramic containers of Roman Period: Evidence from the Peninsular Northwest, en C. Oliveira, R. Morais y Á. Morillo (eds) *Archaeoanalytics: Chromatography and DNA analysis in Archaeology*: 193-212. Esposende, Município de Esposende.

Plinio El Viejo. 2003. *Historia Natural*, Libros VII-XI (Trads. E. Del Barrio et al.), Biblioteca Clásica de Gredos 308. Madrid.

Pérez, G. et al. 2000. La explotación agraria del territorio en época ibérica: Edeta y Kelin, en R. Buxó i Capdevila y E. Pons i Brun (eds) *Els Productes alimentaris d'origen vegetal a l'edat del Ferro de l'Europa occidental, de la producció al consum: actes del XXII Colloqui Internacional per a l'Estudi de l'Edat del Ferro, Girona, 1999*: 151-167. Girona: Museu d'Arqueologia de Catalunya.

Quixal Santos, D. y P. Jardón Giner. 2016. El registro del colmenar ibérico de la Fonteta Ráquia (Riba-Roja, València). *Lvcentum* XXXV: 43-63.

Ribera, A. y M. C. Marín. 2003-2004. Las cerámicas del nivel de destrucción de *Valentia* (75 a.C.) y el final de Azaila. *Kalathos* 22/23: 271-300.

Rosselló Bordoy, G. 1966. *Museo de Mallorca. Sección Etnológica de Muro*. Guía de los Museos de España. XXVIII. Madrid: Ministerio de Educación Nacional.

Daremberg, Ch. y E. Saglio. 1900. *Dictionnaire des antiquités grecques et romaines*, Vol. IV, II. Paris: Librairie Hachette.

Soria Combadiera, L. 2000. Evidencias de producción de miel en la comarca del Júcar (Albacete) en época ibérica, en C. Mata Parreño y G. Pérez Jordà (eds) *Ibers: Agricultors, Artesans I Comerciants. III Reunió sobre Economia en el Món Ibèric (València, 1999), Saguntum. Papeles del Laboratorio de Arqueología, extra* 3: 175-177. València: Universitat de València, Departament de Prehistòria i d'Arqueologia.

Urbina Martínez, D., A. Malalana Ureña y J. Morín De Pablos. 2014. Arroyo Valdespino. Nuevos datos para el estudio de la protohistoria y la época andalusí en la Mancha. Herencia. Ss. V-IV a.C. - XI-XII d.C., en J. Morín de Pablos (coord.) *Los Paisajes Culturales en el Valle del Cigüela*: 593-609. Madrid: AUDEMA.

Vázquez Hoys, A.M. 1991. La miel, alimento de la eternidad, en J. M. Blázquez y S. Montero, S., (coords.) *Aliment. Estudios en Homenaje al Dr. Michel Ponsich,* Gerión Anejos III: 61-93. Madrid: Facultad de Geografía e Historia, Universidad Complutense de Madrid.

Chapter 7

Beekeeping and Problematic Landscapes: Beekeeping and Mining in Roman Spain and North Africa

David Wallace-Hare

University of Exeter (dwallacehare@gmail.com)

Summary: The chapter explores the archaeology of beekeeping through an unusual relationship that developed between beekeeping and mining and quarrying operations in Roman Spain. This relationship can best be characterized as supplementary to these extractive industries. Due to the rather unpredictable nature of where mineral deposits were found, associated communities that formed around them often had to make do with poor or average agricultural lands or otherwise import supplies to sustain the community. Bees, it seems, were used to boost the pollination and yield size of certain crops in such districts. The chapter examines this linkage in the context of beekeeping equipment found at villas in the hinterland of Roman Segobriga, a city that boomed in the early empire thanks in large part to its mineral deposits. The chapter focalizes the economic deployment of bees in this area through the lens of a dossier of land tenure agreements from Roman North Africa which describe the leasing of a class of land known as *subseciva* for keeping bees.

Keywords: Roman Spain, beekeeping, ceramic hives, Latin epigraphy, Roman mining and quarrying, Roman North Africa

Introduction

In the following chapter, I turn our attention to epigraphic and material approaches to the archaeology of beekeeping in one area, central Roman Spain, exploring an unusual relationship that developed between beekeeping and mining and quarrying districts in this area. Roman interest in the Iberian Peninsula during the mid-late Republic seems to have been fed by tales about the wealth of Hispania carried by Phoenician, Greek, Iberian, and later Carthaginian merchants (Diodorus Siculus 5.35.3-4; Pseudo-Aristotle, *de Mirabilibus Auscultationibus* 87 and 135; Herodotus 1.63; Strabo 3.2.11). Italy itself was poor in metals and such interest in external sources was thus unsurprising (Wilson 2002 and 2007). However, the stipulations found in the second treaty between Carthage and Rome struck in 348 BCE forbade Romans from plundering, trading, or colonizing beyond the Fair Promontory in north Africa and Mastia (modern Cartagena) in Spain (Polybius 3.6, 14, and 15). Mastia thus marked the northern boundary of Carthaginian influence in Spain, beyond which Iberian and Greek settlements claimed the rest of the east coast.

The Second Punic War between Rome and Carthage resulted in Rome's acquisition of a sizable portion of the Iberian Peninsula, the southern and eastern parts. The colonization narrative that followed is in broad strokes not debated, with Spain's mineral wealth providing an undeniable attraction to adventurous Roman entrepreneurs. Evan Haley has highlighted

mining as both one of the primary investment opportunities for Roman immigrants to Spain and a source of employment for free migrant labourers within the Peninsula (Haley 1991: 89; cf. most recently Holleran 2016 on this topic). The numerous literary sources referring to pre-Roman and Roman Spain's mines overshadow other attractions through their volume and detail. Polybius, one of our most important sources on the Second Punic War suggested as much in a rather telling statement about past writings on Hispania, and the audience for such accounts. Here he explains how he has written his narrative in a different order than previous treatments of the Peninsula would have led readers to expect:

> Now that I have brought my narrative and the war and the two generals into Italy, I desire, before entering upon the struggle, to say a few words on what I think proper to my method in this work. Some readers will perhaps ask themselves why, since most of what I have said relates to Africa and Spain, I have not said a word more about the mouth of the Mediterranean at the Pillars of Hercules, or about the Outer Sea and its peculiarities, or about the British Isles and the method of obtaining tin, and the gold and silver mines in Spain itself, all matters concerning which authors dispute with each other at great length. I have omitted these subjects not because I think they are foreign to my history, but in the first place because I did not wish to be constantly interrupting the narrative and distracting readers from the actual subject, and

next because I decided not to make scattered and casual allusions to such matters, but assigning the proper place and time to their special treatment to give as true an account of all as is in my power (Polybius 3.57.1-5 (*LCL* 137, transl. Patton).

Previous treatments seem to have prioritized 'the gold and silver mines' of Spain to the degree that Polybius felt he needed to offer an explanation for why he had not yet included these topics in his work. A passage of Strabo provides details from a now lost section of Polybius' work, likely the later section mentioned here above, where the author supplies his audience with some of that detailed information about Spanish mining, including revenue amounts: 'Polybius, in mentioning the silver-mines of New Carthage, says that they are very large; that they are distant from the city about twenty stadia and embrace an area four hundred stadia in circuit; and that forty thousand workmen stay there, who (in his time) bring into the Roman exchequer a daily revenue of twenty-five thousand drachmae' (Strabo 3.2.10 (*LCL* 50)).

Greek and Latin authors during the Principate continued to focalize Roman Spain through the lens of mining, just as they had described its role under the Phoenicians and later under the Carthaginians. For example, the Latin author Solinus (who wrote c. 200 CE) rather colourfully described Hispania as a sort of public treasury: '[i]t abounds in every resource, whatever one might seek that is costly or essential, like gold or silver, it has. It has never lacked iron mines,' (Solinus 23.2 (my translation): *Omni materia affluit, quaecumque aut pretio ambitiosa est aut usu necessaria. argentum uel aurum requiras, habet; ferrariis numquam defecit.*)

The authors of these texts seem to have been writing for audiences hungry for information about one of the most profitable industries of the Iberian Peninsula. Indeed, John Richardson (Richardson 1976: 146) has described the late Republican and early imperial mining activities in southern and eastern Spain as something akin to a 'nineteenth-century gold-rush.' Roman mining and smelting activities during the Republic and Principate had such an environmental impact as to leave detectable levels of lead in ice cores taken in Greenland from pollution produced between 600 BCE and 300 CE. Rosman et al. (1997) showed through isotopic systematics that mining districts in southwest and southeast Spain were the dominant sources of this lead. This was the first study to give quantitative evidence of the importance and scale of these mining districts for both the Carthaginians and Romans. Indeed, lead from the mines of Rio Tinto represented c. 70% of the lead found in Greenland ice between 150 BCE-50 CE.

It seems that due to the rather unpredictable nature of where mineral deposits were found and what must have amounted to a 'gold rush' mentality in the event of new deposits in the Peninsula, associated communities that formed around such deposits often had to make do with poor or average agricultural lands or otherwise import supplies to sustain the community. Beekeeping, I argue in this chapter, seems to have been employed by some Roman villa owners of the early imperial period (first-second century CE) to offset average agricultural lands. To study commercial beekeeping in the Roman period is also to study the ecosystems in which beekeeping was being deployed it turns out.

The best place to examine beekeeping and mining in the Roman Iberian Peninsula is in the territory of Roman Segobriga in the context of beekeeping using ceramic hives in villas belonging to the town's elite. The wealth of this group seems to have been derived from the extraction of *lapis specularis*, a type of translucent gypsum used in Roman windowpane construction and a product for which this region seems to have been famous (Pliny, *Naturalis Historia* 36.160). The unusual concentration and refinements of ceramic hive designs in this area, based on Iberian precursors from eastern Spain, must be understood as a way to boost local agricultural potential. The best 'rule book' for understanding this surprisingly modern way of thinking about how to use bees comes not from Roman Spain itself, but from an adjacent province, Africa Proconsularis. This evidence takes the form of a dossier of land tenure arrangements concerning honey production and hive use at the Villa Magna Variana, an estate owned by the emperor whose lands were leased by his procurators. The arrangements concern the use of beekeeping to extract profits from a certain type of land known as *subseciva*, which referred to lands deemed uncultivable in the past when the original plot division, or centuriation, of a territory was undertaken. The use of bees in such lands is instructive for thinking about mining districts, which sometimes could represent *subseciva* writ large.

Textual and Material Evidence of Apiculture in Pre-Roman and Roman Spain: An Overview

Honey production is one of the few economic activities associated with the pre-Roman Iberian Peninsula, particularly with the eastern and southern Iberian Peninsula. In fact, one recorded king of Tartessus, an ancient southern Spanish kingdom, is said to have invented the art of honey hunting, that is, the collecting of honey from wild beehives (Justin 44.4). While the historicity of Tartessus itself is debated (Aubet 1989; Aguilar 2005), nonetheless, the attribution of honey-hunting to such an obscure figure and far-off corner of the ancient Mediterranean helps argue in favour of its historicity and arguably the importance of bees in the pre-Roman Peninsula. The Greek historian Diodorus Siculus (first century BCE) recorded that the Celtiberians, a large conglomeration of Celtic-speaking peoples to the northeast of the Tartessi, also practiced apiculture before and after the arrival of the Romans (Diodorus Siculus 5.34.2). The association of local populations in central and southern Spain with beekeeping is notable in Greek and Latin sources with few other provincial populations or areas being so closely associated with beekeeping outside Greece or Sicily.

Beekeeping in the Iberian Peninsula is also unique in being included among economic activities practiced *after* Roman occupation in our sources. According to the Greek geographer Strabo (62 BCE-24 CE) the Turdetani, a population in southern Spain, exported substantial amounts of high-quality grain and many types of wine and olive oil (Strabo 3.2.6). They also exported wax, honey, pitch, scarlet dye, and red ochre. While Strabo's language suggests that wax and honey, at least for Turdetania, a region occupying a large part of the Roman province Baetica, were secondary products to grain, wine, and olive oil, nonetheless, they were still regionally important.

Indeed, we even possess an inscription of first century CE date from Roman Corduba marking the transfer of occupancy of an apiary site: 'In the duumvirate of Lucius Valerius Poenus and Lucius Antistius Rusticus, three days before the Kalends of September, Lucius Valerius Kapito, son of Caius, took possession of the apiary site' (*CIL* II 2242 (Córdoba / Corduba, Baetica): *L(ucio) Valerio Poen[o] / L(ucio) Antistio Rustico / IIvir(is) / a(nte) d(iem) III K(alendas) Septembres / L(ucius) Valerius C(ai) f(ilius) Kapi/to alvari locum / occupavit)* (on this inscription see Hanel 2009). Such information is rare in the Roman world, with our only other explicit epigraphic references to participants in commercial beekeeping during the Roman period found at Rome (*AE* 1971, 42; *CIL* VI 9618; cf. Morillo, Morais, and Wallace-Hare 2019). These Roman *mellarii* might have been intermediaries in this trade and not beekeepers themselves, although the possibility remains that they might represent rare examples of urban beekeepers from the ancient world.

Thankfully, such rare but important epigraphic evidence is joined by a wealth of material evidence taking the form of ceramic beehives and honeypots in the Iberian Peninsula. In the 1990s, archaeologists from the Universitat de València, Helena Bonet and Consuelo Mata, uncovered industrial levels of beekeeping in the

Figure 1. Beehives from Puntal dels Llops, Tossal de Sant Miquel, and La Monravana (IV-III century BCE).
(Photo courtesy of the Museo de Prehistoria de Valencia).

form of hundreds of hive fragments in the hinterland of cities belonging to ancient Edetania, an Iberian-speaking zone near modern Valencia. These remains spanned the sixth century BCE to the first century CE and took the form of horizontal ceramic tubes open on both ends with interior striations to aid in comb attachment (Figure 1).

These Iberian ceramic hives may not have been in widespread use even during their period of greatest use, however. Bonet and Mata found that Iberian ceramic hives disappeared in the first century CE. when there seems to have been a move to hives made of lighter and better insulating materials such as esparto grass (Bonet and Mata 1997: 42-43). A similar reduction in ceramic hive usage played out during the Principate. Based on the model established by Bonet and Mata, ceramic hives came to be identified in several other areas of central Spain and northern and southern Portugal (Figure 2).[1] Most notably, Morín and de Almeida discovered the largest number of Roman hives in Central Spain during extensive excavations between 2009-2011 in the territory of Segobriga, west of Edetania (de Almeida and Morín 2012; Morín and de Almeida 2014). Segobriga was home to a Celtiberian population known as the Carpetani or the Olcades (Morín and de Almeida 2014: 290-301). With these later finds from Segobriga, we can see another decline in usage of ceramic hives, just as happened in pre-Roman and late Republican Iberia, with the substitution of hive materials, ceramic for biodegradable materials, occurring during the late second century (Morín and de Almeida 2013: 289-290). The impetus for this change is not yet fully understood. The first century CE Latin agricultural writer Columella of Gades' aversion to ceramic hives may reflect the wider use of organic hives as well.[2]

Other ceramic evidence, which is somewhat unique to the Roman Iberian Peninsula, namely honey pots (sometimes called *vasa mellaria*), were above all storage vessels and seem mostly to indicate the purchase of honey where they occur, not necessarily

sites of production (Figure 3) (see Morais 2006 for the foundational study of these vessels; see also Persano's treatment of these vessels in chapter 4 of this volume). Such vessels can be tested using palynology, or pollen residue analysis of, for instance, residual honey, to determine honey provenance. Melissopalynological analysis has not been applied to beekeeping vessels from the ancient Iberian Peninsula yet. Other testing, using gas chromatography and mass spectrometry of honeypots and hives from the ancient Iberian Peninsula conducted by Oliveira et al. 2013 and 2017, while not designed to reveal floral provenance, nonetheless has been incredibly revealing for understanding techniques of beekeeping (smoking techniques in particular) and the chemical composition of residual honey. Chapter 5 of this volume contains an important new review of the potential of palynology in the archaeology of beekeeping. At present, therefore, honeypots and hives in the Iberian Peninsula, barring such forthcoming pollen residue analyses, cannot necessarily be used to solidify the fuzzy outlines of honey production zones in the pre-Roman and Roman Peninsula.

Lapis Specularis and the Success of Segobriga

As was explained above, various ancient sources point out that beekeeping was a feature of pre-Roman Spain. The distribution of the archaeological remains of beekeeping dated to the Roman period frequently coincides with historically Hispano-Celtic and Iberian zones. This coincidence presents an opportunity to extend our knowledge of local beekeeping regions in Hispania Tarraconensis, one of the three provinces of imperial Roman Spain. Segobriga is a nexus of both Celtiberian and Iberian influences, both of which seem to have informed Roman beekeeping in the area.

Roman Segobriga was recently described by Carlos Noreña (2019) as being a uniquely Romanized city in a provincial backwater. As Noreña put it, the cities of the Mediterranean littoral of the Peninsula were 'plugged in' to the central grid of Roman imperial power and communications. These cities were marked 'by dense populations, high degrees of urbanization, agriculturally productive landscapes, commercial prosperity, concentrations of wealth, and numerous channels of upward mobility for local elites' (Noreña 2019: 3). The central Meseta was not 'plugged in' to this Roman Mediterranean power and culture grid for a variety of reasons, most connected to the Meseta's geographic restraints, with a few notable exceptions. The Meseta is known for its low agricultural productivity, consisting of rocky outcrops with mixed concentrations of oak forests (holm oak especially). In this climate, a continental and not a Mediterranean type, extreme temperatures complicate large-scale planting of Mediterranean staples like olives. The landscape of the Meseta was far better suited to the

---

[1] Ceramic hives (mostly from the Roman period), have been found at Braga, Martinhal (Sagres), Saelices, Las Pedroñeras, and Santa Cruz de La Zarza (Morais 2006; de Almeida and Morín 2012 and 2014; Bernardes et al. 2014), while honey pots (*vasa mellaria*, Port. *meleiros*) have been found at Monte Castêlo (Matosinhos), Chaves, and Conímbriga (Morais 2006; Oliveira, Morais, et al. 2014).

[2] Columella, *de Re Rustica* 9.6.1-2 (my translation): 'After the site has been arranged, you should construct your beehives based on the character of the region, whether that region abounds in cork trees (undoubtedly, we'll make the most advantageous hives out of their bark, seeing as how they don't freeze during the winter, nor blaze with heat during the summer), or if your region is full of giant fennels, which we can also weave hives from because they are similar to the nature of bark. If neither is present, you can weave hives using willow branches in wickerwork fashion. If these too are not at hand, you'll need to construct the hives from the wood of a hollow tree or wood cut into planks. (2) The *worst* hive type is that made of clay because they overheat in the high temperatures of summer and freeze in the cold of winter.'

ranging of livestock and pannage-focused pig farming. By and large, the whole Meseta from the standpoint of imperial communications, density of urbanization, relative wealth, and consistent Roman culture would be defined as a 'hinterland,' to use Curchin's term, or a 'backwater' in Noreña's terminology (Curchin 2004; Noreña 2019).

Backwater cities might escape that status by exploitation of extraordinary local resources, however. In the case of the Central Meseta, this mostly meant tapping into latent mineral resources. This seems to be the story of Roman Segobriga whose unique proximity to the Empire's best supply of *lapis specularis*, a type of selenitic gypsum used for window panes due to its high translucence, elevated it far above its surroundings and led to the creation of a Roman city 'in the middle of nowhere'. The history of the Roman city and the formation of its elite are instructive for thinking about both the integration of local populations in imperial mining and quarrying districts, but also the effects that such mining towns could have in preserving other traditional economic activities in the area.

The site of Segobriga, settled by the fifth century BCE, was perhaps originally chosen because of its landscape, as the city is on an elevation and represented a typical hillfort community or *castro* to use the modern Spanish designation of such sites (Almagro Gorbea 1992: 275–276; Lorrio 2001: 205–207). It is unclear when Segobriga came into Roman possession but we know that by the first century CE the city was called a *civitas stipendiaria,* 'tax-paying town,' by Pliny (*NH* 3.25). Noreña notes that such a designation was characteristic of indigenous settlements in Roman territory (Noreña 2019: 4). Under Augustus, c. 15 BCE, it seems the town became a *municipium iuris Latini*, conferring elevated status on the town and creating new forms of upward mobility among the local elite in the form of municipal office holding. During the Augustan period the town thrived and gained a typical Roman profile with a monumentalized town centre marked by texts commemorating the activities of the local elite and their wider connections. Much of the history of Segobriga during this 'boom' time must be sketched from the archaeological and epigraphic record.

What makes this boom period of Roman Segobriga so fascinating is its local nature. Noreña points out that no Roman *colonia* was planted here, nor do we find clear examples of the immigration of families from the Italian Peninsula (Noreña 2019: 16; cf. Curchin on the unusual absence of much evidence for Italian immigration or military bases in central Spain 2004: 48–50, 84–85, 90–92). According to Noreña, Segobriga's local elite largely consisted of indigenous families. Some individuals seem to have been the descendants of the clients of early governors of Hispania Citerior, a former division of the

Iberian Peninsula during the Roman Republic (Noreña 2019: 15-16). The source of wealth underpinning the rise of a part of the local elite in Roman Segobriga seems to have been exploitation of a mineral for which the town was famous, *lapis specularis* (on the exploitation of sources of *lapis specularis* at Segobriga and the Cuenca region more widely, see Bernárdez Gómez and Guisado di Monti 2002; 2004; 2009; 2010; 2016; and Bernárdez Gómez, Díaz Molina, and Guisado di Monti 2015). Pliny the Elder (*NH* 36.160) says that the best quality *lapis specularis* in the Empire came from mines located within a 150 km radius of Segobriga (Figure 4).

It is clear from Pliny's comment and the rapid monumentalization of this otherwise economically unattractive location in the backwater of central Spain that the export of this resource was the basis of the town's growth and the product which enriched the local elite. Bernárdez Gómez and Guisado di Monti in several publications have demonstrated that exploitation of this stone occurred over the first and second centuries CE and formed the economic focus of Roman Segobriga. These scholars also believe that the Roman *fiscus,* a word denoting alternately the financial administration controlled by the emperor but also the private property of the emperor, directly administered the *lapis specularis* mines and made private concessions to local groups along the lines of the Aljustrel mines (on the *fiscus,* see Brunt 1966). While there is no direct evidence that this was how mines for *lapis specularis* in the area operated, Pliny's detailed knowledge of the stone and its distribution throughout the Empire might indicate state control.

Pliny also tells us that *lapis specularis* was used to make special types of beehives that he must surely have witnessed during his time as procurator of Hispania Tarraconensis. These hives likely originated in the area of Segobriga:

> It is well for the apiaries (*alvaria*) to look due east and to avoid the north wind as well as the west wind. The best hive is made of bark; the next best material is fennel-giant, and the third is osier. Many too have made hives of transparent stone, so that they might look on the bees working inside (*multi et e speculari lapide fecere, ut operantes intus spectarent*), (*NH* 21. 45. (LCL 392)).

What such a hive might have looked like or whether *lapis specularis* would have been used in only one part of the hive is difficult to determine, this being the only reference. The association between the region of Segobriga, *lapis specularis*, and beekeeping is, nonetheless, suggestive of deeper connections between the mining wealth of the city and those who reaped its profits. Some of the mining elite constructed villas in the suburbs of Segobriga and the wider *ager*

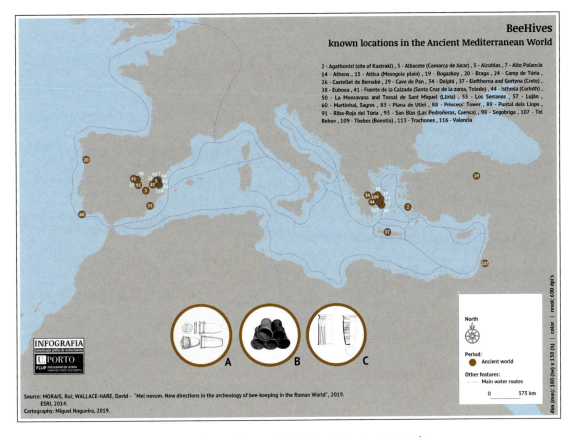

Figure 2. Known Evidence of Ceramic Beehives in the Ancient Mediterranean
(Map: Morais and Wallace-Hare 2019, cartography: Miguel Nogueira).

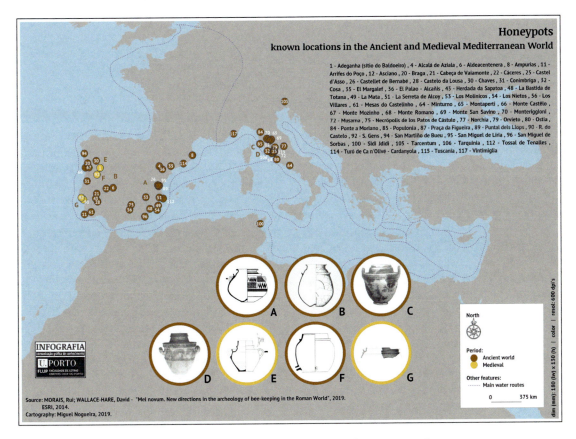

Figure 3. Known Evidence of Ceramic Honeypots in the Ancient Mediterranean
(Map: Morais and Wallace-Hare 2019, cartography: Miguel Nogueira).

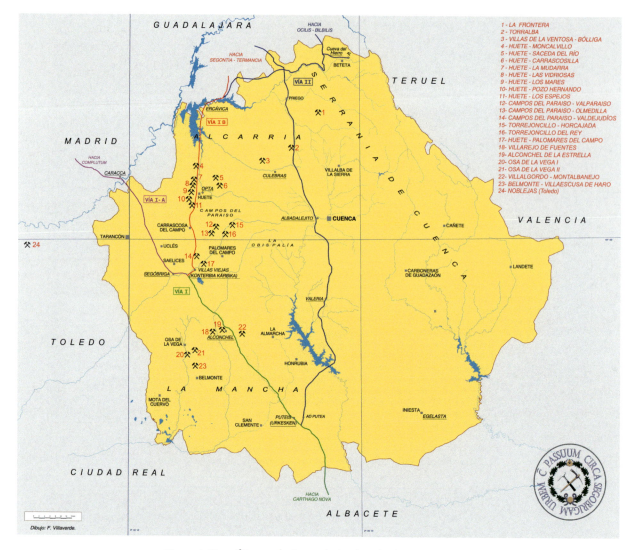

Figure 4. Map of Roman *lapis specularis* mines in the Cuenca area.
Map courtesy of Fernando Villaverde Mora - Equipo de Investigación del lapis specularis.

Segobricensis. These villas were the subject of recent preventive excavations of the *ager Segobricensis* between 2009-2011 during the expansion of a central water pipeline near Cuenca. The archaeological evidence showed that the territory of Segobriga possessed some adequate farmland in certain areas which was used, it seems, for olive and vine production connected with these villas. Importantly, the excavators also uncovered evidence of extensive beekeeping over the first and second centuries in most of these villas, as will become clear in the next section.

Beekeeping at Roman Segobriga

Between 2009 and 2011 Jorge Morín and Rui de Almeida along with several colleagues had the opportunity to undertake preventive excavations of the agricultural hinterland of Segobriga (the *ager*) on the occasion of a large-scale expansion of a water pipeline near Cuenca (de Almeida and Morín 2012). The excavations allowed the team to get a detailed picture of the town's

agricultural basis. In five of the sites excavated (Figure 5: La Peña II, Llanos de Pinilla, Los Vallejos, La Laguna, Casas de Luján II, and Rasero de Luján) the excavators found rims, body fragments, and some bases of ceramic vessels which, thanks to the studies of Morais and Bonet and Mata, could be identified as beehives. Two of the sites where hives were encountered, Llanos de Pinilla and Casas de Luján II, were also sites of olive oil production, with well-preserved olive oil mills, presses, and storage vessels (Urbina, Morín, and Urquijo 2014a-b). Furthermore, three of the sites where hive fragments were encountered, Los Vallejos, Casas de Luján II, and Rasero de Luján, also featured remains of amphoras produced in the region which were indicative of local wine production (Morín and de Almeida 2014: 292). The occurrence of ceramic hives at these sites of agricultural production and the situation of those sites of modest productivity in an area of well-attested mineral output has gone unnoticed but may be significant for understanding the flourishing of beekeeping in this 'backwater' and many others.

Morín and de Almeida's study of beekeeping in the *ager Segobricensis* is important because of the connections they could draw between the Roman period ceramic hive remains found in abundance at five rural sites around the *ager Segobricensis* and the late Iberian Edetanian hives studied by Bonet and Mata (1997). Morín and de Almeida demonstrated clear morphological and technical affinities between the Roman and Iberian hive assemblages (Morín and de Almeida 2014: 302). Equally important, the large number of hive fragments discovered in the territory of Segobriga allowed the excavators to extend the typologies of the Iberian hives examined by Bonet and Mata. They established the existence of a ceramic hive type different from the common 'tubular' Iberian hive type.

This new type, which the excavators dubbed 'troncocónico' ('conical-trunked') was an innovation on the Iberian model which seems to have allowed for 'extension rings,' also called ekes, to be attached to the hive (Figure 6) (Morín and de Almeida 2014: 302).[3] The excavators also emphasized that this conical-trunked hive seems to have been a locally developed form of the second century CE, as this hive was most characteristic of that century and is not found elsewhere in the Peninsula or wider Roman world.

The most important finding of Morín and de Almeida's study was their discovery that imperial apicultural production in the lower region of the Meseta was not the result of new practices introduced by the Romans. Instead, the hive assemblages seemed to indicate the addition of new stimuli and parameters to a long-practiced local tradition. The excavators believed that the hive assemblages and their distribution in the hinterland around Segobriga unequivocally showed 'an intensive production that had no relation to the traditional transhumant practice' (Morín and de Almeida 2014: 284). By 'transhumant practice,' Morín and de Almeida were referring to the itinerant beekeeping alluded to by Pliny the Elder (21.43.7) whereby hives were transported by mule to different forage sites to maximize honey production. The extension rings in the present case would need to have been mounted on stationary hives to work.

The placement of the hives at various villas in the territory of Segobriga is complicated but seems to suggest larger-scale production rather than simple villa subsistence. One nearly complete hive found at Los

Vallejos was located inside a room in the *pars urbana*, or living quarters of the villa, and was next to a Dressel type 2-4 amphora. Both vessels were found beneath the collapsed roof of that room. Morín, de Almeida, and Sánchez in chapter 6 of the present volume suggest that such a space would have been inappropriate for genuine hive use and that it was likely there for storage or cleaning. Other hive fragments were also found in interior contexts at La Peña II, in the *pars rustica / fructuaria* of the villa, in two cases in rooms connected to the vineyard and warehouses, and in another in a large open room of indeterminate agricultural function.

It is only at one site, La Laguna, that a structure (Figure 7) located in an external space appeared to the excavators to represent a site suitable for an apiary (Morín and de Almeida 2014). This space was a high traffic area of the *pars rustica* and may have taken the form of stacked rows of beehives set into some framework accessible from both ends. The proposed reconstruction lends itself heavily to the well-known apiary of Tel Rehov with its rows of stacked hives largely left intact when the apiary was burned c. 900 BCE (Mazar and Panitz Cohen 2007) (Figure 8a-b). This reconstruction is also supported by many similar configurations in the Iberian Peninsula, in particular, traditional apiaries from central and northeastern Spain (see chapters 6, 15, and 16 of the present volume for greater discussion of these structures).

The ceramic beehives were one of the constant elements in Segobrigan ceramic repertoires in the excavated deposits at La Peña II, Llanos de Pinilla, Los Vallejos, La Laguna, Casas de Luján II and Rasero de Luján, making up about 10% of the large vessels (Morín and de Almeida 2014: 292). While a complete hive specimen could not be reconstructed from individual deposits at any one site, a specimen from La Laguna and another from Los Vallejos permitted a reasonable reconstruction due to high levels of preservation among the hive fragments (Morín and de Almeida 2014: 292). These reconstructions are crucial for gauging differences in honey production from late Iberian hives in nearby Edetania. Morín and de Almeida estimated that the approximate capacity of Segobrigan hives over the first-second centuries CE was between 41 and 56 litres, respectively. The hives that could hold 56 liters far exceeded other known Middle (IV-III century BCE) and Late (II-I century BCE) Iberian hive capacities. The 41-litre hive had a slightly greater capacity than other Iberian hives, as well as other hives from the Iberian Peninsula at Braga and Martinhal in Portugal of third-fifth century CE date. (Bonet and Mata 1995: 280; Crane 1983: 17; Morín and de Almeida 2014: 290-292). These two scholars established a clear example of intensive beekeeping which seems to have been a response to stimuli arising from expanded economic opportunities at Segobriga. The basis of that economy was probably

---

[3] On such extension rings, see Harissis 2018 20-21: 'A traditional practice, also known in antiquity, was to elongate horizontal hives by adding a bottomless cylindrical terra-cotta stem ("extension ring"), which was fastened between the lid and the end of the hive, which had projecting rims. With this technique, the beekeeper could easily separate the extension ring from the main hive and harvest part of its crop without disturbing the inner parts; this entailed using less smoke, which was known to harm the taste of honey. Additionally, the extra space provided in the hive prevented swarming.'

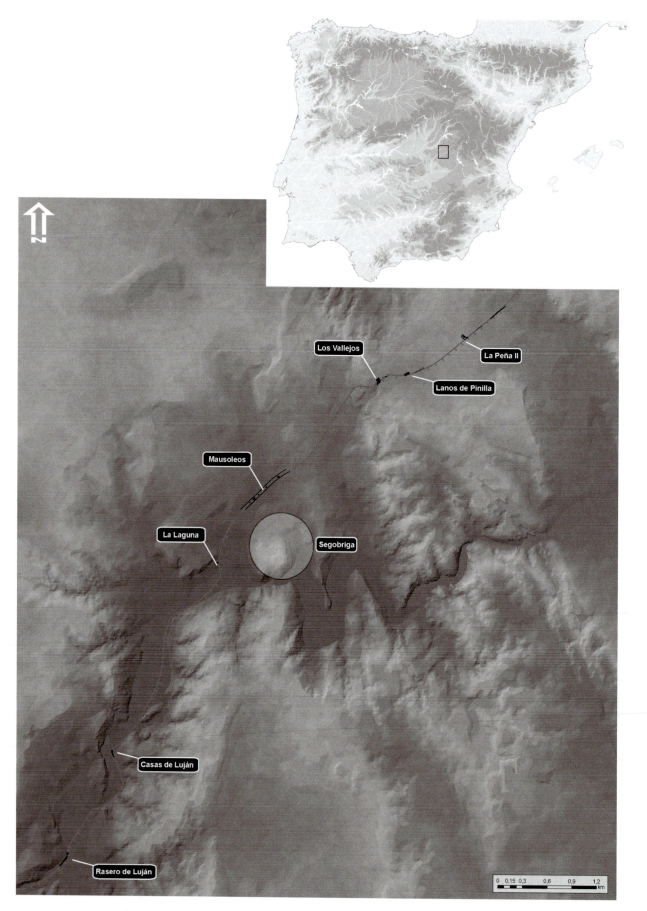

Figure 5. Position of Segobriga in the Iberian Peninsula and Map of the Sites of Excavated Hive Remains (I-II century CE) (Map courtesy of R. de Almeida and J. Morín).

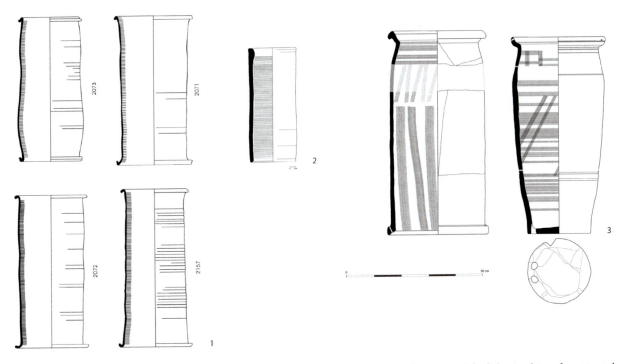

Figure 6. Comparison of Iberian and Segobrigan Ceramic Hives (Drawing courtesy of Jorge Morín): 1) Iberian hives from Puntal des Llops (Valencia region, Spain) (IV-III century BCE) (Bonet and Mata 2002); 2) Roman ceramic hive from the urban *insula* as Carvalheiras, Bracara Augusta (Braga, Portugal) (III century CE) (Morais 2006); 3) left: tube hive from La Laguna (Segóbriga, Cuenca, Spain); right: conical-trunked hive from Los Vallejos and lid (Segóbriga, Cuenca, Spain) (I-II century CE) (de Almeida and Morín 2012).

trade in *lapis specularis* coupled with some olive and vine growing.

I would argue that these hive design changes, suspected tiered frameworks into which hives could be set, and the appearance of ceramic hives across several villas in the same region are to be connected to the benefits bees brought to the farms in which we find them. Bees were also profitable even on small plots of land as we shall see below. The connection between bees and mining in Spain is oblique rather than direct and can be understood best by looking south of the Peninsula to a set of land tenure regulations from the Roman province of Africa Proconsularis dealing with beekeeping on agriculturally problematic lands.

Bees in Problematic Landscapes: A Case Study of Villa Magna Variana in Roman North Africa

An important dossier of land tenure agreements dating to 116-117 CE found at Henchir-Mettich in Africa Proconsularis sheds great light on the potential of beekeeping to Roman villa owners wishing to exploit less productive lands while also improving the success of fruit crops planted on such land. These regulations governed the behaviour of *coloni*, or tenant farmers, on imperial estates in the area (Flach 1978; Kehoe 1988). Among the group of six lengthy inscriptions, one text concerned beekeeping on an estate known as Villa

Magna Variana, or, locally, Mappalia Siga (modern Henchir-Mettich). The land tenure arrangement laid out in the Henchir-Mettich inscription (hereafter the HM inscription) concerned the occupation of unused land called *subseciva* in this area by a group of *coloni*.

According to definitions of this category of land from authors in the *Corpus agrimensorum*, a collection of Roman land surveying manuals from different times compiled c. 450 CE, *subseciva* were lands found outside distributed *centuriae*, or plots of land. Such lands could not be used for agriculture due to their marshy, rocky, or forested nature, or for a number of other reasons and were thus not included in plot divisions of the original Roman territory (for definitions of *subseciva* in the *Corpus Agrimensorum* (ed. and transl. Campbell 2000), see Frontinus 2.24-27; Agennius Urbicus 38. 4-5; *Commentum* 54.21-2, 68.17-18; *De Agris* 272.16-19; see also the discussions of *subseciva* in Kehoe 1988: 37; Roselaar 2010: 140-144; Guillaumin 2007: 157-166; Bluhme, Lachmann, and Rudorff 1848-1852 (v. 2): 390-394, 455-459).

Subseciva re-entered the picture when potential *coloni* sought the right to bring such lands under cultivation. *Subseciva* were said to be in the possession of towns as *ager publicus*, or public land, and could be rented out to *coloni* (Roselaar 2001: 141). In terms of the land conditions addressed in the HM inscription on the estate of Villa

Figure 7. Location of ceramic hives at excavated sites in the territory of Segobriga (L-R, top to bottom):
1: Los Vallejos, 2: La Peña II, 3: La Laguna (suggested apiary location circled), 4: Casas de Luján, 5: Morín and de Almeida's
reconstruction of the hive installations at La Laguna. Photo courtesy of Jorge Morín.

Magna Variana, the *subseciva*, according to Dennis Kehoe (1988: 37), 'must have been largely comprised of hillsides, since the area in which the inscription was discovered is quite hilly with some rather steep mountain slopes.' The HM inscription details the conditions under which *coloni* could occupy *subseciva* under a location- specific application of an earlier land tenure law called the *Lex Mancia*. According to Kehoe, the HM inscription's erection represented the favorable reply by the imperial administration to a petition brought by the *coloni* for

the right to cultivate *subseciva* (Kehoe 1988: 28) (Table 1). This application of the *Lex Mancia* was drafted by the procurators of the estate of the Villa Magna Variana and reflects the imperial administration's interest in gleaning profits from lands formerly unoccupied or deemed uncultivable.

Of interest to us here are those *coloni* who practiced beekeeping on these *subseciva* and the fact that there seems to have been economic incentives to engage in

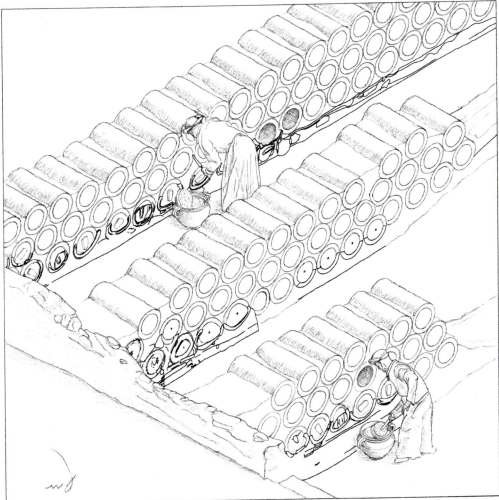

Figure 8a (top): Apiary (c. 900 BCE) at Tel Rehov (Northern Israel)(Photo by Amihai Mazar and courtesy of the Tel Rehov Expedition, The Hebrew University of Jerusalem; Figure 8b (bottom): Reconstruction of the apiary according to Mazar and Panitz-Cohen 2007 (drawing by Ana Iamim; courtesy of Amihai Mazar).

Table 1: The HM Inscription (*CIL* VIII 25902) (Date: 116/117 CE)
(transl. Kehoe 1988): *Subseciva*

Section	Latin	Translation
1.1-10: Preamble and Authorization to Cultivate *Subseciva*	*[Pro sal]ute / [A]ug(usti) n(ostri) Im[p(eratoris)] Caes(aris) Traiani prin[c(ipis)] / totiusqu[e] domus divin(a)e / optimi Germanici Pa[r]thici data a Licinio / [Ma]ximo et Feliciore Aug(usti) lib(ertis) procc(uratoribus) ad exemplu[m] / [leg]is Mancian(a)e. / qui eorum [i]ntra fundo Villae Mag/ n(a)e Varian(a)e id est Mappalia Siga [habitabunt?]* **eis eos agros qui su[b-][c]esiva sunt excolere permittitur lege Manciana / ita ut e<o>s qui excoluerit usum proprium habe/at.** *ex fructibus qui eo loco nati erunt dominis au[t] / conductoribus vilicisve eius f(undi) partes e lege Ma/nciana pr(a)estare debebunt hac cond<i>cione coloni:---*	[On] behalf of the welfare of our Augustus the emperor Caesar Trajan, prince, and his whole divine house, best prince, Germanicus, Parthicus, presented by Licinius Maximus and Felicior, freedmen of Augustus, procurators, based on the example of the law of Mancia. **Those of them <who will have farmsteads> within the estate of Villa Magna or Mappalia Siga are permitted to bring under cultivation, in accordance with the law of Mancia, those fields that are classified as unused [*subseciva*], under the condition that whoever brings them under cultivation will have private use.** Tenants will be obliged to furnish shares in accordance with the law of Mancia from the crops, which will have been raised in that place (sc. *subseciva*), to the landlords or lessees or bailiffs under the following terms: ...

Table 2: The HM Inscription (*CIL* VIII 25902) (Date: 116/117 CE)
(transl. Kehoe 1988): Benefits for Beekeepers

Section	Latin	Translation
1.2—2.6: Rents for Crops Grown on *Subseciva*	*qu[i i]n f(undo) Villae Mag/nae sive Mappali(a) e Siga<e> villas [habe]nt habebunt / domini{ca}s eius f(undi) aut conductoribus vilicisv[e] / eorum in assem partes fructu<u>m et vinea<ru>m ex a/[r] ea[m] partem tertiam, hordei ex area[m]/ partem tertiam fab(a)e ex area{m} [pa]rtem qu/[ar]tam vin<i> de lac<u> partem tertiam oll[e-]/[i co]acti* **partem tertiam mellis in alve/[is] mellari<i>s sextarios singulos qui supra / quinque alveos/ habebit in tempore qu[o vin]/demia mellaria fu[it fuerit],/ dominis aut conducto[ribus vili]/ cisve eius f(undi) qui in assem [6-8] / d(are) d(ebebit).**	Those who have or will have farmsteads within the estate of Villa Magna or Mappalia Siga will be obliged to furnish shares of crops and vines, in accordance with the customary practice of the law of Mancia, to the landlords of this estate or the lessees or their bailiffs, of each crop that he has: a third share of wheat from the threshing floor, a third share of barley from the threshing floor, a third share of wine from the press, a third share of pressed olive oil; **whoever will have over five hives in the time during which the honey harvest has been or [will have been] will be obliged to furnish single sextarii of honey for honey-producing hives to the landlords and the lessees [or] bailiffs of this estate who as a group [---]**

Table 3: The HM Inscription (*CIL* VIII 25902) (Date: 116/117 CE) (transl. Kehoe 1988): Prohibitions Against Removal of Beekeeping Equipment or Bees

Section	Latin	Translation
2.6-13: Penalties for the Fraudulent Removal of Equipment Used in the Production of Honey	*Si quis alveos, examina, apes, [vasa] / mellaria ex f(undo) Villae Magn(a)e sive M/appali(a)e Sig(a)e in octonarium agru[m]/ transtulerit, quo fraus aut dominis au[t]/conductoribus vilicisve ei<u>s <f(undi)> quam <maxime?> fiat, a[lv]/ei{s}, exam<in>a apes vasa mellaria, mel qui in[lati]/ erunt conductor<um> v[ili]corumve in assem e[ius] / f(undi) erunt.*	If anyone will have transported hives, swarms, bees, or honey [vessels] from the estate of Villa Magna or Mappalia Siga into one-eighth land, so that fraud <as much as possible?> is committed against the landlords or lessees or bailiffs of this <estate>, the hives, swarms, bees, honey vessels or honey which will have been brought in will belong to the lessees or bailiffs of this estate as a group.

that activity, incentives ultimately approved by the Emperor through his procurators (Table 2).

These incentives related to how much each tenant-beekeeper would have to pay from the honey and wax produced per hive at Villa Magna Variana. Kehoe explains the situation thus:

The rent for beans was set at either one fourth or one-fifth of the crop, while the rent for honey was calculated on a different basis (1.29-2.6). *Coloni* producing honey could keep all the honey from their first five hives, and then pay one *sextarius* (0.55 litres, or less than one sixth of the typical production of a hive) for each additional one. The procurators thus provided a special incentive to produce honey. On the one hand, the production of honey allowed the *coloni* to increase their overall production without having to use much land; in addition, bees would be useful to an estate by speeding up the pollenization of orchards planned there...The *Fiscus* offered such an incentive to in order to encourage *coloni* to bear the substantial investment needed to raise this crop (Kehoe 1988: 40-41).

By contrast, the *fiscus* could not allow prospective *coloni* on imperial estates to invest their energies in apiculture over high-yield crops simply because it offered more individual profit with minimal rent. Therefore, restrictions were placed on *coloni* preventing them from transferring beekeeping equipment from the *subseciva* of the Villa Magna Variana to land known as *octonarius ager*, 'one-eighth land' (Table 3). *Octonarius ager* was simply private owned land not owned by the emperor which could be leased to *coloni* (Kehoe 1988: 112). For Kehoe, the fact that the *fiscus* had to provide incentives to attract farmers to invest in land on imperial estates reflects a competitive market in which the emperor was not a monopoly landholder but simply the biggest in some areas.

Beekeeping also seems to have been of particular interest to the *fiscus* because of its potential to generate large profits from even small plots of land with relatively minimal subsequent investment after the setting up of an apiary and acquisition of bees. An anecdote from the late Republican Roman agricultural writer Varro makes the interest of the *fiscus* clear (Varro, *de Re Rustica* 3.16.10-11).

Varro tells us that two brothers under his command in Spain, the Veianii, from the region of Falerii, in central Italy, were very well-off but possessed a villa with only a small amount of land no larger than an acre left to them by their father. From this land, however, the brothers yearly received a tidy sum of around 10,000 sesterces, all of it, we are told, from the honey produced from their estate. Their strange set-up involved an apiary built around the whole villa, with a garden at the centre and with bee-friendly plants like thyme, snail-clover, and a plant called honey- or bee-leaf (balm, *Melissa officinalis*) planted around the villa. Varro's *De re rustica* consisted of a variety of conversations about agriculture, with the present anecdote in the third book of that work dealing with other forms of animal husbandry. His anecdote about the Veianii led to the central question of a subsequent conversation in book three about apiculture: 'where should I build an apiary

and of what sort to yield great profits?' (*ubi et cuius modi me facere oporteat alvarium, ut magnos capiam fructus?*) Bees are what Jamie Kreiner (2020), a historian of early medieval pig farming would call 'salvage accumulators.' Bees forage upon natural *and* anthropogenic landscapes in uncontrolled ways, accessing 'hidden' stores of food unused or inaccessible to humans in the form of pollen and nectar and the products of this forage can then be utilized by humans. Interestingly, even this salvage accumulation has unforeseen benefits aside from honey and wax production (on salvage accumulation, see also Tsing 2015).

Kehoe noted that bees would 'be useful in speeding up pollenization of orchards' planted at the Villa Magna Variana (Kehoe 1988: 41), presumably referring to bee-pollinated crops like vetch or fruit trees. Kehoe's hypothesis, harder to illustrate at that time, is now well supported by a string of recent studies on the effect of bee-pollination on crop yields in a variety of climatic conditions, including semi-arid ones (e.g. Al Naggar et al. 2017 (concerning the impact of bee pollination declines on crop yield in Egypt); Gasim and Abdelmula 2018 (on the impact of bee pollination on the yield of Faba Bean (*Vicia faba L.*) in Sudan); Bareke and Addi 2019 (on the impact of honeybee pollination on seed and fruit yield of agricultural crops in Ethiopia)). While we cannot credit the Romans with a comparable level of scientific understanding with respect to how bees could be used to boost crop yields and counterbalance less productive agricultural zones, a process which we are only now coming to fully appreciate, the HM inscription seems to suggest a rough awareness of the positive impact bees could have a) on deriving profit from less productive or complex agricultural zones through honey and wax production, and b) using beekeeping to supplement crop yields.

According to Kehoe, the *fiscus*, 'not only promoted the cultivation of *subseciva*, but desired that the *coloni* might make more intensive use of both new land brought under cultivation and existing fields' (Kehoe 1988: 41).[4] The series of incentives offered by the procurators of Villa Magna Variana to cultivate high-yield crops, in particular, those incentives offered to beekeepers, Kehoe suggested, must be seen in this light. The HM inscription represents the only instance of the selective deployment of beekeeping by the imperial administration to increase agricultural productivity known from the Roman period. Its importance, therefore, is quite significant.

---

[4] Katherine Blouin comes to the same conclusion with respect to agriculturally marginal land in the Nile Delta called 'liminitic land' (*limnitice ge*, 2014: 147-148, 230-233). These limnitic lands (comprising several internal categories) were mostly of an arid type and could be bought land, revenue land, or land in deduction. Importantly, as Blouin demonstrated, all had been plots declared dry (*chersos*) or uninundated (*abrochos*) and subsequently reclaimed for some purpose.

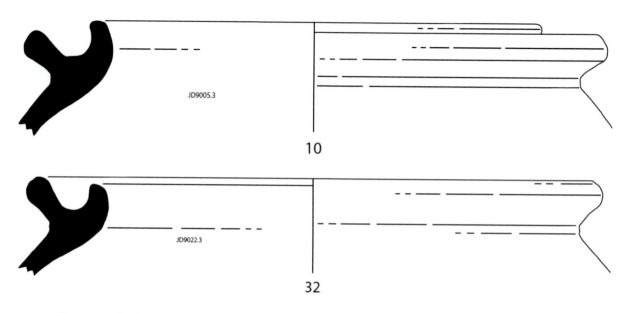

Figure 9a-b. Figure 9a (top): Honey pot (Mukai 2016, Figure119, context 115.10); Figure 9b (bottom) (Mukai 2016, fig, 125, context 118.32) from Aradi (mod. Sidi Jdidi) both dated to the second half of the fifth century CE. Drawings courtesy of Tomoo Mukai.

There is some additional material evidence of beekeeping from Africa proconsularis reminiscent of other areas of intensive apicultural production in the empire, like Segobriga. Material evidence of beekeeping using ceramic hives in the vicinity of the HM inscription is quite recent and has not been connected previously to the efforts of the *fiscus* discussed above. Roughly 100 km to the east of the HM inscription, at Sidi Jdidi, ancient Aradi, excavations have yielded two ceramic vessels we can identify as honeypots from their characteristic recessed rim (Figure 9), and one horizontal ceramic beehive (Figure 10). All three vessels date to the second half of the fifth century CE and are contemporary with similar vessels from Martinhal in southern Portugal and Bracara Augusta in northern Portugal.

Furthermore, we can observe a partial onomastic impact of local beekeeping in the epigraphic record of the area. Under 100 km southwest of Henchir-Mettich, at Sicca Veneria, we know of an epitaph to a Mellaria Prima and further southwest, at Ain el Bey (Numidia), an epitaph to a Iulius Mellarius (*CIL* VIII 16133 (El Kef / Sicca Veneria, Africa Proconsularis): *Mella/ria Pri/ma / v(ixit) a(nnos) L h(ic) / s(ita) e(st)*; *CIL* VIII 5970 (Ain el Bey / Saddar, Numidia): *C. Iulius / Mellari/us v(ixit) a(nnos) LXXXV / o(ssa) t(ibi) b(ene) q(uiescant))*. The creation of a *gentilicium*, or family name, *Mellarius/a* 'honey producer/beekeeper,' and the formation of a *cognomen, Mellarius*, suggest a specialization in this profession at a regional level. The evidence of the HM inscription, associated North African ceramic honeypots and hive, and epigraphic attestations of beekeeping bear some resemblance to a situation observed in the region of Roman Segobriga.

I argue that the ceramic hives found in the territory of Roman Segobriga of roughly contemporary date with the HM inscription parallel in key ways the situation described in that text. The location of the hive deposits and the apiary locations primarily in the agricultural areas (*pars rustica*) of villas in the hinterland of Segobriga suggests a similar rationale as that underpinning the installation of hives on the *subseciva* of the Villa Magna Variana. Unlike Roman North Africa, many ceramic hives have been found in central and eastern Spain spanning the sixth century BCE to the second century CE. This considerable number of ceramic hives allows scholars to document signs of intensification through apicultural technology in ways that suggest innovations. Segobriga is a good example of the supportive role apiculture could play in areas of more complex agricultural potential, especially in areas in which the central commercial focus was not originally agriculture, but connected with mining and quarrying operations. In this sense, such districts might be taken as *subseciva* on a large scale.

Mining Connections?

The presence of apicultural infrastructure in central Spain near a zone of mineral exploitation at Segobriga may reflect a motivation underpinning the use of intensive beekeeping similar to that at Henchir-Mettich. While it is possible that mining simply led to an increase in wealth for members of the local elite who leased exploitation zones, and that this created opportunities for other businesses and livelihoods to flourish, beekeeping in particular may have represented

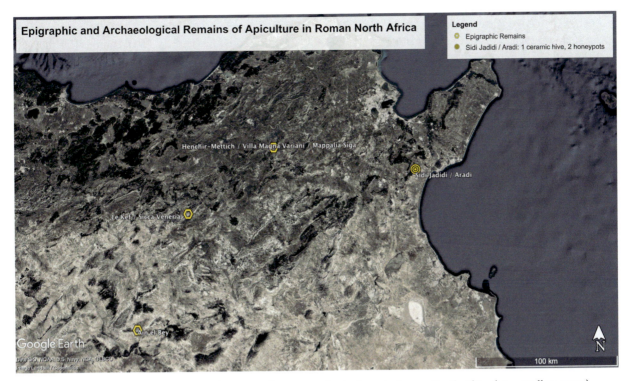

Figure 11. Map of Epigraphic and Archaeological Remains of Apiculture in Roman North Africa (Map: Wallace-Hare).

a welcome supplementary livelihood in areas of central Spain dominated by mining operations. The fact that one could not only increase certain crop yields but also, as Varro highlighted, produce valuable commodities like honey and wax from a small plot of land where easily found bee-friendly plants could be planted to attract bees may have incentivized local beekeepers across the empire.

While honey was always useful, wax especially was a valued commodity for the Romans and was used for important day-to-day functions, as seen, for instance, in the common use of wax tablets for record keeping. In terms of its relevance for mining communities, the use of wax in the lost-wax casting method may have made it productive for metallurgical centres to have apiaries nearby, something discussed in greater detail by Castellano et al. in chapter 5 of this volume. The current chapter has looked more at the usefulness of bees when applied to certain landscapes rather than their potential boon for metallurgy, which represents a fertile area for future investigation. At present, the foregoing interpretation must remain in some ways hypothetical as the hives discovered by Morín and de Almeida have yet to be tested for pollen residue which could enable scholars to understand the provenance of the honey that the hives once contained and thus know better which crops bees may have been made to assist in the Segobrigan villas. Further study is also needed of a diachronic nature, examining the presence of beekeeping in communities built around newly discovered mineral deposits to test the above hypotheses.

Bibliography

Aguilar, M. 2005. *Tarteso: la construcción de un mito en la historiografía Española*. Málaga: CEDMA.

Almagro, M. 1992. La romanización de Segobriga, in F. Coarelli, M. Torelli, and J. Uroz Saez (eds) *Conquista romana y modos de intervención en la organización urbana y territorial, Dialoghi di Archaeologia* 3.10.1–2: 275–88. Rome: Quasar.

Almeida, R. de, and J. Morín. 2012. Colmenas cerámicas en el territorio de Segobriga. Nuevos datos para la apicultura en época romana en Hispania, in D. B. Casasola and A. Ribera I Lacomba (eds), *Cerámicas hispanorromanas II: Producciones regionales*: 725–743. Cádiz: Universidad de Cádiz.

Al Naggar, Y. et al. 2018. Beekeeping and the Need for Pollination from an Agricultural Perspective in Egypt. *Bee World* 95.4: 107-112.

Aubet, M. E. (ed.) 1989. *Tartessos: Arqueología Protohistórica del Bajo Guadalquívir*. Sabadell: AUSA.

Bareke, T. and A. Addi. 2019. Effect of honeybee pollination on seed and fruit yield of agricultural crops in Ethiopia. *MOJ Eco Environ Sci* 4(5): 205-209.

Bernardes, J. P. et al. 2014. Colmeias e outras produções de cerâmica comum do Martinhal (Sagres), in R. Morais, A. Fernández, and M.J.Sousa (eds) *II Congreso Internacional de la SECAH-Ex Officina Hispana. Las producciones cerámicas de imitación en Hispania (Braga, 3 al 6 de Abril de 2013) (tomo I)*: 507–519. Porto: FLUP.

Bernárdez Gómez, M. J. and J.C. Guisado di Monti. 2002. Las explotaciones mineras de *lapis specularis* en

Hispania, in I González Tascón (ed.) *Artifex. Ingeniería romana en España*: 273-298. Madrid: Ministerio de Educación Cultura y Deporte, Secretaría General Técnica.

Bernárdez Gómez, M. J. and J.C. Guisado di Monti. 2004. La minería romana del "Lapis Specularis": Una minería de interior. *Investigaciones arqueológicas en Castilla-La Mancha 1996-2002*: 245-256.

Bernárdez Gómez, M. J. and J.C. Guisado di Monti. 2009. La minería del lapis specularis y su relación con las ciudades romanas de Segóbriga, Ercávica y Valeria, in E. Gonzalbes Cravioto (ed.) *La ciudad romana de Valeria (Cuenca)*: 211-226. Cuenca: Ediciones de la Universidad de Castilla-La Mancha.

Bernárdez Gómez, M. J. and J.C. Guisado di Monti. 2010. La ingeniería minera romana del *lapis specularis* en Hispania, in I. Moreno Gallo (ed.) *Las técnicas y las construcciones en la ingeniería romana, V Congreso de las Obras Públicas*: 405-428. Madrid: Fundación de la Ingeniería Técnica de Obras Públicas.

Bernárdez Gómez, M. J. and J.C. Guisado di Monti. 2016. El comercio del *lapis specularis* y las vías romanas en Castilla-La Mancha, in G. Carrasco Serrano (ed.) *Vías de comunicación romanas en Castilla-La Mancha*: 231-276. Cuenca: Ediciones de la Universidad de Castilla-La Mancha.

Bernárdez Gómez, M. J., M. Díaz Molina and J. C. Guisado di Monti. 2015. Las explotaciones mineras romanas de lapis specularis en la Hispania Citerior y su contexto arqueológico en el Imperio romano, in C. Guarneri (ed.) *Il vetro di pietra. Il lapis specularis nel mondo romano dall'estrazione all'uso*: 19-30. Faenza: Carta Bianca Ed.

Blouin, K. 2014. *Triangular Landscapes: Environment, Society, and the State in the Nile Delta under Roman Rule.* Oxford: Oxford University Press.

Bluhme, F., K. Lachmann and A. A. F. Rudorff. 1848-1852. *Gromatici veteres: die Schriften der römischen Feldmesser.* Berlin: Georg Reimer.

Bonet, H. and C. Mata. 1995. Testimonios de apicultura en época ibérica. *Verdolay* 7: 277-285.

Bonet, H. and C. Mata. 1997. The Archaeology of Beekeeping in Pre-Roman Iberia. *Journal of Mediterranean Archaeology* 10 (1): 33-47.

Brunt, P. 1966. The 'Fiscus' and its Development. *Journal of Roman Studies* 56(1-2): 75-91.

Campbell, J. B. 2000. *The Writings of the Roman land surveyors: introduction, text, translation and commentary.* London: Society for the Promotion of Roman Studies.

Crane, E. 1983. *The Archaeology of beekeeping.* London: Duckworth.

Crane, E. 1999. *The World History of Beekeeping and Honey Hunting.* New York: Routledge.

Curchin, L. 2004. *The Romanization of Central Spain: Complexity, Diversity and Change in a Provincial Hinterland.* New York.

Flach, D. 1978. Inschriftenuntersuchungen zum römischen Kolonat in Nordafrika. *Chiron* 8: 441-492.

Gasim, S. and A. Abdelmula. 2018. Impact of Bee Pollination on Yield of Faba Bean (*Vicia faba* L.) Grown under Semi-Arid Conditions. *Agricultural Sciences* 9: 729-740.

Guillaumin, J.-Y. 2007. *Sur quelques notices des arpenteurs romains.* Franche-Comté: Presses Univ. Franche-Comté.

Haley, E.W. 1991. *Migration and economy in Roman Imperial Spain.* Barcelona: Edicions Universitat Barcelona.

Hanel, N. 2009. Bergbau und Bienenzucht – Zu einer Okkupationsinschrift aus der Umgebung von Córdoba (Spanien). *Der Anschnitt: Zeitschrift für Kunst und Kulture im Bergbau* 61.4: 234-239.

Harissis, H. V. 2018. Beekeeping in prehistoric Greece, in F. Hatjina, G. Mavrofridis and R. Jones (eds) *Beekeeping in the Mediterranean: From Antiquity to the Present*: 18-39. Nea Moudania: Division of Apiculture, Hellenic Agricultural Organization 'Demeter'-Greece; Chamber of Cyclades; Eva Crane Trust - UK.

Holleran, C. 2016. Labour mobility in the Roman world: a case study of mines in Iberia, in L. de Ligt and L. E. Tacoma (eds) *Migration and mobility in the early Roman Empire*: 95–137. Leiden: Brill.

Kehoe, D. P. 1988. *The Economics of Agriculture on Roman Imperial Estates in North Africa.* Göttingen: Vandenhoeck & Ruprecht.

Kreiner, J. 2020. *Legions of Pigs in the Early Medieval West.* New Haven: Yale University Press.

Lorrio, A. 2001. Materiales prerromanos de Segobriga (Cuenca), in F. Villar and M. P. Fernández Álvarez (eds) *Religión, lengua y cultura prerromanas en Hispania*: 199–211. Salamanca: Ediciones Universidad Salamanca.

Mazar, A. and N. Panitz-Cohen. 2007. It is the Land of Honey: Beekeeping at Tel Rehov. *Near Eastern Archaeology* 70 (4): 202–219.

Morais, R. 2006. Potes meleiros e colmeias em cerâmica: uma tradição milenar. *Saguntum* 38: 149–162.

Morillo, Á., R. Morais and D. Wallace-Hare. 2019. Apicultura romana, un nuevo campo en arqueología de la producción. Aportaciones desde el ámbito de la epigrafía y la onomástica, in J. Cabrero Piquero and P. González Serrano (eds.) *PVRPVREA ÆTAS: Estudios sobre el Mundo Antiguo dedicados a la Profesora Pilar Fernández Uriel*: 443–461. Madrid/Salamanca: Signifer Libros.

Morín de Pablos, J. (ed.) 2013. *LA PEÑA II (2009 & 2010): Una unidad de transformación vitivinícola en el territorio de Segóbriga* (MArq Audema 2013: Serie Época Romana/ Antigüedad Tardía). Madrid: AUDEMA.

Morín de Pablos, J. and R. de Almeida. 2014. La apicultura en la Hispania Romana. Producción, consumo y circulación, in M. Bustamente Álvarez and D. Bernal Casasola (eds) *Artifices Idoneos: Artesanos, talleres y manufacturas en Hispania*: 279–305. Mérida: CSIC.

Mukai, T. 2016. *La céramique du groupe épiscopal d'Aradi/ Sidi Jdidi (Tunisie).* Oxford.

Noreña, C. 2019. Romanization in the Middle of Nowhere: The Case of Segobriga. *Fragments:Interdisciplinary Approaches to the Study of Ancient and Medieval Pasts* 8: 1-32.

Oliveira, C. et al. 2014. Análise de fragmentos cerâmicos de potes meleiros e colmeias por cromatografia gasosa acoplada à espectroscopia de massa, in R. Morais, A. Fernández, and M. J. Sousa (eds) *As produções cerâmicas de imitação na Hispania*: 599-610. Porto: FLUP.

Oliveira, C. et al. 2017. Chromatographic analysis of honey ceramic artefacts. *Archaeological and Anthropological Sciences* 11(3): 959-971.

Roselaar, S. 2010. *Public Land in the Roman Republic: A Social and Economic History of Ager Publicus in Italy, 396-89 BC*. Oxford: Oxford University Press.

Rosman, K. J. R. et al. 1997. Lead from Carthaginian and Roman Spanish Mines Isotopically Identified in Greenland Ice Dated from 600 B.C. to 300 A.D. *Environmental Science & Technology 31* (12): 3413-3416.

Tsing, A. 2015. *The Mushroom at the End of the World: On the Possibility of Life in Capitalist Ruins*. Princeton: Princeton University Press.

Urbina, D., J. Morín and C. Urquijo. 2014a. Produccion de vino y aceite en el territorio de Segobriga. Espacios productivos y comercialización. *AnMurcia* 30: 49-70.

Urbina, D., J. Morín and C. Urquijo. 2014b. La producción de aceite en el entorno de Segóbriga (Saelices, Cuenca): almazaras de Casas de Luján y Llano de Pinilla. *AnMurcia* 30: 85-106.

Wilson, A. 2002. Machines, power and the ancient economy. *JRS* 92: 1-32.

Wilson, A. 2007. The Metal Supply of the Roman Empire. In E. Papi (ed.) *Supplying Rome and the Roman Empire*: 109-125. Portsmouth: Journal of Roman Archaeology Supplementary Series.

Chapter 8

Evidence of Dalmatian Beekeeping in Roman Antiquity

Kristina Jelinčić Vučković

Institute of archaeology, Zagreb, Croatia (kristina.jelincic@iarh.hr)

Ivana Ožanić Roguljić

Institute of archaeology, Zagreb, Croatia (ivana.ozanic@iarh.hr)

Emmanuel Botte

Aix Marseille Univ, CNRS, CCJ, Aix-en-Provence, France (emmanuel.botte@cnrs.fr)

Nicolas Garnier

SAS Laboratoire N. Garnier, Vic-le-Comte, France, AOROC laboratory, Ecole Normale Supérieure de Paris, France (Labo.nicolasgarnier@free.fr)

Summary: Honey pots and beehives have not garnered much interest in studies of the history and archaeology of ancient Dalmatia in the past giving the impression of the absence of such evidence. What evidence does exist forms an important addition to the increasingly more perceptible mosaic of ancient Mediterranean apiculture. The current chapter synthesizes what past research exists on this topic and aggregates that with important new research conducted by the authors concerning archaeological evidence of beekeeping in this region.

Keywords: Central Dalmatia, honey, ceramic beehives, stone beehives, Brač, Novo Selo Bunje, chemical and organic residue analysis

Introduction

Apiculture covers a basic human need, consuming sweet food, a necessity as old as human existence itself (Germanidou 2018: 93). But the process of the production and the use of beekeeping products in the Roman Province Dalmatia has not been documented. This chapter will try to fill this gap, focusing on information gathered with new archaeological excavations and field surveys.

Sites with Shards of Ceramic Beehives in Central Dalmatia (Map 1a)

The most common evidence of beekeeping in Central Dalmatia is shards of ceramic beehives. Such ceramic beehives were widespread in Antiquity in the Mediterranean area (Jones 1976; Rosado and Parreño 1995; Crane 1999: 45-46; Anderson-Stojanović and Jones 2002: 345-376; Morín de Pablos and De Almeida 2014). Ceramic beehives are mentioned by Columella who said that they were not favored because during winter,

they froze and in summer, burned (Crane 1983: 52), but maybe this depends on the climate and region.

The first published finds of beehives in Dalmatia are from Stari Grad (*Pharia*) and Stari Grad Plain (Map 1a: 4) on the island Hvar. They are mostly just mentioned in publications without a dating reference or context (Kirigin 2001: 226, T 1: 5; Kirigin 2004: 133-134). A shard is found during the research of a *villa maritima* on the site of Soline, Sv. Klement (Pakleni otoci) (Map 1a: 5) and in Vis (*Issa*) (Map 1a: 6) (Ugarković et al. 2016: 164, Figure 6).

At Vis (*Issa*) along the city walls, architecture has been preserved, but only partially researched. In this context numerous fragments of hives have been found.[1]

Along the coast, most of the beehive shards were found during field surveys. One is located at the late rural Roman site Podgora – Vruje Podspile (Map 1: 7)

---

[1] http://baza.iarh.hr/public/locality/detail/1701 viewed 25 October 2020.

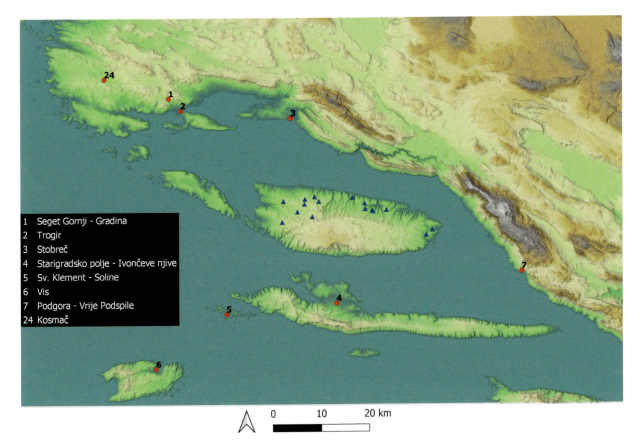

1   Seget Gornji - Gradina
2   Trogir
3   Stobreč
4   Starigradsko polje - Ivončeve njive
5   Sv. Klement - Soline
6   Vis
7   Podgora - Vrije Podspile
24   Kosmač

0   10   20 km

Map 1a

8   Stomorica
9   Podhume - Njivica
10   Dračevica - Ježe
11   Sv. Luka
12   Donji Humac - Gnjilac
13   Nerežišća No.9
14   Škrip - Luke
15   Škrip - Kostirda
16   Postira - Dućac
17   Sv. Juraj na Bračuti
18   Pučišća - Gripe
19   Pučišća - Mladinje brdo
20   Pučišća - Oklade
21   Pučišća - Čod
22   Novo Selo - Bunje
23   Selca - Njivice

0   5   10 km

Map 1b

Njivica Podhume

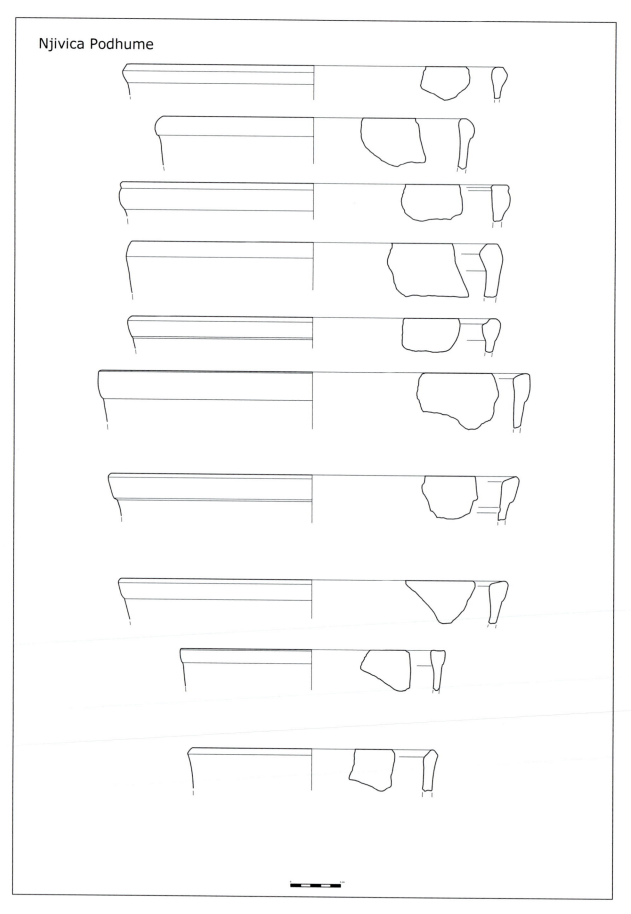

Figure 1. Beehives from the Njivica Podhume site.

(Tomasović 2006: 57, T. XI, 3). There are several finds from hillforts such as Kosmač near Marina (Map 1a: 24), and at the hillfort near hamlet Baradići (Seget Gornji) (Map 1a: 1) (Miletić 2008: 16; Paraman et al. 2020, fig. 33). In Late Iron age hillforts Late Republican and Early Roman amphorae are often found as evidence of trade with the local population. These hives may be part of the same trade and exchange of goods and ideas. Two finds are from urban contexts: Trogir (*Tragurium*) (Map 1a: 2) and Stobreč (*Epetion*) (Map 1a: 3) (Paraman *et al.* 2019; Paraman *et al.* 2020: T. 1:4; Ugarković and Konestra forthcoming T. 5/40).

The island of Brač (see Map 1b) where the largest number of ceramic beehive fragments is found during archaeological excavations and field surveys will form the case study of the present chapter. The finds can be dated, and for part of them, their context is known. Fragments of beehives have been found at 16 sites on the island of Brač. Some of the sites can be dated to the Late Hellenistic period others to the Late Roman period, but some are exclusively Roman as far as the survey results showed. We present the sites with the basic information available from the surveys, excavations, and bibliography.

The most important site is the Novo Selo Bunje site (Map 1b, 22; Figure 5) excavated since 2015, where many fragments of ceramic beehives have been found and where chemical analysis has been conducted (Jelinčić 2005b; Botte *et al.* 2016; Botte *et al.* 2017: 50; Jelinčić Vučković and Botte, 2018: 127-135; Botte *et al.* 2019). This is the largest site on the island where a Roman *villa rustica* was built in the 1st half of the 1st century CE. The site also offers excellent insight into the rural life in Central Roman Dalmatia. This site has been known to researchers since the beginning of the 20th century as parts of the architecture and sarcophagi remains were visible before excavation started. The site is dated from the 1st – 7th century CE. It has a *pars urbana* with small *thermae* or bath and a *pars rustica* with several areas used in the production of wine and olive oil (Bulić 1914: 105–106; Vrsalović 1960: 33–161; Vrsalović 1968; Zaninović 1968: 357–373; Fisković 1981: 105–137; Cicarelli 1982; Zekan 1992: 9–20; Kovačić 1994: 91–97; Stančić *et al.* 2004: 110-112; Jelinčić Vučković 2013: 167–174). The beehive finds from Bunje are found in all areas of the site and can be dated from the 2nd century CE.

Stomorica (Map 1b, 8, Figure 3, Figure 9: 4) is one of the largest sites on Roman Brač. It is presumed that a Roman villa was built here. The main concentration of the finds is to be found in the vicinity of the church of St. Mary. The church was built in the 12th century on the ruins of the Roman architecture still visible on the site. Finds from the prehistoric, late Hellenistic, Early Roman and Late Antique period can be found: *amphorae*,

dolia, tegulae, imbrices, pipes, and pottery (Kovačić and Staničić 1992: 1-2; Stančič *et al.* 1999: 149; Jelinčić 2005a: 97). Several fragments of beehives were found, too, and one of them belongs to the flat base (Figure 3, Figure 9: 4). A flat base made it possible for the hive to stand upright. This suggests the existence of upright beehives but does not exclude the presence other hive types as tubular or horizontal with the rounded base because so far, we have not encountered any extant hives, only hive fragments.

Pučišća Gripe (Map 1b, 18). During surveys, fragments of amphorae, *dolia*, pottery, and *tegulae* were found, mostly on drywalls (Stančič *et al.* 1999: 141) next to the road that probably damaged a part of the site. A small quantity of ceramic beehives was found. Next to the site, there is a location where the remains of stone beehives were found. The lower part of the hive remains without the upper slabs forming a roof which could be removed for the honey collection (Jelinčić 2005a: 101). Finds from Roman Antiquity or any other period were not found among these stone beehives, making the dating impossible. Stone beehives are registered at many sites on the island dating to the late Middle Ages and were in use till the middle of the 20th century. The island of Brač has the most significant stone beehives in the Eastern Adriatic (Škrip, Dol, Blaca...) (Domačinović 1980: 130; Faber and Nikolanci 1985: 4-5; Crane 1983: Figure 131, p. 112-113). They are registered around Omiš and Makarska too. It does not come as a surprise because stone is the primary natural resource of the island, and it was easily accessible for its inhabitants. The hives were built of stone slabs and organized in rows, and hundreds of these beehives existed together.

Selca Njivice (Map 1b, 23; Figure 6: 5-6; Figure 9: 1-2). The site is located in the eastern part of the Selca cemetery. It is assumed that a Roman *villa rustica* was built here. A large pottery concentration, mostly amphorae and bricks, was found (Čače *et al.* 1999: 113). A grave dated to the 4th century CE was discovered in 1921 (Vrsalović 1960: 92-93) as well as part of the inscription (Bulić 1921: 39; Zekan 1992: 14). A substantial quantity of beehive fragments was found on this site, belonging to the body of the beehive.

Postira Dučac (Map 1b, 16): a large pottery concentration was found on the site with some architectural remains. The finds belong mostly to the Roman period and a small quantity to the Late Hellenistic period and pre-Hellenistic period (Stančič *et al.* 1999: 139). Fragments of beehives were found here too.

Donji Humac Gnjilac (Map 1b, 12). On this site, a small quantity of Roman and prehistoric pottery was found. The finds belong to the amphorae, pots, and iron slag (Stančič *et al.* 1999: 158). Beehive fragments were found too.

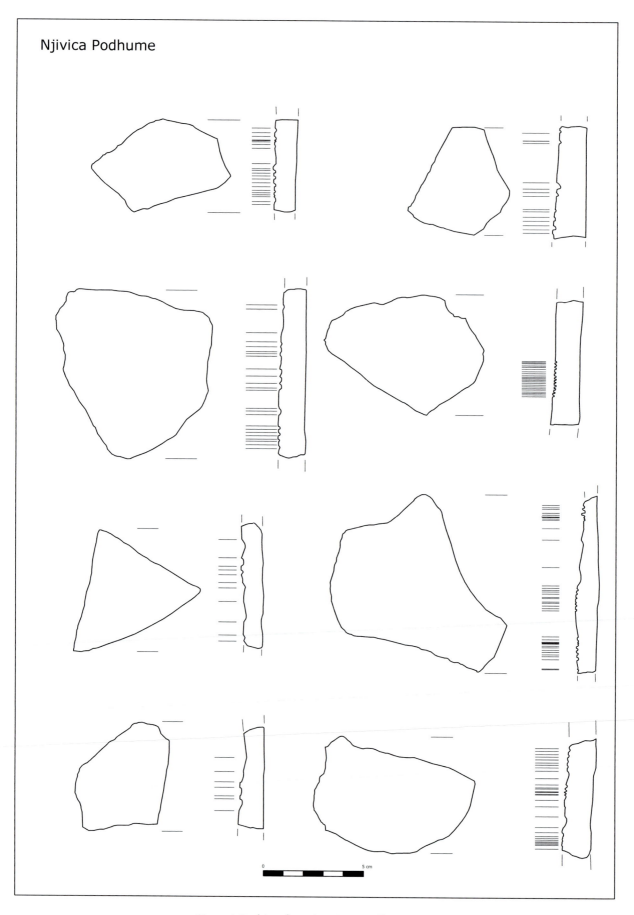

Njivica Podhume

Figure 2. Beehives from the Njivica Podhume site.

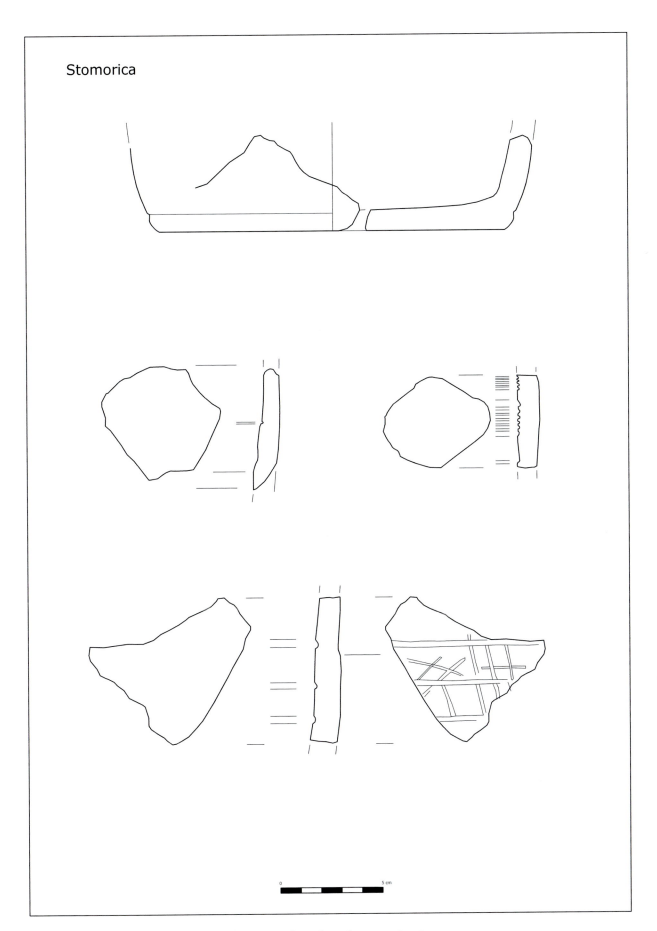

Stomorica

Figure 3. Beehives from the Stomorica site.

Dračevica Ježe (Map 1b, 10): this site was partially damaged during nearby road construction. Some architectural remains can be seen on the site among a large concentration of the pottery, bricks, and mortar (Stančič *et al.* 1999: 147-148). Some of the fragments that were found could have belonged to the beehives mostly because of seemingly shared fabric with the other beehives on the island.

Škrip Kostirda (Map 1, 15). The site is located next to the pond and an old road. A very large concentration of Roman pottery was found (Stančič *et al.* 1999: 168). Some of the pottery of African origin was dated to the 2nd and 3rd century CE. The other pottery fragments belong to the *dolia*, amphorae, and *tegulae* (Jelinčić 2005a: 84-85). An inscription was found here too, and the literature mentions several sarcophagi (Faber and Nikolanci 1985: 24, note. 50, p. IV, 4; Gjurašin 2001: 107; Stančič *et al.* 2004:167). It is assumed that a Roman villa and cemetery existed here. Beehive fragments were found too.

Škrip Luke (Map 1b, 14; Figure 6: 1-2; Figure 7) is one of the largest sites on the island (Stančič 1999: 165; Jelinčić Vučković 2011: 127-149). Some of the drywalls are built with regular stones, and some Roman walls with mortar have been preserved. There are finds of the *tesserae*, mortar, *amphorae*, *tegulae*, pottery, *sarcophagi* remains, coins and an altar with an inscription. The site is dated from the 1st to 5th century CE. Beehive fragments have been found here too.

Pučišća Mladinje brdo (Map 1, 19). A large concentration of finds has been found on this site: *amphorae*, *tegulae*, pottery, glass bracelets, part of a stone column, an altar and *tegulae* with stamps (*Pansiana*) (Jelinčić 2005a: 67-68; Botte *et al.* 2017). Beehive fragments have been found here too.

Nerežišća Ne 9 (Map 1b, 13). The site is located in the interior part of the island next to a valley with fertile soil. A small quantity of finds has been found here (Stančič *et al.* 1999: 119), among them shards of beehives.

Pučišća Oklade PC-PO (Map 1, 20, Figure 6: 6; Figure 9: 3). A large concentration of finds has been found here. They are dated to the late Hellenistic, Early Principate, and Late Antique period (Stančič *et al.* 1999: 145-146). The finds consist mostly of amphorae, but a few beehive shards were found too.

Sv. Juraj Bračuta (Map 1b, 17). This site is dominated by the Early Medieval church surrounded by a large concentration of pottery from Prehistory, the Late Hellenistic, Early Principate and Late Antique periods. Except for the pottery, bricks, glass, and mortar are found (Stančič *et al.* 1999: 139-140) and beehive fragments too.

Pučišća Côd (Map 1b, 21). This site is one of the largest sites where a Roman *villa* was built. Fragments of pottery, sculpture, *tegulae*, *amphorae*, *dolia*, *tubuli* and graves were found. Some architecture remains can be seen too (Stančič *et al.* 1999: 142; Vrsalović 1960: 80; Cambi 2004: 249). Among the finds, beehive fragments were found too.

Sv. Luka (Map 1b, 11). The Early Medieval church dominates the site. Around the church, one Early Christian sarcophagus can be seen, remains of *dolia*, *amphorae*, bricks, pottery plaster, and other finds (Stančič *et al.* 1999: 152; Vrsalović 1960: 102; Fisković 1981: 121) like beehives.

Podhume Njivica (Map 1b, 9, Figure 1-2, 4, 8). The site offers excellent views over the Splitska vrata and the island of Šolta. Some architectural remains are preserved, probably dating to the 1st century BCE. Finds consist of the remains of *dolia*, *amphorae*, mortar, and Roman pottery. Some of the finds belong to prehistory (Stančič *et al.* 1999: 118), and ceramic beehive shards have been found too.

An Attempt at Creating a Typology (Figure 4)

It would be presumptuous to make a typology of the beehives found in Central Dalmatia because thus far, only shards of the body, rims, and bases have been registered. One of the problems is that this type of pottery is not recognized among most of the scholars in Croatia, and as it often happens, the shards lay in the depots unrecognized and unpublished. Thus, we make here a starting point for the creation of a well-established local typology in the hopes that it will help local scholars to recognize this important Dalmatian and Dalmato-Roman economic activity, especially in rural areas. More research is necessary. At present we have seven types of rims with varieties among them (Figure 4). Because we do not possess whole beehives, we cannot attribute these rims to upright or tubular hives. There are at least six identifiable ceramic beehive fabrics (Figure 6). Scoring on these fabrics, a typical characteristic for ceramic beehives in the ancient Mediterranean, when present aids in their recognition of potential hives even from the discovery of a small shard and fabric analysis of those scored varieties aids in identifying those without the typical incised lines.

Njivica Podhume

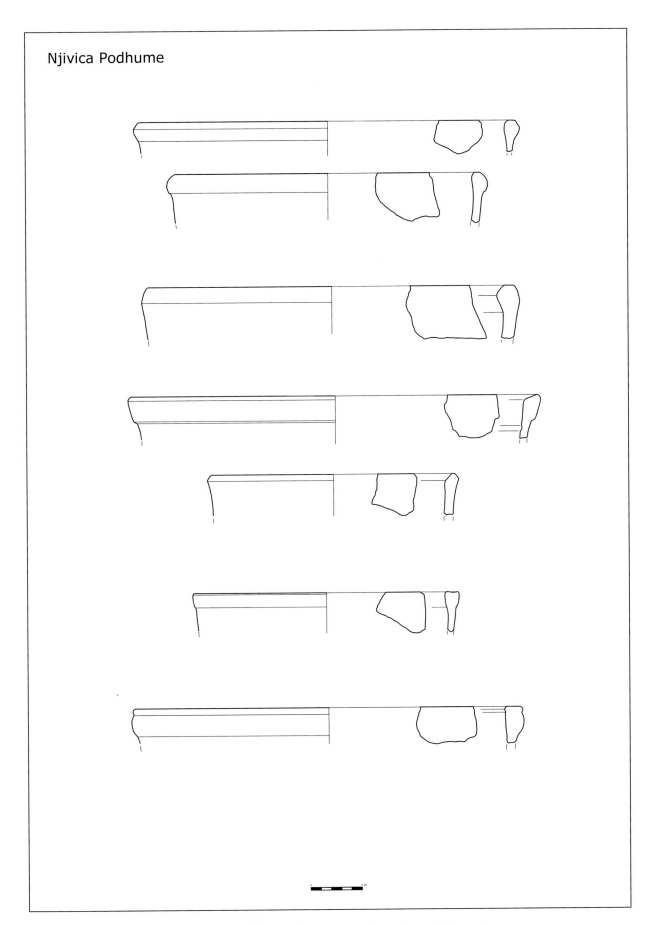

Figure 4. Types of the rims of the beehives from Njivica Podhume site.

The Contribution of Chemical Analyses to Our Understanding of Ancient Honey Production: A Case Study of Novo Selo Bunje

The discoveries made at the Roman villa of Novo Selo Bunje site on the island of Brač, beyond the ceramic forms that had already attracted our attention, were the subject of chemical analyses undertaken to to determine the presence of possible markers of the products of the beehive.

Suppose we can identify beehives by their characteristic shape. In that case, chemical and archaeometric analyses do not allow it to be done for transport containers because honey does not leave chemical markers in the wall (pace Bortolin 2008: 14-16), unlike other foodstuffs, like oil and wine. Indeed, honey is mainly constituted of sugars, highly soluble in water and easily lost after deposition. Some scholars interpret the presence of sugars, mainly disaccharides, as chemical evidence of honey (Oliveira *et al.* 2014). This hypothesis is untenable based on the research in biogeochemistry. By analyzing sediment and aerosols from fields, Rogge *et al.* (2006 and 2007) demonstrated disaccharides and mainly mycose is coming from the cellular membranes of soil microorganisms. Also, levoglucosan has yet to be interpreted as an evidence of smoking a hive before collecting honey, but *is* formed by thermal decomposition of glucose, during cellulose and plant burning. These markers are detected each time the sediment is not removed by washing the potsherd with water before sampling and analysis. Unfortunately, honey does not leave hydrophobic compounds that could be preserved and identified in ceramic vessels. Except low-quality honey that contains some beeswax (hydrophobic and readily identified), well-filtered high-quality honey cannot be detected in ancient ceramics, except in arid contexts such as in Egypt or Saudi Arabia (Garnier 2014).

This approach to chemical analysis on the walls of ceramic containers is not new. These provide valuable information about the product(s) once contained because the latter leaves biochemical markers that are found trapped in the pores of the ceramic. That said, to date, analyses have only revealed wax markers, for example, in Greece (Bortolin 2008: 15). One must take care, however, as mere traces of wax do not necessarily imply a nearby beehive or a link with honey, as wax was also used as a waterproofing agent.

In the specific case of the ceramic fragments from the Roman villa of Bunje (Figure 5), one of the authors (N.G.)

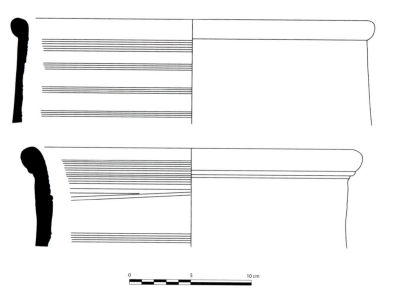

Figure 5. Beehives from the Novo Selo bunje site.

performed the organic residue analysis on some ceramic vessels, used to determine the native organic content from their constitutive chemical markers deposited in the ceramic (Evershed 2008). The chromatographic profile obtained consisted of odd-numbered *n*-alkanes, even-numbered long chain *n*-alcohols, and even-numbered fatty acids (palmitic, oleic, stearic acids and also $C_{24:0}$ to $C_{32:0}$) referred as a distribution characteristic of beeswax (Figure 10) (Garnier et al., 2002; Regert et al. 2003). Monoesters derived from palmitic and oleic acids but also 15- and 16-hydroxy palmitic acids confirm this identification (Dodinet and Garnier, forthcoming). This, therefore, has served to confirm that we are in the presence of beehives.

Conclusion

The number of beehives only on the island of Brač suggests that this particular island was a bee-farming center, but we have to keep in mind that this island was the focus of research only lately, and more attention was given to the pottery finds from its numerous sites. During the survey, prior to the excavation on Bunje, only one small beehive fragment was found, but the excavation itself provided us with a significant number of finds. A situation like this is possible at many other sites, and we need to ask ourselves whether the amount of honey produced in the region was for local, provincial, or reflective of even wider trade. We think that more material from the excavation with precise contexts needs to be published in order to make such comparisons and draw wider conclusions. At the very least, we can assume that the honey that was produced on the island of Brač, for example, was more than enough to fill the needs of the local villas and villages and even more to put on the market at *Salona*.

Figure 6. Main fabrics of the beehives from the island of Brač.

1

2

3

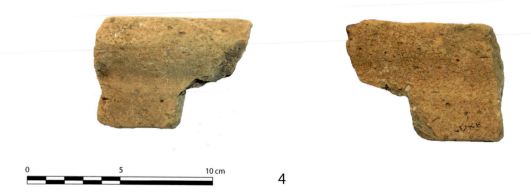

4

Figure 7. Beehives from the Škrip Luke site.

1

2

3

4

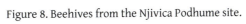

Figure 8. Beehives from the Njivica Podhume site.

Figure 9. Beehives from the Njivice Selca site (1-2), Oklade (3) and Stomorica (4).

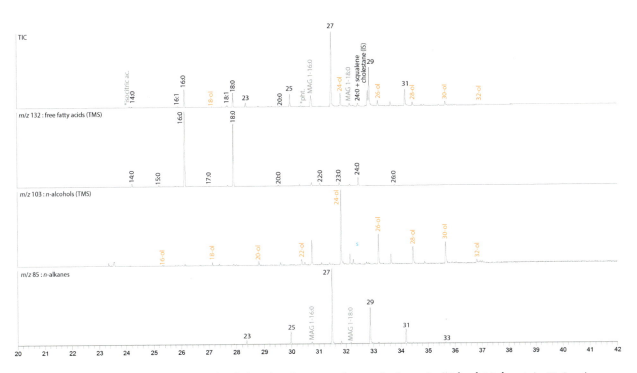

Figure 10. Chromatogram of beehive shards found at the Novo Selo - Bunje site on Brač island © Laboratoire N. Garnier.

It is logical and expected that beehives are found on the Roman rural sites, especially if the area presents a good bee forage. The island of Brač undoubtedly has excellent bee forage, a fact testified by numerous stone beehives used in the last few centuries and toponyms related to bees. Honey was important as a food and as a medicine, and it was necessary to produce it in large amounts to provide for the urban areas or areas without good bee forage. The finds of the hives in the urban centers are more likely to be a testimony of the trade with honey than urban honey production.

Finding beehives fragments on the site of a hillfort dated to the Late Iron Age with finds of Late Republican amphorae so far suggests that the hillfort was part of the trade and not production side. The discoveries of prehistoric beehives so far were not detected in these parts. It is assumed they were made of other materials, perhaps organic material or stone, as presented before. This makes us believe that beehives in the form of a ceramic vessel were not indigenous in Central Dalmatia but brought by Romans or earlier by Greeks.

As for the future, further research in this area should provide more data to see whether some areas were more important in honey production than others and a source of the honey trade. Finding entire hives would give us the possibility to establish a clearer typology. The finds from archaeological contexts and fabric study coupled with typology could help us understand the questions of trade and production of the vessels themselves as an essential part of honey production. We hope that this paper contributed to filling a part of the gap in the knowledge of Central Dalmatian honey production and will make it easier for others interested in the topic in this area to research these hives.

Acknowledgments

The authors of the paper are grateful for the help of Marina Ugarković who shared her unpublished data with us. We are also thankful to Deni Tojčić for helping us with Map 1.

Bibliography

Anderson-Stojanović V.R. and J.E. Jones. 2002. Ancient beehives from Isthmia. *Hesperia* 71: 345-76.

Bonet Rosado, H. and C. Mata Parreño. 1995. Testimonios de apicultura en época ibérica. *Verdolay* 7: 280–281.

Bortolin, R. 2008. *Archeologia del miele*. Mantua: SAP Società Archeologica.

Botte, E., A. Bertrand and K. Jelinčić. 2016. Bunje (Novo Selo, Croatie), Campagne de fouilles 2015. *Cronique des activités archéologiques de l'École française de Rome (online)*. doi: 10.4000/cefr.1519.

Botte, E. et al. 2017. Roman Pottery Finds from the *Villa Rustica* at Novo Selo Bunje site on Brač island, in A. Konestra, G. Lipovac Vrkljan, A. Eterović Borzić, and T. Rosić (eds) *IV. Međunarodni arheološki kolokvij "Rimske keramičarske i staklarske radionice. Proizvodnja i trgovina na jadranskom prostoru i šire," Program i knjiga sažetaka*. Crikvenica: Institut za arheologiju, Muzej Grada Crikvenice 50.

Botte, E. et al. 2019. Bunje (Novo Selo, Croatie), campagnes 2017-2018. *Chronique des activités archéologiques de l'École française de Rome, Balkans, 2018* (online): doi:10.4000/cefr.2419.

Bulić, F. 1914. Trovamenti antichi a Selca. *Bolettino di storia e archeologia Dalmata* 37: 105-6.

Bulić, F. 1921. Starinska iznašašća u Selcima na otoku Braču, *Bolettino di storia e archeologia Dalmata* 44: 39.

Cambi, N. 2004. Kiparstvo na Braču u antičko doba. *Brački zbornik* 21: 239-272.

Ciccarelli, A. 1982. *Zapažanja o otoku Braču*. Beograd: Breviary.

Crane, E. 1983. *The Archeology of Beekeeping*. New York: Cornell University Press.

Crane, E. 1999. *The World History of Beekeeping and Honey Hunting*. New York: Routledge.

Čače, S., T. Podrobnikar and J. Burmaz. 1999. *The Adriatic Island Project (Vol. II): The Archaeological Heritage of the Island of Brac*. Oxford: BAR S803.

Dodinet, E. and N. Garnier. forthcoming. Les analyses organiques en contexte archéologique. Clés d'interprétation croisées de la chimie et de l'ethno-archéobotanique, in C. Pouzadoux, D. Frère, P. Munzi and B. del Mastro (eds) *Les produits biologiques en Italie et Gaule préromaines. Produits alimentaires, médicinaux, magico-religieux, cosmétiques*. Centre Jean Bérard.

Domačinović, V. 1980. Rasprostranjenost pojedinih tipova košnica u Jugoslaviji i pokušaj određivanja njihove relativne starosti. *Etnološka tribina 3, Godišnjak Hrvatskog etnološkog društva*: 129-138.

Evershed, R. P. 2008. Organic residue analysis in archaeology: The archaeological biomarker revolution. *Archaeometry* 50(6): 895–924.

Faber, A. and M. Nikolanci. 1985. Škrip na otoku Braču. *Prilozi* 2: 1-38.

Fisković, I. 1981. Ranokršćanski sarkofazi s otoka Brača. *Vjesnik za arheologiju i historiju dalmatinsku* 75: 105–142.

Garnier, N. et al. 2002. Characterization of Archaeological Beeswax by Electron Ionization and Electrospray Ionization Mass Spectrometry. *Analytical Chemistry* 74: 4868–77.

Garnier, N. 2015. Méthodologies d'analyse chimique organique en archéologie, in C. Oliveira, R. Morais, and Á. M. Cerdán (eds) *ArchaeoAnalytics Chromatography and DNA analysis in archaeology*: 13–39. Esposende: Município de Esposende.

Germanidou, S. 2018. Honey Culture in Byzantium: An Outline of Textual, Iconographic and Archaeological Evidence, in F. Hatjina, G. Mavrofridis, G. R. Jones (eds) *Beekeeping in the Mediterranean - From Antiquity to the present*: 93–104. Nea Moudania: Hellenic Agricultural Organization "Demeter"-Greece.

Gjurašin, H. 2001. Zoran Stančič – Nikša Vujnović – Branko Kirigin – Slobodan Archaeopress, BAR IS (Oxford), 803, 1999, Ocjene i prikazi. *Obavijesti HAD* 33(2): 105-109.

Jelinčić, K. 2005a. *Topografija rustičnih vila na otoku Braču*, Unpublished Master Thesis, University of Zagreb.

Jelinčić, K. 2005b. Rustična vila na Bunjama kod Novog Sela na otoku Braču. *Vjesnik za arheologiju i povijest Dalmatinsku* 98: 121–132.

Jelinčić Vučković, K. 2011. Luke kod Škripa na otoku Braču – novi arheološki nalazi. *Archaeologia Adriatica* 5: 127-149.

Jelinčić Vučković, K. 2013. Terenski pregled lokaliteta Novo Selo Bunje na otoku Braču. *Annales Instituti archaeologici* 9: 167–174.

Jelinčić Vučković, K. and E. Botte. 2018. Arheološko istraživanje na lokalitetu Novo Selo Bunje na otoku Braču, 2017. godine. *Annales Instituti archaeologici* 14: 127-135.

Jones J.E. 1976. Hives and Honey of Hymettus: beekeeping in Ancient Greece *Archaeology* 29.2: 80-91.

Kirigin, B. 2001. Zaštitna arheološka iskopavanja u okolici Starog Grada na otoku Hvaru godine 1984. i 1985. *Diadora 20*: 209-255.

Kirigin, B. 2004. Faros, parska naseobina. Prilog proučavanju grčke civilizacije u Dalmaciji. *Vjesnik za arheologiju i historiju dalmatinsku* 96: 9-301.

Kovačić, V. and Z. Staničič. 1992. Stomorica ranosrednjovjekovna crkva kod Ložišća, *Konzervatorski bilten Regionalnog zavoda za zaštitu spomenika. kulture u Splitu* 11: 1-2.

Kovačić, V. 1994. Topografija pojedinačnih nalaza, in J. Belamarić, R. Bužančić, D. Domančić, J. Jeličić-Radonić, V. Kovačić (eds) *Ranokršćanski spomenici otoka Brača*, 91–97.

Miletić, A. 2008 [2009]. Saltus tariotarum, *Opuscula archeologica* 32: 7-20.

Morín de Pablos, J. and R. de Almeida. 2014. La apicultura en la Hispania romana: producción, consume y circulación, in M. Bustamante and D. Bernal (eds) *Artífices Idóneos. Artesanos, talleresy manufacturas en Hispania, Anejos Archivo Español de Arqueología* 61: 290-302. Mérida: CSIC.

Oliveira, C. et al. 2014. Análise química cromatográfica a fragmentos cerâmicos de potes meleiros e colmeias, in R. Morais, A. Fernández and M. J. Sousa (eds) *As produções cerâmicas de imitação na Hispania. Braga, Monografias Ex officina Hispana II* : 599-610. Porto: Faculdade de Letras da Universidade do Porto.

Paraman, L. et al. 2019. Report on New Excavations in Ancient Trogir. The 2018 Croatian and Austrian Mission, *Jahreshefte des Österreichischen Archäologischen Institutes* 88.

Paraman, L., M. Ugarković and M. Steskal. 2020. Terenski pregled i dokumentiranje gradinskih nalazišta na širem trogirskom području u 2019. godini kao uvod u sustavno istraživanje Hiličkog poluotoka, *Annales Instituti Archaeologici XVI*, forthcoming.

Regert, M. et al. 2003. Structural characterization of lipid constituents from natural substances preserved in archaeological environments. *Measurement Science and Technology* 14: 1620–30.

Rogge, W. F., P. M. Medeiros and B. R. T. Simoneit. 2006. Organic marker compounds for surface soil and fugitive dust from open lot dairies and cattle feedlots. *Atmospheric Environment*, 40(1): 27–49.

Rogge, W. F., P. M. Medeiros and B. R. T. Simoneit. 2007. Organic marker compounds in surface soils of crop fields from the San Joaquin Valley fugitive dust characterization study. *Atmospheric Environment*, 41(37): 8183–8204.

Stančič, Z. et al. 1999. *The Archaeological Heritage of the Island of Brač*, Croatia, The Adriatic Islands Project, Vol 2, BAR International Series 803. Oxford: BAR.

Stančič, Z. et al. 2004. Arheološka baština otoka Brača. *Brački zbornik* 21: 3–238.

Tomasović, M. 2006. Ostaci iz prapovijesnog i antičkog razdoblja na primorskoj strani Podgore, Arheološka slika Podgore, in A. Kunac, K. Mucić and M. Tomasović (eds) *Podgora od prapovijesti do srednjeg vijeka- kulturno- topografska razmatranja (obalni dio):* 56-62. Makarska: Gradski muze.

Ugarković, M. et al. 2016. Arheološka istraživanja rimske vile u uvali Soline na otoku Sveti Klement (Pakleni otoci, Hvar), lipanj 2015. godine. *Annales Instituti Archaeologici* XII: 160-165.

Ugarković, M. and A. Konestra. 2020. Stobreč; bedemi: pregled keramičkih nalaza s istraživanja 2012. godine, in: *Okolica kaštelanskog zaljeva u prošlosti, Izdanja Hrvatskog arheološkog društva*, forthcoming.

Vrsalović, D. 1960. Kulturno-povijesni spomenici otoka Brača. *Brački zbornik* 4: 33–161.

Vrsalović, D.1968. Povijest otoka Brača. *Brački zbornik* 6.

Zaninović, M.1968. Neki primjeri smještaja antičkih gospodarskih zgrada. *Arheološki radovi i rasprave* 4-5: 357–373.

Zekan, M. 1992. Istočni dio Brača-povijesni pregled do utemeljenja Sumartina. *Zbornik radova:* 9–20.

Baza antičkih lokaliteta, Institute of archaeology, Zagreb, Croatia, viewed 3 November 2020 http://baza.iarh.hr/public/locality/map , http://baza.iarh.hr/public/locality/detail/1701

Chapter 9

Ancient Rock-Cut Apiaries in the Mediterranean Area:
Some Case-Studies

Roberto Bixio

Hon. Inspector for Archaeology, sector Artificial Cavities, Italian Ministry of Cultural Heritage, Italy; Dr.h.c.
National University of Architecture and Construction of Armenia (roberto.bixio@gmail.com)

Andrea Bixio

Centro Studi Sotterranei - Genoa, Italy (andrea_bixio@yahoo.it)

Andrea De Pascale

Museo Archeologico del Finale/Istituto Internazionale di Studi Liguri, sezione Finalese - Finale Ligure/Savona,
Italy (depascale@museoarcheofinale.it)

Summary: A particular category of ancient and medieval apiary, consisting of structures carved into rock, is one encountered at many sites in the Mediterranean. Three types have been identified: 'shelf or cell' apiaries, where each hive is contained in individual cavities carved into a rocky face; 'open chamber' apiaries, consisting of wider rooms carved into cliffs, used simply as shelter for hives built on various types of support placed in the back wall; 'enclosed chamber' apiaries, in which the rooms are totally carved out of the rock, and the hives are leaning against the wall that closes the outer side. In this case the beehives can be unmovable, dug directly into the rock, communicating externally through individual small flight holes, or horizontal mobile tubular hives corresponding to vertical slits. According to various historical sources, which report use of bees as 'biological weapons', their use is considered plausible also in Cappadocia where the rock-cut apiaries both most prolific and at times were associated with underground shelters, clearly used as defences from raids.

Keywords: Rock-cut apiaries, carved hives, weapon-like bees, flight holes and slits, Cappadocia, Malta

Introduction

Since 1991, the *Centro Studi Sotterranei* (Centre for Underground Studies) of Genoa (Italy), in collaboration with several Turkish and Italian Universities, and under the aegis of the Turkish Ministry of Culture, has been leading a series of speleo-archaeological wide-ranging missions in various areas of Anatolia. The specific target of these surveys is to locate, explore and document the ancient artificial cavities, i.e. carved by man in the natural rocky cliffs and in the subsoil in order to realize rock-cut and underground structures intended for habitation or worship, for defence and military purposes, for water collection or regulation, or for utilitarian and agricultural purposes, such as warehouses, oil crushers, winepresses, stables, pigeon houses and, last but not least, apiaries, i.e. structures specifically excavated for beekeeping.

The investigations in Cappadocia have obviously led us to consider and compare other ancient rock-cut sites devoted to beekeeping, located in Malta and in various other areas of the Mediterranean. In the following chapter, we consider their different characteristics, knowing that the documentation so far gathered is certainly not exhaustive and the topic can be further developed.

The Rock-cut Apiaries of Cappadocia

Cappadocia, a historic Roman and then Byzantine Empire region, is located in central Turkey (Figure 1). With its 20,000 sq.Km of volcanic rocks, mainly consisting of soft tuff strongly shaped by meteoric agents in spectacular morphologies, Cappadocia features one of the rockiest landscapes in the world. In its subsoil are enclosed at least 1000 Christian churches (Ousterhout 2017: 5), hundreds of shelters, thousands of tombs and pigeon houses, countless dwellings and real rock-cut villages, carved in the cliffs and pinnacles of buttes and plateaus, at an average altitude of 1200 m a.s.l.(Bixio R., Castellani *et al.* 2002; Bixio R. 2012). In

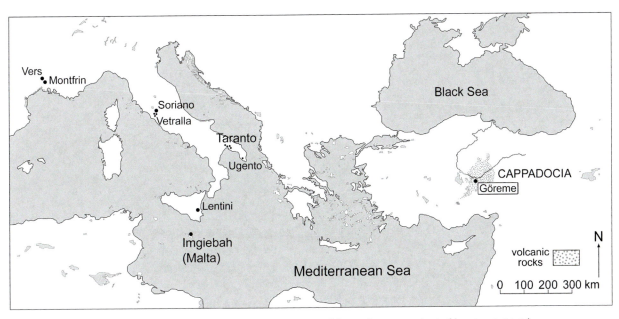

Figure. 1. Rock-cut apiaries documented in the area of the Mediterranean basin (drawing R. Bixio).

most cases, these hypogeal structures date back to the Byzantine period, between the 6th and 11th century AD. In later centuries, following the conquest of the Seljuks, and then of the Ottomans, many sites were reused, reworked or increased, but still in the presence of a strong component of Greek people. This influence ended only in 1923 as the result of the population exchange following the Lausanne treaty. However, even today several rock-cut structures are partly used by the current Turkish people for agricultural or tourist purposes.

Here, in 2001, for the first time we located two rock-cut apiaries (Bixio R., Bologna *et al.* 2004). One of these, which better preserves remains of the past and is still partially in use (Figures 2, 3), allowed us to understand

the criteria and functions of all the other apiaries discovered later in the region, many of which have been long since abandoned and partly destroyed.

Between 2006 and 2007, Gaby Roussel (2006; 2008) of the Apistoria Association of Dijon, identified and catalogued 56 cave-apiaries in the region but there are likely many more still to be documented. During the 2012-2014 surveys of the *Centro Studi Sotterranei*, within the mission of the University of Tuscia of Viterbo, directed by Professor Maria Andaloro, and thanks to other researchers ((Bobrovskyy, Grek, Lucas), at the site of Göreme, the Byzantine *Korama* (Figure 4), in a very small area of only four square kilometres, another 12 new rock-cut apiaries were identified (Bixio R. and Germanidou 2019).

Figure. 2. Göreme. Niketas rock-cut apiary, likely Byzantine in date (Ousterhout 2017: 406). On the left, cells with immovable hives, still in use. In the background, the mobile tubular hives, at present disused (photo G. Bologna).

Figure. 3. Göreme. Old cylindrical basket beehive, covered with dried animal excrements, in the rock-cut apiary next to the hermitage of the stylite Niketas, in Kızılçukur/Meskendir valley (photo G. Bologna).

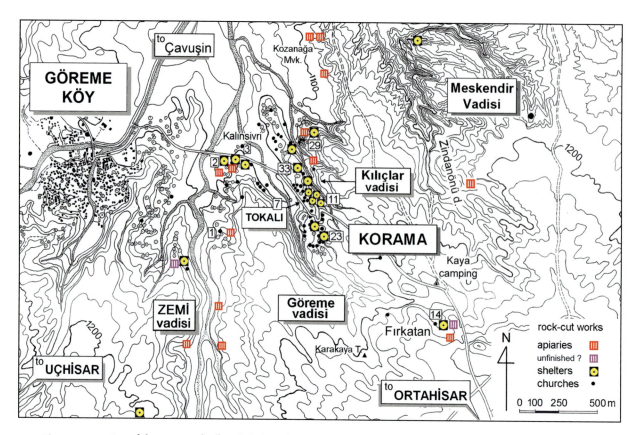

Figure. 4. Location of the apiaries (red) and shelters (yellow) identified in the *Korama*/Göreme area (drawing R. Bixio).

Bees Like 'Biological Weapons'?

It is worth remembering that, thanks to broader explorations, in the same restricted area, where at least 86 rock-cut churches and monasteries and 49 carved refectories are concentrated (Andaloro et al. 2017; De Jerphanion 1925-1942; Jolivet-Levy 2015 and Ousterhout 2017), we identified the remains of 17 underground shelters, 14 of which were previously completely unknown (Bixio A. *et al.* 2018a and 2018b).

Underground shelters are peculiar and widely spread facilities in Cappadocia consisting of structures fully excavated inside the rocks, for tens or hundreds and, sometimes, thousands of metres, equipped with passive and active defensive devices, like heavy millstone-doors, traps, redoubts and escape ways, devoted to hiding a population during raids and war (Bixio R. and De Pascale 2015).

The apiaries so far localized at Göreme are therefore adjacent, or in any case, very close not only to the monastic or residential installations, but also to the aforementioned shelters (Figure 4). By linking these discoveries with the studies of Sophia Germanidou (Germanidou 2013; 2015 and 2016), we assumed that bees here, in addition to producing honey and wax, could also have been used in shelters as a defensive weapon (Bixio R. and Germanidou, in press). During

the frequent Arab incursions, historically taking place between the 7th and the end of the 10th century, or further combat scenarios, such as the arrival of Seljuks, it is in fact plausible that the mobile hives, tubular type (see below), were cast by the defenders into the shaft-traps to scare off, or even kill would-be attackers (Figure 5).

This assumption is supported by sources that, in different contexts, attest the use of bees in antiquity for such purposes. In the case of besieged cities, bees were in fact introduced into the so-called 'counter-mine tunnels,' together with wasps, scorpions, vipers, or even bears or other wild beasts, as an alternative to the use of fire or flooding. Aeneas Tacticus, 360 BCE, in the *Poliorketikà* (treatise on the 'Conquest of Cities'), speaks of it, and the Roman commander Lucullus and his men were said to have experienced such measures in 72 BCE, in his siege of Themiscyra (Asia Minor), during the Mithridatic wars (Bettalli 1990; Bonetto 1997: 356-392). Even in the Byzantine treatises of military tactics, in chapters on naval battles there is talk of launching '...venomous creatures. When the pots are shattered, the animals bite and by their poison wipe out the enemy' (Dain 1937: 83-84 and Dennis 2010: 527). Eva Crane adds several other ancient and recently documented examples: around AD 500 to defend an Irish convent by raiders; in AD 908, in the siege of the English Chester by the Danes; in AD 940, during a siege

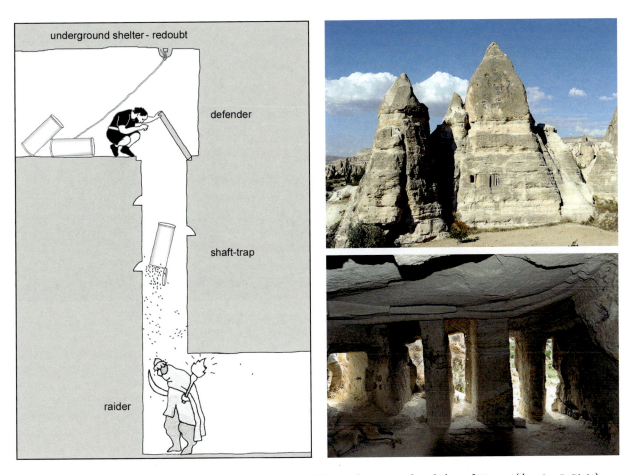

Figure. 5. Left - Reconstruction of a possible use of bees in defence of a trap inside a shelter of Göreme (drawing R. Bixio).

Figure. 6. Right, above - Göreme. Pensile rock-cut apiary (photo A.Bixio).

Figure. 7. Right, below - Göreme. Remains of a rock-cut apiary at ground level (photo A. Bixio).

of a Saxon stronghold by the Duke of Lorraine; and so on (Crane 2003). During the Crusades, Christians and Muslims attacked each other by hurling hives, in addition to stones, fire and quicklime. It is reported that Richard the Lionheart carried a very large number of beehives on ships just to be used in the field. In Crane's article there is also reproduced an illustration of a manuscript of 1326, preserved at Oxford, in which a machine 'for hurling bees into a besieged fortress' is even represented: the device consists of a kind of windmill on the rotating blades of which the hives were placed and launched using centrifugal force. Among the most recent testimonies is mentioned that of 1862, concerning the American Civil War; the clashes in East Africa between British and Germans during World War I (1914-18); and during the Vietnam War in traps, between 1965 and 1975.

Peculiarity of the Cappadocian Rock-Cut Apiaries

According to our surveys in Cappadocia, all the rock-cut apiaries, even if each with its own peculiarities, have a similar basic conformation, consisting of rooms entirely excavated in the tuff in which they were or, in

a few cases, are still housed the hives (Figure 2) (Bixio and Germanidou 2019). In other words, the cavities housing the hives are closed on all sides. We have named this type 'enclosed chamber rupestrian apiaries' (see below).

Like the numerous rock-cut pigeon houses in the same areas, the apiaries are generally excavated at several metres above the ground, in the rock faces of cliffs or pinnacles (Figure 6 and Figure 8, centre), but in some cases they are placed at ground level (Figure 7).

From the outside they are identified by the presence of ordered rows of small round holes (flight holes), often combined, sideways, with parallel series of vertical slits for the bees' entrance (Figure 8). The beekeeper can access from a small front or side entrance, originally closed by a wooden door. In the inner chamber, the flight holes, as a rule, correspond to fixed hives. These accommodations consist of vertical compartments subdivided into cells by horizontal slabs inserted in grooves (Figure 8, left), each with a movable closure on the back (Figure 2, left side, in the foreground), generally no longer existing: they were evidently

Figure. 8. Interior and exterior of a rock-cut apiary in the area of Göreme (photos M. Traverso, A. Carpignano, A. Bixio).

functional to a sedentary beekeeping. Behind the vertical slits (Figure 8, right) were presumably placed mobile tubular hives, corresponding to cylindrical baskets covered with dried animal excrement (Figure 2, right side, in background, and Figure 3), originally lying on each other. It is likely that the tubular beehives were used to practice the so-called nomadic beekeeping. Moreover, their characteristics (shape, size, mobility) fit well with the above-cited hypothesis for defensive use in underground shelters with vertical traps (Figure 5).

The Rock-cut Apiary at Ugento in South Italy

In the case of the rock-cut village of Ugento (province of Lecce, Apulia - Italy), a structure carved inside a rocky ridge of an ancient abandoned quarry and covered with a slightly pitched roof made of blocks of stone, has strong similarities with some Cappadocian rock-cut apiaries.

Initially this structure was considered a funeral *columbarium*, perhaps dating back to Roman times. Recently it has been reconsidered as an apiary because each of the 60 square niches, carved in five horizontal rows, 50 cm deep, in the lower part has a hole of about 4 cm in diameter, communicating with the outside. The shape, size and position of these holes are very similar to those of the 'flight holes' of the Cappadocian apiaries. Thus, the comparison of these elements now allows us to hypothesize the purpose of the Ugento structure as related to beekeeping (Calò and Santucci 2017).

The Rock-Cut Apiaries at Vers in South France

It would be appropriate to investigate more thoroughly some of the apiaries called by Masetti *ruches-placards*,

i.e. cupboard-hives, carved in the rock face of an abandoned quarry (*Carrière de l'Estel*) - next to several 'niche apiaries' (type a.3), mentioned below - the latter, from the short description, seem quite similar to the Cappadocian ones (Masetti 2000: 359-361). In the quarry an inner chamber would be excavated, while the cells for the hives would be realized in the thickness of the rock curtain. The hives would have the entrance for bees (flight holes) on the outside, and the mouth for inspection and taking honey on the opposite side, inside the chamber.

Mixed Rock-cut Apiaries: The Case of Imgiebah (Malta)

In 2002, together with Raffaele Cirone of the Italian Federation of Beekeepers of Rome, a short investigation was conducted by the *Centro Studi Sotterranei* in Malta to get the documentation for some (already known) ancient rupestrian apiaries (Bixio, Traverso and Cirone 2002; Cirone 2001).

The site of Imgiebah (the name means 'apiary') lies on the hill overlooking the innermost point of the bay of St. Paul, behind the last houses of Xemxjia, on the northeastern coast of Malta. The slopes are characterized by limestone outcrops where shallow caves open, and by terraces bordered with extensive dry-stone walls. The surrounding area is rich in ancient remains. The road climbing toward the summit, partly carved into the rock, is attributed to the Roman period, as the remains of a 'villa,' with a Punic tomb identified nearby.

At a sharp bend in the road there are three apiaries, located on contiguous and overlooking terraces. The lower one extends along a front of twelve metres (Figure

Figure. 9. Interior and exterior of the lower rock-cut apiary of Imgiebah at Malta (photos M. Traverso).

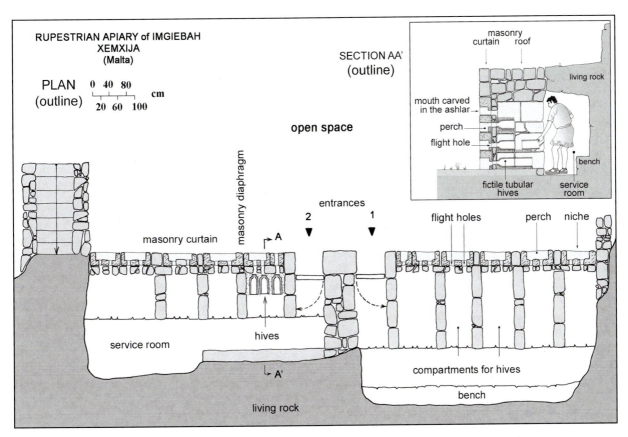

Figure. 10. Sketch of the lower rupestrian apiary of Imgiebah at Malta (drawing R. Bixio).

9). It consists of a wall about three metres high, with exposed squared off stones, carefully restored in recent times with mortar. It has three rows of overlapping niches, for a total of forty-seven mouths of various sizes. It is divided into two sectors that are accessed through two low small doors placed in the centre (points 1 and 2 of the plan in Figure 10).

After entering the inner rooms, we realize that the structure consists of a natural shallow cave, against which a composite masonry structure was placed. The cave irregularly extends along the whole length of the front, with a small depth of about one or two metres: a kind of room, long and narrow, sheltered by a roof partly of stones and partly of living rock, jutting out, and closed at both ends by walls. Its genesis is due, in origin, to natural erosion of the limestone, without human intervention. Localized artificial excavations are still visible today, to obtain the space for bench and small shelves, to lay oil lamps and tools suitable for the apiary management.

Figure. 11. Exterior of the upper rock-cut apiary of Imgiebah at Malta (photo M. Traverso).

Figure. 12. Interior of the upper rock-cut apiary of Imgiebah at Malta (photo M. Traverso).

The masonry part consists of ashlars of various sizes, i.e. rock nearby quarried and purposely squared. It is globigerina limestone, very soft to cut, which then hardens on air exposure. A lengthwise wall, advanced a few metres with respect to the natural room, thus encloses the natural rock shelter. The inside is covered by a second wall, leaning against the first, of rougher dry stones. The two spaces thus obtained (service rooms) are in turn divided into parallel compartments by orthogonal partitions formed by rows of ashlar and covered with masonry supported by horizontal slabs or opposing stones (cappuccina covering), arranged between the partitions (Figure 9).

In some ashlars of the outer wall there are carved niches, arranged on three, sometimes four superimposed horizontal rows starting from the ground level. The openings, except for some rectangular ones, have the appearance of oven mouths with the upper profile like a flat arch (Figure 9, central). They penetrate the limestone ashlar for about 30 cm forming shelves that could serve as perches for the bees coming from the outside. On the rear wall, closing each niche, two or more squared holes (flight holes) were carved, piercing the diaphragm of stone; the mouths of the hives were set against the inner side of each hole. The hives are made of cylinders of terracotta, open at one end and closed on the side of the short narrowing that forms the stumpy neck (Figure 9, right). A single mouth or five small holes were present here to allow the entrance of bees (flight counter-holes). The tubular hives were horizontally placed in the compartments defined by internal partition walls, in superimposed rows resting on movable horizontal stone shelves, supported by vertical side slabs and sealed with clay. Each row could contain, depending on the number of outer holes, two or more hives side by side. The open side of the cylinder (back mouth of the hive) was facing the interior (service room). A wood stopper, now missing, closed it, sealed with propolis (by bees) and wax (by man). The stopper

was removed during the honey collection. Sometimes, extensions (*chambre au miel*) were added, then removed at the time of honey collection (Masetti 2002).

Technical operations (inspections, fumigation and honey collection) took place in the service room with all comfort for the beekeeper (and for the bees): above the hives larger openings allowed the bees to flee outside when the honey was harvested (Masetti 2002).

The scheme just described is essentially the same in the other two hives, although some structural changes are detected. For example, the upper apiary, which appears more archaic, consists of a single room housed in a much larger rock shelter (more than three meters wide), entirely covered by the overhanging roof of living stone. Inside there are no orthogonal partitions, or supporting slabs for the natural cover, but only the longitudinal closing wall in dry stones. Here, the hives were resting on shelves constructed in the wall itself, in arched niches, instead of movables slabs (Figure 12). The openings in the outer ashlars, are rectangular niches rather than oven-like mouth, but always with two flight holes each (Figure 11). There are no external niches inside the middle apiary, but the flight holes are constructed directly in the ashlars of the drywall, at the horizontal joints of the two intermediate rows. According to the survey of Masetti in the archipelago there are at least 17 apiaries similar to those described above (Masetti 2002).

Proposal for a Classification of the Rock-cut Apiaries

In addition to the apiaries described above, at several sites in the Mediterranean basin various forms of rock-carved apiaries are known. Thanks to the documentation of several researchers, a wide variety of characteristics have been synthesized. Those known to us and cited herein are indicated on the map of figure 1. We think it useful, therefore, to report the updated classification already proposed for the 2012 International Workshop on Speleology in Artificial Cavities (Bixio and De Pascale 2013).

Schematic representations of several types of rock-cut apiaries are shown in plates 1 and 2, synthesizing the characteristics of the beekeeping structures documented in Cappadocia and Malta with those existing in the territory of southern France and central-southern Italian regions: Latium, Sicily and, especially, Apulia.[1] So far, three general categories of rupestrian apiaries have been identified: wall apiaries (category a), open chamber apiaries (category b), and enclosed chamber apiaries (category c). Each of them is, in turn, split into several sub-types. However, we have to point out that every apiary has its own peculiarities, often associated with masonry.

Rock-cut wall apiaries (category a)

Rock-cut wall apiaries are those structures in which the housing for the hives was achieved by excavating a vertical cliff. The rock face may be completely natural or achieved by rectifying the roughness of the stone. According to the shape of the structure, we classify the rock-cut wall apiaries in four types.

a.1) Shelf apiaries

It is the simplest housing type achieved by excavating a horizontal parallelepiped in the rock, with one of the long sides open toward the outside. In this sort of shelf, the hives are located side by side. The apiary can consist of a single shelf, or more shelves superimposed and/or staggered. Some examples have been found in the following localities in the province of Taranto (Apulia): Fantiano and Malabarba (Grottaglie), Triglie (Crispiano), and S. Vito (Mottola).

a.2) Cell apiaries

It took considerably more skill to realize an apiary with cells. These are constructed from the excavation of single and/or overlapping parallelepipeds, similar to stacked boxes, separated by thin curtains obtained by saving the rock from excavation. Since it is clear that producing a continuous shelf (a.1 type), would be easier and faster than carving the same volume divided into several units, we assumed that the housing in cells perhaps had the advantage of imparting a greater thermal inertia to beehives. Furthermore, each cell, closed by a

wicket, could be used directly as a hive without further containers of wood or other material, and the combs are attached directly to the rock. Some examples have been documented in the locality of Crispiano in the province of Taranto (Apulia - Italy).

a.3) Niche apiaries

We will call 'niches' the parallelepipeds vertically carved out, and shallow. Those in the province of Viterbo are between 76 and 84 cm high, between 37 and 51 cm wide and between 38 and 43 cm deep (Bortolin 2008: 88-9). The rock-cut apiaries with niches consist of individual cavities, excavated side by side on vertical natural rock walls. In each one, a single hive is vertically housed. In France there are examples of apiaries with niches obtained on regularized faces of old abandoned stone quarries. The apiary at Vers dates back to 1793. From the description, the niche itself, closed by a small door, provided with flight holes and hinged on the rocky frame, served as a hive, without the need of additional containers (Masetti 2000: 361). As previously said, close to this a housing system is briefly described, which seems more likely of the full carved chamber type: see below 'Enclosed chamber rupestrian apiaries (category c)' (Masetti 2000: 359-361). Some examples have been documented in the following localities: in France, Gard region, Quarry of Estel (Vers), and the Quarry of Montfrin. In Italy, Viterbo province (Latium), Castello di Bolsignano, Pian Castagno at Soriano nel Cimino; Casale Mangani, Casalino, Querceto Nardini, Poraglie, and Pian della Noce at Vetralla (documented by De Minicis 2018). We should point out that the housing system with niches is considerably more common in masonry structures and is very popular in England (Crane 1983: 118-150).

a.4) Compartment and cell apiaries

This is a more complex structure than the previous ones: it consists of compartments, i.e. high vertical niches, adjacent to each other, separated by rock diaphragms obtained by saving the rock from excavation of a short space slightly behind the vertical of the outer wall of the cliff. The diaphragms are also provided with parallel grooves (runners), carved on both sides, so that horizontal slabs can be inserted to divide into multiple overlapped cells each compartment for the hives housing.

This type of installation, as well as representing a combination and evolution of types a.2 and a.3, can be considered a form of transition from wall rock-cut apiaries to those with open chamber (type b.2), and with some similarities to those with full enclosed chamber (type c.2). Some examples are in Masseria S. Angelo (Massafra), Taranto province (Apulia).

---

[1] For the latter see www.perieghesis.it/Atlante degli avucchiari del tarantino.

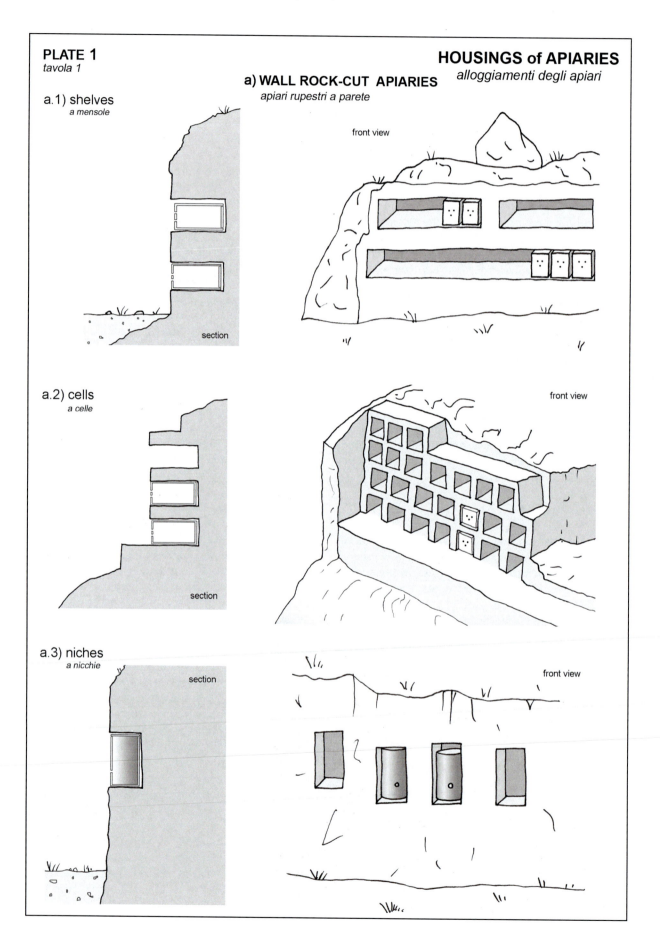

PLATE 1
tavola 1

HOUSINGS of APIARIES
alloggiamenti degli apiari

a) WALL ROCK-CUT APIARIES
apiari rupestri a parete

a.1) shelves
a mensole

front view

section

a.2) cells
a celle

front view

section

a.3) niches
a nicchie

section

front view

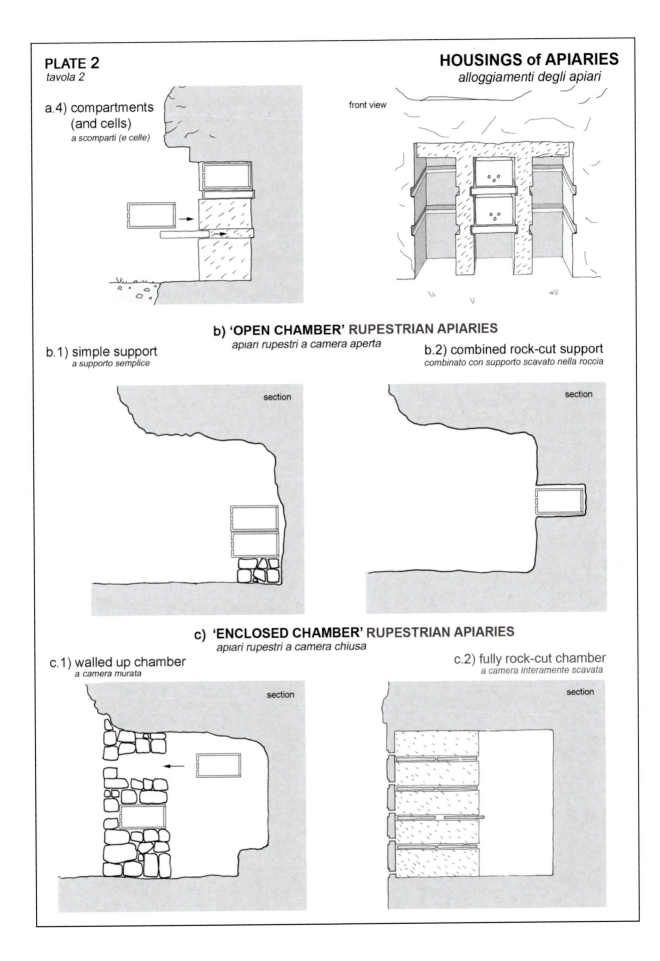

PLATE 2
tavola 2

HOUSINGS of APIARIES
alloggiamenti degli apiari

a.4) compartments
(and cells)
a scomparti (e celle)

front view

b) 'OPEN CHAMBER' RUPESTRIAN APIARIES
apiari rupestri a camera aperta

b.1) simple support
a supporto semplice

b.2) combined rock-cut support
combinato con supporto scavato nella roccia

section

section

c) 'ENCLOSED CHAMBER' RUPESTRIAN APIARIES
apiari rupestri a camera chiusa

c.1) walled up chamber
a camera murata

c.2) fully rock-cut chamber
a camera interamente scavata

section

section

Open chamber rupestrian apiaries (category b)

These are installations in which the housing for the hives are no longer constructed in the external walls of the cliffs, but placed inside an underground cavity in the rock mass, with the side walls and its own rock coverage defining the upper part of the apiary, 5-6 ft in height or more. This space can consist of a cave, i.e. a pre-existing natural cavity, or may have been purposely excavated, in part or totally, by the beekeeper to obtain a service area for the beehives' handling.

We define 'open chamber' as a cavity the entrance of which does not have any type of closure. Thus, it is completely open at the front, and the room is merely an additional shelter for the hives which are housed on various types of supports placed inside, leaning against the back wall. We can subdivide open chamber apiaries into two types:

b.1) Simple support apiary

The hives simply lie if necessary on superimposed rows, on the floor of the cavity, isolated by a simple plank or raised off the ground by a dry stone wall, without additional infrastructures. Some examples are known from the following localities: Madonna della Scala and Masseria S. Angelo (Massafra, Taranto province, Apulia), and S. Lania (Lentini, Catania province, Sicily) (Franco Dell'Aquila, personal communication).

b.2) Combined rock-cut support apiary

In this case, on the back wall of the cavity, or on the side walls, some housings for hives have been excavated. The accommodations are usually horizontal shelves very similar to those described above for the rock-cut wall apiaries. Some examples are known from the following localities, in the Taranto province (Apulia): Masseria Torretta (Massafra), and Masseria Vicentino (Grottaglie). A special case is described as type a.4 at Masseria S. Angelo (Massafra, Taranto province, Apulia), consisting of a shallow antechamber with more complex compartments and cells. As already mentioned, it can be considered a transitional form.

'Enclosed chamber' rupestrian apiaries (category c)

In this case the cavities housing the hives are closed on all sides, with the entrance to the chamber in the form of a door. We have chosen to distinguish these apiaries in a new category, rather than simply aggregate them with the 'open chamber' type, because this type of structure requires a substantial change for the beehive handling. In fact, the hives are no longer leaning against the walls of the cavity as in the previous examples (Figure 13), but are inserted in the curtain (masonry or rocky) that closes the external front of the chamber (Figure 14). In this way the beekeeper can directly access the rear part of the hives without having to remove them from their accommodation and without disturbing the flight of the bees (Figure 2), unlike what happens in the wall apiaries or open chamber apiaries described above, where the beekeeper must work standing in front of the flight holes and / or must remove the hive.

We distinguish two types of enclosed chamber apiaries:

c.1) 'Walled up chamber' rupestrian apiary

This type is composed of a cave (a rocky natural cavity) closed on the outer side by dry stone walls, and inside

Figure. 13. Hives leaning against the walls of the cavity with flight holes on the same side of the beekeeper (drawing R. Bixio).

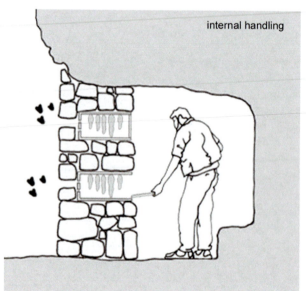

Figure. 14. Hives housed in the outer wall of the cavity, with access for inspection on the back, on the opposite side of the flight holes (drawing R. Bixio).

divided in compartments, built with dry stones too. This is the model documented at Imgiebak, on Malta (see above). Of course, there are also apiaries entirely built in masonry, with the same functions as the rupestrian ones.

c.2) 'Fully rock-cut chamber' apiary

This model, largely present in Cappadocia (and, perhaps, at one site in France, at Vars), is conceptually equivalent to those of Malta. The main difference lies in the fact that the Cappadocian apiaries do not exploit existing natural cavities, but are fully artificially excavated inside the tuff faces of cliffs and pinnacles. The chambers are closed by a natural curtain of rock also on the external side and are internally divided by vertical compartments achieved by preserving the rock from excavation.

Another peculiarity, so far noticed only in Cappadocia, are the series of vertical slits carved alongside the rows of flight holes. We note that they had the same function as flight holes, but were destined for a different type of hive (horizontal baskets), likely used for nomadic beekeeping.

Bibliography

Andaloro, M., V. Valentini and P. Pogliani. 2017. Painting on Rock Settements in Cappadocia. A Database and a Virtual Museum for the Enhancement of the Area, in Parise et al. (eds) *Cappadocia-Hypogea 2017: Proceedings of the International Congress of Speleology in Artificial* Cappadocia: 493-502. Istanbul: Ege Yayinlari.

Bettalli, M. (ed.) 1990. *Enea Tattico, Poliorketkà, La difesa di una città assediata*. Studi e testi di storia antica. Pisa: Ediz. ET.

Bixio, A. et al. 2018a. Kiliçlar Kalesi: una fortezza 'lineare' a Göreme (Cappadocia, Turchia). *Opera Ipogea* 2/2018: 109-126.

Bixio, A., R. Bixio and A. De Pascale. 2018b. Rock-cut Shelters in Göreme. Where Geomorphology, Hypogeal Architecture and Hydrology Meet, in M. Şener (ed) *Proceedings of the Cappadocia Geosciences Symposium*, Niğde, October 2018: 80-91. Niğde (Turkey): Ōmer Halisdemir University.

Bixio, R. (ed.) 2012. *Cappadocia. Records of the Underground Sites*. British Archaeological Reports-BAR, International Series 2413. Oxford: Archaeopress.

Bixio, R., G. Bologna and M. Traverso. 2004. Cappadocia 2003. Gli apiari rupestri dell'Altopiano Centrale Anatolico (Turchia). *Opera Ipogea* 1/2004: 3-18.

Bixio, R., V. Castellani and C. Succhiarelli (eds) 2002. *Cappadocia, le città sotterranee*. Rome: Istituto Poligrafico e Zecca dello Stato.

Bixio, R. and A. De Pascale. 2013. A New Type of Rock-cut Works: The Apiaries, in M. Parise (ed) *Proceedings of the International Workshop on Speleology in Artificial*

Cavities 'Classification of the typologies of artificial cavities in the world.' Torino/Italy - May 2012. *Opera Ipogea* 1/2013: 59-74.

Bixio, R. and A. De Pascale. 2015. Defensive Devices in Ancient Underground Shelters. Comparison among the sites of Aydıntepe, Ani, Ahlat and Cappadocia, in Turkey. Proceedings *Symposium on East Anatolia-South Caucasus Cultures, Erzurum, october 2012, Atatürk Üniversitesi Edebiyat Fakültesi Erzurum*, vol. II: 461-480. Cambridge: Cambridge Scholars Publishing.

Bixio, R. and S. Germanidou. 2019. Rock-cut apiaries and underground shelters in Göreme: a link? *Opera Ipogea* 2: 23-34.

Bixio, R., M. Traverso and R. Cirone. 2002. Apicoltura rupestre a Malta. *Opera Ipogea* 3/2002: 19-26.

Bonetto, J. 1997. La guerra sotterranea, in M. Busana (ed) *Via per montes excisa*: 337-398. Rome: L'Erma di Bretschneider.

Bortolin, R. 2008. *Archeologia del miele*. Mantova: SAP.

Calò, S. and E. Santucci. 2017. Hypogea with niches of southern Apulia. *Proceedings of the International Congress of Speleology in Artificial Cavities 'Hypogea 2017'*, Cappadocia, March 6/8 2017: 20-29. Istanbul: Ege Yayinlari.

Cirone, R. 2001. Un apiario in pietra riaffiora dopo 3000 anni nel cuore del mare. *Apitalia* 6: 10-17.

Crane, E. 1983. *The Archaeology of Beekeeping*. London: Duckworth.

Crane, E. 2003. The Use of Bees and their Products in Warfare. *Bee World* 84.2: 94-97.

Dain, A. 1937. *La 'Tactique' de Nicéphore Ouranos*. Paris: Les Belles Lettres.

De Jerphanion G. 1925-1942. *Une nouvelle province de l'art byzantine. Les églises rupestres de Cappadoce*. Bibliothèque Archéologique et Historique. Paris: Librairie Orientaliste Paul Geuthner.

De Minicis, E. 2018. L'apicoltura rupestre nella Tuscia. In R.M. Carra Boncasa and E. Vitale (eds), *Quaderni Digitali di Archeologia Postclassica* 11: 93-119. Palermo: Antipodes.

Dennis, G. 2010. *The Taktika of Leo VI*, Corpus fontium historiae Byzantinae, vol. 49. Washington, D.C.: Dumbarton Oaks Texts (12).

Germanidou, S. 2013. Evidence from written sources and Archaeological finds for a type of "Biological War" using bees in Byzantium. *Byzantina Symmeikta* 23: 91-104.

Germanidou, S. 2015. Dovecotes from the Roman and Byzantine periods. *Herom-Hellenistic and Roman Material Culture* 4.1: 33-52.

Germanidou, S. 2016. Byzantine honey culture. *Byzantium Today* 7. Athens: National Hellenic Research Foundation, Institute of Historical Research, Section of Byzantine Research.

Greco, V.A. n.d. *Perieghesis. Viaggio nella storia del paesaggio agrario del tarantino - Atlante degli avucchiari del tarantino*. Access 07/2019: www.perieghesis.it.

Jolivet-Lévy, C. 2015. *La Cappadoce.Un siècle après G. De Jerphanion.* Paris: Geuthner.

Masetti, L. 2000. La carrière de l'Estel. *La santè de l'abeille* 180 11-12, pp. 358-361.

Masetti, L. 2000. Malte: l'île des chevaliers et des abeilles. *L'Abeille de France* no. 887/2002. Paris.

Ousterhout, R. 2017. *Visualizing Community.* Dumbarton Oaks Studies XLVI. Washington D.C.: Dumbarton Oaks Research Library.

Roussel, G. 2006. Découverte de vieux ruchers en Cappadoce. *Cahiers d'Apistoria* 5 A: 39-46.

Roussel, G. 2008. Ruchers de Turquie. *Cahiers d'Apistoria* 7 A: 37-44.

Chapter 10

Appiaria vel in civitate vel in villa: Apiculture in the Early Medieval West

Javier Martínez Jiménez

Faculty of Classics, Churchill College, University of Cambridge (jmj50@cam.ac.uk)

Summary: This chapter explores the socio-economic role and nature of apiculture in the post-Roman Latin West, assessing the scarce archaeological evidence, the anecdotal mentions in literary sources, and the unexpectedly thorough legislation of the successor kingdoms. With this scant information it is possible to reach three preliminary conclusions. First, that honey and wax were probably more relevant in the early Medieval centuries than in the preceding Roman centuries. Second, that there was a shift in the legal status of bees from wild to domesticated animals, and thus considered private property of the beekeeper. Third, that while beekeeping and honey hunting remained a mostly rural activity, it is possible that beekeeping flourished in the less-dense urban contexts of the post-Roman centuries, perhaps playing a role in the emerging garden economy of the period.

Keywords: Apiculture; Latin West; early Middle Ages; urban economies; beekeeping legislation

In *The Simpson*'s episode no. 104 (first aired in 1994), a swarm of bees abandoned its hive in order to sit on top of Homer's pile of sugar. The original owners of the hive followed the swarm and tried to buy the bees back from Homer, offering him 2000 dollars. Before they could pay him, however, the rain washed the sugar away and the swarm flew off, leaving Homer crying on a sugary puddle. Homer is unaware of the fact, but claiming property of a swarm that had landed on his property is a modern-day premise which goes all the way back to Roman times because in Roman law bees are seen as wild animals (*apes ... quarum constat feram esse naturam*).[1] This is a principle, however, that did not apply in the early Middle Ages. During those centuries, bees (even swarming ones) were the property of a beekeeper and could be claimed back without the need to buy them back from a second party. This level of protection appears to be parallel to the perceived social and economic importance of apiculture in the post-Roman period.

Even if it is generally accepted (but not usually acknowledged) that apiculture played an essential role in ancient economies, the study of beekeeping is limited in comparison. Apiculture's apparent marginality, its (almost total) archaeological invisibility, and its scarce presence in the sources perhaps can explain the lack of attention it has received in the non-specialised literature. As with many other aspects, this is especially true for the post-Roman period and the early Middle Ages (for our purposes, here defined broadly between the fifth and tenth centuries). The two main, thousand-page, reference works for the economy of this period (McCormick 2001; Wickham 2005) have only three entries in total (all in McCormick's book).

This is not to say that there has been no interest in early Medieval beekeeping, but this has focused on the Greek East (Germanidou 2018), and there is virtually nothing in the literature for the Latin West – or at least nothing which is not part of a longer narrative (e.g., Crane 1999; Kritsky 2017; Mané 1991). For these purposes, I want to present in this paper a (very) summarised overview about the relevance of beekeeping in the West in this period, proposing also that there might have been an increasing demand at the time, and introducing the possibility of a post-Roman urban beekeeping tradition.

Bees, Honey, and Wax in the Post-Roman Centuries

Honey and wax played a major role in ancient economies in ways and forms which are familiar to us nowadays (e.g., Allsop and Miller 1996; Balandier 2006; Fernández Uriel 2011), and the early Middle Ages were no different.

Honey, for example, is described in various degrees of detail by a number of early Medieval authors, like Anthimus (fl. 510; *De obs. cib.*), Isidore of Seville (d.

---

* This chapter has been written during my position as a PDRA within the Impact of the Ancient City Project led by Prof. Andrew Wallace-Hadrill. This project has received funding from the European Research Council (ERC) under the European Union's Horizon 2020 research and innovation programme (grant agreement n° 693418). I would also like to thank the GEMS of the History Faculty at Cambridge, where this idea began to brew, and to Andrew Wallace-Hadrill, Andrew Dalby, Caroline Goodson, Sam Ottewill-Soulsby, Manuel Moreno, Maria Duggan, Cameron Moffett and the anonymous reviewers for their help and useful comments.

[1] 'Bees... whose nature is agreed to be wild', *Dig.* XLI.1.5.5. This and all other translations are by the author, unless specified.

636; *Etym.*, XX.2.18, 36) and Hrabanus Maurus (d. 856; *De Univ.*, XXII.1.7). They all mention how honey itself could be eaten on its own or from the comb, used for cooking in pastries and roasts, or as an essential part of medicines. Similarly, honey also appears in other texts both as a prized commodity and a staple good. For instance, in the sixth century, honey was one of the products which the bishops of Mérida distributed to the poor and needy (*VPE*, V.3.7).[2] At a later date, in seventh-century Wessex, ten jugs of honey could pay a year's rent over ten hides of land (*LI*, LXX.1) and, in the eight century, there are mentions of taxes and tributes paid in honey both in Carolingian France (Kritsky 2017: 253) and early Umayyad Spain (Carmona González 1992).

Honey played an essential role also in drinks, many of which were mixed (or were brewed) with it (*Etym.*, XX.3.10-3). Mead (*medus* in Latin) is the most obvious example, having a significant role in society, becoming in British and Scandinavian contexts the strong alcoholic drink of preference – especially after wine ceased to be as easily available as it had been in the Roman period.[3] Mead became a high-status drink linked to prestige feasting and aristocratic drinking, with mead-halls appearing prominently in texts such as Welsh bardic poetry and *Beowulf* (Moffett 2017; Osborn 2006: 272; Pollington 2011). But there were many other drinks which were usually mixed with honey, like wine and must (*mulsum, oenomelon, melicratum*), cider (*hydromelum*), vinegar (*oximelum*) or essence of roses (*rhodomelum*). Amongst Franks and Burgundians, absinth wine with honey was highly regarded (Greg. Tur., *Hist.*, VIII.31; cf. Ven. Fort., *VSM*, I.39). Even water simply sweetened with honey was a common beverage (Greg. Tur., *Hist.*, V.10).

Despite these evident examples of the high regard of honey, wax might have been the primary reason why bees were kept (Kritsky 2017: 254). Wax was, after all, used for many different purposes because of its characteristic chemical and physical properties (insoluble, malleable solid with a low melting point), which suggest a generalised and daily consumption. These uses include dedicatory *ex-votos*, writing tablets, document sealing, water-proofing, and an essential part of bronze-sculpting. Sidonius Apollinaris (d. 485), furthermore, also mentions wax as an essential component of pictures and paintings (Sid. Ap., *Ep.*, VII.14.5).[4] But in the early Medieval world wax was notably used for candles.

Artificial lighting in the ancient world was achieved in many ways, including torches and candles, but oil lamps appear very clearly to have been predominant. These lamps (regardless of the material) were typologically very similar: an open flat form with a spout for the wick. Olive oil was the most prestigious fuel for these lamps (Kimpe, et al. 2001), but other types of fat (resin, pitch, whale blubber) were also burned in lamps to light, especially when olive oil was not available (Fouracre 1995: 70). Lamps of this type continued to exist into the early Middle Ages, especially in those areas where olive oil was produced, like Umayyad Spain (Carvajal López 2009; cf. Karivieri 1996: 58-9), and the sources mention lamps also in Merovingian and Carolingian Francia (Fouracre 2018); a sharp contrast with Britain, where they vanish after the second century (Eckardt 2000; Allason-Jones 2005: 80-1).

We know that oil was still shipped beyond the Mediterranean coastline into the Carolingian period (and also in Britain, where it was still imported during the sixth century – Hemer, et al. 2013), but it is clear that wax candles were becoming increasingly important.[5] This is certain by the ninth century, when we know of tithes and taxes paid in candles levied in Carolingian Francia and Anglo-Saxon England (Fouracre 2018: 313-4; Gittos 2013: 46, 110-22), showing a preference of the candle over the lamp. Wax candles, of course, appear long before that, especially in religious contexts, where light has a very important symbolic significance (cf. Dell'Elicine 2008). Wax candles, with their smell-less and smoke-less bright flame (as opposed to tallow), already appear prominently in fifth- and sixth-century Christian texts, linked to cemetery vigils and personal offerings (e.g., Sid. Ap., *Ep.*, II.10; Ven. Fort., *VSR*, 8 and 47; *Carm.*, V.2.125). Candles also appeared in evening processions (Romano 2016), and they were the focus of the feast of Candlemas. It should be mentioned that these liturgical candles could be quite large, some as tall as a person (Ven. Fort., *VSR*, 74), overall suggesting an increasing demand for beeswax candles.

A final element that can further underline the perceived importance of beekeeping in post-Roman societies is the attention paid to it in the law codes of the successor kingdoms (table 1). Of course, beekeeping legislation is as ancient as legislation itself, and specific laws on honey theft and beekeeping appear already in Hittite texts (Lewis and Llewelyn-Jones 2018: 281).

---

[2] *Si quis enim ... ad atrium ob necessitatem accessisset, liquorem olei, vini vel mellis a dispensantibus poposcisset* = 'If anyone indeed ... approached the atrium [of the xenodochium] out of need [and] asked from the dispensers a measure of oil, wine or honey [Bishop Masona would give them more than they asked for].' For other late antique bishops and the distribution of foods, see Dunn 2014 and Izdebski 2012.
[3] Although it was still present (Hemer et al. 2013).
[4] *[Q]uem crebro melius infigit et argilla simulacris et cera picturis* = 'Why, clay often puts these [lines] more successfully into images or wax into

pictures.' This technique goes all the way to ancient Egypt (Kristky 2015: 104-10).
[5] Perhaps because of the perceived sacred view of bees (Fife 1939).

LAW	LOCATION	APPROX. DATE	COVERS	EFFECT
PLS VIII.1-2	Gaul	500	Theft of locked hives	fine: 45 *solidi* and return of property
PLS VIII.3	Gaul	500	Theft of fewer than 6 non-enclosed hives	fine: 15 *solidi* and return of property
PLS VIII.4	Gaul	500	Theft of more than 7 non-enclosed hives	fine: 45 *solidi* and return of property
LV VIII.6.1	Hispania	570? (*antiqua*)	Claiming property of wild bees and provisions for the theft thereof	Property ascribed by marking location of the hive three times; fraud or theft punished with 20 lashes and twofold compensation
LV VIII.6.2	Hispania	570? (*antiqua*)	Damage caused by bees	Beekeeper to move hives or face fine of 5 *solidi* plus compensations
LR XLVIII.6	Italy	630	Theft of hives	fine: 12 *solidi* and triple compensation
LLP 318	Italy	643	Theft of hives in an apiary	fine: 12 *solidi*
LLP 319	Italy	643	Theft of hives in a tree	fine: 6 *solidi*
LV VIII.6.3	Hispania	650	Trespassing into apiary with intent of theft	3 *solidi* and 50 lashes for trespassing, if freeman or 100 lashes if unfree; ninefold compensation and 50 lashes for theft if freeman, or sixfold if unfree
LB XXII.8	Bavaria	740	Claiming back one's swarm from someone's tree	Previous owner can have three goes at scaring the swarm back
LB XXII.9	Bavaria	740	Claiming back one's swarm from someone's hives	Previous owner can have three goes at scaring the swarm back
LB XXII.10	Bavaria	740	Dispute over property of hives	Judge's rule unless accused produces 6 oathtakers
LA IX.2	Wessex	890	Theft of hives	fine: 120 shillings [per hive?]
LHB II.37.113	Wales	940	Theft of honey and damage to trees	fine: 24 *denarii* per damaged tree and 24 per stolen hive and honey
LHB II.45.1-9	Wales	940	Price of swarms	Price of different types of swarms.

These law codes were heavily based in pre-existing Roman legislation, and focused mainly on the fines to be paid by those caught stealing hives.[6] The laws tend to specify fines in coins and the amount varies depending on the number of hives involved or the circumstances of the theft. For example, hives stolen from a locked and roofed apiary were fined at 45 *solidi* plus the return of the bees in early Frankish law (*PLS*, VIII.1, 2), while the fine was only a third of that amount (15 *solidi*) if the apiary was not roofed and only fewer than six hives went missing (*PLS*, VIII.3).[7] Similar thefts in later Frankish law were fined with fewer *solidi* (12), but with an extra compensation that tripled the value of the stolen goods (*LR*, XLVII.6), a practice mirrored in the seventh-century Lombard Edict of Rothair (*LLP*, 318). Even in Anglo-Saxon law (*LA*, IX.2), the theft of hives (*beoðeofe*) was punishable at 120 shillings, which may reflect the 12 *solidi* of the Ripuarian and Lombard fines. Visigothic law was even harsher, because the

theft was also punishable with lashes (*LV*, VIII.6.3), even if the monetary fine was smaller (no more than 3 *solidi*).[8] These punishments were severe, but they pale in comparison with the death sentence given by count Ranulf to the thief who stole a jar of honey (*vas cum melle repositum*) from a priest. While the subject of the theft (in this case, a holy man) is important, the actual stolen item (the honey jar) is what earned the death penalty (*Vita Eparchii* 10).[9]

There was, however, another set of laws that diverted clearly from preceding Roman legislation, and those are the laws that regulate the property of the hives. Under Roman law (Frier 1983), bees were wild animals, so no one could claim property over them. Indeed, if the bees swarmed away the beekeeper lost all rights over them

---

[6] The laws are quoted and translated in the appendix.

[7] If more than six hives were stolen, then the fine was the full 45 *solidi* (*PLS*, VIII.4).

[8] 50 lashes (100 if it was a slave) simply for trespassing into an apiary and, if theft was proven, the compensation was ninefold the value stolen.

[9] This anecdote raises issues: the story is most probably an exaggeration if not a complete fabrication, but it shows an ignorance of laws on bees, which might have been intentional or a reflection of contemporary legal practices.

(*Instit.*, II.68 = *Dig.*, XLI.1.5.5).[10] Moreover, harvesting honey from wild bees in an estate was not theft *per se*, although trespassing into someone's estate was (*Dig.*, XLVII.2.26). Only after bees had been 'housed' in an artificial hive (*alveo concludere*) was it legal to obtain usufruct from them (*Dig.*, VII.1.9.1; X.2.8.1; XXXIII.7.10; XLI.2.3.16).

Under the laws of the successor kingdoms, beekeepers could claim the property of a swarm and were even responsible for the damage caused by their bees, which could be seen as a way of protecting beekeepers and minimise conflict over a valuable resource (a contrast with modern practices and beekeepers offering Homer Simpson $2,000 for a swarm that had been theirs). A Lombard law regarding the theft of bees in trees (*LLP*, 319) echoes an earlier Visigothic law (*LV*, VIII.6.1), by which it was possible to claim property of bees that lived in a tree or rock on your land by marking it with three 'characters'. This practice is different from earlier Roman laws (esp. *Dig.*, XLI.1.5.2),[11] where the bees had to be housed in a hive in order to claim them. Similarly, in Bavarian law, bees that land on a tree belong to the owner of the tree, although the previous owner of the swarm can try to claim them back (*LB*, XXII.8). Other laws (*LB*, XXII.9-10) also cover the possibility of people luring their neighbour's bees by placing more welcoming hives, in which case they will become the new owners of the bees unless the original owner can scare them away. This perceived private property of the hive, completely different from Roman practices, is still echoed in medieval Welsh laws, where hives in trees are considered private property (*LHB*, II.37.113; *CC*, VI.42; cf. Crane and Walker 1984: 24-5).[12]

This necessarily short overview should be enough to highlight the apparent importance of bees and apiculture in the post-Roman centuries. However, the question remains as to where the honey and wax that was in such high demand came from. It will be unsurprising to see that, as in many other aspects, early Medieval patterns of beekeeping mirror Roman practices, although it is only possible that, against general perceptions, early Medieval apiculture also took place in urban contexts.

Rural Beekeeping and Honey Hunting

It is safe to assume that apiculture was an eminently rural activity in the early Medieval period, much in the way that it had been in the Roman centuries (Crane 1999; Fernández Uriel 2017: 924), despite the scarce positive evidence. Roman authors, at least, would have us believe that bees belonged solely in the countryside and in the world of agriculture even if the correlation between bees and pollination was not yet known (Grant 1949). Virgil (d. 19 BCE; *Georg.* IV), Pliny (d. 97 CE; *Nat. Hist.* XI, XXI, XXII), Columella (d. 70 CE; IX, X), and Varro (d. 27 BCE; III.16), dedicated many pages to describe the best practices on beekeeping. This was a tradition built on the Hellenistic treatises on apiculture like those of Aristomachus of Soli and Philistus of Tarsus (*Nat. Hist.* XI.19; Lewis and Llewelyn-Jones 2018: 284), which always linked bees to villas and rural estates as part of the *pastio villatica*, and highlighted apiculture as a very profitable enterprise (Balandier 2006: 186).[13] This view is supported by the archaeology of the period: the apiaries identified near Segobriga in Spain (Almeida and Morin 2012; Morín and Almeida 2014) and Eastern Portugal (Wallace-Hare 2019) are good examples of large-scale rural beekeeping in the Roman period. In this period we know that the exploitation of beehives was leased, both in public (like the *alvari locus* occupied by Lucius Valerius Kapito – *CIL* II.2242) and private land (Varro III.16.10; cf. Fernández Uriel 2017: 932), usually taking advantage of those areas which were not good enough for intensive agriculture (e.g., Strabo III.2.6; *CIL* VII.25902).

If we look now at the early Medieval evidence, we find that there is very little, but it nevertheless supports this continuity of Roman practices. Of the laws listed above, many specify regulations on honey hunting in forests (*LV*, VII.6.1; *LLP*, 319; *LB*, XXII.10; *LHB*, II.37.113). Similarly, the descriptions of enclosed apiaries (like those specified in *PLS*, VIII.1-2) are similar to those rural apiaries identified in Portugal (Wallace-Hare 2019). Passing anecdotal mentions in the literary sources also agree that bees belong in the *rus*: Sidonius Apollinaris certainly depicts bees as part of the idealised rural countryside (Sid. Ap., *Carm.*, II.3),[14] while in the ninth century Eulogius of Córdoba describes the large amount of hives that are located at *Penna Mellaria* (literally, the 'honey-producing mountain'; *Mem.*, III.11.2).[15]

---

[10] *(...) nam et apes idem faciunt, quarum constat feram esse naturam (...) donec revertendi animum habeant, quod si desierint revertendi animum habere, desinant nostra esse et fiant occupantium* = 'and bees, whose nature is agreed to be wild, do the same (...) that if these animals should not have the intention to return, they also cease to be ours and become the property of the first occupant.'

[11] *Apium quoque natura fera est: itaque quae in arbore nostra consederint, antequam a nobis alveo concludantur, non magis nostrae esse intelleguntur quam volucres, quae in nostra arbore nidum fecerint. ideo si alius eas incluserit, earum dominus erit* = 'Also, the nature of bees is wild, thus those that may settle in a tree of our property, before they are housed in a hive, it is understood are not our property any more than the birds that nest in our tree; so, if anyone captures them, he shall be their owner.'

[12] The law, however, states that if the owner of the trees is notified of the intention of the hunter, the honey hunter may keep it, suggesting that beekeepers were employed to capture wild swarms, just like it happens today.

[13] It is debatable whether Roman beekeeping was a complementary activity to farming (Fernández Uriel 2017: 933) or not (Almeida and Morín 2012).

[14] *Quid faceret laetas segetes, quod tempus amandum/ messibus et gregibus, vitibus atque apibus* = 'What makes the fields happy? What time is dear to the crops and the flocks, and the vineyards and the bees?'

[15] *[A]ntiquitus congestos in illa celsiore rupe ap<i>um industria favos*

Archaeologically, however, there is no comparable evidence to the known Roman rural apiaries.[16] If the Bavarian laws are any indication of early Medieval practices (*LB*, XXII.9), then we should assume that beehives must have been mostly made out of perishable materials, like bark (*cortices*), wicker (*surculli*), or wood or hollowed logs (*lignum*),[17] making them hard to identify. The wicker 'skep' of Anglo-Saxon England is, in this aspect, perhaps the most famous type in the literature (Osborn 2006). It is also possible that ceramic hives existed in the early Middle Ages: the constant use of *vas* (*vasculum, vascellum*) in the texts alongside *alveum* and *favus* could indicate that hives were conceived still as pots. Some Roman ceramic beehives might have continued in use after Antiquity (as could be the case of the ceramic hive of Braga), but no specific typologies have been identified yet, and the claims that some of the truncated-cone vessels from early Umayyad Spain were Attic-style beehives are very debatable (Gutiérrez Lloret 1994: 145; Morín and Almeida 2014: 290).

This being said, the most interesting thing about the beehive from Braga is that it appeared in an *urban* context, which would defy all preconceptions about Roman beekeeping practices. Maybe the chronology of the context (fifth century) is key to understand this (so-far) unique find. After all, there is currently some discussion about the possibility of urban apiculture in Antiquity (Mavrofridis 2018), and the post-Roman urban environment (which was less dense than its earlier counterpart), might have made it more welcoming for bees and less problematic for beekeepers.

The Possibility of Hives and Beekeeping in Urban Contexts

Nowadays we see a proliferation of urban beehives (e.g., Cadwaller, et al. 2011; Lorenz and Stark 2015; Moore and Kosut 2013), but beekeeping in Antiquity is not usually associated with urban spaces. This could be for the sole reason that bees are seen as a potential danger to neighbours and animals, and may unexpectedly and uncontrollably swarm into buildings (cf. Mavrofridis 2018).

In Roman cities, this might have been a concern because of their dense habitation pattern. Furthermore, environmental studies of Romano-British towns have shown no traces of honeybees in urban areas (Kenward 2005; Smith 2012: 50-1),[18] but bees in city contexts are

known for other periods of Antiquity. We first encounter them in Iron-Age Mesopotamia, where it appears that urban beekeeping was not uncommon. The Akkadian inscription of the Assyrian governor Shamash-resh-assur (c. 1100-700 BC) indicates that bees and gardens went hand in hand.[19] Perhaps the local circumstances (where open cities and the abundance of gardens made these the only places where bees could thrive) explain this pattern (Lewis and Llewelyn-Jones 2018: 285-6). Excavations from Isthmia (a sanctuary city without a dense built environment) show that beekeeping at a large scale (beyond self-supply) was already part of urban economies in the Hellenistic period (Anderson-Stojanović and Jones 2002). It is also clear that in the Iberian peninsula during the Iron Age apiaries existed inside settled areas (Bonet and Mata 1997; Morín and Almeida 2014: 283; Quixal and Jardón 2016). Lastly and chronologically closer, there are documents from Byzantine Egypt that describe people keeping hives in the roofs of their houses in settled areas (Mavrofridis 2018). These exceptional examples underline the fact that it is not impossible to combine dwellings and beekeeping, or to discard the possibility of pre-Modern urban beekeeping.

If we now look in detail at the post-Roman period, we will find that there is some evidence to suggest that urban beekeeping existed and that it was, at least, conceivable.

First of all, we encounter one of the Visigothic laws. This *antiqua* law in Recceswinth's code (*LV*, VIII.6.2) covered apiaries in urban and rural areas: *Si quis appiaria in civitate aut in villa forsitan construxerit.*[20] In this case, the word *civitas* unequivocally refers to a city as it is opposed to *villa*. The rest of the law, regulating the damage caused by bees, covers the possibility of bees located in areas where they were likely to affect neighbours, a reasonable consequence of keeping bees in close proximity to humans and animals. Of course, this law has to be taken with a pinch of salt, and we should keep in mind that this type of legislation is universal in its scope; it is more *normative* than *practical* in its approach (Rio 2009: 200-10). The *aut civitate aut in villa* clause was, perhaps, more rhetoric device than legal specification but, in either case, it opens the possibility that urban apiaries were conceivable and not actively prohibited.

A second instance of beehives kept in urban contexts appears much later, during the Mercian defence of Chester against the Viking siege of AD 907 (Kritsky

---

maiores nostri viderunt, Pinna mellaria vocatus est = 'From Antiquity our elders saw fit that beehives [*favos*] were gathered in that most lofty rock, which is called Pinna Mellaria because of the diligence of the bees.' The site of Peñamelaria still exists outside Córdoba.

[16] But when it appears, as it will be evident below, beekeeping appears in urban contexts.

[17] These are, obviously, very common practices (Crane 1999).

[18] Of course, this is only a small selection of sites, and does not necessarily reflect Empire-wide patterns.

[19] 'I, Shamash-resh-assur, governor of Sukhi and Ma'er, have brought from the mountains of the people of Kabka into the land [of] Sukhi the bees that collect honey (...) I settled them in the gardens of the city of Gabbarini, so as they could collect honey and wax...' (transl. in Lewis and Llewelyn-Jones 2018).

[20] 'Should anyone build an apiary in a town or village'.

2017: 253). As recorded in the *Fragmentary Annals of Ireland*, the Norwegians were mining under the walls in order to collapse them, and they set up timber frames, covered in skins to protect the diggers from boulders and boiling water (and ale). To counter this, '[t]he Saxons then scattered all the beehives there were in the town on top of the besiegers' (transl. Randler 1978: 173). If this account were true, then it would be certain evidence for the existence of beehives inside the walls of Chester, which as a main Mercian centre would be consuming wax, honey and (especially) mead. These beehives, however, might have been transported into the city before the siege, and not represent necessarily urban apiaries.

Lastly, we find an anecdotal mention of bees in the seventh-century *Vita* of Saint Medardus of Noyon (in Neustria), where a thief attempted to steal the honey from the bishop's apiary only to be attacked by an angry swarm of bees (*VSMed*, V.15-6). Even if the location of this event is not specified (beyond the 'house' of the saint: *vasa eius domum nocturna obreptione praeveniens distenta melle furaverat*),[21] it refers to the miracles of a bishop in his own house, which for Gaul in this period we can assume was located in an urban site (Loseby, 2020).

I will be the first to admit that these anecdotes provide no conclusive evidence to prove the existence of urban beekeeping in the early Middle Ages, although they highlight that the presence of bees in urban areas was not unthinkable. If we now turn to archaeology, we can find more supportive evidence.

The most unequivocal piece of evidence supporting the existence of urban apiaries comes from the finds of bees and hives in Middle Saxon layers of London and the Coppergate excavation in York (Kenward 2005). In the latter, the archaeologists identified a structure with so many preserved remains of bees that it might have been an urban bee shelter: perhaps permanent, perhaps to keep the hives protected over winter, something that echoes the account of the siege of Chester. Less conclusive are the sixth-century remains identified at Tintagel castle, where there is a structure (Site C in the middle terrace) that could have been either a mead processing area (Moffett 2017) or, perhaps, an apiary for skeps. While the urban nature of Tintagel is debatable, it certainly acted as a main central place (cf. Martínez and Tejerizo 2015), and still would show the presence of bees in inhabited areas in this period.

Another problematic example comes from the city of Braga in Portugal, already mentioned above. There, one of the large excavations inside the old Roman city (the Insula das Carvalheiras site) unearthed the remains of what originally was taken to be a water pipe, but more recent examination has proven it to have been a ceramic hive (Morais 2006: 157; Oliveira, et al. 2017). Ceramic hives are more commonly found in Mediterranean areas (Francis 2009; 2012), but the appearance of this type of hive in Atlantic contexts should make us reconsider earlier assumptions on the use and distribution of ceramic hives in the Roman period.[22] More importantly, the hive is, typologically, datable to the third century even if it was obtained from fifth century abandonment layers. On the same site, second and third century honey pots ('potes meleiros') have been excavated in situ (Delgado 1997). While it is more usual to find urban residue in rural contexts (a result of waste disposal and manuring), it is not common for the opposite to happen. As a result, it should be assumed that the hive was originally located in the urban area, either in relation to the second/third-century honey pots or, more likely, in a fifth-century environment (cf. Wallace-Hare 2019: 44).

Overall, the evidence is thin at best, and I would not advocate for a widespread tradition of urban beekeeping in the early Middle Ages (cf. Mavrofridis 2018). But the evidence is, perhaps, substantial enough. If these hints are considered alongside the type of urban environment that characterised the early Middle Ages, there may be a stronger case to defend the habitual presence of hives in urban or suburban areas.

Apiculture and the Early Medieval Urban Garden Economy

Traditional narratives of decline and fall of the Roman city have underlined the increasing number of 'empty' areas and production sites in areas formerly occupied by monumental or residential structures (cf. Ward-Perkins 2005). While the processes of ruralisation are now being assessed under a different light, the fact remains that many post-Roman cities had an agricultural component that had not existed before.

Recent studies on Italy (Goodson 2018) have shown that gardens in eighth- and ninth-century urban contexts were valuable possessions, and that urban garden production played an essential role in local economies. These gardens provided staple vegetables and legumes, and even fruit for self-consumption and small-scale market sales. Even if Goodson does not include beekeeping within her analysis of the garden economy, it is easy to see how apiculture might have

---

[21] '[The thief] came into the house with the cover of darkness so he might steal his [Saint Medard's] hives and honey.' The beehive is in this text referred to poetically as *castrorum habitacula*: 'cells of the fort' rather than the more mundane *apiarium, alveum* or *vas apium*. Also, this story must be read with the same caveats as that of the thief who was executed for stealing honey in the life of Eparchius mentioned above.

[22] In fact, pottery hives from Britain are also known from the Rockbourne villa (Allason-Jones 2005: 100).

fitted within this garden economy. After all, wax and honey were valuable staples that anyone could obtain 'freely' from their own yards. In fact, apiculture could be a very profitable enterprise within this garden economy for three main reasons. Firstly, because a hive can be harvested more than once a year;[23] secondly, because the bees' natural tendency to swarm means that an apiary could potentially double its output every year; and lastly, beekeeping requires a minimal input in terms of working hours – hours that could be spent in any other activity, garden-related or not (Allsop and Miller 1995). A dedicated garden owner with any number of hives could certainly make a sweet profit on the side.[24]

The lack of specific references to urban gardens in the rest of the West (especially for the earlier periods, where gardens appear only sporadically and not necessarily in urban contexts),[25] is certainly a problem, but archaeology shows, however, that there were many open spaces in old Roman cities which could (and might) have been devoted to agriculture.

The so-called 'dark/black earths/soils' are dark, organic layers, largely devoid of material culture first identified in British archaeology between clearly datable Roman layers and clearly datable High Medieval ones (Yule 1990). This usage soon spread into continental contexts, and became a characteristic element of the post-Roman city. Our knowledge of 'dark earths' has improved in the recent years, and they cannot anymore be seen as large-scale phases of abandonment, but rather as the even formation of new soil as a result of middening, domestic dumping, timber-building debris, and other aeolic and colluvial siltation processes (Macphail 2010; Macphail, et al. 2010; Nicosia, et al. 2013). The high organic content, however, would suggest a direct link with gardens or animals (Christie 2006: 261-2), even if the specifics need to be studied on a case-by-case basis.

Many cities across Gaul and Spain have these open, unbuilt spaces in area formerly occupied by houses and monuments (Esmonde Cleary 2013; Martínez, et al. 2018: 102-5), a situation much more evident in Britain, where cities are virtually abandoned by the sixth century (and only later re-emerging as urban nuclei; Crabtree 2018; Speed 2014). These towns were less dense because of the expansion of suburbs (around Christian shrines or harbours) and the concentration of habitat around the new foci of power (episcopal and civil palaces). This created a settlement model where traditional 'urban' and 'rural' zones were not as clearly demarcated and separated as they might have been in the past.[26] In fact, the reduction of wall enclosures left many 'urban' areas outside the fortifications. Dark earths are found in the 'gaps' between these multi-focal urban spreads. If we now remember the examples mentioned earlier of pre-Roman urban beekeeping, we will find that this post-Roman model shares with the former the same combination of spread urbanism mixed with gardens and horticulture.

There is, of course, a great degree of chronological and regional variation. Main Mediterranean economic and political centres (like Mérida, Arles, Córdoba, Tarragona, Seville, or Marseilles) kept their town centres 'urban' for longer, with even phases of revitalised urbanism (Loseby 2020; Martínez Jiménez 2013), so archaeological evidence for 'dark earths' or empty plots is scarce until the seventh or eighth centuries. Minor centres, secondary nuclei, and British towns, on the other hand, show these patterns much earlier, clearly noticeable during the sixth century. But, overall, cities from the fifth century onwards show degrees of 'ruralisation' that could (in those early stages) be linked to the well-defined garden economies of Italy of the eighth. Furthermore, new urban foundations of the early Middle Ages, like Hamwic in Britain (Brisbane 1994; Jervis 2011), Dorestad in Francia (Pestell 2011), or the earlier Reccopolis in Spain (Henning, et al. 2019) are characterised by this open urbanism which combined dwellings and production areas. This would show that this model was the way in which cities and urban spaces were conceptualised from the very beginning.

While the extent, depth and nature of the ruralisation of the Roman city in the West varies from case to case, it is clear that domestic agricultural outputs must have been important for urban communities. This is especially true after the fifth century, when the volume and intensity of bulk transport of daily commodities across the Mediterranean all but collapsed (Loseby 2012). Honey and wax can certainly qualify as bulk daily commodities and, as a result, it is probable that their long-distance trade declined in this period, meaning that the demand had to be met locally.[27] In this context, ceramic (and, to a lesser extent, glass) table wares are a good point of comparison. In the Roman period, table wares (especially red-gloss wares like *Terra Sigillata* and ARS) had been widely traded across the Mediterranean and beyond, but in the early Medieval period the widespread distribution of these items stopped. As a result, production became increasingly regional and, in

---

[23] Which was a well-known fact in Antiquity, so that even Petronius put hives as examples of fast growth (*Satyr*. 76).

[24] There were, of course, beekeepers who managed large numbers of hives (cf. Aparici Martí 1998: 37; Carmona Ruiz 2000).

[25] Laws, for instance, are vague when describing location of gardens: e.g., *LV*, VIII.3.2, 7 and 15; *PLS*, XXVII.7-11. A more detailed analysis of the documentary evidence could definitely cast more light on this issue.

[26] Of course, Roman cities had been centres of skilled production (Erdkamp 2012), but not usually of agricultural produce.

[27] It is difficult to tell, as there are hardly any notices of honey trade in the West for this early period, which coincides with the end of honey-related amphora typologies.

many cases, local, domestic and unskilled in order to keep up the supply (Esmonde Cleary 2013: 325-9; Jervis, et al. 2016; Martínez, et al. 2018: 119-20).

All things considered, the changes in the long-distance trade of daily commodities, together with the evident increase of localised urban production, create an economic environment in which it would make sense to have urban or suburban apiculture. Urban gardens, which are becoming increasingly visible in the archaeology, *could* have been perfect locations for these apiaries – acknowledging, of course, that this is a situation yet unconfirmed by the archaeology.

Conclusions

Apiculture has always been a key element in pre-Modern economies, and the early Middle Ages are no exception. While the collapse of the villa economy during the fifth century might have affected the inter-regional transportation of bulk agricultural commodities (including wax and honey), local and domestic production could have soon met the demand once cheap imports were not available anymore. This is, at least, what is clearly happening with other better-known products, like pottery or glass.

Furthermore, during this period we see cultural changes that might have prompted an increasing demand for wax and to a lesser extent, honey – both of which were most likely supplied locally. The importance of lighting in Christian rites went beyond the symbolic relevance of light, and churches demanded a constant supply of lamps, candles and fuel. In regions where oil for lamps was not available (or had to be imported), we see candles soon taking precedence. Similarly, and especially in Britain, the increasing importance of mead as a strong alcohol with the decline of wine (or, rather, the confining of wine to Christian contexts) could have implied a growing demand for honey for drinking and feasting.

It is clear that Roman rural patterns of apiculture and honey hunting continued in the West after the fifth century (the legal and literary sources are very clear about it), although beekeepers were given, in these legal contexts, more rights over wild bees than they had enjoyed in the Roman past (or than they would enjoy today, going back to Homer Simpson).

What is more surprising, however, is that the transformation of the Roman city and of long-distance supply patterns open the door to the possibility of urban beekeeping in these centuries. The evidence is scarce and unevenly distributed, with clear indications for urban hives in insular contexts from the eighth century onwards and anecdotal evidence for the continent in the fifth and sixth. The geographical

and chronological gaps make it difficult to propose a general tradition of urban beekeeping, but the local importance of garden economies (as identified in Italy and extrapolated to other sites where 'dark earths' are present) could further support this proposal. Of course, the very specific type of post-Roman urban landscape that develops from the fifth century onwards is key – especially when compared to previous historical periods when urban beekeeping is recorded.

The first municipal statutes of the city of Seville after the Christian conquest of 1248 regulated beekeeping. In this late medieval context, the municipal statutes specifically ruled that apiaries had to be located away from the city, in isolated and less productive areas, and with a given distance between them (Carmona Ruiz 2000). These laws show how serious a business apiculture had become by that period (honey and wax had become the second largest export of the Kingdom of Castile, second only to wool), and seems to involve large numbers of hives per owner. But it also shows that by then honey and wax had become a large-scale business, that the products were traded and exported, and that the densely built urban environments had pushed hives firmly to the countryside.

Appendix – Early Medieval Laws on Beekeeping

- ***PLS* VIII.1**: *Si quis unam apem hoc es unum vasum deintro clavem furaverit et tectum super habuerit cui fuerit adprobatum, mallobergo 'antedio olechardis' hoc est, MDCC denarios qui faciunt solidos XLV culpabilis iudicetur excepto capitale et dilatura* = Should anyone steal one swarm of bees, i.e., one hive, kept locked under a roof and it is proved against him (this is an 'antedio olechardis' in the Malberg gloss), it is judged [that he shall pay] 1800 *denarii* which make 45 *solidi* in addition to return of the swarm plus compensation.
- ***PLS* VIII.2**: *Si quis unam apem hoc est unum vascellum, ubi amplius non fuerit, furaverit et ei fuerit adprobatum, causam superius conpraehensam convenit onservare* = Should anyone steal one swarm, which is one hive, where there were no more, and it is proved against him, he shall be held liable as in the above case.
- ***PLS* VIII.3**: *Si amplius vero usque ad sex apes foris tectum furaverit, ut aliquid exinde remaneant et ei fuerit adprobatum, mallobergo 'leodardi' hoc est, DC denarius qui faciunt XV solidos culpabilis iudicetur excepto capitale et dilatura* = Should someone steal up to six swarms outside a roof where more remain and it is proved against him (called 'leodardi' in the Mallberg gloss) it is judged [that he shall be liable to pay] 600 *denarii* which make

15 *solidi* in addition for return of those stolen plus a compensation.

- **PLS VIII.4**: *Si vero septem aut amplius furaverit et si adhuc aliquid exinde remaneant et retro clavem fuerint cui fuerit adprobatum, mallobergo 'antedio texaga olechardis' hoc est, MDCCC denarios qui faciunt solidos XLV culpabilis iudicetur excepto capitale et dilatura* = If truly someone steals seven or more hives and still others remain there under lock and it is proved against him (called 'antedio texaga olechardis' in the Mallberg gloss) it is judged [that he shall be liable to pay] 1800 *denarii* which make 45 *solidi* in addition to returning them plus a compensation.

- **LV VIII.6.1**: *Si quis apes in silva sua aut in rupibus vel in saxo aut in arboribus invenierit, faciat tres decurias qu<a>e vocantur caracteres; unde potius non per unum caracterem fraus nascatur. Et si quis contra hoc fecerit adque alienum signatum inruperit, duplum restituat illi, cui fraus inlata est, et pretera XX flagella suscipiat* = Should anyone find bees in his forest, or in cliffs or in a rock, may he make three marks which are called characters; so that from one single character may fraud not arise. And should anyone act against this and break into someone else's marked tree, may he repay double to whom the fraud was caused, and receive 20 lashes.

- **LV VIII.6.2**: *Si quis appiaria in civitate aut in villa forsitan construexerit, et alii damnum intulerit, statim moneatur, ut eas in abditis locis transferre debear, ne forsitan in eodem loco hominibus aut animalibus damnum inferant. Et qui haec precept aut testationem neclexerit et damnum subfocationis in quatrupes intulerit, quod mortuum fuerit, duplum restituat; quod vero debilitatum, ille obtineat et simile somino reddat et pro iudicis contestatione, quam audire neclexit, V solidos coactos exolvat* = Should anyone build an apiary in a town or village, and any damage should result to others thereby, he must straightway be notified to move it elsewhere, that the bees may not inflict further injury upon men or animals in that locality; and if, after such notice, the owner should neglect to move said apiary, and any quadruped should be injured by the bees, the owner of the latter shall give two animals for every one that is killed, and one for each that is crippled, to the owner thereof; and shall be entitled to keep said injured animals; and shall be compelled to pay five *solidi* for neglecting to heed the warning of the judge.

- **LV VIII.6.3**: *Si quis ingenuus in appiaria furti causa fuerit conprehensus, si nihil exinde abstulerit, propter hoc, quod ibidem conprehensus est, III solidos solvat et L flagella suscipiat. Ceterum si abstulerit, novecuplum cogatur exolvere et predictum numerum flagellorum excipiat. Servus vero, si ingressis nihil abstulerit, C*

ververibus addicatur. Quod si abstulerit, sexcuplo reddere conpellatur; pro quo si dominus satisfacere noluerit, eum serviturum illi, qui damnim pertulit, tradat = Should any freeborn man enter an apiary for the purpose of theft, and is caught there, even if he takes nothing, for the mere fact of being arrested in such a place, he shall pay three *solidi* and receive fifty lashes. However, if he stole anything, he shall be compelled to pay ninefold its value, and shall receive the aforesaid number of lashes. If a slave should trespass in an apiary, without stealing anything, he shall receive a hundred lashes; and if he stole anything, he shall be compelled to restore sixfold the value of the same; and if his master is unwilling to render satisfaction for his act, he must deliver said slave to him who suffered the loss.

- **LR XLVIII.6**: *Si quis de appiario apem furaverit, XII solidos conponat, et quod furaverit in triplo conponat* = Should anyone steal a hive from an apiary, may he pay 12 *solidi*, and pay in triple what he stole.

- **LLP 318**: *Si quis de apiculari vas cum apibus furatus fuerit, unum aut plura, componat solidos XII* = Should anyone steal one or many hives from an apiary, may he pay 12 *solidi*.

- **LLP 319**: *Si quis de arbore signata in silva alterius apes tulerit, componat solidos 6. Nam si signata non fuerit, tunc quicumque invenerit iure naturali habeat sibi, excepto in gagio regis. Et si contingerit, ut dominus cuius silva est supervenerit, tollat mel et amplius culpa non requiratur* = Should anyone steal hives from a marked tree in someone else's forest, may he pay 6 *solidi*. But if it is not marked, then whoever may find them shall have them as per natural law, excepting in the gagius [??] of the king. And if necessary, should the master of the forest come, may he take the honey without requiring further guilt.

- **LB XXII.8**: *Si apes, id est examen alicuius ex apile elapsus fuerit et in alterius nemoris arborem intraverit et ille consecutus fuerit, tunc interpellat eum cuius arbor est, et cum fumo et percussionibus ternis de traverse secure, si potest, suum eiciat examen; verumtamen ita, it arborem non laedatur. Et quod remanserint, huius sint, cuius arbor est* = If bees, that is, someone's swarm, escape from the beehive, and land in a tree in another's grove, and [their owner] follows them, then let him call him to whom the tree belongs, and let him drive out his swarm with smoke and three blows of a cross-axe, if he can, so that the tree is not injured. And let what remains belong to the owner of the tree.

- **LB XXII.9**: *Si autem in capturis quae ad capiendas apes ponuntur, id est vascula apium, simile modo interpellat eum cuius vasculum est, et studeat suum eicere examen. Verumtamen vasculum non aperietur* *nec ledetur: si ligneum est, ternis vicibus lidat eum terries; si ex corticibus aut ex surculis conpositum fuerit, cum pugillo ternis vicibus percutiatur vasculum et non amplius, et quos eicerit suae erunt, et quae remanserint ipsius erunt cuius vasculum est* = If, however, bees enter receptacles placed for the capture of bees, that is, beehives, in a similar way, let him call him to whom the hive belongs, and let him strive to drive out his swarm. Nevertheless, let him not open or injure the hive. If it is wood, let him throw it three times on the ground; if it is held together by bark or by wicker wood, let him strike the hive three times with his fist and no more. And the bees that are driven out are his, and those that remain belong to him who owns the hive.

- **LB XXII.10**: *Si autem dominum arboris vel vasculi non interpellaverit et sine illius conscientiam eiectum domini restituerit et ille cuius vasculum fuerat, eum conpellaverit, ut ex suo opera vel arbore retulisset et restituendi conpellaverit, quod untprunt vocant, et ille alius, si negare voluerit et dicit suum consecutus fuisse: tunc cum sex sacramentalibus iuret, quod ex suo opera ipsum examen iniuste non tulisset nec illud ad iudicium restituere deberet* = If, however, he does not inform the owner of the trees or hives, and without his [the latter's] knowledge brings back the swarm to his own property, and the owner of the hive compels him to return them from his own hive or tree and to make restoration, which they call untprut, and if that other one wishes to deny it and says that he followed his own swarm, then let him swear with six oathtakers that he did not unjustly bring his swarm from his [the other's] hive, nor ought he to compensate for this according to the judgment.

- **LA IX.2**: *Geo was golðeofe ⌐ stodðeofe ⌐ beoðeofe ⌐ manig witu maran ðonne opru; nu sint eal gelic buton manðeofe: CXX scill* = Formerly the fines to be paid by those who stole gold and horses and bees, and many other fines, were greater than the rest. Now all fines, with the exception of stealing men, are alike: 120 shillings. [transl. Attenborough, 1922].

- **LHB II.37.113**: *Quidam dicunt, si perforetur quercus causa mellis, quod pro apertione quercus XXIIII denarii, et pro melle et apibus XXIIII denarii, et pro unoquoque illorum XXIIII denarii redduntur* = It is said, if an oak is drilled into for the sake of the honey; for breaking the oak 24 *denarii* are to be paid, and for the honey and the hive 24 denarii, and for only one of those 24 *denarii*.

- **CC, VI.42**: *O deruyd dyuot gwenyn y lestyr dyn arall ac yssu y mel a mynnv or neb pieiffo y mel y dieissywaw or dyn bioed y gwenyn alewssei y mel; kyfreith adyweit na diwygir dim idaw, sef achaws yw, py aniueil bynnac ny dylyo eu perchennawc y bygeilaeth nae gadw, ny dyly ynteu difwyn y llwgyr nar gweithret, sef anyueileit yw y*

rei hynny gwenyn, neu letuegin adofer o fwystuil gwyllt, neu ederyn = If bees come to another person's skep and eat the honey, and the owner of the honey wills to be compensated by the owner of the bees which devoured the honey, the law says: that he is to have no compensation because, whatever animal is not to be attended or guarded by the owner (such as bees, or a wild animal, or a bird), [said owner] is not to compensate for damage or deed [caused by his animals]. [amended from the transl. by Owen (1841)].

Primary Sources

CC = *Cyfreithiau Cymru*. Ed. A. Owen, 1841. *Ancient Laws and Institutes of Wales*. London, Commission on the Public Documents of the Kingdom.

Columella = Columella, *De re rustica*. Edd. E. Forster and E. Heffner, 1954. Loeb Classical Library 407. Cambridge, Harvard University Press).

De obs. cib. = Anthimus, *De observatione ciborum epistula ad Theudericum regem Francorum*. Ed. E. Liechtenhan, 1963. Corpus Medicorum Latinorum 8. Berlin, Academia Scientiarum.

De Univ. = Hrabanus Maurus, *De Universo*. Ed. J.-P. Migne, 1851. PL 108. Paris, Sirou.

Dig. = Justinian, *Digestum*. Ed. P. Krueger, 1915. Corpus Iuris Civilis, v. 2. Berlin, Weidmann.

Etym. = Isidore of Seville, *Etymologiae*. Ed. W. Lindsay, 1911. Oxford. Clarendon.

Georg. = Virgil, *Georgics*. Ed. H. Rushton Fairclough, 1916. Loeb Classical Library 63: 97-260. Cambridge, Harvard University Press.

Greg. Tur., *Hist.* = Gregory of Tours, *Historiae*. Edd. B. Krusch and W. Levison, 1951. MGH SS. rer. Merov. 1. Hannover, Hahnan.

Instit. = Gaius, *Institutiones*. Transl. E. Poste, 1904. Oxford, Clarendon Press.

LA = *Laws of Alfred*. Ed. F. Attenborough, 1922. The Laws of the Earliest English Kings: 33-94. Cambridge, Cambridge University Press.

LB = *Lex Baiuwariorum*. Ed. E. Heyman. 1952. MGH LL nat. Germ. 5.2. Hannover, Hahnan.

LHB = *Leges Howeli Boni*. Ed. A. Owen, 1841. Ancient Laws and Institutes of Wales: 749-903. London, Commission on the Public Documents of the Kingdom.

LI = *Laws of Ine*. Ed. F. Attenborough, 1922. The Laws of the Earliest English Kings: 33-94. Cambridge, Cambridge University Press.

LLP = *Liber Legis Langobardorum Papiensis*. Ed. G. Pertz, 1968. MGH LL 4: 291-606. Hannover, Hahnan.

LR = *Lex Ripuaria*. Edd. F. Beyerle and R. Buchner, 1954. MGH LL nat. Germ. 3.2. Hannover, Hahnan.

LV = *Lex Visigothorum*. Ed. K. Zeumer, 1902. MGH LL nat. Germ. 1. Hannover, Hahnan.

Mem. = Eulogius of Toledo, *Memorialis Sanctorum Libri Tres*. Ed. J.-P. Migne, 1851. PL 115. Paris, Sirou.

Nat. Hist. = Pliny the Elder, *Naturalis Historia*. Edd. VV.AA., 1938-62. Loeb Classical Library. Cambridge, Harvard University Press.

PLS = *Pactus Legis* Salicae. Ed. K. Edhart, 1962. MGH LL nat. Germ. 4.1. Hannover, Hahnan.

Satyr. = Petronius, *Satyricon*. Ed. M. Heseltine, and W. Rouse, 1913. Loeb Classical Library 15. Cambridge, Harvard University Press.

Sid. Ap., *Carm.* = Sidonius Apollinaris, *Carmina*. Ed. W. Anderson, 1936. Loeb Classical Library 296. Cambridge, Harvard University Press.

Sid. Ap., *Ep.* = Sidonius Apollinaris, *Epistolae*. Ed. W. Anderson, 1936. Loeb Classical Library 296. Cambridge, Harvard University Press.

Strabo = Strabo, *Geographiká*. Ed. H. Jones, 1923. Loeb Classical Library 50. Cambridge, Harvard University Press.

Varro = Varro, *De re rustica*. Edd. W. Hooper and H. Ash, 1934. Loeb Classical Library 283. Cambridge, Harvard University Press.

Ven. Fort., *Carm* = Venantius Fortunatus, *Carmina*. Ed. F. Leo, 1881. MGH AA 4.1: 1-270. Berlin, Weidmann.

Ven. Fort., *VSM* = Venantius Fortunatus, *Vita Sancti Martini*. Ed. F. Leo, 1881. MGH AA 4.1: 294-370. Berlin, Weidmann.

Ven. Fort., *VSR* = Venantius Fortunatus, *Vita Sanctae Radegundis*. Ed. B. Brusch, 1885. MGH AA 4.2: 38-49. Berlin, Weidmann.

Vita Eparchii = *Vita et virtutes Eparchii reclusi eclosimensis*. Ed. B. Krusch, 1896. MGH SS rer Merov 3: 550-64. Hannover, Hahnan.

VPE = *Vitae Patrum Emeritensium*. Ed. A. Maya Sánchez, 1992. CCSL 116. Turnhout, Brepols.

VSMed = Pseudo-Venantius, *Vita Sancti Medardi*. Ed. B. Brusch, 1885. MGH AA 4.2: 67-73. Berlin, Weidmann.

Bibliography

Allason-Jones, L. 2005 [1989]. *Women in Roman Britain*. York: Council for British Archaeology.

Allsop, K. and J. Miller. 1996. Honey revisited: a reappraisal of honey in pre-industrial diets. *British Journal of Nutrition* 75: 513-20.

Almeida, R. R. de, and J. Morín de Pablos. 2012. Colmenas cerámicas en el territorio de Segobriga. Nuevos datos para la apicultura en época romana en Hispania, in D. Bernal Casasola and A. Ribera Lacomba (eds) *Cerámicas hispanorromanas II. Producciones regionales*: 725-43. Cádiz: University of Cádiz.

Aparici Martí, J. 1998. De la apicultura a la obtención de la cera. Las "otras manufacturas" medievales de Segorbe y Castellón. *Millars: Espai i Història* 21: 31-50.

Anderson-Stojanović, V. and E. Jones. 2002. Ancient beehives from Isthmia. *Hesperia* 71.4: 345-76.

Balandier, C. 2006. L'importance de la production du miel dans l'économie gréco-romaine. *Pallas* 64: 183-96.

Bonet Rosado, H. and C. Mata Parreño. 1997. The archaeology of beekeeping in pre-Roman Iberia. *Journal of Mediterranean Archaeology* 10.1: 33-47.

Brisbane, M. 1994. Hamwic (Saxon Southampton): the origin and development of an eighth century port and production centre, in *Archéologie des villes dans le Nord-Ouest de l'Europe (VIIe-XIIIe siècle). Actes du IVe Congrès International d'Archéologie Médiévale (Douai, 26, 27, 28 septembre 1991)*: 27-34. Caen: Société d'Archéologie Médiévale.

Cadwaller, A. et al. 2011. Supporting Urban Beekeeping Livelihood Strategies in Cape Town. Unpublished BSc thesis, Worcester Polytechnic Institute.

Carmona González, A. 1992. Una cuarta version de la capitulación de Tudmir. *Sharq al-Andalus* 9: 11-7.

Carmona Ruiz, Mª A. 2000. La apicultura sevillana a fines de la Edad Media. *Anuario de Estudios Medievales* 30.1: 387-421.

Carvajal López, J. C. 2009. Pottery production and Islam in south east Spain. A social model. *Antiquity* 83: 388-98.

Christie, N. 2006. *From Constantine to Charlemagne. An Archaeology of Italy AD 300-800*. Aldershot: Ashgate.

Clarke, G. 1942. Bees in Antiquity. *Antiquity* 16: 208-15.

Crabtree, P. 2018. *Early Medieval Britain: The Rebirth of Towns in the Post-Roman West*. Cambridge: Cambridge University Press.

Crane, E. 1999. *A World History of Beekeeping and Honey Hunting*. London: Duckworth.

Crane, E. and P. Walker. 1984. Evidence on Welsh beekeeping in the past. *Folk Life* 23.1: 24-48.

Delatouche, R. 1977. Regards sur l'agriculture aux temps carolingiens. *Journal des savants* 2.1: 73-100.

Dell'Elicine, E. 2008. Discurso, gesto y comunicación en la liturgia visigoda (589-711). *Bulletin du Centre d'études Médiévales d'Auxerre* 2 [online].

Delgado, M. 1997. Potes meleiros de Bracara Augusta. *Portugallia* 17-18: 149-65.

Dunn, G. 2014. Episcopal crisis management in Late Antique Gaul: The example of Exsuperius of Toulouse. *Antichthon* 48: 126-43.

Eckardt, H. 2000. Illuminating Roman Britain, in G. Fincham et al. (eds) *TRAC99: Proceedings of the Ninth Annual Theoretical Roman Archaeology Conference, Durham 1999*: 8-21. Oxford: Oxbow.

Erdkamp, P. 2012. Urbanism, in W. Scheidel (ed.) *The Cambridge Companion to the Roman Economy*: 241-65. Cambridge: Cambridge University Press.

Esmonde Cleary, S. 2013. *The Roman West, AD 200-500: An Archaeological Study*. Cambridge: Cambridge University Press.

Fernández Uriel, P. 2011. *Dones del cielo: Abeja y miel en el Mediterráneo antiguo*. Madrid: UNED.

Fernández Uriel, P. 2017. Productos de la Hispania romana: miel y púrpura. *Gerión* 35: 925-43.

Fife, A. E. 1939. *The Concept of the Sacredness of Bees, Honey and Wax in Christian Popular Tradition*. Stanford: Leland Stanford Junior University.

Fouracre, P. 1995. Eternal light and earthly needs: practical aspects of the development of Frankish immunities, in W. Davies, and P. Fouracre (eds) *Property and Power in the Early Middle Ages*: 53-81. Cambridge: Cambridge University Press.

Fouracre, P. 2018. 'Framing' and lighting: another angle of transmission, in R. Balzanetti, J. Barrow and P. Skinner (eds) *Italy and Early Medieval Europe: Papers for Chris Wickham*: 307-15. Oxford: Oxford University Press.

Francis, J. 2009. Ancient Greek ceramic beekeeping equipment at the University of Ottawa. *Museion* II 9: 159-70.

Francis, J. 2012. Experiments with an old ceramic beehive. *Oxford Journal of Archaeology* 31.2: 143-59.

Frier, B. 1983. Bees and lawyers. *The Classical Journal* 78.2: 105-14.

Germanidou, S. 2018. Honey culture in Byzantium: an outline of textual, iconographic and archaeological evidence, in F. Hatjina, G. Mavrfridis and R. Jones (eds), *Beekeeping in the Mediterranean From Antiquity to the Present*: 93-104. Nea Moudania: Division of Apiculture, Hellenic Agricultural Organization 'Demeter'-Greece; Chamber of Cyclades; Eva Crane Trust - UK.

Gittos, H. 2013. *Liturgy, Architecture, and the Sacred Places in Anglo-Saxon England*. Oxford: Oxford University Press.

Goodson, C. 2018. Garden cities in early medieval Italy, in R. Balzanetti, J. Barrow and P. Skinner (eds) *Italy and Early Medieval Europe: Papers for Chris Wickham*: 339-55. Oxford: Oxford University Press.

Grant, V. 1949. Arthur Dobbs (1750) and the discovery of the pollination of flowers by insects. *Bulletin of the Torrey Botanical Club* 76.3: 217-9.

Gutiérrez Lloret, S. 1994. *La cora de Tudmir de la Antigüedad al mundo islámico: Poblamiento y cultura material*. Madrid: Casa de Velázquez.

Hemer, K. et al. 2013. Evidence of early medieval trade and migration between Wales and the Mediterranean Sea region. *Journal of Archaeological Science* 40: 2352-9.

Henning, J. et al. 2019. Reccopolis revealed: first geomagnetic mapping of the early medieval Visigothic royal town. *Antiquity*, 93.369: 735-51.

Izdebski, A. 2012. Bishops in Late Antique Italy: social importance vs. political power. *Phoenix* 66.1: 158-75.

Jervis, B. 2011. A patchwork of people, pots and places: Material engagements and the construction of 'the social' in Hamwic (Anglo-Saxon Southampton), UK. *Journal of Social Archaeology* 11.3: 239-65.

Jervis, B, et al. 2016. Early Anglo-Saxon pottery in South East England: recent work and a research framework for the future. *Medieval Ceramics* 36.

Karivieri, A. 1996. *The Athenian Lamp Industry in Late Antiquity* (Papers and Monographs of the Finnish Institute in Athens 5). Helsinki: Finnish Institute in Athens.

Kenward, H. 2005. Honeybees (Apis mellifera Linnaeus) from archaeological deposits in Britain, in D. Smith, M. Brickley and W. Smith (eds) *Fertile Ground: Papers in Honour of Susan Limbrey* (Symposia of the Association for Environmental Archaeology 22): 97-107. Oxford: Oxbow.

Kimpe, K., P. Jacobs and M. Waelkens. 2001. Analysis of oil used in late Roman oil lamps with different mass spectrometric techniques revealed the presence of predominantly olive oil together with traces of animal fat. *Journal of Chromatography* A 937: 87-95.

Kritsky, G. 2015. *The Tears of Re: Beekeeping in Ancient Egypt*. Oxford: Oxford University Press.

Kritsky, G. 2017. Beekeeping from Antiquity through the Middle Ages. *Annual Review of Entomology* 62: 249–64.

Lewis, S. and L. Llewelyn-Jones. 2018. *The Culture of Animals in Antiquity: A Sourcebook with Commentaries*. London: Routledge.

Lorenz, S. and K. Stark. 2015. Saving the honeybees in Berlin? A case study of the urban beekeeping boom. *Environmental Sociology* 1.2: 116-26.

Loseby, S. 2012. Post-Roman economies, in W. Scheidel (ed.) *The Cambridge Companion to the Roman Economy*: 334-60. Cambridge: Cambridge University Press.

Loseby, S. 2020. The role of the city in Merovingian Francia, in B. Effros and I. Moreira (eds) *The Oxford Handbook of the Merovingian World*: 583-611. Oxford: Oxford University Press.

Macphail, R. I. 2010. Dark earths and insights into changing land use or urban areas, in D. Sami and G. Sped (eds) *Debating Urbanism Within and Beyond the Walls AD 300-700* (Leicester Archaeology Monograph 17): 145-66. Bristol: University of Leicester.

Macphail, R. I., H. Galinié and F. Verhaeghe. 2003. A future for Dark Earth? *Antiquity* 77: 349-358.

Mané, P. 1991. Abeilles et apiculture dans l'iconographie médiévale. *Anthropozoologica* 14-15: 25-48.

Martínez Jiménez, J. 2013. Crisis or crises? The End of Roman towns in Iberia between the late Roman and the early Umayyad periods, in E. van der Wilt and J. Martínez Jiménez (eds) *Tough Times: The Archaeology of Crisis and Recovery* (BAR IS 2478): 77-90. Oxford: Archaeopress.

Martínez Jiménez, J., I. Sastre de Diego and C. Tejerizo García. 2018. *The Iberian Peninsula AD 300-850. An Archaeological Perspective* (Late Antique and Early Medieval Iberia 6). Amsterdam: Amsterdam University Press.

Martínez Jiménez, J. and C. Tejerizo García. 2015. Central places in the post-Roman Mediterranean: regional models for the Iberian Peninsula. *Journal of Mediterranean Archaeology* 28.1: 81-103.

Mavrofridis, G. 2018. Urban beekeeping in Antiquity. *Ethnoentomology* 2: 52-61.

McCormick, M. 2001. *The Origins of the European Economy: Communications and Commerce AD 300-900*. Cambridge: Cambridge University Press.

Moffett, C. 2017. Slate discs at Tintagel castle: evidence for post-Roman mead production? *The Antiquaries Journal* 97: 119-43.

Moore, L. J., and M. Kosut. 2013. *Buzz: Urban Beekeeping and the Power of the Bee*. New York: New York University Press.

Morais, R. 2006. Potes meleiros e colmeias em cerâmica: uma tradição milenar. *Saguntum* 38: 149-61.

Morín de Pablos, J. and R. de Almeida. 2014. La apicultura en la Hispania romana: producción, consumo y circulación, in M. Bustamante Álvarez and D. Bernal Casasola (eds) *Artífices idóneos: artesanos, talleres y manufacturas en Hispania*: 279-305. (Anejos a Archivo Español de Arqueología 71). Madrid: CSIC.

Nicosia, C., Y. Devos and Q. Borderie. 2013. The contribution of geosciences to the study of European Dark Earths: a review. *Post-Classical Archaeologies* 3: 145-170.

Oliveira, C. et al. 2017. Chromatographic analysis of honey ceramic artifacts. *Archaeological and Anthropological Sciences* 10.

Osborn, M. 2006. Anglo-Saxon tame bees: some evidence for beekeeping from riddles and charms. *Neuphilologische Mitteilunden* 107.3: 271-83.

Pestell, Y. 2011. Markets, emporia, wics, and 'productive' sites: Pre-Viking trade centres in Anglo-Saxon England, in D. Hinton, S. Crawford, and H. Hamerow (eds) *The Oxford Handbook of Anglo-Saxon Archaeology*: 557-76. Oxford: Oxford University Press.

Pollington, S. 2011. The mead-hall community. *Journal of Medieval History* 37.1: 19-33.

Quixal Santos, D. and P. Jardón Giner, P. 2016. El registro material del colmenar ibérico de la Fonteta Ràquia (Riba-roja, València). *Lucentum* 35: 43-63.

Radner, J. (ed.) 1978. *The Fragmentary Annals of Ireland*. Dublin: Dublin Institute for Advanced Studies.

Rio, A. 2009. *Legal Practice and the Written Word in the Early Middle Ages: Frankish Formulae c. 500-1000*. Cambridge: Cambridge University Press.

Romano, J. 2016 [2014]. *Liturgy and Society in Early Medieval Rome*. Abingdon, Routledge.

Smith, D. 2012. *Insects in the City: An Archaeological Perspective on London's Past*. BAR BS 561. Oxford, Archaeopress.

Speed, G. 2014. *Urban Transformations from Late Roman Britain to Anglo-Saxon England*. Oxford: Archaeopress.

Wallace-Hare, D. 2019. Civitas mellifica? New research on Roman (and pre-Roman) beekeeping among the Igaeditani, in I. Sánchez and J. Morín (eds) *De Civitas Igaeditanorum a Laydaniyya: Paisajes Urbanos de Indanha-a-Velha (Portugal) en épocas tardoantigua y medieval* (BAR IS 2943): 55-71. Oxford: Archaeopress.

Ward-Perkins, B. 2005. *The Fall of Rome and the End of Civilization*. Oxford: Oxford University Press.

Werner, J. 1964. Frankish royal tombs in the cathedrals of Cologne and Saint-Denis. *Antiquity* 38: 201-16.

Wickham, C. 2005. *Framing the Early Middle Ages: Europe and the Mediterranean, 400-800*. Oxford: Oxford University Press.

Yule, B. 1990. The dark earth and late Roman London. *Antiquity* 64: 620-8.

Chapter 11

The Production and Trade of Wax in North-Eastern Iberia, XIV-XVI c.: The Case of Catalonia

Lluís Sales i Favà

King's College London (lluis.sales_fava@kcl.ac.uk)

Alexandra Sapoznik

King's College London (alexandra.sapoznik@kcl.ac.uk)

Summary: Combined analysis of the accounts of the Lleuda of Mediona, a tax on trade coming into Barcelona, and of the Lluminària of the Cathedral of Barcelona offers new insight into the trade in beeswax in late medieval Catalonia. As one of the great cities of the Mediterranean, Barcelona had constant access to the ubiquitous North African wax shipped via Mallorca and Valencia. Yet despite this, it is clear that locally produced wax remained an important and desirable commodity throughout the period. Both the higher price of this *cera de la terra* compared with Maghrebi wax and the effort the Cathedral put into sourcing it, suggest that domestic wax was a high-quality product. This was the result of widespread and productive apiculture in the region, consideration of which also sheds light on the wider context of competing demands within the medieval environment.

Keywords: Apiculture; bees; wax; Barcelona; trade

Beeswax had many uses in medieval Europe: as a source of luxury lighting, for sealing documents, waterproofing cargo for transport, casting bronze. Above all, beeswax was necessary for Christian religious ritual.[1] Beeswax candles, believed to reflect the virginity and chastity of Christ and Mary, were necessary for the proper conduct of the Mass, and wax candles burned on altars and tombs, before images, shrines, baptismal fonts and crosses. The stages of a Christian's life were marked by wax candles, their memory commemorated with wax candles, their sins atoned for through gifts of wax candles. The Purification of the Virgin, was marked by blessing the wax for the year, hence its popular vernacular name, Candlemas. Easter, the high point of the Christian liturgical year, was celebrated through the symbolic recreation of Christ's life through the use of wax candles of which the large Paschal candle, often weighing more than a hundred kilograms, was the most spectacular. Constant demand for wax drove an extensive trade encompassing the whole of the Mediterranean. This wax was the product of many millions of bees foraging among billions of flowering plants. As such, its supply was also susceptible to environmental change and fluctuations in climate and patterns of vegetation.

This paper will analyse the tax record of the Lleuda of Mediona from 1434-1572, and the accounts of the Lluminària of the Cathedral of Barcelona from 1374-1589. Together, these series offer a new insight into trade in beeswax in late medieval Barcelona. These documents also provide evidence of extensive beekeeping across Catalonia, Rousillon and Cerdanya, a feature of the rural economy of this region which deserves further study. The ecological factors which facilitated widespread and productive apiculture will be considered, placing beekeeping within the wider context of competing demands within the medieval environment.

I.

This section will analyse the trade in, and consumption of, beeswax in late medieval Barcelona. This will be done by assessing two unpublished series of accounts: the Lleuda of Mediona, a tax on local trade, and the accounts of the *Lluminària del Cos Preciós de Jesucrist*, an office in the Cathedral of Barcelona. In total, 1,371 entries related to the trade in wax have been collected from 31 different books of the Lleuda of Mediona covering, with interruptions, the period 1434 to 1572. Information from this series will be discussed in connection with a further 3,409 entries regarding incomes and expenses from 30 books of the Lluminària covering the years 1374 to 1589, again with interruptions. By assessing

---

[1] The research for this paper was funded by the Leverhulme Trust and is part of the project 'Bees in the Medieval World: Economic, Environmental and Cultural Perspectives' (RPG-2018-080). For more on this and what follows, see Sapoznik 2019.

the quantity of wax traded into the city, its origins and quality, particularly that of domestically produced Catalan wax, this study offers insight into the role of beeswax in the economic and religious life of the city.

Neither the accounts of the Lleuda of Mediona nor the Lluminària have received historical attention, despite being rich sources for the history of the city. As such, they merit description and explanation. It is well known that the Cathedral of Barcelona was a powerful institution with a wide territorial domain, both within the city and outside it, where it held key jurisdictional lordships. By the 1580s, the end of the period under study here, the chapter of the cathedral was administered by forty canons and about a hundred beneficed clergy who held offices with specific revenues (Fàbrega i Grau 1978: 26-27). The administrator of the Lluminària was one of these canons. Far from being a unified institution, the Cathedral of Barcelona, like any pre-modern cathedral, was a compound of semi-independent entities (Conesa Soriano 2015: 40). These *obres*, *pabordies*, sacristies or chapels were managed separately and, in different contexts, could compete for revenues or make economic transactions with each other (Conesa Soriano 2015: 44). This seems to have been the case for the relations established between the main Sacristy (*Sagristia Major*) and the Lluminària, who in given periods would, for example, exchange wax and candles.[2] As yet there has been no research into the Lluminària as an institution or its role within the Cathedral (on the cathedral of Barcelona, see Torres i Ferrer 2001; Conesa Soriano 2017; Fatjó Goméz 1999). It was, however, likely to have been the greatest wax consumer within the Cathedral.

The Lluminària was chiefly in charge of managing the celebrations related to Corpus Christi and to taking care of the *Custòdia* or Ostensorium, the golden structure in which the Holy Body was exposed, relic-like, to worshippers.[3] It also devoted part of its efforts and funds to the *enramada*, the spectacular floral decoration of the whole Cathedral during the feast of Corpus Christi. The Lluminària was in charge of acquiring all of the materials needed for these tasks, and would split the cost with the *bací de la Custòdia*, an institution

closely related to the Lluminària.[4] The account books of the Lluminària reveal regular purchases of wax and candles, mostly for the liturgical ceremonies related to the Holy Body, and to a lesser extent celebrations such as Candlemas.[5] Another interesting source of income for the institution was the leasing of the wooden crosses placed on top of the coffins during the funerals, as well as numerous revenues derived from pious donations (*bací*), the burials themselves (*vasser*), estate rents (*censos, morabatins*) and the installments of long-term or even semi-perpetual credits (*censals*).[6]

The Administrator or *Protector* of the Lluminària—considered a minor office within the structure of the chapter—supervised the salaries of the scribe of the cathedral and numerous aides (porters, candle-lighters, messengers) as well as the purchases made, including those of already made candles from the external chandlers (*candelers*) over the fourteenth and fifteenth centuries, and then of mostly raw wax during the sixteenth century.[7] These chandlers were granted a monopoly over provisioning the Lluminària with wax, in a complex arrangement by which wax that had burned in the cathedral but could still be reused was resold to the same chandler at a cheaper price.

Lleuda Reial i de Mediona was a compound of taxes on the trade (*lleudes*) and circulation (*passatge*) of all commodities in Barcelona. In 1222 James I and the noble Guillem de Mediona agreed on consolidating the group of taxes which the latter had already been collecting in the market of the city (the *lleuda de Mediona*), and those which would remain in the hands of the monarch (the *lleuda reial*) (see Gual Camarena 1976: 56-65 (doc.1)).[8] The fraction known as the Lleuda of Mediona in its biggest part had been transferred to the Pia Almoina of the Cathedral by the fourteenth century. This institution was in charge of the levying

---

[2] In 1582 the Lluminària bought candles, in bulk, from the Sacristy, worth 2 s. See Arxiu de la Catedral de Barcelona (ACB), Administració de la Lluminària, 1581-1582, s.f. This is repeated over the following years. Institutional history of the Cathedral of Barcelona is meager: a greater analysis of the abundant sources in its archive is needed.

[3] An example of one of the many duties in relation to this, in August 1374 the Lluminària paid for the arrangement of the silk cover that protected the Ostensorium: *adobar la cuberta de seda de la custòdia (...) ço és entre tatxes daurades e claus, e de sos tribaylls*. See ACB, Altres llibres de la Lluminària, Rebuda de l'Acapte, 1374-1413, s.f. On the Custòdia, see Mas 1916 and also Dalmases 1992: 118. For a recent perspective of these structures, focused on northern Europe, see Timmermann 2009. For the introduction of the eucharistic feast in the late middle ages, see Rubin 1991.

[4] See a couple of examples, early and late, of these expenses in ACB, Administració de la Lluminària, 1427-1428, f.10r (for 1427) and in ACB, Administració de la Lluminària, 1584-1586, f.25v (for 1586).

[5] As an example, on June 19 of 1377 the accounts state a payment of 58 s. to *Francesc Cortós, candeler, per les segones candeles qui serveixen a les segones vespres de Corpus Cristi el dicmenge en les vytaves; la profesó, porten lo bisbe, canonges, beneficiats, clergues, scolans, lechs, tothom qui sia el cor*. See ACB, Altres llibres de la Lluminària, Rebuda de l'Acapte, 1374-1413, s.f. (19-VI-1377). For Candlemas, see ACB, Administració de la Lluminària, 1407-1411, f. 16r (26-I-1411). The Lluminària would also purchase wax on a regular basis for the consumption of the intervening clerics at the *cor* (the central choir) and at the main altar. For the former, see ACB, Administració de la Lluminària, 1433-1435, f. 24v (31-VIII-1434) and the for the latter, ACB, Administració de la Lluminària, 1442-1447, f. 14r (19-IV-1443).

[6] These crosses often had to be repaired and repainted. See for instance ACB, Administració de la Lluminària, 1531-1539, f. 14v (1532).

[7] ACB, Liber vulgariter de la Crehueta Nuncupatus (1455-1495). The commercial relationship between the Lluminària and these chandlers, to whom the institution granted a monopoly, will be the subject of further study. For this, see ACB, Administració de la Lluminària, 1442-1447, f. 56r (1445).

[8] Pere Orti Gost has studied the origin and structure of the tax in great detail (Orti Gost 2000: 397-433).

the tax. The whole compound of taxes continued to be collected in Barcelona until the liberal revolution in the nineteenth century.

The Lleuda of Mediona was a complicated tax, and a specific tariff was applied depending on commodity. Wax was taxed by weight at 2s. 8d. per càrrega (about 124 kg). For wax weighing less than half a càrrega, the tariff was retained by Mediona; if more, the latter collected one third of the tax and the remaining two thirds went to the king.[9] The way in which the tariff was applied to wax demonstrates that both the buyer and the seller were obliged to pay. The existing registers, however, list only the name of the person being taxed, the commodities on which the tax was levied, and the amount owed. They do not specify whether the person paying the tax was the buyer or the seller. It is possible, then, that some of the commodities listed have been double counted, if both the buyer and seller were listed as separate entries in the register. A further frustration arises from the number of exemptions of the tax granted to various groups—most importantly, the general exemption granted to the citizens of Barcelona in 1232. This means that the accounts do not record purchases or sales made by locals (De Capmany y de Monpalau 1962: 15-16 (doc. 7)).[10] Nonetheless, even with these limitations, it is clear that the Lleuda of Mediona is an invaluable source for studying the commerce of the city.

With these caveats in mind, taking a closer look at wax, a single commodity among the wide variety of items traded through Barcelona, offers the opportunity to consider trade into the city with some precision. It seems likely that, because wax was a product of mostly rural origin, the great majority of it was imported raw, and the entries recorded were largely sales made by foreigners to locals. Therefore, there would have been few double entries in the source for this commodity.[11]

Wax was traded along with at least one other product in 30% of the entries analyzed (417 out of 1,371). Small amounts of cotton appear in over a quarter of these, from which it may be suggested that the chandlers or institutions that needed candles could be buying the material for the wicks along with the wax from one single provider.

Bearing in mind the discussion above, Table 1 demonstrates that the trade in wax fluctuated dramatically over our period, although with a general upward trend in the first decades of the sixteenth century.[12] The oscillations likely correspond to either climatic or political circumstances which impacted the supply of wax into the city. For example, political factors certainly explain the relatively low figures for 1463 and 1465, corresponding with the Catalan Civil War (1462-1472), in which some parts of Catalonia had fallen in the hands of the royalists and thus the trade with the city – controlled by the Consell de Cent and the Diputació – was blocked.[13] It is well known that during this troubled period masses and religious ceremonies were reduced to the minimum level of observance, and this would have had a significant impact on demand for wax and its consumption (Torres i Ferrer 1997: 126).

The Barcelona wax trade was also characterized by a marked seasonality. Over 50% of the transactions took place during the first three months of the year, with January alone accounting for 40.1% of the registered entries in the Lleuda.[14] Yet intriguingly, the Lluminària seems to have spent most of the money on wax and candles a bit later in the year: March to May accounted for 37.1% of wax purchases by the institution, coinciding with the preparation for the feasts of Easter and Corpus Christi.[15]

Over the period analyzed, the Lluminària purchased on average 700 libras of wax (or about 280 kg). The accounts indicate a steep decline in consumption over the later fifteenth century, from an average of 464 kg per year between 1427-1447, to only 210 kg per year from 1479-1491, the next period for which there is surviving data.[16] Wax purchases fell even further, averaging only 121 kg per year from 1532-1553, and recovery is only

---

[9] In the case of the passatge of wax, the tariff was of 10 d. per each càrrega. See Orti Gost 2000: 411.

[10] For a broader reflection on the nature of the registers, see the introduction to the edition of the first available book of the Lleuda by Salicrú 1995: 20.

[11] For example, in 1487, out of 69 entries concerning wax, in 60 cases this was the raw product; either with no other specification (57 cases), either white (1) or either white and yellow (3). Cera obrada (likely, rendered candles) appears in 7 entries, along with 1 entry of candeles de cera. One entry notes both wax obrada and not obrada.

[12] The trade of wax alone represented up to 1.8% of the total tax revenues of the Lleuda de Mediona in 1526 (8,639 d. of 476.976 d.), whilst in preceding and succeeding years that figure seems to be around 0.6% (in 1488, wax levies were 1,641 d. of a total 281,672 d. -0.58%; in 1555, 3,306.5 d. of 517,969 d. -0.64%; in 1572, 2,167 d. of 341,522 d. -0.63%).

[13] See the case, for instance, of the shortage of meat analyzed by Banegas López 2017: 83-87 and 238-239 or of wood claimed by the local council in Arxiu Històric de la Ciutat de Barcelona (AHCB), 1B. II-17 (1465-1467), f. 192r-192v (5-III-1467). We thank Laura Miquel Milian for this reference.

[14] Considering only the years for which we have the 12 months available (1487, 1488, 1491, 1492, 1555, 1572), the sequence is the following: Jan., 40.1%; Feb, 2.3%; March, 10.8%; April, 4.3%; May, 9.6%; June, 5.9%; July, 4.8%; Aug., 4.7%; Sep., 3.6%; Oct., 5.5%; Nov., 3%; Dec., 5.3%.

[15] Considering only the years for which we have the 12 months available (1377, 1378, 1408-1410, 1428, 1432-1438, 1441-1446, 1480-1482, 1488-1490, 1523-1525, 1532-1560, 1562-1575, 1578-1585, 1588), the sequence is the following: Jan., 2.6%; Feb, 4.8%; March, 11.7%; April, 13.6%; May, 11.9%; June, 8.4%; July, 4.3%; Aug., 7.6%; Sep., 8.2%; Oct., 9.4%; Nov., 4.6%; Dec., 2.8%; Unsp., 10.1%.

[16] Without data for 1430. Peak points in this period are 1429 (739.6 kg.) and 1445 (759.8 kg.). Data is also missing for 1484-1486.

Table 1. Entries of Wax in the Market of Barcelona
(Accounts of the Lleuda de Mediona)

Year (available months)	Weight (kg.) (entries n.)	Amended weight (kg.) (entries n.)[a]
1434 (1)	3,897.2 (8)	-
1463 (3)	166 (7)	1,372 (58)
1465 (3)	417.5 (20)	1,690 (81)
1486 (7)	6,128.2 (48)	18,684 (146)
1487 (12)	9,454.9 (69)	9,454.9 (69)
1488 (12)	13,285.6 (97)	13,285.6 (97)
1489 (5)	7,649.2 (70)	11,383 (104)
1490 (7)	6,178 (23)	18,835 (70)
1491 (12)	4,715.6 (78)	4,715.6 (78)
1492 (12)	11,719.4 (127)	11,719.4 (127)
1493 (5)	7,671.2 (64)	11,418 (95)
1510 (7)	8,741.6 (78)	26,641 (238)
1511 (5)	2,168.2 (102)	3,227 (152)
1526 (10)	27,526.4 (299)	30,031 (326)
1538 (3)	3,445.4 (47)	24,967 (341)
1539 (8)	6,198.6 (136)	7,504 (164)
1555 (12)	3,982.2 (49)	3,982.2 (49)
1572 (12)	2,593.6 (49)	2,593.6 (49)
Total	125,938.8	471,503

a. Taking into account the months with data for each year, and the period of the year for which data is available. *Sources*: ACB, Llibre d'Administració de la Lleuda de Mediona, 1434; 1463; 1465; 1486-1487; 1487-1488; 1488-1489; 1490-1491; 1491-1492; 1492-1493; 1510-1511; 1526 (Jan.-Feb.); 1526 (March-May); 1526 (June-July); 1526 (Aug.-Oct.); 1538-1539 (Oct.-Feb.); 1539 (March-Aug.); 1555 ('21', Jan.-Feb.); 1555 ('36', March-April); 1555 ('19' May); 1555 ('30' June); 1555 ('45' July); 1555 ('20' Aug.); 1555 ('35' Sept.); 1555 ('33' Oct.); 1555 ('23' Nov.); 1555 ('31' Dec.); 1571-1573.

noted from the mid-1550s, when on average 280 kg of wax were purchased from 1554-1589.[17]

Taking into account only the full years for which evidence survives from both sources, 1487-1491, 1538, 1539, 1572, wax purchased by the Lluminària represents between 0.4% to 7.6% of the total wax sold by foreigners in Barcelona.[18] This reinforces the suggestion that

the Lluminària was an important actor in the wax market in the city, and its records may therefore help in determining the origins of the wax consumed in Barcelona. Again, the combination of both sources, will provide a much clearer image of this trade.

II.

Both the Lleuda de Mediona and the accounts of the Lluminària provide tantalizing evidence as to the origins of the beeswax being bought and sold. The accounts of the Lleuda de Mediona are not a conclusive source for assessing the origins of the wax traded in Barcelona, since not all of the entries include the origins of the wax. Of the total 312,787 kg. of wax on which the tax was levied during the years under scrutiny, 205,739 kg. (65.8%) were traded by someone whose origin is known.[19] The origin of the person taxed should, in general, be a reliable indicator of where the wax was heading and specially, as mentioned above, from where it came from (see Figure 1).[20]

Wax remittance from Mallorca and Valencia was largely of Maghrebi origin.[21] Ongoing studies will describe this key trade.[22] But for the moment we will focus here on Catalonia and the so-called *Comtats* (Rousillon and Cerdanya), the individuals from which handled at least 21.8% of the total wax that was traded during the years analyzed (27,231 kg. out of a total 125,114.9 kg.). Table 2 zooms into detail by showing the traders of wax in Barcelona that were declared inhabitants of these regions (the 21.8% of the given total). From this, several zones stand out as particularly important.

The first important region, measured in terms of weight, is the Roussillon.[23] The trade between the region and the Principality of Catalonia had been active before the last decades of the fifteenth century, a period

---

[17] Data is lacking for 1561 and 1587. In this latter period some years present figures above the general mean of 280 kg.: 1554, 1555, 1562, 1565, 1574, 1576, 1577 and 1586.

[18] In 1487 the Lleuda levied 9,459.9 kg. of wax and the Lluminària acquired 39.98 kg (0.42%); 1488, Lleuda 13,285.6 kg. – Lluminària 187.33 kg. (1.41%); 1489, Lleuda 18,358.1 kg. – Lluminària 181 kg. (0.98%); 1490, Lleuda 10,590.9 kg.- Lluminària 210.2 kg. (1.98%); 1491, Lleuda 4,715.6 kg – Lluminària 348.37 kg. (7.38%); 1538, Lleuda 13,781 kg. – Lluminària 64.4 kg. (0.46%); 1539, Lleuda 9,297.7 kg – Lluminària 217.03 kg. (2.3%); 1572, Lleuda 2,593.6 kg – Lluminària 197.33 kg. (7.6%).

[19] On the contrary, if we take into account the entries but not the total weights, origins of traders are only stated in one third of the cases (405 out of 1,371).

[20] In relation to specific exemptions, Salicrú (1995: 8-13) discussed why some origins were mentioned, but not others.

[21] As an example, the 2 quintars, 1 arrova and 18 lliures (≈100,8 kg.) of *cera berberesca* sold by Tomàs Gil, merchant of Valencia, to the Lluminària in 1555 [ACB, Administració de la Lluminària, 1551-1567, f. 35r (6-IX-1555)]. See also the evidence provided by Macaire 1986: 328 and 399, and also by Benassar Vaquer 2001: 72-82 and 376-378, for Mallorca; and by Guiral 1974: 99-131, and also by Salvador Esteban 1990: 19-49, for Valencia. The traders identified as moors (only between 1487 and 1492) in Figure 1 were likely also commercializing this barbaresque wax. According to Salicrú (1995: 10), those labeled as foreigners (*forahabitants*) could be individuals that had no special fiscal consideration but were non-Genoese Italians, Germans, Castilians or Valencians.

[22] This is currently under study as part of the project 'Bees in the Medieval World': see fn.1.

[23] The records likely describe all individuals from Rousillon and Cerdanya as such in order to keep track of their special fiscal status. These merchants had to pay the totality of their taxes for the wax trade to Mediona, and not to the King, even if the weight exceeded 0.5 càrrega. See Orti Gost 2000: 411.

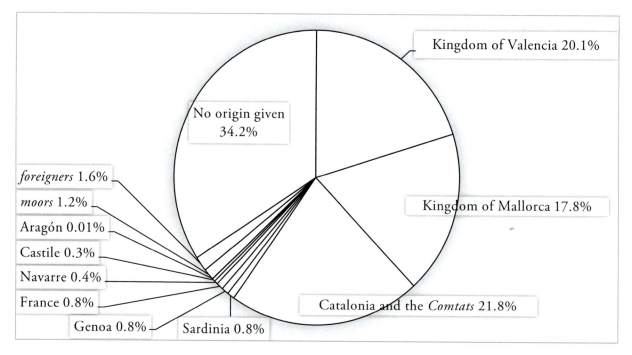

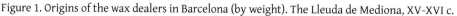

Figure 1. Origins of the wax dealers in Barcelona (by weight). The Lleuda de Mediona, XV-XVI c.

Sources: ACB, Llibre d'Administració de la Lleuda de Mediona, 1434; 1463; 1465; 1486-1487; 1487-1488; 1488-1489; 1490-1491; 1491-1492; 1492-1493; 1510-1511; 1526 (Jan-Feb.); 1526 (March-May); 1526 (June-July); 1526 (Aug.-Oct.); 1538-1539 (Oct.-Feb.); 1539 (March-Aug.); 1555 ('21', Jan.-Feb.); 1555 ('36', March-April); 1555 ('19' May); 1555 ('30' June); 1555 ('45' July); 1555 ('20' Aug.); 1555 ('35' Sept.); 1555 ('33' Oct.); 1555 ('23' Nov.); 1555 ('31' Dec.); 1571-1573.

when the Crown of Aragon and the Kingdom of France escalated tensions, and would actually resume in the XVI century. [24] The Roussillon was a hinge through which the southern regions of the Crown of Aragon traded foodstuffs and agricultural produce towards the Kingdom of France (Serra and Sanllehy 2005: 486-489). But turmoil between the two states and the cession first (1462) and conquest afterward (1463) of the *Comtats* by Louis XI had a visible impact on trade, including wax. This situation was not reversed until the Treaty of Barcelona of 1493, by which the *Comtats* were returned to Aragon with the lavish entry of Isabel and Ferdinand II into the city of Perpignan.

The Roussillon is not known to have been a wax-producing region during the late middle ages.[25] In fact, most fiscal and notarial sources at our disposal show that, on the contrary, wax would have moved from the south (Catalonia) to the Roussillon, either as its final

stage, or as an entrepot to reach further north (See Riera Melis 1986: 113; Riera Melis 2003: 80-81; Ferrer i Mallol 2008: 120; Riera Melis 2017).[26] We may hypothesize that the 4,717 kg. traded by merchants from Perpignan would have been mostly from purchases made by them in Barcelona to be taken to their homeland. This impression is supported by the fact that this appears to be a strongly specialized trade, dealt with chiefly by three individuals (Pere Gomar and Gabriel Gomar, from Perpignan, between 1526 and 1539 and Joan Oliveres, also from Perpignan, in 1555) who, along with wax, were in charge of transporting commodities that traditionally flowed from south to north: sugar, cumin, saffron, pepper, cinnamon, licorice, 'Greek glue' and cotton.

This could also be the case of the Vallès, from which we detect three main traders: Francí Valls (1486-1493), Gilabert Valls (1526), likely a successor of the latter, and Bernardí Pujalt (1489), all from the town of Granollers.

---

[24] As has been assessed by several historians, for example Catafau 2012: 27 and also Puig 2005: 23-40.

[25] Although it did have certain reputation for the honey, which was even shipped to the Eastern Mediterranean. See for example Archives Départementales des Pyrénées-Orientales (ADPO), B.139, f. 93r (6-VII-1381) and also ADPO, B.250, Notularum (1352-1438), s.f. (3-IV-1397). In 1617, Miquel Agustí (1616: 188v) would highlight the quality of the *white honey* produced in Ópol and Salses, in the Roussillon. Still in 1857 Antoine Siau organized a survey of the bee harvesting in the Department of the Pyrénées-Orientales which ascertained that, out of 229 municipalities, in 209 -including many highland towns of the Pyrenees- there was some extension of beehives. See Lemeunier 2004: l: 387-404.

[26] The Lleuda of Cotlliure charged 2 s. for each càrrega of wax that touched the port, generally transiting from south to north. See Sobrequés 1969-1970: 65-84; Riera Melis 1980: 91-125 and also Gual Camarena 1976: 102-107 (doc.9). The accounts of another less-well known toll, the Lleuda of Cadaqués (South of Cotlliure and collected by the Counts of Empúries) show that between July and November of 1391, of 53 shipments of wax which touched this port, only 3 carried wax from the north to Barcelona. On the contrary, most shipments (49) show the transit of the product from Barcelona, the Balearic Islands and Valencia toward Cotlliure, Canet, Aigües-Mortes or Avignon. See Archivo Ducal de Medinaceli (ADM), Arxiu Comtal d'Empúries, ref. 4740, rt. 279.

Table 2. The Origins of the Catalan Wax Traders in Barcelona, in Weight Supplied (kg.). Accounts of the Lleuda De Mediona

Year	Catalan South		Roussillon	Vegueria of Girona	Vallès	Cerdanya	Osona	Region of Lleida[c]	Bages	Anoia-Penedès	Baix Llobregat	Barcelonès
	Camp de Tarragona[a]	Catalan Ebro[b]										
1434												
1463				43.2								
1465				8.2								
1486				249.6	119.6							
1487				135.2								
1488	278.4	1,216.8		355.2	722.8					228.8		
1489	41.6	135.2		270.8	748.8				624.0			
1490				64	111.2							
1491	83.2			145.6	639.6		41.6	4.8	166.4			
1492	603.2	1,298.8	194.8	275.6	962.8	1,861.6		1,412.4	374.4			
1493	910.6	260		64	436.8	124.8		135.2		197.6		
1510				873.6			1,206.4				10.4	
1511	124.8		4.8					10.4		74.4		
1526	270.4		1,147.1	1,243.4	296	41.6	451.2		9.6		140	63
1538			1,912.4			104						
1539	1,185.6		896	621.2								
1555			468	83.2				116.4	64.8	83.2		
1572	36.4		93.6						20			72.8
Total	3,534.2	2,910.8	4,716.7	4,368.8	4,037.6	2,132	1,699.2	1,679.2	1,259.2	584	150.4	135.8
%	23.7		17.3	16.1	14.8	7.8	6.2	6.2	4.6	2.1	0.6	0.5

a. Includes the modern comarques of El Tarragonès, Alt Camp and Baix Camp; b. Includes the modern comarques of Baix Ebre, Montsià, Terra Alta and those merchants identified as inhabitants of the Templar territories; c. Includes the modern comarques of El Segrià, La Noguera and La Segarra. Sources: ACB, Llibre d'Administració de la Lleuda de Mediona, 1434; 1463; 1465; 1486-1487; 1487-1488; 1488-1489; 1490-1491; 1491-1492; 1492-1493; 1510-1511; 1526 (Jan-Feb.); 1526 (March-May); 1526 (June-July); 1526 (Aug.-Oct.); 1538-1539 (Oct.-Feb.); 1539 (March-Aug.); 1555 ('21', Jan.-Feb.); 1555 ('36', March-April); 1555 ('19' May); 1555 ('30' June); 1555 ('45' July); 1555 ('20' Aug.); 1555 ('35' Sept.); 1555 ('33' Oct.); 1555 ('23' Nov.); 1555 ('31' Dec.); 1571-1573.

Table 3. Identified Wax Purchased by the *Lluminària* of the Cathedral of Barcelona (in Kg.)

years	Anoia-Penedès	domestic[a]	Vegueria of Girona	Catalan Ebro[b]	Maresme	Camp de Tarragona[c]	Bages	Maghreb	Flanders
1480-1500	-	-	-	-	-	19.2	4.2	-	-
1501-1520	-	-	-	-	-	-	-	-	-
1521-1540	55	60.6	-	-	-	-	-	204.4	-
1541-1559	232.7	76.1	-	-	-	-	-	610.1	172.4
1560-1589	526.8	141.5	232.1	140.2	116	80.4	43.9	847.2	-
Total	814.5	278.2	232.1	140.2	116	99.6	48.1	1,661.7	172.4

a. Wax acquired from a peasant (*pagès*) or identified as *cera de la terra*; b. It only includes El Priorat. c. It only includes El Baix Camp. *Sources*: ACB, Administració de la Lluminària, 1479-1481; 1481-1483; 1487-1491; 1523-1524; 1523-1525; 1531-1539; 1539-1545; 1547-1551; 1551-1567; 1565-1571; 1571-1575; 1575-1576; 1577-1579; 1578; 1579-1581; 1581; 1582-1584; 1584-1586; 1588-1589.

These merchants could have been using Barcelona as an entrepot for the wax they bought abroad, since in most of their entries they paid the right of circulation (*passatge*).[27]

The rest of the origins stated in the accounts are plausible provenances for domestic wax, especially in light of the allusions made in the accounts of the Lluminària. Table 3 confirms purchases from zones such as the city of Girona and its territory (*vegueria*), the Catalan Ebro region, the Camp de Tarragona and specially the Anoia-Penedès lands.

Little is known yet about the production of wax in the area of Girona during pre-modern times. Nonetheless, there is evidence of individuals or institutions in possession of large numbers of beehives, which could have been for market-oriented production. This appears to have been the case of the monastery of Santa Maria de Roca Rossa, in Maçanet de la Selva, which kept at least 25 beehives during the fourteenth century, or of some farmsteads in early seventeenth-century Empordà, that hosted as many as 95.[28]

From a contemporary perspective, the appearance of the Catalan Ebro and the area of Tarragona as regions of wax provision is not surprising. Along with the north of the Kingdom of Valencia and Teruel, the area presented an ideal ecological habitat for bee harvesting and even today is reputed for its honey.[29] The profusion of references to beehives in the local ordinances of the villages of La Conca de Barberà during the later middle ages serve as evidence (See Martínez i Garcia 2010: 16-17; see also

assessments of wax, honey and beehives in the personal income taxes in places such as Valls (1394), Alcover (1393) and Reus (1445) in Morelló Baget 1993: 370 and in Morelló Baget 1997: 936.). The river Ebro itself served as the main conductor through which apicultural products were shipped toward the consumption centres.[30]

Other origins might seem unexpected from a contemporary perspective. The Anoia-Penedès appears to have been the most solicited zone for the provision of wax by the Lluminària, with inland towns as Piera and Vilafranca acting as nodes of distribution.[31] A hypothesis still to be verified is that the agrarian expansion in the late sixteenth century could have expelled bee harvesting to mostly marginal lands.[32]

The elevated county of Cerdanya, in the Comtats and above 1,000 mamsl., could have also provided the city of Barcelona with wax obtained under particular conditions, for example the harsh winters that imposed transhumance between the valleys and the mountains, and only one annual harvest.[33]

The accounts of the Lluminària and the Lleuda give evidence of a wide availability of wax in Catalonia, with the produce flowing into the city of Barcelona mostly from the deforested southern Catalonia-wide region

---

[27] About Granollers and also the aforenamed Valls as a key town and lineage of the late sixteenth century candle production in the region, see García Espuche 1998: 269-271.

[28] For the first case, see ADM, Arxiu Vescomtal de Cabrera (Arxiu Històric d'Hostalric), doc. 3730 (28-II-1338) (rt. 0986, foto. 0516-0519) and also Arxiu Diocesà de Girona (ADG), U-31 f. 222v- 223v (7-VIII-1357); for the latter, see Gifre Ribas 2012: 374.

[29] This survey, undertaken by Pascual Madoz between 1846 and 1850 in his *Diccionario Geográfico*, assessed a wide presence of honey and wax production in the Ebro region. See for instance the entries of towns such as El Perelló, Rasquera and Tivissa in Madoz 1985: 211, 252 and 413 respectively.

[30] As it is assessed by the 1252 toll of the Ribera of the Ebro, which charged 10 s. for each càrrega of wax in the Catalan towns along the fluvial shore. See Gual Camarena 1976: 107-110 (doc.10).

[31] Beekeeping is a well-documented occupation in the region in premodern times, both in the local regulations (Sans i Travé and Guasch i Dalmau 1979: 221-246, specifically articles 23, 24, 25, 30, 55 and 56), and in the activities of the peasantry, which were authenticated by notaries and the jurisdictional courts. See Gual i Vilà and Jorba i Serra. 2013: 111-113 and also Gaya Catasús 2016: 1107-1114.

[32] Where it could have coexisted with widely documented lime kilns. Some examples of the latter for the town of Piera are to be found in Arxiu de la Corona d'Aragó (ACA), Notarials, Ig-875, s.f. (12-I-1560) and s.f (8-VII-1560). For this agrarian expansion, see, for instance, Valls Junyent 1990: 99-136.

[33] Wax harvesting is not unknown for the Pyrenees. Contemporary evidence is provided for the Pallars Sobirà (Violant i Simorra 1935: 222-228), Berga (Serra i Coma and Ferrer i Alòs 1985: 184-185) and Tremp (Madoz 1985: 447-448).

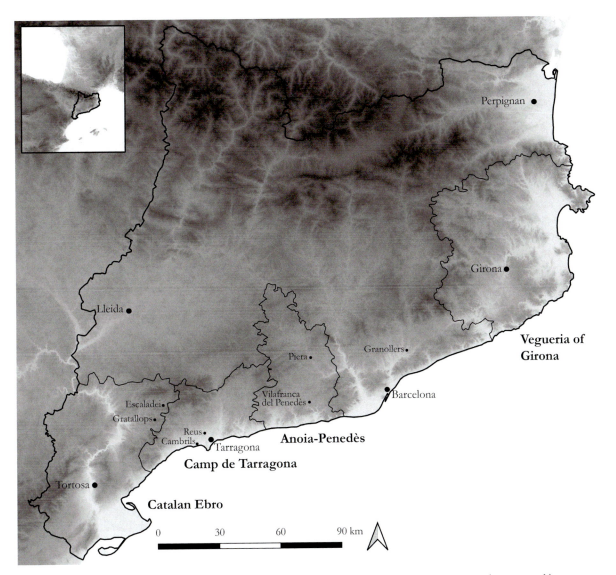

Figure 2. The wax-producing regions in Catalonia, XIV-XVI c. (Map: L. Sales i Favà and A. Sapoznik)

(the Ebro and Tarragona), but also from zones as Anoia-Penedès - preferred by the Cathedral - Girona and even the Pyrenees highlands. Individuals from these regions were involved in the wax trade. Their profiles and the means of transportation are approached in the following paragraphs.

In 1487 the accounts of the Lleuda show that out of 56 individuals that dealt with domestic wax trade, presumed to be mostly sellers, only five did more than one single operation (in all cases, 2). The pattern is similar to 1572, when out of 40 individuals, only three were involved in more than one operation (one Joan Baptista Gori sold a total of 63 kg. of wax twice in January and once in June in the city; the two others were only assessed twice). This could suggest a limited professionalization of the trade and production of wax in Catalonia during the late middle ages.[34]

The accounts of the Lluminària provide some insight into these issues. Sometime between 1491 and 1518 the institution modified its management of wax. After a long period during which the Cathedral had been relying on an external chandler, for reasons not yet known it built a workshop under the vaults of the temple (*la cambra de obrar cera*) where, from then on, wax would be rendered into candles.[35] Not only does this imply the organization of a new body of wax specialists hired by the Cathedral, but also the centralization of its provision. The canons would fetch the produce directly from the providers. This is the reason why the scribes started giving details about the origins of wax, which were usually from petty traders or producers.

---

[34] Although candles could only be sold by chandlers, *especiers* and given intermediaries, in Barcelona the trade of raw wax was not

regulated by specific professional ordinances. See Vela 2003: 1065-1084.

[35] A first reference to this chamber is in ACB, Llibre d'Alabarans 1511-1519, f. 39v (24-VII-1518). We have only one inventory of the wax, both raw and rendered into candles, kept by the Lluminària in 1563. See ACB, Administració de la Lluminària, 1551-1567, f. 80v-81r (12-VIII-1563).

According to the statutes of the Lluminària only found for 1565 onward, the canon-*baciner* of the Cathedral was in charge of the purchases, under the supervision of the Protector of the Lluminària. If the former were to leave the city in his quest for wax, all expenses made were to be covered by the Lluminària.[36]

This wax could sometimes be purchased directly from peasants from places nearby Barcelona. Although one Joan from the Feixes farmstead of Cabrera d'Anoia received as much as 56 l. 14 s. for an undetermined, but clearly very large, amount of wax in 1562, most transactions of this kind were like the 3.3 kg. sold from a peasant surnamed Terrades in 1572.[37]

We also have evidence of wax bought on-site, especially in the Anoia-Penedès area. This was generally conducted by those responsible in the Cathedral, although in certain periods a broker could assume the duty. Such is the case of Marc Bas from the town of Capellades, who purchased *cera de la terra* for the Lluminària between 1545 and 1555.[38] Another broker, Geronim Balaguer, acquired 182 kg. of yellow wax from *diverses persones* in the fair in Vilafranca del Penedès in 1562.[39] In this case, though, the Protector himself had gone personally to the fair to supervise the process and returned to Barcelona with the wax loaded on a mule.

The most eloquent reference found so far about the exploration of the domestic wax market by the Cathedral of Barcelona dates to 1574. The accounts describe how the chapter allowed the Protector of the Lluminària, Cristòfol Borrell, to take 100 l. from the main Sacristy (*traguessen del·la caxa del·la sacristia*) - he was thus not dealing with their own funds - and hand them to one canon Balsells in order to travel either to Tarragona or the Escaladei, a Cistercian monastery in the Priorat, to buy wax.[40] The product would be whitened only once it arrived in Barcelona.[41] The outcome of this operation is revealing of how the institution monitored the market. The whole amount of money was replaced in September of 1574 because no wax had been found where they had judged there would be some (*per tant que no se trobà çera hont pensàvem*).[42]

It has been mentioned that wax was transported to the city on mule-back from the regional fairs.[43] The accounts of the Lluminària also show that for longer distances domestic wax could be carried by sea, and that the institution itself arranged its own shipment contracts (*nòlits*).[44] This is the case of the wax shipped from the seaport of Cambrils to Barcelona in 1573, which we know had been harvested in inland Gratallops, in the Priorat.[45] Before that, in the latter village, a local muleteer had been hired with his two donkeys. The wax had been transported by him to Reus – which appears as a regional node in the Camp de Tarragona for this product– and afterward to the harbor of Cambrils to be shipped by sea to Barcelona. Once in the city, the Lluminària generally hired porters to carry the produce from the harbor (*de mar*) to the nave of the Cathedral, and from there up to the chamber.[46]

The logistical effort that the Lluminària was making during the sixteenth century with domestic wax is clear. The product was sold by petty traders who could have been producers themselves. These purchases were sometimes made in the market in Barcelona, but others were the result of forays into remote regions of inland Catalonia. To our knowledge, during the fifteenth and sixteenth centuries the Cathedral had within its reach a constant supply of Maghrebi wax which, considering the weights given by the Lleuda, would have been enough to sustain its needs entirely.

The recourse to this *cera de la terra* is rather more paradoxical once we realize that domestic wax was slightly more expensive, over time, than the *cera berberesca*. In March and April of 1532 two remittances of yellow wax from local peasants gave a rate of 28.5 d. for each lliura (d./l.), whilst on December of that same year, Barbary wax (which was also usually yellow) was taxed at 27 d./l.[47] Several years after that, in 1569,

---

36 See ACB, Administració de la Lluminària, 1565-1571, s.f. (10-I-1565) and also ACB, Carpeta miscel·lània, segles XVI-XVII (31-X-1581).

37 See respectively, ACB, Administració de la Lluminària, 1551-1567, f. 77r (10-IX-1562); ACB, Administració de la Lluminària, 1571-1575, f. 16r (16-III-1572). A certain amount of wax was also sold by confraternities of peasants. Such is the case of the 58 kg. beared by the *confraria de Sant Dionís i Senén* of the town of Piera in 1550. See ACB, Administració de la Lluminària, 1547-1551, f. 32v (11-XI-1550). Many sales of this type appear in the accounts of the Lleuda. Groups of men from several towns paid their taxes for transactions of petty amounts of wax, likely sold to ecclesiastical institutions. One single example refers to the 2 kg. of wax sold by the men of suburban Sant Andreu del Palomar in 1526, that contributed to a certain *lluminària*, likely of a parish of the city. See ACB, Llibre d'Administració de la Lleuda de Mediona, 1526 June-July, 7, s. f. (30-VI-1526).

38 See ACB, Administració de la Lluminària, 1539-1545, f. 40v (30-IV-1545) and ACB, Administració de la Lluminària, 1551-1567, f. 35r (10-VI-1555).

39 See ACB, Administració de la Lluminària, 1551-1567, f. 77v (18-IX-1562).

40 The honey made in the charterhouse was also praised, as was asserted by Gil 1600: 54r-56r.

41 ACB, Administració de la Lluminària, 1571-1575, f. 13r (IX-1574).

42 Just one year before, in 1573, the Lluminària paid 2 s. per day to a man for conducting a mule *i altres bèsties*, with which wax was transported, and also for looking for wax (*sercar dita cera*) around the area of Reus. See ACB, Administració de la Lluminària, 1571-1575, f. 64r (1573).

43 Wax was carried by the same means from the fair of Santa Coloma de Queralt in 1559. Generally, the animals were rented from the muleteers (*traginers*), a growing sector in the sixteenth century. See ACB, Administració de la Lluminària, 1551-1567, f. 58r (1559).

44 See ACB, Administració dela Lluminària, 1571-1575, f. 66r (15-VI-1573). For a previous *nòlit* for barbaresque wax, see ACB, Administració de la Lluminària, 1565-1571, f. 49r (1569).

45 ACB, Administració de la Lluminària, 1571-1575, f. 64r (1573).

46 See for instance ACB, Administració de la Lluminària, 1565-1571, f. 50v (16-X-1570).

47 See ACB, Administració de la Lluminària, 1531-1539, f. 11v (5-III-1532); f. 11v (15-IV-1532) and f.12v (23-XIII-1532).

two different purchases of *cera de la terra* – one from a chandler and another from a peasant – gave an average price of 49 d./l.[48] That same year, North African wax was found by the Lluminària in the seaport of Sitges at a lower 44 d./l.[49]

The higher prices of domestic produce might be explained by a smaller supply, and maybe also by its overland transportation costs. However, we may also speculate about how this wax was judged by contemporaries, who potentially considered it to be a better product than its Northern African counterpart. Although the sources are silent on this, it seems likely that *cera de la terra* was recognized on the market as a high quality product, possibly used in circumstances where a better performance of the candle was required.[50]

Analysis of the Lleuda of Mediona clearly demonstrates the importance of local production to the trade in wax and provides a crucial indication of the prevalence of beekeeping in this region during the later middle ages. In particular, it appears that the area of Tortosa and the lower Ebro and inner Catalonia, stretching north toward the Camp de Tarragona, Conca de Barberà and Anoia-Penedès formed an especially important wax-producing zone. The reasons for this merit further consideration and may provide a new means by which to consider the late medieval environment, offering insight into the changing landscapes of the pre-modern Mediterranean.

III.

Left to its own devices and without human intervention, the vegetation of the Mediterranean basin is characterized by the presence of pine forest and trees such as the cork and holm oak. Centuries of human activity have profoundly altered this landscape. Indeed, deforestation was probably the most important environmental change in the Mediterranean from Antiquity through the Middle Ages. This was caused by a variety of factors, including Roman shipbuilding, the Muslim expansion, and high medieval population growth and urbanization. Within this context, it is

of note that Tortosa was created through extensive woodland clearance, and the city strictly maintained its remaining forest through careful fire management (Houston 1964: 144; Bolòs 2004: 312-15; Lloret and Marí 2001: 155-163). The loss of tree cover through deforestation allowed greater light to reach the subsequent layers of vegetation, but removing the protective layer of shade and dense tree roots increased evaporation and promoted soil erosion. The plants which thrive in the subsequent complex shrubland associations, the soft-leaved garrigue, which grows over calcareous soils, and the denser maquis which grows over silicaceous ones, can withstand this dry climate and thin soils. Additionally, the maquis has two layers or heights: the tall maquis of three to five metres, and the low maquis of about 1.5 metres of shrubs such as cistus, often brought about by deforestation and degradation of the high maquis. At its most degraded, the plantlife of the maquis overlaps with many of the plants of the garrigue, such as kermes oak, rosemary and lavender, while garrigue is also associated with thyme, white cistus and sage (See e.g. Houston 1964: 86-88).

Above all, in the middle ages deforestation was driven by extensive pastoralism and the need to graze large flocks and herds across vast tracts of land. One of the more remarkable differences in medieval agricultural regimes in southern Europe compared with those of the north, are the large regions in which animal and cereal husbandry were relatively poorly integrated. These were areas of transhumance, in which livestock foraged at great distance from settlements.[51] The importance of this for the vegetation of the Mediterranean is clear. Pastoralists made extensive use of fire to clear swathes of land of mature vegetation in order to increase the growth of younger, more palatable and easier to access forage. The plants which survive in such conditions are able either to withstand fires or to regenerate quickly thereafter (see e.g. Moreno and Oechel 1994). This is a manmade landscape, and successive fires allowed for the dominance of certain dense associations of plants such as rosemary, heather and cistus. In Catalonia, these systems of transhumance could also be integrated into arable systems, as in the seasonal movement of herds onto the fields of the *masos*, farmsteads which often had pastures and significant animal holdings of their own (Congost 2015: 15; Farías 2009: 37-42). The prevalence of sheep, which eat vegetation down almost to the ground, and voracious goats is of particular importance in understanding the degradation of vegetation and landscape.

---

[48] ACB, Administració de la Lluminària, 1565-1571, f. 49r (1569).

[49] ACB, Administració de la Lluminària, 1565-1571, f. 49r (29-VIII-1569). Still other examples are to be found for 1539 and for 1547. In the former year, *cera nova de la terra* (yellow) was purchased at 31 d./l. while North African produce was acquired at 26.5 d./l. See ACB, Administració de la Lluminària, 1539-1545, f. 11r (19-V-1539) and f. 11v (14-X-1539). In 1547, one same chandler (Joan Oliveres) sold white domestic and barbaresque wax within two months. The former cost 38 d./l. and the latter 29.5 d./l. See ACB, Administració de la Lluminària. 1547-1551. f. 16r (15-VI-1547) and f.16r (18-VIII-1547).

[50] This would coincide with the impression given by the chronicler Agustín Horozco, from Cádiz, in 1598. Comparing it with the production from North Africa, he judged the domestic Valencian wax (along with the one from Venice) as the most esteemed (*tienen más primor y nombre*). See Horozco 1845: 174.

[51] Due to its economic and political importance, the mesta of Castile is the most well-known and well-documented of these systems, but transhumance was also prevalent throughout the Crown of Aragon, such as the lligallos of Aragon and Valencia.

Yet the degraded shrublands of the maquis and garrigue provide abundant bee forage. When Braudel wrote derisorily of the 'monotonous wilderness of heath and rosemary' which characterized stretches of Aragon in the hinterland of the Ebro, he was describing a region rich in flowering plantlife, well-suited to beekeeping with long and varied flowering periods (Braudel 1972: 399). Additionally, the types of plants which flower in these shrublands provide forage for particularly palatable honeys. It was surely the early flowering rosemary among other spring flowers from which the much-prized first honey harvest derived (Canova 1999: 22; Vila 2003: 199).[52] Since wax is produced to contain honey, it seems likely that the degraded and forest-cleared landscapes created especially suitable bee habitats, which also produced high volumes of honey and wax that could become the products of regional and even long-distance trade.

While it appears that beekeeping was related generally to agriculturally marginal land, it seems possible that it was even more specifically associated with fire-forged landscapes. For example, in the 1340s Bernat Blandric of Lloret, just south of Girona, was a charcoal burner who also kept beehives. So closely intertwined were these activities that he even promised his guarantor in a credit transaction for an advance purchase of coal, that if he did not supply the goods he would pay him five hives (Sales i Favà 2019: 397). The combination of beekeeping and charcoal burning is striking, for this was a period of increased pressure on woodland resources to make charcoal for smelting iron. The expansion of charcoal burning during this period, necessary to make agricultural implements and weaponry, also added to the deforestation caused by pastoralism. Although in managed woodlands coppicing would have staved off absolute tree loss, the loss of dense canopy would have facilitated the growth of lower layers of vegetation.

This is not to say that beekeeping was only practiced alongside burning. A multiplicity of land uses is evinced in a temporary lease in the Puig Rodó de Romanyà in 1368, in which the lessor retained the right to the beehives on the land, in addition to being able to harvest acorns and take cork. The paltry 3s. annual rent may indicate that, despite the preponderance of usable products, the land was nonetheless considered marginal for other purposes once the rights to these were removed (Sales i Favà 2019: 397, and more generally: 395-404).

The hives in question were most likely made of cork where it was available, or other abundant wood as the log hives of the Sierra de Albarracín (Lemeunier 2004:

394-402). These were well-suited to the intense heat of summer and as well as light and easy to move (see e.g. Carmona Ruiz 1999: 394).[53] The latter was an important feature because beehives, much like sheep, were moved between summer and winter foraging areas in a kind of bee transhumance (Lemeunier 2004: 387-404; for central Castile see also Sanchez Benito 1987: 99-104; Sanchez Benito 1988: 95-100; Sanchez Benito 1989:19). Transhumant apiculture was particularly important at medium altitudes. For example, traditionally at Pena-roja (Teruel, Aragon), cold winters and a total absence of flowering plants necessitated either leaving the bees significant amounts of honey to feed them through the winter—a practice common throughout northern Europe—or, preferably, moving the hives down to sites such as Horta (Terra Alta, Tarragona), Aldover (Baix Ebre, Tarragona) or the plain of Castelló (Lombarte Arrufat and Quintana i Font 1989: 77, which also gives an indication of the annual cycle of flowering plants and their respective honeys; medium altitudes are noted also in Lemeunier 2004: 402). The easy-to-move nature of Iberian hives is most clearly demonstrated by the preoccupation of the law codes with thefts of hives themselves, rather than only the swarms (e.g. Ortega Gil 1995: 45-71). The placement of the hives was typically highly regulated, and bees were to be kept away from vineyards, where it was thought bees could injure the grapes, and away from livestock which could knock over hives and otherwise interfere with the bees (Ortega Gil 1995: 51-2; Sanchez Benito 1989: 15, 18; Sanchez Benito 1987: 100-3; Carmona Ruiz 1999: 394-97; Aparici Martí 1999: 37; Aparisi Romero: forthcoming).

Certainly the documentary evidence supports the overall impression that the majority of beekeeping took place away from arable agriculture and viticulture. But despite medieval beliefs about keeping pastoral husbandry and apiculture separate, it would appear that there was in fact a symbiotic relationship between the two activities. Part of this may have been due to social, economic and occupational structures built around peripatetic employments and land use. But a hitherto unconsidered aspect of this is that the purposefully altered landscapes provided plentiful bee forage.

Accepting that fires are 'recognized as one of the main ecological drivers of landscape and vegetation changes within the Mediterranean Basin', can help explain the association of beekeeping with pastoralism (Lloret and Mari 2001:155). It may be tentatively suggested that extensive beekeeping may be a sign of a particular type of landscape degradation, and in this way has the

---

[52] Although it must be noted that not all of the plants of the maquis produced palatable honey. Honey made from the highly prevalent arbutus (strawberry tree) is notably bitter.

[53] See also Llibre d'Administració de la Lleuda de Mediona, 1501-1502, s.f. and Llibre d'Administració de la Lleuda de Mediona, 1544 (XI) - 1545 (IV), s.f. which records honey sold in Barcelona directly with their 'bornieta' (i.e. inside the cork hives).

potential to tell us much about human intervention on the medieval environment.

The period under study also covers a well-known period of climate change, with the end of the medieval warm period and the onset of the little ice age. The effect of this climatological change within the Mediterranean requires further detailed study. However, particular periods of increased flooding within the Crown of Argon are well attested to through notarial registers and, intriguingly, records of rogation ceremonies. The latter occurred in times of heavy rains and floods ('pro serenitate' rogations) and in time of drought ('pro pluvia' rogations) in five levels of increasing severity (Martin-Vide and Barriendos 1995: 201-221; Reixach Sala 2018: 321-41). These show a decrease in 'pro pluvia' rogations from the late 16th to mid-17th centuries, which corresponds with increasing records of flooding from notarial remarks (Barriendos i Vallvé 1995: 53-70, esp. 56; Peris i Albentosa 2004: 679). In the valley of the Ebro, where floods caused by heavy rains were also compounded by snowmelt from the Pyrenees, spring floods were particularly common (Barriendos and Martin-Vide 1998: 473-491, esp. 478). The consequences of the particularly high incidence of spring floods in this region, given the importance of spring forage for bees and the physical inability of bees to collect nectar and pollen in rain and high winds, needs to be considered in further depth to place long-term changes in apiculture and wax production within this important zone.

IV.

Combined analysis of the accounts of the Lleuda of Mediona and the Lluminària of the Cathedral of Barcelona offers a new insight into the trade in beeswax in late medieval Barcelona. It particular, these records demonstrate the as yet underexplored importance of domestically produced wax in this trade. As one of the great cities of the Mediterranean, Barcelona had easy access to the ubiquitous North African wax shipped via Mallorca and Valencia. Yet despite this, it is clear that locally produced wax remained an important and desirable commodity throughout our period. Both the higher price of this *cera de la terra* compared with Maghrebi wax and the effort the Cathedral put into sourcing it, suggest that domestic wax was a high-quality product.

This wax was the product of a particular kind of manmade landscape, one in which large-scale pastoralism, expansion of settlement and intensification of certain types of land use, led to deforestation and radically altered the plantlife of the region. Yet the vegetation of the maquis and garrigue provided an abundance of bee forage. Favourable bee habitats, combined with high demand for wax, no doubt made beekeeping an attractive and lucrative by-employment

for local producers. The prevalence of domestic wax, and the indications as to its origins within the records, shows the importance of apiculture in regions of Catalonia and the Comtats. It is clear the beekeeping was common across southern Catalonia, especially along the Ebro and in Tarragona, into Anoia-Penedès, north into the region of Girona and even into the highlands of the Pyrenees. In highly productive zones, wax and honey were harvested two or even three times a year, traditionally at the beginning and end of summer (Vila 2003: 199; Gómez Vozmediano, Sánchez Gonzále and Priego Rodríguez 2005: 151). The lag between when wax was harvested and when it entered the market suggests that producers held wax back until demand increased in preparation for important wax-consuming festivals such as Candlemas (2 February)—this can explain the very high proportion of wax entering the city in January, visible in the Lleuda of Mediona, which does not correspond with harvest times.

Domestic wax, like that from North Africa, arrived raw into the city, and was then cleaned and whitened before being made into candles. That this was a major operation is indicated by the development of wax-processing workrooms within the Cathedral itself—further demonstration of the importance of wax candles for religious ritual. Every step of this process bore a cost, and the intricate accounting systems which developed around the production of wax candles and the recycling of used wax are indicative of the fiscal resources needed to keep the Cathedral supplied with adequate amounts of wax.

Constant need for wax to celebrate the mass, religious festivals and life-cycle events, in addition to providing a luxury lighting source, made wax a sought after commodity, demand for which was met through both long-distance trade and local production. Close examination of the trade in, and consumption of, beeswax offers insight into a range of economic, cultural and environmental factors in late medieval Catalonia.

Bibliography

Agustí, M. 1616. *Llibre dels Secrets de Agricultura, casa rústica y pastoral*. Barcelona: Esteve Libreros.

Aparici Martí, J. 1999. De la apicultura a la obtención de la cera. Las 'otras manufacturas' medievales de Segorbe y Castellón. *Millars: Espais i Història* 21: 31-50.

Aparisi Romero, F. forthcoming. *L'apicultura a la València medieval*.

Banegas López, R.A. 2017. *Sangre, dinero y poder. El negocio de la carne en la Barcelona medieval*. Lleida: Editorial Milenio.

Barriendos, M. 1995. La dinámica climática de Tortosa (S. XIV-XIX). *Recerca* 1: 53-70.

Barriendos, M. and J. Martin-Vide. 1998. Secular climatic oscillations as indicated by catastrophic floods in the Spanish Mediterranean coastal area (14th-19th centuries). *Climatic Change* 38: 473-491.

Benassar Vaquer, O. 2001. *El comerç marítim de Mallorca: 1448-1531*. Palma: El Tall.

Bolòs, J. 2004. *Els orígens medievals del paisatge català. L'arqueologia del paisatge com a font per a conèixer la història de Catalunya*. Barcelona: Institut d'Estudis Catalans.

Braudel, F. 1972. *The Mediterranean and the Mediterranean World in the Age of Philip II. Vol.1* (trans. S. Reynolds). London: Collins.

De Capmany y de Monpalau, A. 1962. *Memorias históricas sobre la marina, comercio y artes de la antigua ciudad de Barcelona. Reedición anotada*. Barcelona: Cámara Oficial de Comercio y Navegación, 3 v.

Canova, G. 1999. Api e miele tra sapere empirico, tradizione e conoscenza scientifica nel mondo arabo-islamico, in G. Canova (ed) *Scienza e Islam*: 69-92. Rome: Herder.

Carmona Ruiz, Ma. A. 1999. La apicultura sevillana a fines de la Edad Media. *Revista española de estudios agrosociales y pesqueros* 185: 131-154.

Catafau, A. 2012. La frontière, une "calamité féconde" pour les Comtés nord-catalans au Moyen Âge, in F. Corrons, E. Trenc and G. Dorel-Ferré (dir.) *"Comme une étoffe déchirée" Les Catalognes avant et après le traité des Pyrénées*: 19-30. Canet: Trabucaire.

Conesa Soriano, J. 2015. La gestion d'un patrimoine ecclésiastique urbain à la fin du Moyen Âge: l'inscription du chapitre cathédral de Barcelone dans la ville. *Histoire Urbaine* 42: 37-55.

Conesa Soriano, J. 2017. Entre l'Église et la ville: le chapitre et les chanoines a Barcelone au sortir de la guerre civile catalane (1472-1500). Unpublished PhD thesis, Université Paris-Sorbonne.

Congost, R. 2015. The Catalonia of the *mas*: The old Catalonia, in R. Congost (ed.) *The Catalan Mas: Origins, transformations and the end of an agrarian system*: 7-24. Girona: Associació d'Historia Rural, Centre de Recerca d'Historia Rural (Institut de Recerca Històrica) de la Universitat de Girona and Documenta Universitària.

Dalmases, N. 1992. *Orfebreria catalana medieval: Barcelona 1300-1500: aproximació a l'estudi*. Barcelona: Institut d'Estudis Catalans, 2 v.

Fàbrega i Grau, A. 1978. *La vida quotidiana a la catedral de Barcelona en declinar el Renaixement: any 1580*. Barcelona: Real Academia de Belles Arts de Sant Jordi.

Farías, V. 2009. *El mas i la vila a la Catalunya medieval: Els fonaments d'una societat senyorialitzada (segles XI-XIV)*. Valencia: Universitat de Valencia.

Fatjó Goméz, P. 1999. La Catedral de Barcelona en el siglo XVII: las estructuras y los hombres. Unpublished PhD thesis, Universitat de Barcelona.

Ferrer i Mallol, M. T. 2008. Navegació, ports i comerç a la mediterrània de la Baixa Edat Mitjana, in J. Pérez Ballester and G. Pascual (eds) *Actas V Jornadas internacionales de Arqueología subacuática. Comercio,*

redistribución y fondeaderos. La navegación a vela en el Mediterráneo (Gandia, 8-10 novembre 2006): 113-166. Valencia: Universitat de València.

García Espuche, A. 1998. *Un siglo decisivo: Barcelona y Cataluña, 1550-1640*. Madrid: Alianza.

Gaya Catasús, J. 2016. Societat i economia agrària al Penedès, segles XVI-XVII. El terme de Subirats, la parròquia de Sant Pere de Lavern. Unpublished PhD thesis, Universitat Pompeu Fabra, 2 v.

Gifre Ribas, P. 2012. *Els senyors útils i propietaris de mas. La formació històrica d'un grup social pagès (vegueria de Girona, 1468-1730)*. Barcelona: Fundació Noguera.

Gil, P. 1600. *Libre primer de la historia Cathalana en lo qual se tracta de Historia o descripció natural, ço es de cosas naturals de Cathaluña*.

Gómez Vozmediano, M.F., R. Sánchez González and A. Priego Rodríguez. 2005. *La apicultura en los Montes de Toledo*. Toledo: Asociación para el Desarrollo Integrado del Territorio Montes de Toledo.

Gual Camarena, M. 1976. *Vocabulario del Comercio Medieval. Colección de aranceles aduaneros de la Corona de Aragón (siglos XIII y XIV)*. Barcelona: El Albir.

Gual i Vilà, V. and X. Jorba i Serra. 2013. Conflictivitat rural a la Catalunya Nova (Concà d'Òdena i senyoria de Poblet) a l'època moderna. *Estudis d'Història Agrària* 25: 99-124.

Guiral, J. 1974. Les relations commerciales du Royaume de Valence avec la Berbérie au XVe siècle. *Mélanges de la Casa de Velázquez* 10: 99-131.

Horozco, A. 1845. *Historia de la ciudad de Cádiz*. Madrid: Imprenta de Don Manuel Bosch.

Houston, J.M. 1964. *The Western Mediterranean World: An Introduction to its Regional Landscapes*. Aylesbury: Longman.

Lemeunier, G. 2004. Viajes de abejas. La trashumancia apícola en la Cataluña norte (s. XIX), in J. L. Castán Esteban and C. Serrano Lacarra (coord.) *La trashumancia en la España mediterránea: historia, antropología, medio natural, desarrollo rural*: 387-404. Zaragoza: Centro de Estudios sobre Despoblación y Desarrollo de Áreas Rurales.

Lloret, F. and G. Marí. 2001. A comparison of the medieval and current fire regimes in managed pine forests of Catalonia (NE Spain). *Forest Ecology and Management* 151: 155-163.

Lombarte Arrufat, D. and A. Quintana i Font. 1989. L'apicultura tradicional a Pena-roja. *Alazet* 1: 73-98.

Macaire, P. 1986. *Majorque et le commerce international: 1400-1450 environ*. Lille: Atelier Reproduction des theses.

Madoz, P. 1985. *El Principat de Catalunya, Andorra i zona de parla catalana del Regne d'Aragó: al "Diccionario geográfico-estadístico-histórico de España y sus posesiones de Ultramar"*. **Barcelona: Curial, 2 v.**

Martínez i Garcia, M. 2010. *Apicultura tradicional a la Conca de Barberà*. Valls: Cossetània.

Martin-Vide, J. and M. Barriendos. 1995. The use of rogation ceremony records in climatic reconstruction: A case study from Catalonia (Spain). *Climatic Change* 30: 201-221.

Mas, J. 1916. *La custòdia de la Sèu de Barcelona en l'any 1522*. Barcelona: La Renaixensa.

Morelló Baget, J. 1993. Consideracions al voltant d'una font de tipus fiscal: els Llibres d'Estimes de Reus, in M. Sánchez Martínez (comp.), *Estudios sobre renta, fiscalidad y finanzas en la Cataluña bajomedieval*: 349-380. Barcelona: CSIC.

Morelló Baget, J. 1997. Els impostos sobre la renda a Catalunya: redelmes, onzens i similars. *Anuario de Estudios Medievales* 27/2: 903-967.

Moreno, J.M. and W. C. Oechel (eds). 1994. *The Role of Fire in Mediterranean-Type Ecosystems*. London: Springer.

Ortega Gil, P. 1995. Hurtos de colmenas: Apuntes históricos. *Cuadernos de Historia del Derecho* 22: 45-71.

Orti Gost, P. 2000. *Renda i fiscalitat en una ciutat medieval: Barcelona, segles XII-XIV*. Barcelona: CSIC.

Peris i Albentosa, T. 2004. Calamitats climàtiques i economia agrària a la ribera del Xúquer entre els segles XV i XIX. *Estudis d'Història Agrària* 17: 51-70.

Puig, C. 2005. À l'origine des premières taxes douanières: les leudaires en Roussillon et en Cerdagne (XIIIe - milieu du XIVe siècle), in G. Larguier (dir.), *Douanes, États et Frontières dans l'est des Pyrénées de l'Antiquiteé à nos jours*: 23-40. Perpignan: PUP-AHAD.

Reixach Sala, A. 2018. La percepción de las inundaciones en la Cataluña nororiental entre los siglos XIV y XVII. De las notas crísticas a las autobiografías populares. *Revista de História da Sociedade e da Cultura* 18: 321-41.

Riera Melis, A. 1980. La lezda de Colliure bajo la Administración mallorquina. 1: La reforma de aranceles de finales del siglo XIII. *Acta Historica et Archaeologica Mediaevalia* 1: 91-125.

Riera Melis, A. 1986. *La Corona de Aragón y el reino de Mallorca en el primer cuarto del siglo XIV*. Barcelona: CSIC.

Riera Melis, A. 2003. Barcelona en els segles XIV i XV, un mercat internacional a escala mediterrània. *Barcelona Quaderns d'Història* 8: 65-83.

Riera Melis, A. 2017. El regne de Mallorca, la corona d'Aragó i França al començament del segle XIV. *E-Spania: Revue électronique d'études hispaniques* 28.

Rubin, M. 1991. *Corpus Christi: The Eucharist in Late Medieval Culture*. Cambridge: Cambridge University Press.

Sales i Favà, L. 2019. Crèdit i morositat a la Catalunya del segle XIV. El cas de la baronia de Llagostera. Unpublished PhD thesis, Universitat de Girona.

Salicrú, R. 1995. *El tràfic de mercaderies a Barcelona segons els comptes de la Lleuda de Mediona (febrer de 1434)*. Barcelona: CSIC.

Salvador Esteban, E. 1990. El tráfico marítimo Barcelona-Valencia durante los siglos XVI y XVII. Su significado en el conjunto del comercio importador valenciano. *Pedralbes* 10: 19-49.

Sanchez Benito, J. M. 1987. Aproximación al estudio de un sector económico en Castilla a fines de la edad media: La explotación colmenera, in *Hernán Cortés y Su Tiempo. Actas de Congreso Hernán Cortés y su tiempo, V centenario (1485-1985)*: 99-104. Cáceres: Mérida Editora Regional de Extremedura.

Sanchez Benito, J. M. 1988. Poder y propiedad: Los hermanos de la Santa Hermandad Vieja de Toledo. *I Congreso de Historia de Castilla-La Mancha* 6: 95-100.

Sanchez Benito, J. M. 1989. Datos sobre la organización de la producción apícola castellana en la baja Edad Media. *Estudis d'Historia Economica* 1:11-25.

Sans i Travé, J. M. and D. Guasch i Dalmau. 1979. Les ordinacions del Montmell (segle XV). *Miscel·lània Penedesenca* 2: 221-246.

Sapoznik, A. 2019. Bees in the Bees in the medieval economy: Religious observance and the production, trade, and consumption of wax in England, c.1300-1555. *Economic History Review* 72: 1152-1174.

Serra, E. and M. A. Sanllehy. 2005. Comerç transpirinenc a Catalunya segons la documentació de la Generalitat de Catalunya (s. XVI-XVII), in J-M. Minovez and P. Poujade (eds) *Circulation des marchandises et réseaux commerciaux dans les Pyrénées, XIII-XIXe siècle / Circulació de mercaderies i xarxes comercials als Pirineus (segles XIII-XIX)*: 473-522. Toulouse: CNRS and Université de Toulouse-le Mirail.

Serra i Coma, R. and L. Ferrer i Alòs. 1985. Un qüestionari de Francisco de Zamora (1789). *Estudis d'Història Agrària* 5: 159-207.

Sobrequés, J. 1969-1970. La Lleuda de Cotlliure de 1317. *Cuadernos de Historia Económica de Cataluña* 21: 65-84.

Timmermann, A. 2009. *Real Presence: Sacrament Houses and the Body of Christ, c. 1270-1600*. Turnhout: Brepols.

Torres i Ferrer, M. J. 1997. La catedral de Barcelona y la Guerra Civil Catalana (1462-1472). *Medievalismo* 7: 99-138.

Torres i Ferrer, M. J. 2001. Catedral de Barcelona en el siglo XV: gestión económica del patrimonio y proyección social y política. Unpublished PhD thesis, Universitat de Barcelona.

Valls Junyent, F. 1990. Creixement agrari i diferenciació social pagesa a la comarca de l'Anoia entre començaments del segle XVI i mitjans del XVII. *Pedralbes* 10: 99-136.

Vela, C. 2003. El control de la candeleria de cera a Barcelona. Una visió diacrònica (s. XIV-XVI), in S. Claramunt (coord.), *El Món urbà a la Corona d'Aragó del 1137 als decrets de Nova planta : Barcelona, Poblet, Lleida, 7 al 12 de desembre de 2000. Actes*: 1065-1084. Barcelona: Publicacions de la Universitat de Barcelona.

Vila, P. 2003. *Resum de geografia de Catalunya*. Barcelona: Institut d'Estudis Catalans.

Violant i Simorra, R. 1935. El cultiu tradicional de les abelles, al Pallars Sobirà. *Agricultura i ramaderia* 12: 222-228.

Chapter 12

Del panal a la mesa: La miel en la Corona de Aragón (siglos XIV-XV)

Pablo José Alcover Cateura

Food Observatory (ODELA)—Universitat de Barcelona (palcovca@gmail.com)

Resumen: La miel fue el principal edulcorante de las clases populares en la Europa medieval. Su uso como alimento y medicina fue generalizado en Occidente. El presente trabajo es la primera monografía sobre la miel en la población cristiana de la Corona de Aragón. A través de este estudio que analiza desde la producción al consumo del edulcorante se intenta demostrar que es un elemento ilustrativo para comprobar las diferencias sociales en la Edad Media. Este trabajo forma parte de una investigación en curso sobre la miel en la Corona de Aragón durante los siglos XIV y XV.

Palabras clave: Historia de la alimentación; Corona de Aragón; miel; dulces; colmenas

Summary: Honey was the principal sweetener of the common people in medieval Europe. Its use as a food and medicine was a general feature in the West. The present contribution is the first study concering honey in the Christian population of the Crown of Aragon. This study, which analyzes this foodstuff from production to consumption, is meant to demonstrate that honey is an illustrative element for showing class differences in the medieval period. This work forms part of an ongoing investigation into honey in the Crown of Aragon during the 14th and 15th centuries.

Keywords: food history; Crown of Aragon; honey; confectionery; hives

> 'Abejas benditas,
> santos abejares,
> dan miel a los hombres
> y cera a los altares.'
> Adagio apícola castellano

Introducción

La explotación económica del bosque medieval se resumía en que todo se podía vender si alguien lo compraba. Se vendían productos con los siguientes orígenes: Primero, mineral (azabache, nitrato potásico y substratos de varios tipos, entre otros). Segundo, vegetal (frutos frescos y secos, madera, leña, pez, trementina, aceite de enebro, alquitrán, carbón y ceniza). Tercero, animal (animales y pieles de caza, colmenas, miel, cera y palomina). Cuarto, productos de artesanía y manufacturados para mercados próximos (tablas de madera ya talladas para elaborar muebles concretos, paños, prendas y piezas de azabache) (Sesma Muñoz 2001: 204-215).

Por lo tanto, la miel viene normalmente del bosque. Allí estaba preferentemente su zona de producción. El capítulo comienza con la producción y recolección del alimento. No se puede comprender en profundidad su comercio y usos sino se sabe de dónde viene. Posteriormente, se analiza su comercialización. Ésta permite conocer cómo y en qué cantidades se compraba, realidades importantes para entender su consumo. Más

adelante y para finalizar se detallan sus usos. A través de ellos, se aprecian los principales indicadores que demuestran que el edulcorante es un objeto de estudio con el que se pueden distinguir los estamentos.

Producción y recolección

Las dos razas autóctonas de abeja, la española (*Apis mellifera iberica*) y la negra (*Apis mellifera mellifera*) fueron probablemente las mayores productoras del edulcorante. Por ahora, la documentación consultada no ofrece más datos al respecto.

Las áreas de actividad apícola eran amplias. Se aprovechaban todas las tierras posibles. Las áreas estaban normalmente situadas en terrenos elevados que no se empleaban para el cultivo. Asimismo, estaban compuestas de prados y/o bosques que presentaban la mayor biodiversidad disponible y estaban cerca de ríos (Sesma Muñoz 2001: 204-215). Todas las características de las zonas de producción del edulcorante responden, como es lógico, a las necesidades propias de los enjambres (Càrrere 1977: 331-332). La quema de rozas era un peligro para los panales. Los señores jurisdiccionales prohibirían seguramente quemarlas cerca de su ubicación (Carmona Ruiz 2000: 401). Tres de los grandes depredadores para los enjambres eran el oso ibérico (*Ursus arctos pyrenaicus*), la garduña (*Martes foina*) y la comadreja (*Mustela nivalis*). Los tres, especialmente el oso, ingerían notables cantidades del edulcorante o de las larvas de abejas, ricas en proteínas. Su actividad podría haber tenido un control, pero la falta de fuentes impide confirmarlo.

Se compara ahora la miel con un producto análogo, el azúcar. La implantación del azúcar fue tardía. Llegó a finales del siglo XIV a la Corona de Aragón. La producción local de este nunca cubrió la demanda y por ello la gran mayoría se importaba. Las principales áreas de producción azucarera estaban en Oriente, en Siria, Palestina y Egipto. Esto incrementaba notablemente su precio (Riera Melis 2013a: 126). En cambio, la miel era siempre producto local. No se han documentado importaciones notables de miel seguramente porque no hacían falta.

La actividad melera histórica dejó testimonios en la toponimia. Cercano a *Puigpunyent* (Sierra de Tramontana, Mallorca) está el *Penyal de sa Mel* ('el Peñasco de la Miel'); a escasos kilómetros de Berga (*Berguedà*, Cataluña) se halla el *Coll de la Mel* ('la Colina de la Miel'); y en la *Seu d'Urgell* (*Alt Urgell*, Cataluña) aún es transitable el *Camí de la Mel* ('el Camino de la Miel').

A las áreas apícolas habituales, se añadía un caso particular: los panales dentro de oquedades en ribazos (muros perimetrales) de viñedos. Había dos opciones: por un lado, la 'natural', los enjambres construyeron sus hogares en agujeros presentes en un muro. Por otro, la 'artificial', un agujero se hizo a propósito dentro del muro para facilitar la instalación de los insectos.

A la viña le basta el aire para polinizarse ¿Entonces qué aportaron los enjambres? El propietario del viñedo obtuvo dos mejoras importantes: un aumento de la calidad del racimo y una limpieza de la podredumbre de la cosecha. Ambas mejoras se producen porque las abejas liban el jugo de las uvas dañadas. La primera incrementó la valoración del producto. La segunda permitió una cobertura sanitaria natural en un periodo histórico carente de tratamientos fitosanitarios. Esta técnica se documenta a partir del siglo XIX, aunque su origen se remontaría probablemente a la Edad Media.[1]

Los apicultores de los siglos XIV y XV fabricaron colmenas 'fijistas', es decir, sin cuadros móviles como

las actuales. Su proceso de fabricación se desconoce por falta de fuentes (Crane 1999: 214-216).

Había dos tiempos de recolección anual de miel: primero, en primavera, a partir de finales de abril. Segundo, a inicios de otoño, a partir del día de san Miguel (29 de setiembre). Sin duda, la mejor miel era la otoñal. Los refranes lo recuerdan: *la bona mel, per Sant Miquel* ('la buena miel, por San Miguel') y *per Sant Miquel hi ha la millor mel* ('por San Miguel, hay la mejor miel'). En ambos casos, los panales se cortaban con un tempanador, utensilio de hierro ligero y manejable. Al cortar los panales los animales sufrían mucho estrés. También causaba muchas muertes, por la defensa del panal y por estar en proceso de cría.

Los panales se cortaban e introducían en una caja. Allí, se separaban las calidades: los que tenían la mejor miel iban a una cesta de mimbre. Aquí se obtenía el mejor edulcorante que era de color claro. La llamaron la 'buena miel'. La otra selección era de los panales carentes de miel. Éstos eran aplastados. Al hacerlo, se iba formando una bola que se destinaba para fabricar cera. Los panales restantes eran prensados. El resultado del proceso era una miel de peor calidad que la clara. La nombraron 'miel negra' por su tonalidad más oscura que la de mejor sabor. En resumen, la recolección aportaba 'buena miel', 'miel negra' y cera (Serra de Manresa 2019: 48-50).[2]

Posteriormente, la 'buena miel' era almacenada en jarras cerámicas, barnizadas o no, y en botes de vidrio. La 'negra' era guardada en otras jarras. Los materiales y recipientes permitían distinguir los productos fácilmente. Todo se dejaba reposar unos días para separar impurezas. Finalizado el reposo, una parte de la producción se calentaba a cuarenta grados. Esta técnica permitía más tiempo de conservación en estado líquido. El resultado era la conocida como 'miel cocida'. Por último, debido a la pérdida del olor al calentarse, se perfumaba poniendo ramitas de las plantas nectaríferas correspondientes en la última capa. La parte restante, la 'miel cruda', era envasada directamente y duraba pocos días. No obstante, era muy apreciada por su sabor potente y aroma intenso (Serra de Manresa 2019: 48-50).

Comercio

Mi tesis doctoral recientemente leída es un estudio comparado de las ordenanzas municipales bajomedievales relacionadas con el cargo de *mostassaf*, un inspector del mercado. Se han analizado 35

---

[1] Agradezco esta información sobre la polinización de abejas en viñedos a Albert Roig Deulofeu, arqueólogo del paisaje e historiador y a Lluís Busom, apicultor, considerado uno de los mayores expertos en esta técnica hoy en desuso.

[2] La economía de alcance es un concepto que podría ser aplicado en el caso del negocio del apicultor. Es decir, producción conjunta de varios productos para optimizar recursos y abaratar el precio de las unidades.

Figura 1. *Rusc* (colmena 'fijista') conservado en el Museu Comarcal de l'Horta Sud (Valencia). Se desconoce su datación. Imagen realizada con el permiso de Clara Pérez, gerente del Museu.

El comercio de miel circulaba por el mercado interior y exterior. Al principio, circuló sólo por el mercado interior. Posteriormente, las amplias zonas de producción local fueron una de las razones que explican su exportación en grandes cantidades. De nuevo, se diferenciaba con el azúcar que se importaba generalmente de Chipre, del Reino Armenio de Cilicia y de Creta (Riera Melis 2013a: 126).

Segundo, es vital conocer las medidas metrológicas. Las unidades ponderales más usual eran la arroba (7.8 kg. aprox.) y el quintar (cuatro arrobas, 31.2 kg. aprox). Los recipientes de venta más populares eran los cántaros de dos arrobas, medio quintar, todos de cerámica barnizada o no. Todos los cántaros se cerraban con una pleita de esparto, un trenzado resistente, con tres anillas (Bajet Royo 1994: 579; Bertran de Heredia Bercero 1998:180).

Explicados los dos elementos de análisis del comercio, analicemos el transporte. Para el transporte marítimo se utilizaban jarras grandes de hasta unos 60kg. cubiertas con un tejido de estopa: la arpillera (Càrrere 1977: 332). También iban con tres anillas hechas de un tejido grueso: la driza de bandera. En los dos casos, los tejidos eran impermeabilizados encerándolos probablemente con trementina, un aceite esencial obtenido al destilar al vapor la resina de pino (Gómez Urdáñez 2019: 58, López Pizcueta 1992: 43, Sesma Muñoz 2001: 207).[4]

recopilaciones de normativas locales de los principales municipios de los reinos y territorios de la Corona de Aragón (siglos XIV-XV). Este trabajo es el punto de partida de este apartado (Alcover Cateura 2019: 201-203).

Primero, es esencial saber qué productos se vendían. Había miel de distintos tipos que recibía nombres según la planta melífera (Serra de Manresa 2019: 65- 82).[3] Cada región de la Corona de Aragón producía variedades propias. Por ejemplo, en la marina de Llucmajor (Mallorca) se elaboraba de romaní (Aguiló 2002: 125).

Al llegar a puerto, una parte de las jarras se cargaba directamente para su transporte. Otras se vaciaban y su contenido se distribuía en recipientes más pequeños. En cualquier caso, el transporte lo realizaría un animal de carga, mulo o asno, o un carro tirado por estos. Los animales llevaban en su lomo una caja de madera, la aportadera, dónde dentro iban los recipientes más pequeños (Ollé Albiol 1996: 45). Ésta era resistente para evitar daños por el traqueteo del viaje. No había restricciones para su entrada. Por ejemplo, en una ordenanza de Igualada las recuas de animales de carga y filas de carros entraban por cualquier puerta de la muralla de una población y por el puerto (Castellà Raich 1954: 43).[5]

---

[3] Sigue una lista de las variedades principales: de abejera, de acacia blanca, de meliloto, de encina, de roble, de almendro, de madroño, de abeto, de alhucema, de espliego, de gayuba, de boja, de brezo, de brecina, de castaño, de jara, de algarrobo, de limonero, de naranjo, de malva, de mejorana, de orégano, de toronjil, de arándano, de nabo, de olivarda, de manzano, de poleo, de rabaniza blanca, de zarza, de romero, de ajedrea, de salvia, de tilo, de tomillo, de cantueso, de esparceta, de alfalfa, de veza y de milflores.

[4] El uso de trementina se documenta en un encargo de telas para la Seo de Zaragoza del año 1431. Es posible que también se usara para el transporte de mercancías.

[5] Las ordenanzas de venta de miel no mencionan el paso obligado por una de las puertas o caminos hacia la ciudad o

La venta

La miel se vendía en mercados municipales. En general, su origen era variado: podía ser local, llegando del mismo lugar, villa, ciudad y/o de municipios vecinos. También podía ser regional, llegando de poblaciones que distaban a varios días por transporte fluvial, marítimo o terrestre, del mismo reino o de otro cercano (Batlle Gallart 1994: 533; Càrrere 1977: 332; Salicrú Lluch 1995: 112). Además, la miel del campo de Tarragona, de Tortosa y Mequinenza, las de mayor prestigio, se exportaba en grandes cantidades por todo el Mar Mediterráneo (Càrrere 1977: 331-332).

Las cargas eran importantes y a veces a gran escala. Podían llegar hasta 41t. y arribaban a los puertos de Alguero, Cagliari, Bugía, Túnez, Sicilia, Rodas, Esmirna, *Altotoch*, Beirut, Alejandría, Quíos y Constantinopla (Càrrere 1977: 331-332; Del Treppo, 1975: 71, 150; Giménez Soler 1910: 192; Peláez 1981: 216).[6] Los duques de Borgoña podrían haber consumido miel de la Corona de Aragón. Esta es una de las hipótesis a partir del estudio de los restos de polen hallados en pozos negros del palacio de los duques de Borgoña en Brujas (Deforce 2010: 337-342).

Las ordenanzas sobre la venta de miel son poco numerosas en las recopilaciones de normativas locales en comparación a las de otros alimentos, como pan y carne. Esto no es porque fueran un edulcorante poco importante sino porque había una escasa presión fiscal sobre su comercialización. Además, las normas locales sólo legislan la comercialización del edulcorante al por mayor, al ser el único que estaba fiscalizado (Alcover Cateura 2019: 201-203).

En general, la normativa municipal menciona que cualquier vecino, hombre o mujer, cristiano, judío, musulmán o converso puede ser vendedor de miel. Se señala a veces un tipo concreto de vendedor, un campesino apicultor, como es el caso de Igualada (Cataluña) e Ibiza (Islas Baleares) (Castellà Raich 1954: 43; Ferrer Abárzuza 2002: 290). Por otro lado, las fuentes notariales recogen otro tipo de vendedor, un mercader, especializado o no en este producto (Càrrere 1977: 331; Cerdà Mellado 2012: 87).

Según las ordenanzas, las tiendas de los vendedores se debían localizar en un rincón de una plaza pública. De nuevo, Igualada e Ibiza nos ofrecen ejemplos de esta normativa (Castellà Raich 1954: 43; Ferrer Abárzuza, 2002: 290). En el Principado de Cataluña, la parada consistía en una mesa rudimentaria, formada por

dos cajas de madera que hacían de patas y uno o dos tablones que hacían de tablero. Encima de la mesa iban los recipientes, separados por tamaño y tipo de miel. En caso de inclemencias climáticas, el vendedor montaba un toldo con unos palos clavados en el suelo y una tela. Todo era fácilmente desmontable porque debía quedar despejado al final del día. Las paradas del resto de territorios y reinos de la Corona de Aragón serían similares (Soler Sala, 2006: 520-527).

La normativa municipal de Barcelona y Igualada detalla que en el mismo sitio dónde se vendía miel, se vendía aceite de oliva (Bajet Royo 1994: 578-580; Castellà Raich 1954: 43). Esto sería por dos motivos: el uso de los mismos recipientes y facilitar al comprador la adquisición de productos usados para la conservación (Bajet Royo 1994: 578-580).

Por otra parte, las tiendas de los mercaderes eran plantas bajas o casas-tienda. La ubicación de estos locales podría estar en cualquier vía pública o plaza del municipio. También habría jarras y cántaros grandes para la venta al por mayor. Los cántaros se forraban de mimbre para protegerlos de los golpes (Càrrere 1977: 331; Casas Homs 1970: 35, 37).

Los recipientes para aceite de oliva y miel contaban con dos sellos en todos los establecimientos de los mercados de los diversos territorios de la Corona de Aragón. El primero era el de su fabricante, para poder identificarle si la pieza presentaba taras. Este se situaba en un lugar visible fácilmente. Esta marca se documenta especialmente en el Principado de Cataluña (Bajet Royo 1994: 579; Olivar Daydí 1952: 94; Batllori Munné y Llubià Munné 1949: 109).

El segundo sello era del *mostassaf*, un inspector del mercado que confirmaba así la calidad del contenido de la jarra (Arxiu Municipal de Girona, Fons Ajuntament de Girona, Llibre del mostassaf, RG 17490, f. 2v-6v; Bajet Royo 2002: 140-142; Cabanes 1989: 77; Caria 1994: 57; Falcón Pérez 1978: 116; Guinot Rodríguez 2006: 97; Sevillano Colom 1957: 175). El sellado implicaba el cobro de un impuesto indirecto. Dicha marca se realizaba consistía en el escudo del propio inspector, del municipio, de un argentero o una cruz (Arxiu Municipal de Girona, Fons Ajuntament de Girona, Llibre del mostassaf, RG 17490, f. 2v; Bajet Royo 1994: 265-268, 578-580; Ferrer Abárzuza 2002: 276).[7] Esta técnica de sellado se ha documentado arqueológicamente en

---

villa. Por ejemplo, una ordenanza del libro del *mostassaf* de Igualada sólo se menciona que en un lugar de la plaza del grano se tiene que vender todo el aceite y miel.

[6] *Altotoch* es probablemente la ciudad turca de Alaçati.

[7] En Zaragoza se conservan numerosos pregones del siglo XV que muestran la actividad del *mostassaf* durante la inspección de objetos de la metrología usados por los vendedores, Archivo Municipal de Zaragoza, Serie Libros de Cridas o pregones, PRE. 2, Cuadernillo 1, ff. 1v-3v; Cuadernillo 2, ff. 2r-3v; Cuadernillo 4, ff. 1r-3v. Este es el primer libreo de pregones zaragozanos. Hay dieciséis más con documentación similar del mismo inspector.

Santa Creu de Rodes (Cataluña) (Mataró Pladelasala, Ollich y Puig Griessenberg 2010-2012: 90-99). Es posible que alguno de los recipientes cerámicos tuviera miel, pero no se tendrán datos hasta que no se realicen estudios al respecto. Seguramente ocasiones en que los mercaderes y apicultores utilizaban circuitos de venta paralelos al oficial. Así se libraban de pagar el tributo (Càrrere 1977: 331-332).

Las estafas más comunes recogidas en las ordenanzas consistían en venderla endurecida y añadir agua. Por ejemplo, en el reino de Mallorca, la multa por vender edulcorante al por mayor en mal estado eran habitualmente 10 *lliures* y la pérdida del producto (Pons, 1949: 58). En el reino de Valencia, no se podía reutilizar ningún recipiente de miel, al no estar barnizado, y la pena por hacerlo eran 5 *sous* (Furió y García-Oliver 2007: 99-100; Guinot Rodríguez 2006: 281).

Por ahora, no se ha encontrado ordenanzas municipales sobre la venta de jalea real[8] y propóleo.[9] El hecho no implica la inexistencia de ambos productos, sino que no se fiscalizaban, por lo que no generaron un registro.

La compra y el almacenaje doméstico

Los compradores al menudeo iban habitualmente a la tienda con sus propios recipientes, *alfabions*[10] y jarras pequeñas (Ferrer Abárzuza 2002: 187; Rotger Capllonch 1969: 147; Sastre Moll 1997: 136, 159).[11] Seguramente, los clientes al por mayor usaban sus recipientes propios, pero no se tiene información al respecto.

En general, el edulcorante se guardaba sobre todo en jarras y cántaros en las casas, y también en otros recipientes como *pimenteres*[12], botes y ollas (Casas Homs 1970: 35, 37; Santanach Soler 1998: 242; Vinyolas Vidal 2001: 485, 513). La tipología de la *pimentera*, cacharro de origen medieval, se mantuvo prácticamente inalterada hasta, como mínimo, finales del siglo XVIII.[13] En este siglo, este cacharro tenía una forma cóncava en su interior para que fuera fácil sacar miel con una

Figura 2. *Pimentera* de la colección Marroig conservada en las dependencias del Servicio de Patrimonio Histórico del Consejo de Mallorca. La pieza se ha datado en el siglo XVIII. La decoración consiste en unos riñones. El taller es Manises (Valencia). La imagen es del fotógrafo Juan Ramon Bonet. Se tiene su permiso de reproducción para esta publicación. La fotografía original se encuentra en el catálogo de la cerámica de la colección Marroig (Cantarellas, C. *et al.* 2006: 121).

cuchara. Además, se decoraba a menudo con riñones (Cantarellas et al. 2006: 121). Quizá esta decoración sea porque la miel se empleaba en fórmulas magistrales para tratar las piedras en el riñón. Es posible que estas piezas en la Baja Edad Media estuviesen decoradas con los mismos motivos, pero no se tiene información al respecto. En el inventario del domicilio de Benet Martí Roig, apotecario de Barcelona, datado en el año 1560 se mencionan *pimenteres* con o sin boca de bronce (Cerdà Mellado 2012: 93). Es muy posible que estas dos tipologías existieran en el medievo. El bronce no tenía ningún uso técnico específico: una loza o madera tenía

---

[8] Complemento alimenticio lo producen las abejas obreras, que transforman el polen de las flores dentro de su organismo y secretándola por sus glándulas. Cabe destacar que la cosecha de la jalea real se realiza de las celdas de las abejas reinas directamente, ya que son las que tienen la mayor cantidad de sustancia. Debido al hecho de que la jalea real es perecedera, a veces se añade miel o cera de abeja para prolongar su preservación.

[9] Substancia que las abejas recolectan y elaboran a partir de resinas vegetales para arreglar o fortalecer la colmena y para desinfectar sus celdas. Es ingrediente en pomadas, jabones y productos farmacéuticos que apoyan al sistema inmunológico.

[10] Tinajas de cerámica barnizada pequeñas de cuatro nansas.

[11] Los *alfabions* fueron usados hasta época reciente en Ibiza para contener miel.

[12] Vaso pequeño de cerámica barnizada que contenía pimienta, otras especias y miel.

[13] Agradezco esta información al personal de restauración y conservación del Museo de Mallorca.

la misma utilidad ¿Entonces, porqué lo pusieron? Por razones mágicas porque se inscribían exclusivamente en bronce fórmulas mágicas para evitar que fueran allá ratones y moscas (Page 2019: 454). Los cántaros en las casas iban a veces forrados de hojas de palmera para proteger el recipiente de golpes (Casas Homs 1970: 37).

Usos

Cañamiel, melar, meloso, melaza, melón, melocotón, entre otros, comparten el mismo origen etimológico ¿Porqué sucede esto? La miel fue habitualmente el principal edulcorante de la población medieval (Riera Melis 2013a: 74).[14] Por este motivo, era la referencia a lo dulce en la cultura colectiva. Así se fomentó la aparición de dicciones de uso corriente vinculadas a su agradable sabor. *Això és mel* (esto es miel) o *mel* (miel), se utilizan todavía en catalán para referirse a que un alimento o situación es agradable.

Los alimentos básicos de la población medieval de la Corona de Aragón eran pan, carne, pescado y vino (Alcover Cateura 2019: 131). Estos eran los que tenían el circuito comercial más controlado por las autoridades. Además, eran los alimentos más fiscalizados por los gobiernos municipales y los señores jurisdiccionales (Alcover Cateura 2019: 140-190). La miel no era esencial en la alimentación de la población, pero sí tenía muchos usos a lo largo del año.

Las recetas de cocina y la elaboración de medicamentos eran sus usos principales. En la cocina, superaba al azúcar en dulzura, contundencia y potencia gustativa (Carré Pons 2017: 255). El producto tenía dos usos principales en cocina: edulcorante y conservante. Uno no tenía porqué excluir al otro. De hecho, en numerosos platos y productos, ambos usos eran inseparables.

Como edulcorante mejoraba el sabor del vino *piment*, bebida consumida por amplios sectores sociales.[15] Asimismo, dulcificaban *sèmolas* y *farines*, una especie de polentas. Estos platos tenían origen popular, pero eran apreciados por las clases altas. Para mejorar su sabor cocían los granos en leche de almendra y no en agua, como su receta original. Para seguir mejorándolos y espesándolos, le añadían miel o azúcar. Su consumo estaba ampliamente extendido

entre todos los estamentos. Durante Cuaresma y otros períodos de guardar, se consumían sin endulzar, salvo si estabas enfermo (Riera Melis 2017: 407). Además, confería dulzor a la *salsa salvatgina*, de sabor fuerte. Su potencia gustativa se lograba con una base de vinagre. Se utilizaba para acompañar carnes de caza y volatería (Banegas López 2014: 158). También quitaba el sabor amargo de la mantequilla cruda, alimento escasamente consumido en las tierras al sur de los Pirineos.

Como conservante se utilizó con rábanos dulces y nabos tiernos. La elaboración consistía en mezclar las verduras en una olla hirviendo con caldo de carne y el edulcorante (Carré Pons 2017: 23). Un caso particular son los rábanos *galesc*. Primero se cortaban en trozos pequeños y después se cocían en agua con sal. Cuando ya estaban bien cocidos, se retiraban y se metían en agua fría ocho días. El agua se tenía que cambiar a diario. Más adelante, se hervía en *exerop,* es decir, miel con agua, a razón de media escudilla de agua por libra de miel. Cuando la mezcla estaba tibia, se metían los rábanos dentro. La cantidad debía ser la misma de *exerop* y de rábanos. En ambos casos, se guardaban en cacharros para su conservación, normalmente en botes o ollas (Grewe, Soberanas y Santanach 2004: 274, 279). Hasta aquí las recetas dónde la miel especialmente edulcora o conserva.

En el resto de casos, la miel tanto dulcifica como conserva. Se utilizaba como ingrediente imprescindible en los postres de las mesas populares, como el turrón, (Colesanti, 2008: 148; Grewe, Soberanas y Santanach 2004: 274, 294). Todos los dulces se elaboraban habitualmente en los hogares. Los turrones se comían en Navidad. Por lo tanto, los dulces fueron consumidos normalmente durante fiestas religiosas. De aquí, el uso de expresiones como *per Tots Sants castanyes, i per Nadal torrons* ('por Todos los Santos castañas, y por Navidad turrones'). Los turrones y otros dulces, excepto *confits*, se guardaban en las *caixas torroneres* ('cajas para turrón'), que iban cerradas con llave para evitar tentaciones (Vinyoles Vidal 2001: 485, Villalmanzo Cameno 1999: 252).

A lo largo del año se comían *confits*, excepto durante la Cuaresma y otros períodos de abstinencia. *Confit* es un vocablo que englobaba a caramelos, frutas frescas y secas, condimentos y especias.[16] Estos dulces se

---

[14] De hecho, según Riera, ya era así desde la Alta Edad Media.

[15] Vino aromatizado con miel, jengibre, canela, clavo y pimienta. Su composición variaba según la región y las capacidades económicas de cada individuo o comunidad. Es el heredero del vino *conditum paradoxum* de la antigua Roma, que contenía vino, miel, pimienta, azafrán, lentisco, laurel, dátiles asados y pasas. En la Edad Media, se consumía frío en verano, caliente en invierno y por Navidad, Adviento y otras fiestas del calendario litúrgico. Variedades de esta bebida se consumen todavía en el centro de Europa. Por ejemplo, el *glühwein* alemán, el *glögg* suevo, *grzane wino* polaco. En todos los casos coinciden en ser bebidas típicas de Adviento y Navidad.

[16] Esta es una lista de alimentos confitados: nuez (tres plantas del género *juglans regia*), avellana (siete especies del género *corylus avellana*), almendra (cuarenta y tres plantas del género *prunus dulcis*), piñón (*pinus pinea*), pistacho (*pistacia vera*), calabaza (veintiún plantas de los géneros *cucurbita* y *lagenaria*), naranja (veinte especies del género *citrus sinensis*), membrillo (*cydonia oblonga*), cidro (*citrus medica*), oliva (setenta especies del género *olea europea*), jengibre (dos especies del género *zingiber officinale*) y anís (*pimpinella anisum*). Además, el agua de cebada (antiguo preparado alimentario para elaborar horchata) y la *espongea* (tipo pasta real esponjosa, de aquí su

elaboraban también en casa (Vinyoles Vidal 1988: 153-154, 165; Gimeno Blay et al. 2009: 73, 77). Las familias los guardaban normalmente en *pots*, botes de cerámica barnizada (Batlle Gallart 2007: 850; Sastre Moll 1997: 136; López Pizcueta 1992: 32). En Jueves Lardero, el jueves que da comienzo al Carnaval y en el día de Carnaval, fue tradicional consumirlos (Vinyoles Vidal 1988: 153-154, 165).

Durante la Baja Edad Media, el azúcar substituyó progresivamente a la miel tanto en la elaboración de dulces como de medicamentos (Bajet Royo 1994: 493; Casas Homs 1970: 37; López Pizcueta 1992: 38, 43-45; Vela Aulesa 2003: 49).[17] Primero desde su introducción a mediados del siglo XII hasta la primera mitad del siglo XIV tuvo casi exclusivamente usos medicinales. Posteriormente, se extendió como alimento de lujo en las mesas privilegiadas (Riera Melis 2013b: 383-384). Este proceso de substitución de uno por otro se vinculó especialmente a las clases adineradas (Pérez Samper 1998: 54-56, 136). La adopción progresiva del azúcar por parte de las élites urbanas se debe diversos factores: los médicos consideraban al azúcar blanco el edulcorante más sano; era una forma de marcar las diferencias económicas con las clases populares; la ostentación implícita de permitirse estar al día de las corrientes culinarias; y la simbología positiva del blanco que representaba la pureza e inocencia (Riera Melis 2013a:74). A finales del siglo XIV hubo una cierta 'democratización' de su consumo, pero el alcance de este fenómeno es complicado (Coulon Damien 2001: 742; Ouerfelli 2008: 337).

En los recetarios medievales, libros propios de las bibliotecas privadas de miembros de élites urbanas y de órdenes regulares poderosas, la miel siempre tiene un papel menor que el azúcar. Por ejemplo, en el *Llibre de Sent Soví* (siglo XIV), el más conocido recetario del medievo catalán, hay un total de seis recetas con miel frente a las veintiséis con azúcar. La miel, excepto en dos recetas del *Llibre*, es sustituto barato del azúcar (Santanach Soler y Barrieras 2016: 174-239).

En las ordenanzas de la corte del rey Pedro el Ceremonioso, este especifica que el *museu*, un criado de su máxima confianza, debía tener siempre miel para el consumo personal de la familia real. Seguramente era para elaborar polentas, dulces y salsas. El rey señala la necesidad de comprar azúcar constantemente para tenerlo todo el año en la despensa. Especifica su uso en todos los *confits* de consumo propio (Gimeno Blay et al. 2009: 73, 77).[18] Por tanto, el rey usaba preferentemente el azúcar, aunque no abandonaba el consumo de miel.

Los médicos de la Corona de Aragón recomendaron comer miel con moderación. El alimento lo consideraban astringente, razón añadida para pautarlo en pequeñas cantidades. Si era posible, el azúcar blanco se sugería emplear como su substituto. Dos recetas con miel tenían efectos adversos. El vi *piment* inflamaba el estómago y las *sèmolas* y *farines* causaban la multiplicación de gusanos intestinales. Estos gusanos generaban obstrucciones en la vejiga y los riñón (Carré Pons 2017: 229, 250, 255-256, 264; Riera Melis 2017: 407). En cambio, el rábano *galesc* con o sin miel tenía efectos positivos y adversos. Los positivos eran principalmente orinar con más facilidad, tratar el dolor de riñones, vejiga, la carraspea, aumentar la leche materna y la fogosidad en el hombre. Los adversos eran sobretodo dolores de ojos, boca y aumento de las ventosidades (Faraudo de Saint-Germain 1943: 138-139).

La *Concordie Apothecariorum Barchinone* recoge numerosas fórmulas magistrales con miel. Con ella, se elaboraron jarabes simples y compuestos, electuarios,[19] emplastos,[20] ungüentos, aceites y colirios (Ballester y Castelló 1944: 31-59; Vela Aulesa 2003: 186, 225). Dichos fármacos ayudaron a prevenir y/o a combatir el vértigo, la gota, las piedras en el riñón, los tapones de cera, los tumores, el vómito, las úlceras, la indigestión y los ronquidos. También se usó como pomada para limpiar y curar heridas (Montero Hernández y Gutiérrez Urbón 2016: 52). El médico Arnau de Vilanova la recomendó para ancianos y enfermos por ser un alimento caliente y reconfortante (Gil-Sostres 1996: 383).

Según un tratado de veterinaria para caballos del siglo XV, el edulcorante popular se utilizó junto a otras substancias para tratar enfermedades como bubones en el cuello, problemas de corazón y heridas en el lomo (Gili 1985: 56, 58, 87). En la misma línea, en un tratado de cetrería del siglo XIV, al ave de caza azor (*accipiter gentilis*) se aconsejó darle carne embadurnada de miel cuando enfermaba (García Sempere 2002: 104). Las colmenas y su preciado líquido sirvieron también para elaborar trampas para cazar osos. Esta práctica se documenta en el siglo XVIII, pero es posible que su origen sea anterior (Torrente Sánchez-Guisande 1999: 134).[21]

---

nombre, típica de la repostería sefardí) se endulzaban.
[17] Este cambio de aprecia, por ejemplo, en las ordenanzas referidas a confiteros de Barcelona del año 1448.
[18] Ibídem, loc. cit.

[19] Medicamento de consistencia líquida, pastosa o sólida, compuesto de varios ingredientes, casi siempre vegetales, y de cierta cantidad de miel, jarabe o azúcar, que en sus composiciones más sencillas tiene la consideración de golosina. Esta consideración explica su origen etimológico del griego ἐκλείχειν (*ekleikhein*) 'lamer'.
[20] Preparado farmacéutico de uso tópico, sólido, moldeable y adhesivo.
[21] La literatura sobre veterinaria y cetrería medieval es amplísima. Conviene destacar algunas obras, las del Dr. Fradejas Rueda: Fradejas Rueda 1999; Fradejas Rueda 2004; Fradejas Rueda 2010; Fradejas Rueda 2017.

Finalmente, la miel se utilizaba para untar el cuerpo de los criminales condenados a la picota para provocarles picaduras (Ferrer Abárzuza 2002: 275).[22] Esta práctica se ha documentado en Ibiza, pero es probable se extendiese por otros municipios de la Corona de Aragón. Este es posiblemente el origen del *moixó foguer*, personaje del carnaval catalán con probables orígenes medievales y que hoy en día pervive en Vilanova i la Gertrú (Garraf) y Valls (Alt Camp) (Massip 2008:138).[23] El personaje va completamente desnudo y se cubre el cuerpo entero con miel y plumas. Se esconde dentro de una caja llena de plumas de volatería y le acompaña una comitiva. Ocasionalmente, el *moixó* sale de la caja arrojando plumas por todos lados. Hasta inicios del siglo XX, el hombre escogido para este papel era pobre y de mala reputación.

Conclusiones

La miel tuvo numerosos usos en la Corona de Aragón durante la Edad Media. El primero y más importante, es que fue uno de los principales edulcorantes y conservantes utilizados por la gran mayoría de la población. Esto contribuyó a que se elaborasen y vendiesen muchas variedades. Cada región contaba con sus variedades propias, tanto de 'cruda' como de 'cocida'. La 'cruda' era la más valorada y la 'cocida' la más utilizada.

Los *confits* con miel se consumía a lo largo del año, excepto en tiempos penitenciales como la Cuaresma. También tenía un papel importante en la elaboración de dulces durante las fiestas del calendario litúrgico. Además, se utilizaba en la elaboración de salsas de sabor potente para las carnes más finas y caras, las de caza y de volatería. Otro uso habitual de la miel era como conservante de verduras.

Asimismo, sus usos medicinales eran variados y apreciados por médicos y pacientes. No se tomaba normalmente sola sino como componente de una fórmula magistral. Dicho esto, se recomendaba a los débiles, por la edad o la enfermedad, para reconfortarse. Caballos y aves de caza también se aprovechaban de las curas con miel para tratar diversas enfermedades.

Estos usos principales de la miel se relacionan con los tres tipos de vendedores: campesinos apicultores, mercaderes y apotecarios. Seguramente, la miel empleada para el consumo sería comprada preferentemente a los dos primeros. En cambio,

la utilizada para fines medicinales, se adquiría probablemente en la botica.

Por otra parte, una utilización tenebrosa era untar el cuerpo de criminales para causarles daño mediante las picaduras. Para finalizar, al no ser un alimento básico en la alimentación, su producción, distribución y venta no tenía normativas locales muy restrictivas ni padecía mucha presión fiscal por medio de impuestos indirectos.

A través de la miel se puede estudiar la sociedad urbana bajomedieval de la Corona de Aragón. Su mayor uso es un indicativo de pertenencia a las clases populares. En cambio, su menor consumo es típico de las clases más adineradas.

Sin duda, el azúcar tuvo mayor estima entre las clases sociales más ricas. Estas lo veían como un alimento más sano que la miel, siguiendo la opinión de los médicos de la época. Asimismo, era un signo de riqueza y una forma fácil de diferenciar una mesa opulenta de una popular. También la simbología del color jugaba a favor del maná de la caña: el color amarillento de la miel se vinculaba a la riqueza, la codicia y a los judíos. Por contra, la blancura del azúcar lo convertía en un alimento puro e inocente, fuera de toda connotación negativa. Aún así, la miel fue el edulcorante que marcó la mentalidad, la consciencia colectiva en el medievo, como atestiguan las dicciones, expresiones y refranes todavía en uso.

Bibliografía

Aguiló, C. 2002. *Toponímia i etimología*. Barcelona: Abadia de Montserrat.

Alcover Cateura, P. J. 2019. Els mercats alimentaris de la Corona d'Aragó a través de la documentació municipal, Tesis doctoral inédita, Universitat de Barcelona, Facultat de Geografia i Història.

Bajet Royo, M. 1994. El Mostassaf de Barcelona i les seves funcions en el segle XVI: edició del "Llibre de les ordinations." Barcelona: Fundació Noguera.

Bajet Royo, M. 2002. Policia de mercat a l'època medieval. *Revista de Dret Històric Català* 2: 121-143.

Banegas López, R. A. 2014. Una anàlisi dels productes i tècniques de cuina al Llibre de Sent Soví, en J. Santanach i Suñol (coord.), *Llibre de Sent Soví* (Col·l. 7 Portes): 129-166. Barcelona: Editorial Barcino.

Batlle Gallart, C. 1994. Francesc Ferrer, apotecari de Barcelona vers el 1400, i el seu obrador. *Miscel·lània de Textos Medievals* 7: 499-547.

Batlle Gallart, C. 2007. Guillem Eimeric, jurista d'una família patrícia de Barcelona (+ 1301). *Anuario de estudios medievales* 37/2: 823-866.

Batllori Munné, A. y L.M. Llubià Munné. 1949. *Ceràmica catalana decorada*. Barcelona: Llibreria Tuebols.

Bertran de Heredia Bercero, J. 1998. Tipologia de la producció barcelonina de ceràmica comuna baix

---

[22] Se usa la versión de imprenta de este libro, que me facilitó el mismo autor. No coincide la paginación con el libro acabado, de difícil consulta debido a una tirada de pocos ejemplares.

[23] Agradezco la información sobre el personaje a la Dra. Natalia Moragas Segura, profesora Serra Hunter en el Departamento de Historia y Archeología de la Universidad de Barcelona.

medieval: una proposta de sistematizació, Ceràmica medieval i postmedieval: circuits productius i seqüències culturals, en J. I. Padilla Lapuente y J. M. Vila Carabas (eds) *Ceràmica medieval i postmedieval. Circuits productius i seqüències culturals:* 177-206. Barcelona: Publicacions Universitat de Barcelona.

Cabanes, M. L. (ed.) 1989. *Almotacen. el llibre del mustaçaf de la ciutat d'alacant.* Alicante: Departamento de Publicaciones e Imagen Ayuntamiento de Alicante.

Cantarellas, C. et al. 2006. *La cerámica de la col.lecció Marroig.* Palma: Consell de Mallorca.

Caria, R. 1994. Les ordinacions municipals de l'Alguer (1526). *Revista d'Història del Dret* 22: 45-70.

Carré Pons, A. (ed.) 2017. *Regiment de sanitat per al rei d'Aragó. Aforismes de la memòria. Arnau de Vilanova.* Barcelona: Universitat de Barcelona.

Càrrere, C., 1977. *Barcelona 1380-1462. Un centre econòmic en època de crisi.* Barcelona: Curial, vol. 1.

Carmona Ruiz, M. A. 2000. La apicultura sevillana a fines de la Edad Media. *Anuario de Estudios Medievales* 20/1: 387-421.

Casas Homs, J. M. 1970. L'heretatge d'un mercader barceloní. Darreries del catorzèn segle. *Cuadernos de Historia Económica de Cataluña:* 9-112.

Cerdà Mellado, J. A. 2012. *La loza catalana de la colección Mascort.* Torroella de Montgrí: Fundació Mascort.

Castellà Raich, G. 1954. *Libre de la mostaçaferia. Ordinacions de la vila d'Igualada.* Igualada: Centro de Estudios Comarcales de Igualada.

Crane, E. 1999. *The World History of Beekeeping and Honey Hunting.* New York: Routledge.

Colesanti, G.T. 2008. *Una mujer de negocios catalana en la Sicilia del siglo XV: Caterina LLull i Sabastida. Edición y estudio de su libro maestro 1472-1479.* Barcelona: CSIC.

Coulon, D. 2001. El comercio catalán de azúcar en el siglo XIV. *Anuario de Estudios Medievales* 31 2: 727-756.

Deforce, K. 2010. Pollen analysis of 15th century cesspits from the palace of the dukes of Burgundy in Bruges (Belgium): evidence for the use of honey from the western Mediterranean. *Journal of Archaeological Science* 37/ 2: 337-342.

Del Treppo, M. *Els Mercaders catalans i l'expansió de la corona catalano-aragonesa al segle XV.* Barcelona: Curial.

Falcón Pérez, M. I. 1978. *Organización municipal de Zaragoza en el siglo XV: con notas acerca de los orígenes del régimen municipal en Zaragoza.* Zaragoza: Departamento de Historia Medieval de la Facultad de Filosofía y Letras.

Faraudo de Saint-Germain, L. 1943 (ed.) *El «Libre de les medicines particulars». Versión catalana trecentista del texto árabe del tratado de medicamentos simples de Ibn Wáfid, autor médico toledano del siglo XI.* Barcelona: Real Academia de Buenas Letras de Barcelona.

Ferrer Abárzuza, A. 2002. *El Llibre del mostassaf d'Eivissa. La vila d'Eivissa a la baixa edat mitjana.* Ibiza: l'Editorial Mediterrània-Eivissa.

Fradejas Rueda, J. M. (ed.) 1999. *Textos clásicos de cetrería, montería y caza.* Madrid: Mapfre.

Fradejas Rueda, J. M. 2004. *El arte de cetrería de Federico II.* Ciudad del Vaticano: Biblioteca Apostólica Vaticana.

Fradejas Rueda, J.M. 2010. La versión castellana del "Livro de falcoaria" de Pedro Menino de Gonzalo Rodríguez de Escobar. *Incipit* 3: 85-100.

Fradejas Rueda, J.M. 2017. Los libros de caza medievales y su interés para la historia natural. *Arbor: Ciencia, pensamiento y cultura* 5/2: 1-9.

Furió, A. y F. Garcia-Oliver. 2007. *Llibre d'establiments i ordenacions de la ciutat de València. I.: (1296-1345).* Valencia: Universitat de València,

García Ramón, J. L. et al. 2016. *Estrabón. Geografía.* Madrid: Gredos, 2 vols.

García Sempere, M. 2002. *Quan d'ombri Déu sa curatura, aproximación a un tratado catalán de cetrería en verso.* En *La caza en la Edad Media* (José Manuel Fradejas Rueda ed.). Tordesillas: Universidad de Valladolid.

Gil-Sostres, P. (ed.) 1996. *Arnau de Vilanova, Opera Medica Omnia.* Barcelona: Universitat de Barcelona,

Gili, J. Lo cavall. 1985.*Tractat de manescalia del segle XV.* Oxford: The Dolphin Book Co.

Giménez Soler, A. 1910. El comercio en tierra de infieles durante la Edad Media. *Boletín de la Real Academia de Buenas Letras de Barcelona:* 171-199.

Gómez Urdáñez, C. 2019. *Iluminaciones naturales y revestimientos cromáticos: Historia de los acabados de la Catedral de Santa María de la Huerta de Tarazona (siglos XIII-XXI).* Zaragoza: Prensas de la Universidad de Zaragoza.

Grewe, R., A. J. Soberanas y J. Santanach. 2004. *Llibre de Sent Soví. Llibre de totes maneres de menjar. Llibre de totes maneres de confits.* Barcelona, Editorial Barcino.

Guinot Rodríguez, E. 2006. Establiments municipals del Maestrat, els Ports de Morella i Llucena (segles XIV-XVIII). Valencia: Universitat de València.

De Osma, J. y G. Scull 1908. *Apuntes sobre cerámica morisca: Los maestros alfareros de Manises, Paterna y Valencia; contratos y ordenanzas de los siglos XIV, XV y XVI.* Valencia: Hijos de M.G. Hernández,

Jordi González, R. 1997. *Aportació a la història de la farmacia catalana.* Barcelona: Fundació Uriach 1838.

López Pizcueta, T. 1992. Los bienes de un farmacéutico barcelonés del siglo XIV: Francesc de Camp. *Acta historica et archeologica medievalia* 13: 17-73.

Massip, F. 2008. Rei d'innocents, Bisbes de burles: rialla i transgressió en temps de Nadal, en J. L. Sirera Turó (ed.) *Estudios sobre teatro medieval:* 131-146. Valencia, Publicacions de la Universitat de València.

Mataró Pladelasala, M., I. Ollich y A. M. Puig Griessenberger. 2010-2012. Unes mesures de ceràmica trobades a Santa Creu de Rodes (el Port de la Selva, Alt Empordà). *Arqueologia Medieval* 6-7: 90-99.

Montero Hernández, A. M. y J. M. Gutiérrez Urbón. 2016. Miel sobre piel. "Proyecto Lumbre". *Revista*

Multidisciplinar de Insuficiencia Cutánea Aguda 11: 52-57.

Ollé Albiol, M. 1996. *El llibre de les abelles (De setis, lligallos i abellers)*. Barcelona: Abadia de Montserrat.

Olivar Daydí, M. 1952. *La ceramica trecetista en los países de la corona de Aragón*. Barcelona: Editorial Seix Barral.

Ouerfelli, M. 2008. *Le sucre: production, commercialisation et usages dans la Méditerranée médiévale*. Leiden: Brill.

Page, S. 2019. Medieval magical figures. Betweem image and text, en S. Page y C. Rider (eds) *The Routledge History of Medieval Magic*: 432-457. Oxford: Routledge.

Peláez, M. J. 1981. *Catalunya després de la guerra civil del segle XV*. Barcelona: Curial.

Pérez Samper, M. A. 1998. *La alimentación en la España del Siglo de Oro*, Huesca: La Val de Onsera.

Riera Melis, A. 2013a. El azúcar en la farmacopea y la alta cocina árabes medievales. en F. Sabaté (ed.) *El sucre en la Història. Alimentació, quotidianitat i economia*: 21-90. Lleida: Universitat de Lleida.

Riera Melis, A. 2013b. Sucre per a després d'una pesta: Barcelona, 1349-1350, en M. Sánchez Martínez et al. (ed.) *A l'entorn de la Barcelona medieval. Estudis dedicats a la doctora Josefina Mutgé i Vives*: 367-386. Barcelona: CSIC.

Riera Melis, A. 2017. *Els cereals i el pa en els països de llengua catalana a la baixa edat mitjana*. Barcelona: Institut d'Estudis Catalans.

Rotger Capllonch, M. 1969. *Historia de Pollensa*. Palma de Mallorca: Imp. Sagrados Corazones, vol. 2.

Santanach Soler, J. 1998. Ceràmica comuna d'època moderna, en J. I. Padilla (ed.) *Ceràmica medieval i postmedieval: circuits productius i seqüències culturals*: 225-272. Barcelona: Universitat de Barcelona.

Santanach Soler, J. y M. Barrieras. 2016. *Llibre de Sent Soví*, Barcelona: Barcino.

Salicrú Lluch, R. 1995. *El tràfic de mercaderies a Barcelona segons els comptes de la lleuda de Mediona (febrer de 1434)*. Barcelona: CSIC.

Sastre Moll, J. 1997. *Alguns aspectes de la vida quotidiana a la Ciutat de Mallorca: època medieval*. Palma de Mallorca: Institut d'Estudis Baleàrics.

Sesma Muñoz, J. A. 2001. El bosque y su explotación económica para el mercado en el sur de Aragón en la Baja Edad Media, en J. Clemente Ramos (ed.) *El medio natural en la España medieval. Actas del I Congreso sobre ecohistoria e historia medieval*: 195-216. Cáceres: Universidad de Extremadura.

Torrente Sánchez-Guisande, J. P. 1999. *Osos y otras fieras en el pasado de Asturias, 1700-1860*. Proaza: Fundación Oso de Asturias.

Sevillano Colom, F. 1957. *Valencia urbana medieval a través del oficio de Mustaçaf*. Valencia: Instituto Valenciano de Estudios Históricos, Institución Alfonso el Magnánimo, Diputación Provincial de Valencia.

Vela Aulesa, C. 2003. *L'Obrador d'un apotecari medieval segons el llibre de comptes de Francesc Ses Canes (Barcelona, 1378-1381)*. Barcelona: CSIC.

Vinyoles Vidal, M. T. 1988. El rebost, la taula i la cuina dels frares barcelonins al 1400. En *Alimentació i societat a la Catalunya Medieval*: 137-166. Barcelona: CSIC.

Vinyoles Vidal, T. 2001. La qualitat de vida en un mas del vallès al segle XV. Estudi de l'inventari del mas Canals de Rubí (Can Rosés), en M. T. Ferrer Mallol, J. Mutgé Vives y M. Riu Riu (eds) *El mas català durant l'Edat Mitjana i la Moderna (segles IX-XVIII). Apectes arqueològics, històrics, arquitectònics i antropològics. Actes del Col.loqui celebrat a Barcelona del 3 al 5 de novembre de 1999*: 479-520. Barcelona, CSIC.

Villalmanzo Cameno, J. 1999. *Ausiàs March: colección documental*. Valencia: Institució Alfons el Magnànim, Diputació de València.

Bibliografía de manuscritos

Arxiu Municipal de Girona, Fons Ajuntament de Girona, *Llibre del mostassaf*, RG 17490.

Chapter 13

Honey and Wax in Medieval Tyrol on the Basis of Tyrolean Land Registers (*Urbaria*) and Books of Accounts[1]

Barbara Denicolò

University of Salzburg (barbara.denicolo@sbg.ac.at)

Summary: The chapter is an outline or first conclusion of a more extensive study on the history of honey and wax in medieval Tyrol. It discusses references to the production or importation of bee products in Tyrolean land registers (*urbaria*) and account books. The relevant literature on medieval food history argues that honey was widely used in the Middle Ages as a substitute for high-priced sugar. The small and limited area of Tyrol never really appears, because current research is mainly of a more general character. It remains unclear to what extent these studies can be applied to the situation in Tyrol. A specific investigation of the conditions in Tyrol, which occupies a special position due to political, climatic, geographical and cultural conditions, is therefore necessary. The chapter investigates whether and under what circumstances honey was produced in Tyrol or whether honey and wax were bought from outside.

Keywords: Middle Ages; Tyrol; account books; *urbaria*; honey; wax; oil

Preface

The following chapter of regional history focuses on the Tyrol region in the middle of the Central Alps, which is now divided between Austria and Italy. It is a small, distinct area with geographical and climatic peculiarities. Tyrol lies on the temperate border area of continental, Atlantic and Mediterranean influences and, due to its mountain structure, features many different subclimates. In the higher regions there is a subcontinental alpine mountain climate with low temperatures.

The area, which encompasses the north and south sides of the main Alpine ridge, was for a long time under common rule as the county of Tyrol. In the Tyrolean territory there is the Brenner Pass, the lowest of the Alpine crossings and therefore in the Middle Ages one of the most common travel and trade routes over the mountains, connecting north and south. The Tyrolean sovereigns ensured proper roads and the safety of travelers and took in turn customs duties and charges. The growing salt and silver mining made the territory and its rulers even more prosperous (Hödl 1998: 147-156; Hörmann 2007: 208-212; Maleczek 2009: 954; Mayer 1919/20: 113-116, 120-125, 164-65; Riedmann 1985: 431-437).

In the 12th century, the former part of the duchy of Bavaria grew steadily into an autonomous and prosperous domain, under the rule of the Bavarian aristocratic family called the Counts of Tyrol. In 1253 they were succeeded by the Meinhardiner Line, which quickly gained power and wealth, but died out in the male line already by 1335. The last surviving daughter Margarete, called Maultasch, handed the land over to her closest relatives from the Habsburg family, for whom Tyrol was of particular strategic importance (Brandstätter 2007; Hödl 1998: 147-156; Hörmann 2007: 208-212; Maleczek 2009: 954; Mayer 1919/20: 113-116, 120-125, 164-65; Riedmann 1985: 431-437). Until 1379 no sovereign was residing in Tyrol, which had a massive effect on the availability of historical sources from that time. In the 1420s Friedrich IV moved his residence from Meran in South Tyrol to Innsbruck in North Tyrol. After Friedrich's son, Siegmund, had retired from power in 1490, Innsbruck became the residence of the former Roman-German king and later Emperor Maximilian I. The last Habsburg ruler to reside in Tyrol was Ferdinand II and his wife Philippine Welser (Hödl 1998: 147-156; Hörmann 2007: 208-212; Maleczek 2009: 954; Mayer 1919/20: 113-116, 120-125, 164-65; Riedmann 1985: 431-437).

This chapter is an excerpt from a more complex study on the history of honey and wax in Tyrol, which is still in progress. In the work *Honig und Wachs im Mittelalter* I discuss references to honey and wax through different source genres. Currently, honey is an important touristic and marketing attraction in Tyrol, which is portrayed as a genuinely local product with a long and venerable historical tradition. The relevant literature on medieval food history argues that honey was widely used in the Middle Ages as a substitute for high-priced

---

[1] I thank Dr Armin Torggler for the successful initial spark for this study and the many other helpful suggestions.

sugar (e.g. Hundsbichler 1985: 205-206; Rösener 1985: 150; Schubert 2010: 164; Schulz 2011: 418, 541).

In comparison to sugar and its history, honey and wax are relatively ignored in the literature. Sugar, as the wickedly expensive and exotic spice that changed culinary tastes, has obviously fascinated the more and encouraged more research. However, even publications on the history of sugar can provide a good deal of information on honey, since honey and sugar have always been closely linked and the history of honey is often understood teleologically as the prehistory of sugar. The small and limited area of Tyrol never really appears in the literature, because research on this topic hitherto has mainly been of a more general character. It remains unclear to what extent these studies can be applied to the situation in Tyrol. A specific investigation of the conditions in Tyrol, which occupies a special position due to political, climatic, geographical and cultural conditions, is therefore necessary.

For this study I have used the most important genres of the Tyrolean written tradition of the 13th to 16th centuries, such as cookbooks and rent registers (*urbaria*), *Weistümer* (collections of local law traditions, awards or by-laws from central Europe), customs tariffs, notarial registers (*Notariatsimbreviaturen*), various public documents and account books related to Tyrol.

Urbaria list the possessions of individual aristocratic or ecclesiastical lordships, urban cooperative ventures or foundations and bourgeois companies as well as the levies flowing from them in the form of natural products or money. Books of accounts document the expenses for food and other luxury goods of these individuals or lordships. Cookbooks and collections of medical and dietary recipes, on the other hand, show the role honey played in the kitchen, dietetics and medicine, as well as the dishes and medicines in which it was employed. Cookbooks, however, primarily reflect the dietary choices of the wealthy, the main readership of this genre. They provide little information about the diet of the urban and rural population (cf. several cookbooks with a relation to Tyrol: Aichholzer 1999: 183-244; Danner 1970; Lemmer 1983; Prentki 1989: 199-222; Thurnher and Zimmermann 1979; Weiss Adamson 1996).

Furthermore, there are more normative sources in our possession that report on target states that should naturally follow on these prescriptions, allowing for fewer statements about how these targets would be implemented in practice. Thus, the *Capitulare de villis*, the regulations for managing Charlemagne's rural estates from the end of the 8th century, prescribes which food should be produced on Carolingian estates, but reports nothing about their implementation (Lerner 1984: 87-98).

Weistümer, on the other hand, are oral principles of law which were later written down and which regulated communal living and the relationship between subjects and lordship. These collections provide a good insight into the everyday life of the peasant class in Tyrol (e.g. Grass and Faussner 1994a, 1994b; Grass and Finsterwalder, 1966; Zingerle and Egger 1888, 1891; Zingerle and Inama-Sternegg 1875, 1877, 1880).

In the German-speaking countries, private and public legal transactions of various kinds were primarily carried out and documented in the form of charters certified with seals, whereas in the South they were performed and documented by so-called notarial deeds written by a public notary, who also recorded all notarized transactions in his personal registers. Since Tyrol was a border region, both source genres are widespread. As sealing was an aristocratic privilege for a long time, these early Tyrolean notarial registers mainly reflect the bourgeois urban environment. They provide many insights into everyday life through the marriage contracts, neighborhood disputes and wills they contain (e. g. Bitschnau and Obermair 2009, 2012; Huter 1937, 1949, 1957; Voltelini 1899; Voltelini and Huter 1951).

Finally, customs regulations, like other commercial documents, provide information about honey and wax as goods, as well as about the value attributed to them (Bohn 2008: 123-124; Denicolò 2013: 83-85; Ertl 2008: 73-74; Esch 2010: 249; Heyd 1879: 550-557; Hödl 1998: 141-142; Riedmann 1985: 504-510; Simonsfeld 1968: 55; Stolz 1942: 276-279, 287-289; Stolz 1953: 17-44, 49-50, 59-77, 207-218, 232-234, 239-259; Stolz 1955: 33, 53-54, 66-74, 77, 80-81, 85, 97, 100, 112, 227-239; Stromer 1995: 139).

With the help of the sources presented here, which complement each other very well in their different priorities, I would like to find out whether and under what circumstances honey was produced in Tyrol or whether honey and wax were bought from outside.

This chapter discusses only the results from the two most important source groups outlined before due to space and time limitations: Account books and *urbaria*. After a description of the two source genres it reviews the references found to apicultural products therein and attempts to synthesize some initial impressions about the above research question. Finally, it discusses the limitations of these sources.

The current chapter is based on extensive archival research and contains numerous source quotations in German, which, for the purposes of the chapter are paraphrased or translated into English. The current chapter is not an exhaustive evaluation of all the Tyrolean account books and *urbaria* of the Middle

Ages, which are generally not edited and scattered in numerous archives. For this reason, I only present a few selected sources from different contexts, periods and regions, which, however, enable a representative overview.

The chapter does not treat aspects of medieval beekeeping or techniques and equipment for the production and processing of wax and honey, nor does it aim to explain in detail where the imported honey came from and how it arrived in Tyrol.

Introduction

Bees and honey play a prominent role in (South) Tyrolean tourism advertising campaigns, and various interest groups and distributors also emphasize the close connection between honey or fragrant beeswax and closeness to nature or the rural, native peasant culture. The population associates rural beekeeping and its products with primeval landscape conservation, a high-quality and healthy, balanced nutrition, with well-being, tradition and (culinary) delights. Similar associations evoke cosmetic products 'with milk and honey' which promise rich and natural care.

Because of the great importance of bees, honey, and wax in the Levant in general and in the Old Testament in particular - the Promised Land was the land where milk and honey flow - honey and wax as well as their producers played an essential role in Christianity. Early Christians consecrated milk and honey at Easter and praised the bees because they produced the wax for the Easter candle. The bee and its qualities became a symbol of Mary, Jesus and the religious. There are legends of bee miracles and several saints are venerated in this context (Crane 1989: 106-108, 136-141, 147-149; Dutli 2012: 10-11, 87-109; Lerner 1984: 9-20, 29-41, 48-72, 75-81, 87-109; Makarovič 1989: 157-158; Soler 1996: 73-74). Already in pre-Christian antiquity honey was regarded as gods' food and was a popular offering during the cult of the dead; bees were seen as birds of the soul and the Muses. Numerous dietetic-medical texts underline the positive qualities of honey, agricultural tracts deal with beekeeping and were spread further. In order to satisfy the great demand, Rome imported large quantities of honey from Greece, Spain and North Africa (Amouretti 1996: 218; Kritsky 2017: 250-253). Besides the taste and symbolism of honey, wax was also an important commodity in antiquity, which was used as sealing wax or writing material, as well as for craftwork, medicine, fine arts, *inter alia* (Makarovič, 1989: 159).

The value of honey and wax continued into the medieval period throughout Europe. Penitential books (*libri poenitentiales*) and *Weistümer* prescribed compensations and punishments for killing bees or destroying any combs and honey (Crane 1989: 117; Lerner 1984: 87-

98). Finally, in the older historical literature from the medieval period, interest and tithes in the form of honey, wax or beehives appear commonplace. Lerner even considers them as the first and most original form of taxation in monastic contexts (Crane 1989: 115; Lerner 1984: 48-72, 87-98; Lippmann 1929: 367-368; Rösener 1985: 45-47, 150; Warnke 2009b).

Tyrol as a special case?

In (scientific) literature honey appears as the most important ingredient in former times to bring sweetness into the human diet. In less nuanced contexts, it may even seem that, in contrast to the extremely precious sugar, honey was relatively widespread and accessible (Hundsbichler 1985: 205-206; Rösener 1985: 150; Schubert 2010: 164; Schulz 2011: 418, 541).

'North of the Alps, honey was the sweetener of choice until modern times. It was not only used for sweetening food, pastries and drinks, in the production of mead and sometimes even for preserving food, but was also particularly important for nutrition because of its richness in indispensable carbohydrates. Honey also has the advantage that it can be stored for a long time without any additives,' (Schulz 2011: 661-662).

Therefore, honey and wax were 'highly sought-after commodities. [...] The extraordinarily large demand for honey led to the intensification of bee hunting and keeping, to the collection of honey levies and to a lively honey trade' (Warnke 2009a: 117-118). Honey could thus be won 'without big effort through adequately extended beekeeping in the house-gardens or in the forests' (Hundsbichler 1985: 205-206). 'Beehives stood on many farms and were regularly exposed in heath and forest areas' (Rösener 1985: 150). Beehives often changed hands as a dowry, payout or wage (Crane 1989: 115; Lippmann 1929: 46-47). And 'together with the honey produced in considerably larger quantities, wax also played a major role as a seigniorial tax' (Warnke 2009b: 1889). Also among merchants and at customs and toll stations wax and honey were regarded as full-fledged forms of payment in many areas (Pechhacker 2003: 15-17). Honey therefore, seems to be at least until the late 15th or 16th centuries a very important and appreciated food ingredient, a popular remedy and preservative, and was also used in agriculture. Wax was required in churches and monasteries, but also in private homes.

The available Tyrolean written sources do not confirm this impression, or at least only to a limited extent. References to honey and beekeeping are surprisingly rare: Is there a lack of traditions or sources here? Was beekeeping also practiced in Tyrol and if so, to what extent? To what extent is modern marketed honey *really* connected to a medieval tradition of apiculture in Tyrol?

Instead of a complete analysis of all possible types of sources, in the following two sections of the chapter two sources respectively are examined in greater detail: firstly *Urbaria,* and secondly account books, which alternately document ownership and production on the one hand, and acquisitions of medieval households on the other. By analyzing a selection of lower aristocratic, knightly, baronial, monastic, and princely books of accounts dating from the 13th to the 16th centuries and originating from various regions of the *Hochstift Brixen* and the County of Tyrol, I will adumbrate why honey is rarely present in the sources and particularly in the *urbaria,* and what conclusions can be made from this about demand and supply. At present, I have only employed print editions with a subject index for the following review, without considering the mostly unexplored sources in the various archives, which remain likely an important source. Therefore, only some highlights from the sources will be presented, which should illustrate fundamental aspects and problems, and point out new directions for further research.

Urbaria or land registers

Urbaria contain all the property rights of the individual manors together with the tithes and interests resulting from them. They generally reflect up to a certain point the agricultural production of the valleys and districts, which means that the levies generally correspond to the main production goods of the farms, i.e. in Tyrol mainly livestock, eggs, cereals and meat, and from the very high altitude farms called *Schwaighöfe* primarily cheese. However, the most important agricultural product in South Tyrol was wine. Many Austrian and Bavarian monasteries and estates owned land in South Tyrol, from which they received natural taxes in the form of wine. It is notable that in Tyrol neither honey nor wax nor even beehives belonged to the traditionally taxed and tithed property (Klos-Buzek 1956: 50-52).

The examined medieval *urbaria* of different dominions in the territory of today's North and South Tyrol show a similar situation. There are no mentions of bees, honey or wax, either in the *urbarium* of Count Meinhard II of Tyrol from 1288, in the oldest complete *urbarium* of the outer County of Gorizia from 1299 (Zingerle 1890; Klos-Buzek 1956: 21-42), or in the two *urbaria* of the County of Tyrol from the time of Friedrich IV at the beginning of the 15th century (Archival Document: Tiroler Landesarchiv TLA, Urbar 1. 2, ,*Alt Urbar der gantzen Grafschafft Tirol Ambter 1406 1412',* TLA Urbar 1.3 ,*1410/1423/1450'*). The *urbarium* called *Rattenberger Salbuch* contains in the section '*Es ist zu mercken wez mon nicht in dem gericht zu Ratenberg hat und dem chasten nicht dient'* (things which are not available in the district of Rattenberg and therefore do not generate income) the note '*Item ymppen hat mon nicht'* ('people don't have bees') (Bachmann 1970: 11).

The land registers of ecclesiastical manors in Tyrol also contain only a few references to wax and honey levies from their own estates, although secondary literature regularly mentions ecclesiastical manors having their own apiaries due to their higher wax consumption. However, the few wax tithes recorded in the *urbaria* cannot have covered the actual demand. While there is neither wax nor honey mentioned in the extensive 1420 *urbarium* of the Hospital of the Holy Spirit, the largest institution of that kind in Bozen (Schneider 2003), the *urbarium* of Stams Abbey for the years 1306-33 records a pound of wax per year: '*Item mediam curiam habebimus [...] et agros tres [...] de quibus dat annuatim libra cere'* (Köfler 1978: 73). The oldest land register of the Benedictine abbey St. Georgenberg in Fiecht from 1361-70 demands from the farm Niederhof in Vomp among other duties also a *scutellam mellin,* a bowl of honey of unknown size (Bachmann 1970: 134, 158). The land register of Marienberg Abbey of 1390 does not record any honey, but according to the chapter *Piper et cera* (!) at least four people pay for a field, a garden or a grain aid between one and four pounds of wax (a total of ten pounds per year) or once also a candle. Note the mention of wax in an equivalent position to pepper, which gives an indication of its appreciation at the time (Schwitzer 1891:118-199). It is not always clear whether the individuals themselves kept bees or bought the wax. In the cases mentioned before the wax is obviously not a tax in kind related to the assets they held.

At least for (South)-Tyrol oil as an alternative for wax candles seemed to be quite important, also in the religious sphere. A series of so-called oil rates are listed, for example, in the 1398 *urbarium* of the parish church of St. Nikolaus in Meran. A separate chapter lists 36 rents from one to four *galetae* of oil for houses or plots of land in the city, once also decidedly intended for the eternal light of a civic association (Schwitzer 1891: 335-343).

A *Gelte* (medieval Latin *galeta*) is a common measure for wine, oil, grain and honey, especially in southern Germany. In the literature there are different conversion keys. While Walter Schneider calculates 5.718 kg c. for a *galeta* as an oil measure, Cristina Belloni states approximately 34 liters for a *galeta* of oil and 31.7 liters c. for a *galeta* of grain, based on older research undertaken by Martini, Schneller and Rottleuthner (Belloni 2004: LXXII; Belloni 2009: XXV; Martini 1883: 793; Schneider 2003: L; Schneller 1898: 140-162). The five kilos are rather little, especially since literature and customs regulations often state that honey was traded in barrels. 30 liters would fit much better there, especially if we take a look at the representation of the *Tacuinum Sanitatis,* where honey is just taken out of a barrel (Figure 1).

However, it remains unclear which type of oil was meant, whether it was purchased or self-produced and

Figure 1. Anonymous, Italy, first half of the 15th century. BnF, NAL 1673, fol. 82r. The dietary guide, called *Tacuinum sanitatis*, describes various foods and their health effects, including honey. The picture shows in the background several basket-shaped beehives, which were placed in a corner, protected from the weather. In the foreground a man is taking honey from a large wooden barrel.

what the final intended use was. Several formulations show, however, that the so-called eternal lights of various civic associations were supported by oil and not by wax donations (Schwitzer 1891: 335-343). In the renewed and updated *urbarium* of 1424 the number of oil rents is even significantly higher (Schwitzer 1891: 344-360).

The only example of a honey and wax fief known is provided by the land register of the Benedictine convent Sonnenburg in the Puster Valley (1296 and 1315-1335, respectively), which owned extensive land properties in the region and collected not only money but also the common natural taxes of the region (Wolfsgruber 1968: 10-13, 31-36). Among the *feuda officiorum*, the so-called *Amtlehen*, which the monastery could give to *ministeriales*, servile unfree nobles, for their services, we find not only a carpenter's, charcoal burner's or dyer's fiefdom, but also a *Honiglehen* (*feudum mellis*), which had to pay interest on various properties in the form of wax and honey: '*1 curia ze Wise ad mel et ceram and 1 feudum*', and '*Item dominus Rudegerus habet feudum unde mel solvere tenetur, Her Rudeger hat 1 lehen, da geit er honich von*' (Wolfsgruber 1968: no. 596 and 634 on p. 111 and 114; cf. also table XII–XIII on p. LXXVI and LXXIX). Unfortunately, the nature of this feudal relationship, whether the owner of this *feudum* was really a beekeeper or *Zeidler*, or whether he could also buy the honey or procure it in some other way does not follow from this information.

These occasional glimpses into the various *urbaria* of spiritual and lay institutions provide surprisingly rare mentions of honey and wax in comparison with wine, grain, cheese or meat. Apparently, to judge from these texts, beekeeping played only a minor role in the area. Perhaps it is just a problem of sources, however. Agricultural products that do not belong to the interest-bearing goods (*Zinsgut*) may be marginal and economically uninteresting, and so do not appear in the written sources. It is possible that beekeeping was thus quite widespread in Tyrol, but remained entirely in the private sector. Perhaps people kept bee colonies for domestic use and harvested honey and wax, or exploited at least the dwellings of wild bees in the forest.

Account books

Despite sporadic evidence of wax and honey levies in land registers, especially in those of spiritual provenance it can be assumed that honey and wax were hardly or at least only in very small quantities produced or gathered in Tyrol and mainly imported or bought at markets and fairs in order to cover the supposedly high demand. Books of accounts of different persons, families or dominions can yield information about various kinds of purchases and expenses (Fouquet

2000: 15-18; Mersiowsky 2000). Here, too, the Tyrolean princes provide important sources. They are discussed here first because of their comprehensiveness and continuity, but also due to the outstanding social and economic importance of the Tyrolean *Landesfürsten*.

The account books examined for this chapter are the accounts of the Prince's Central Administration and mainly contain the settlements of the individual *Amtleute* (civil servants) with the Chamber as well as the purchases and expenditures of the court. They allow conclusions about financial development and outline the nature of the princely incomes and outgoings (Brandstätter 2005: 126; Fouquet 2000: 15-18; Kaltenbacher 2006: 49-55; Kogler 1901: 549-551; Torggler 2018). What have survived to the present day are first the oldest series, the so-called *Ältere Tiroler Rechnungsbücher* out of the Meinhardinian period (1280s to the 1360s). Of those, the first three (until c. 1300) are available as a printed edition and indexed by subject. Then follow, secondly, the ten surviving account books of Friedrich IV, which cover the first thirty years of the 15th century.[2] Thirdly, there follow the so-called younger *Tiroler Rechnungsbücher*, seamless annual accounts under Archduke Sigismund of Austria, called *der Münzreiche* ('rich in coins') starting from the 1460s, and finally from 1493 to 1519 the *oberösterreichische Kammerraitbücher* of king and emperor Maximilian I (Riedmann 1984: 315-317; Wiesflecker 1986: XVIII-XIX, 2). All except those three already mentioned are unedited. There are few studies of the material culture to which these account books speak, one is the study by Otto Stolz, which does not go beyond the mere mention of an occurrence of sugar, honey and wax or oilcloth (Stolz 1957).

On the one hand, the listed expenses in these account books can simply represent official expenses. On the other hand, the accounts also include expenses in the name of the duke or one of his representatives: cost of living, deliveries of goods to the ducal court, purchase of clothing, spices, medicines, jewelry and imported foodstuffs, payment of all kinds of court debts etc. In most cases - the examples are from the books of Friedrich IV - the Duke himself appears as the issuer of these instructions ('*für meins herrn zergaden gen Insprugg*' or '*ainen mains herren briefe*'), as does the Duchess ('*meiner frawn gnadn zu notdurft irer chuchn*'), as well as her closer associates, such as the Kitchen Master or Chamberlain (Kaltenbacher 2006: 71-74, 79-85, 88-93, 98-100; Mayer 1919/20: 123-127, 129-139; Stolz 1957).

---

[2] The account books of Friedrich IV were previously examined in the context of the author's MA thesis (Denicolò 2013) and are not very helpful overall, since during this period more and more contributions in kind were replaced by simple cash payments and, on the other hand, several expenses were only summarily accounted for.

In the first volume of the edition of the older Tyrolean account books procured by Christoph Haidacher, the manuscripts A (IC 277, 1288-90) and B (MC 8, 1305-1310) of the Tyrolean Regional Archives contain a total of 11 mentions of wax or wax lights, one mention of honey (but none of sugar or confectionery, which can consist of sugar or honey) and 20 mentions of oil (cf. the index of Haidacher 1993). Wax, between one pound and two *centenarios* or centner, with one exception, always appears among the expenses; it was purchased by various officials, often as a donation to religious institutions (*ad capellam*), for the Chamber (*ad cameram*) or for the Prince himself (*ad expensam domini*) (Haidacher 1993: 90, 92, 119, 129, 180, 251, 266). Eberhard Jäger, provost of Sterzing, charged in 1291 expenses probably for wax lights: '*Item pro 2 cereis lb 2 ad ecclesiam in Strazperch*' as well as oil for the lights: '*oleo ad luminaria ecclesie pro galetis 4 lb 5 g*' (Haidacher 1993: 235-236, 253).

Oil was also used for lighting in the southern Tyrol, but in many cases probably more as a food for Lent. This is suggested by oil donations to hospitals or vineyard workers as well as the frequent mention in the context of other *Quadragesimalia* or *Fastenspeis*, i.e. Lenting foods. Some districts (Ämter) in the south regularly recorded olive oil rents of up to 20 *galetas* per year. These were mainly local authorities from the area of Lake Garda or Trento and customs officers along the Brenner route (Haidacher 1993: 183-185; 220-222).

While sugar and confectionery do not appear in these two manuscripts, there is a mention of honey: the judge Gerold von Gries charged in 1291 expenses for an unspecified quantity of honey, which he sent to the residence Castle Tyrol, among other things: '*pro melle et rebuis aliis illuc missis m 16 lb 1 ½*' (Haidacher 1993: 340).

The second volume contains three other account books (IC. 278, IC 279 and records concerning the siege of Weineck) from the years 1293 to 1305, and lists wax 25 times, oil 24 times, twice each sugar and confectionery, but no honey (cf. the index of Haidacher 1998). Here, too, wax is only purchased, either in the form of honeycombs (*rauba cerae*) (Haidacher 1998: 137) or in quantities ranging from a few *libras* to six *centenarios*. These purchases were usually made by officials along the Brenner route or from the southern Tyrol, either for the Chamber or by request of the duke (Haidacher 1998: 137-139, 154, 167-169, 262, 281-282, 321, 355, 392-393, 428, 432-434). Some entries of wax were made without further explanation or were intended for individuals. Other purchases of wax, approximately for 40 pounds per centner, went *ad capellam* (e.g. for Candlemas or for a funeral) or to the *ballista* (catapultier or archer) together with strings and red earth (Haidacher 1998: 195-197, 228-229, 239-241, 289-291, 311, 320, 326-328, 356, 363-365, 378, 413-415).

Although the amounts of money were usually always given, it is difficult to make significant comparisons. This is either because the amounts are collective sums or because different measurements and currency units make comparisons difficult. Wax, however, definitely belonged to the higher-priced goods, as not only the rather high amounts but also its inclusion in the context of spices and other imported products show. In 1294, the customs officer of Klausen reported 500 pounds of wax for more than sixteen *marks* '*ad rationem lb 8 rauba computata*' (at the cost of 8 pounds per comb) (Haidacher 1998: 355), and the provost of Friedberg bought 283 pounds of wax for 14 *marks* and 30 *solidi* in 1294 (Haidacher 1998: 320). Finally, the report given by Heinrich Zelner, customs officer of Klausen, on spending for the prince in 1295 is particularly informative. Together with sugar, almonds and figs, he charged '*centenaries 4 cere et libras 5, qui valent ver m 16 lb 2, ad rationem lb 40 pro centenario*'. A centner of figs cost '*4 lb pro centenario*' in the same settlement (Haidacher 1998: 413-415). Furthermore, the provost of Gufidaun charged 1296 in the name of the duke a large amount of wax '*ad expensarum dominorum [...] cere libras 107 pro lb 43 minus s 7*' and '*dedit ad cameram [...] zukari libras 142 ½ pro m 14 s 50 [...] cere centenarios 10 libras 74 ½*' (Haidacher 1998: 281). Oil, finally, again appears in this volume both as income (including the cost of pressing oil, '*pro pressure olei*') and as expenditure (Haidacher 1998: 326-328, 359, 443-445, 261-262), especially by officials north of the Brenner Pass and in connection with fasting food (Haidacher 1998: 363-365, 368-370, 379-381, 428). The more extensive manuscript (IC 280, 1295-1301) edited in volume no. 3 offers a similar result: 16 mentions of wax and 21 of oil, two mentions of honey and four of sugar (cf. the index of Haidacher 2008). Here, too, oil appears among both income, and expenditure, and was sent to the residences in Innsbruck and Castle Tyrol, in some cases dedicated to be used as a food for lent (*Fastenspeis* or *ad coquinam*) (Haidacher 2008: 54-56, 75, 85-87, 154, 204-206, 277, 287, 289).

Wax in quantities of up to eight centner, sometimes in the form of honeycombs, goes again *ad cameram*, ad *expensas dominorum*, or *ad capellam* especially for Candlemas (Haidacher 2008: 54-56, 74-76, 80-82, 84-86, 98-100, 182-183, 204-206, 227, 241, 246, 255, 285). The frequent appearance of wax among purchases in the context of high-priced goods such as saffron, pepper, rice, sugar or almonds is again striking. The accounts of Heinrich Zelner in 1297 also reveal the proportion of prices between cotton, wax, saffron and sugar: '*Item cere libras 53 pro lb 21 s 4*; *item croci libras 11 pro lb 50 minus s 10*; *item zukari libras 35 ½ pro lb 35,5*; *item bombicis libras 50 pro 12, 5*; *item pro saccis involturis, funibus, et stora s 45*' (Haidacher 2008: 229-230). Finally, the account that the tailor Werner of Meran made in 1298 for himself and Götschlin of Bozen concerning the purchases for the sovereign in Venice are particularly impressive. There they bought almonds, jewelry, rice, cotton, all kinds of fabrics, stoats, dates, ginger, pine nuts, medicines, 25 pounds of grains of paradise, 360 pounds of raisins, 15 pounds of cinnamon, 50 pounds of saffron, 5 pounds of nutmeg, 12.5 pounds of galanga and 12.5 pounds of cloves. Furthermore, they spent '*pro 105 libris zukari gross lb 3 s 12 d 8*' and '*per 14 centenariis cere gross lb 10 s 15*' (Haidacher 2008: 228)·

The question arises why the wax was bought in Venice and then brought to Tyrol with all the other overseas trade goods. Possibly because wax was a long-distance good, too. It could either come from the Mediterranean area itself, or from the large honey exporting territories in today's Germany or Eastern Europe (for greater detail on the wax trade in the western medieval Mediterranean, see chapter 11 of the present volume). The forest-covered regions of Eastern Europe, Poland, Lithuania, Slovenia, the Czech Republic and Russia, where beekeeping was an important economic factor, were renowned for their abundance of bees and honey already in the Middle Ages. Venice also maintained regular ship connections to the North Sea together with the Hanseatic League. The Hanse brought the wax and honey to the west and south and supplied the large honey markets in Augsburg, Nuremberg, Frankfurt, Cologne, Vienna, Wroclaw, Prague, Bruges, Narbonne or St. Denis or even the trading metropolis of Venice (Kritsky 2017: 255-257; Lerner 1984: 82-86, 124, 132-142; Lippmann 1929: 44-45, 48, 50, 354-356, 358, 360, 367-368; Mihelič 1989: 36; Schulz 2011: 663-664; Stolz 1955:59-60; Warnke 2009b: 1888-1890).

Amongst expenses for purchasing honey is only found twice, while the purchase of sugar was found four times. Interestingly enough, these four times South Tyrolean officials bought about a hundred pounds of sugar and sent it to Tyrol. These were always either customs officers who were in contact with these goods by profession, or persons who were known to have travelled to Venice (Haidacher 2008: 54-56, 98-100, 228-230). In 1296 Planco of Meran accounted expenses for honey as a fasting food: '*pro mellus galetis 5 in Quadragesima lb 5*' (Haidacher 2008: 89-91). And in 1306 the following expenses were recorded in a calculation for a lunch at the beginning of January: '*in prandio pro sagimine* [i.e. fat] *lb 4 s 5; item pro unzia 1 croci s 10; item per 1 galeta mellis g 13; item portitoribus g 2*' (Haidacher 2008: 299).

The account books of Friedrich IV from the years 1413 to 1436[3] are by far not as extensive and detailed as those of the older and younger series. This is probably

one of the reasons why they offer only a little relevant information to the questions asked in this chapter. Honey is neither included in income nor expenditure; wax, however, is purchased several times. Sugar is bought twice by the official called Kellner auf Tirol or the customs officer in Bolzano, and once even a sugar loaf is reported as a gift (Denicolò 2013: 115-116). Possibly only the confectionery and the preserved nuts could contain honey, which Andre von Caldinetsch charged in 1432/33: a payment to the '*spezkger hie vmb ingemacht nussen von Venedig raus und ander confect xx lb pn*' (Denicolò 2013: 118, cites TLA, Cod. 136 fol. 81v).

However, the same picture is not only found in the princely account books but also in the account book of the Künigl family of Ehrenburg from 1494 to 1519 and in those of the Vilanders family from 1415 to 1417. Honey, wax candles and the like are only found, if at all, in the form of purchases (Egg 2004: 22-23; Goller 2007). The Vilanders account book does not mention wax and only once the purchase of wicks is mentioned, possibly to make candles within the household (Goller 2007: 82). Oil, on the other hand, was purchased more frequently: According to a Künigl's calculation once four *galetas* of tree oil (olive oil) clearly for the church: '*gen Sandt Petter to Kyenss geben zu dem Ewigen Liecht und soll da mitt dye Kirchen ain jar beleichten*' (Egg 2004: 29, 46, 78-79). Also in the accounts of 1497, there is a record of honey bought from a local butcher, who maybe was also the producer of the honey: '*Mer hab ich auss geben dem Cristan Lechner umb iii Pf honyg vi kr und hab im geben für alles schlachtigen so er gethan hatt im Herbst hünz auff Weinachtn xviiii kr*' (Egg 2004: 97, 99). For comparison, a day's work then cost six kreutzer and half a *Star* (a Southern German, Swiss and italian measure for grain with different sizes from 30 liters up to 80 liters) of lentils cost seven kreutzer (Wiesflecker 1986: 144). The situation is similar with religious institutions. For example, the accounts book of the parish church of St. Mary's in Gries near Bozen-Bolzano for the years 1422 to 1440 lists no entries for honey but 14 mentions of wax among the purchases (Stamm and Obermair 2011).

But even books of account are not sources that depict the situation truthfully, i.e. record all expenditures and revenues meticulously and offset them against each other. The accounting books, for example, usually do not report the ostensibly large part of the food produced from one's own farms. They only show what was *additionally* purchased or acquired from a different district. Especially (fresh) basic foods are not displayed in contrast to exotic spices and fruits, southern wines and the like, which were purchased through long-distance trade. Similarly, generalizing designations such as *kuchlspeis* i.e. food and *gwurz* for spice remain just as unclear as the lump-sum sums of money which went directly to the prince at his personal disposal to buy often luxury food. Gifts that regularly changed

---

[3] Preserved in the Regional Archive in Innsbruck: Tiroler Landesarchiv Cod. 206 (1413/14-1415), 130 (1413-1415/16), 207 (1415-16), 114 (1416-17), 132 (1416-1419), 133 (1424/25), 134 (1425/26), 135 (1426-1428), 136 (1432/33), 137 (1433–1436).

hands, even in the form of natural products, are also not recorded (Haidaicher 1983: 27-30, 102-103, 211-219, 231-234; Rösener 2008: 95, 117; Spieß 2008: 66-73; Wiesflecker 1986: 70, 91).

Conclusion

Honey was apparently not present at all as a tithe product, and wax only played a negligible role compared to meat, grain and cheese. Interests paid as wax were limited to spiritual institutions. The certainly large demand for lighting and liturgical-religious purposes could certainly not be met by this. By contrast, the account books record much larger quantities of wax, which was purchased mainly in the south of Tyrol. The mentions of wax far exceed those of honey, which is also remarkably rare. Honey appears only as a purchased commodity, but is less present in relation to the more prestigious sugar (*pace* Schulz 2011: 337-338, 639-640). On the basis of the sources investigated to this point, the demand for honey and wax in Tyrol was apparently not covered by local beekeeping, but by imports or purchases. In this context, the relevant literature always mentions the large exporting centers in the Nuremberg Reichswald or in Eastern Europe, whose products were also negotiated in the trading metropolis of Venice (Bohn 2008: 123-124; Denicolò 2013: 83-85; Ertl 2008: 73-74; Esch 2010: 249; Heyd 1879: 550-557; Hödl 1998: 141-142; Lerner 1984: 82-86, 124, 132-142; Lippmann 1929: 44-50, 354-360, 367-368; Mihelič 1989: 36; Riedmann 1985: 504-510; Schulz 2011: 663-664; Simonsfeld 1968: 55, 103; Stolz 1942: 276-279, 286-289; Stolz 1953: 17-44, 49-50, 62-66, 71-76, 207-218, 232-234, 239-359; Stolz 1955: 33, 59-60, 66-74, 227-239; Stromer 1995: 139).

Nevertheless, it cannot be assumed that in Tyrol no hive or log beekeeping was practiced simply because honey and wax did not belong to the regular rent and tithe system and therefore appear only sporadically in the written sources. It can therefore be supposed that the inhabitants of individual farmsteads and households outside the landlord's sphere of influence kept bee colonies close to their homes to obtain honey and wax for their own use, or discovered and exploited wild beehives in the surrounding forests. To date, there is no evidence of systematic honey gathering in the forest, nor is there any evidence of organized and licensed commercial beekeeping (*Zeidlerei*) in Tyrol (Pechhacker 2003: 15-17; Mihelič 1989: 33; Schulz 2011: 660; Lerner 1984: 110-123). It was unlikely that larger quantities of honey and wax could be produced in Tyrol at all due to the technical and natural conditions of the time. The alpine climate, the altitude, the high proportion of coniferous forest and the lack of nectar and pollen-rich plants might certainly have been a factor (Bucher et al. 2004; Glaser 2012).

A significant part (over 50%) of the country is still covered with primarily coniferous forest (88%). The honey extracted from these forests is in fact not real honey, but dark brown to reddish honeydew produced by treelice. In the Middle Ages, this honeydew was probably considered inferior due to its colour, as very light to transparent honey types were considered to be of higher quality and were therefore particularly appreciated. In addition to forest honey, chestnut honey, which is very typical of the natural environment of South Tyrol, also has a particularly dark colour. Light honey is obtained mainly from oaks and lime trees, which are only rarely found in Tyrol, in contrast to some areas in Bavaria, Thuringia, Saxony and Eastern Europe. In Tyrol, the entire accessible and woodless and thus cultivable area may have been used for grain production or livestock farming. Cereal fields, which incidentally do not depend on bee pollination at all, are not interesting as nectar sources for bees, nor are grasses. The possibility of collecting nectar in Tyrol in the Middle Ages was probably much smaller in earlier centuries than it is today. A possible forest beekeeping would have been possible only under difficult conditions due to the altitude and especially the steepness of the Tyrolean forests and meadows. Indeed, 60% of the forest is located in an inclination class between 41% and 80% (http://www.provinz.bz.it/land-forstwirtschaft/wald-holz-almen/wald-in-suedtirol.asp; https://www.tirol.gv.at/umwelt/wald/datenundfakten/).

It is also possible that the passion for sweets was not as pronounced in Tyrol or the Alpine region as in other regions of Europe. According to Bruno Laurioux, culturally influenced tastes play a major role in the preference for certain foods. English cuisine, for example, appears to be very sweet based on traditional recipes, whereas hardly any sweeteners were used in French cuisine. He places the preferences of Italian cookbooks in the center, but makes no mention of the German language area (Laurioux 2005: 293-298, 315, 340-342; Laurioux 1996: 467).

A particularity of (southern) Tyrol is the importance of northern Italian oil, especially for liturgical lighting. Honey appears as a purchased product, but is less present in relation to the more distinguished and dietetically more powerful sugar (Schulz 2011: 337-338, 639-640).

The end of this short history of honey in Tyrol finally marks the rise of sugar, beginning in the 14th century, which led to a marginalization of honey; in dietetics, sugar was considered more nutritious, digestible and healing than honey. Sugar was many times more expensive and could not be produced by the people themselves. It was therefore more valuable, more exclusive and socially more distinctive than the already precious and expensive honey. Honey thus became a cheaper substitute for sugar, which was either unavailable or unaffordable, and was able to maintain this position until the 18th century. Then, due to the rapid fall in prices, sugar finally gained the upper hand and honey almost completely lost its importance until recent times.

Epilogue

Whoever wants to undertake the task of writing a history of honey and wax in Tyrol and the surrounding area has to deal with broader questions, such as the yield ratio of honey and wax or of *Zeidlerei* and domestic beekeeping or how the relationship between supply and demand changed in Central Europe and the Central Alpine region over time. About these questions there are only contradictory statements so far in scholarship. There is also a lack of reliable information on specific aspects of beekeeping, such as construction and material of beehives, apiaries, bee gardens, or the techniques for extracting and processing honey and wax. Contemporary pictorial sources and ethnological findings could provide information on this, since beekeeping and honey harvesting have hardly changed at all until late modern times (Crane 1980: 112-118; Crane 1990: 312-325, 329-394, 483-552; Schulz 2011: 639-640, 662-663, 666; Warnke 2009b: 1888-1889).

In such a broader investigation, other types of sources that shed light on other areas of life and thus provide different types of information can help. In addition to archaeological, ethnological and pictorial sources, countless other *urbaria* and accounting books, especially from monasteries, convents, hospitals, parishes and dioceses, which certainly had an increased need for wax, offer rich source material. Further material, especially for the Tyrolean region, also features notarial instruments and registers of their imbreviatures, customs regulations, the so called *Weistümer* and *Traditionsbücher*, inventories and wills, provisioning lists and regulations, supply contracts, food orders and stocktaking, private household books, administrative and normative documents of various kinds, diaries and chronicles, dietary and medical texts, as well as cookbooks and the so-called *Kunstbücher*, which contain recipes for different aspects of daily life (Kritsky 2017: 255-257; cf. also Blockmans 1999: 2-9; Felgenhauer-Schmiedt 1993:108; Grass and Faussner 1994a, 1994b; Grass and Finsterwalder 1966; Kühnel 1977a: 53-55; Kühnel 1977b: 5; Santifaller 1929, 1941; Voltelini 1899; Voltelini and Huter 1951; Zingerle and Egger 1888, 1891; Zingerle and Inama-Sternegg 1875, 1877, 1880).

Bibliography

Aichholzer, D. 1999. „*Wildu machen ayn guet essen...*": *Drei mittelhochdeutsche Kochbücher: Erstedition, Übersetzung, Kommentar.* Bern et. al.: Lang.

Amouretti, J. M.-C. 1996. Villes et campagnes grecques, in J.-L. Flandrin and M. Montanari (eds) *Histoire de l'alimentation*: 133-150. Paris: Fayard.

Bachmann, H. (ed.) 1970. *Das Rattenberger Salbuch von 1416.* Innsbruck: Wagner.

Belloni, C. (ed.) 2004. *Documenti trentini negli archivi di Innsbruck (1145-1284).* Trento: Provincia autonoma, Soprintendenza per i beni librari e archivistici.

Belloni, C. (ed.) 2009. *Documenti trentini nel Tiroler Landesarchiv di Innsbruck (1285-1310).* Trento: Provincia autonoma, Soprintendenza per i beni librari e archivistici.

Bitschnau, M. and H. Obermair (eds) 2009. *Tiroler Urkundenbuch: Abt. 2, Die Urkunden zur Geschichte des Inn-, Eisack- und Pustertals: Bd. 1., Bis zum Jahr 1140.* Innsbruck: Wagner.

Bitschnau, M. and H. Obermair (eds) 2012. *Tiroler Urkundenbuch: Abt. 2, Die Urkunden zur Geschichte des Inn-, Eisack- und Pustertals: Bd. 2., 1140 bis 1200.* Innsbruck: Wagner.

Blockmans, W. 1999. The Feeling of Being Oneself, in W. Blockmans and A. Janse (eds) *Showing status: Representation of Social Positions in the Late Middle Ages*: 1-16. Turnhout: Brepols.

Bohn, R. 2008. Handelsmacht im Norden Europas: Die Hanse – eine Interessensgemeinschaft von Fernhändlern, in R. Bohn et al. (eds) *Fernhandel in Antike und Mittelalter*: 111-127. Stuttgart: Theiss.

Brandstätter, K. 2005. Der Hof unterwegs: Zum Aufenthalt Herzog Friedrichs IV. von Österreich in Wiener Neustadt 1412/1413. In K. Brandstätter and J. Hörmann (eds), *Tirol-Österreich-Italien: Festschrift für Josef Riedmann zum 65. Geburtstag*: 125-139. Innsbruck, Wagner.

Brandstätter, K. 2007. Zur Entwicklung der Finanzen unter Herzog Friedrich IV. In G. Mühlberger and M. Blaas (eds), *Grafschaft Tirol – Terra Venusta: Studien zur Geschichte Tirols, insbesondere des Vintschgaus*: 219-235. Innsbruck, Wagner.

Bucher, E. et al. 2004. *Das Pollenbild der Südtiroler Honige.* Bozen, Landesagentur für Umwelt und Arbeitsschutz.

Crane, E. 1980. *A Book of Honey.* Oxford: Oxford University Press.

Crane, E. 1999. *The World History of Beekeeping and Honey Hunting.* New York: Routledge.

Danner, B. 1970. Alte Kochrezepte aus dem bayrischen Inntal. *Ostbairische Grenzmarken* 12: 118-128.

Denicolò, B. 2013. Essen, Trinken und Kleidung am Hof Friedrich IV. von Tirol 1413-1436. Unpublished Magister thesis, University of Innsbruck.

Dutli, R. 2012. *Das Lied vom Honig: Eine Kulturgeschichte der Biene*. Göttingen: Wallstein.

Egg, I. M. B. 2004. Das Raitbuch der Familie Künigl zu Ehrenburg (1494-1500): Kurzkommentar und Tranksription. Unpublished Magister thesis, University of Innsbruck.

Ertl, T. 2008. *Seide, Pfeffer und Kanonen: Globalisierung im Mittelalter*. Darmstadt: Wissenschaftliche Buchgesellschaft.

Esch, A. 2010. Italienische Kaufleute in Brügge, flandrisch-niederländische Kaufleute in Rom, in G. Fouquet and H.-J. Gilomen (eds) *Netzwerke im europäischen Handel des Mittelalters*: 245-261. Ostfildern: Thorbecke.

Felgenhauer-Schmiedt, S. 1993. *Die Sachkultur des Mittelalters im Lichte der archäologischen Funde*. Frankfurt am Main et al.: Lang.

Fouquet, G. 2000. Adel und Zahl – es sy umb klein oder groß: Bemerkungen zu einem Forschungsgebiet vornehmlich im Reich des Spätmittelalters, in H. von Seggern and G. Fouquet (eds) *Adel und Zahl: Studien zum adeligen Rechnen und Haushalten in Spätmittelalter und früher Neuzeit*: 3-24. Ubstadt-Weiher: Verlag Regionalkultur.

Glaser, R. 2012. Historische Klimatologie Mitteleuropas, in *Europäische Geschichte Online* (EGO) 2012: http://www.ieg-ego.eu/glaserr-2012-de.

Goller, J. 2007. Das Rechnungsbuch der Herren von Vilanders von 1415-1417 (Tiroler Landesarchiv, Handschrift 523): Auswertung und Transkription. Unpublished Magister thesis, University of Innsbruck.

Grass, N. and H. C. Faussner (eds) 1994a. *Österreichische Weistümer 19: Die Tirolischen Weisthümer 6 Erg.-Bd. 2: Oberinntal, Gerichte Hörtenberg und St. Petersberg*. Vienna: Braumüller.

Grass, N. and H. C. Faussner (eds) 1994b. *Österreichische Weistümer 20: Die Tirolischen Weisthümer 7 Erg.-Bd. 3: Oberinntal, Gerichte Imst, Landeck, Laudeck und Pfunds*. Vienna: Braumüller.

Grass, N. and K. Finsterwalder (eds) 1966. *Österreichische Weistümer 17: Die Tirolischen Weisthümer 5 Erg.-Bd. 1: Unterinntal*. Vienna: Braumüller.

Haidacher, C. (ed.) 1993. *Die älteren Tiroler Rechnungsbücher: Analyse und Edition 1: (IC. 277, MC. 8)*. Innsbruck: Tiroler Landesarchiv.

Haidacher, C. (ed.) 1998. *Die älteren Tiroler Rechnungsbücher: Analyse und Edition 2: (IC. 278, IC 279 und Belagerung von Weineck)*. Innsbruck: Tiroler Landesarchiv.

Haidacher, C. (ed.) 2008. *Die älteren Tiroler Rechnungsbücher: Analyse und Edition 3: (IC. 280)*. Innsbruck: Tiroler Landesarchiv.

Haidacher, C. 1983. Beiträge zur Bevölkerungs- und Sozialstruktur der Stadt Innsbruck im Mittelalter und in der beginnenden Neuzeit. Unpublished PhD thesis, University of Innsbruck.

Heyd, W. 1879. *Geschichte des Levantehandels im Mittelalter 2*. Stuttgart: Cotta.

Hödl, G. 1998. *Habsburg und Österreich 1273-1493: Gestalten und Gestalt des österreichischen Spätmittelalters*. Vienna et al.: Böhlau.

Hörmann, J. 2007. Kanzlei und Registerwesen der Tiroler Landesfürsten bis 1361: Ein Überblick, in G. Mühlberger and M. Blaas (eds) *Grafschaft Tirol – Terra Venusta: Studien zur Geschichte Tirols, insbesondere des Vintschgaus*: 207-218. Innsbruck: Wagner.

Hundsbichler, H. 1985. Nahrung, in H. Kühnel (ed.) *Alltag im Spätmittelalter*: 196-231. Vienna et al.: Edition Kaleidoskop.

Huter, F. (ed.) 1937. *Tiroler Urkundenbuch: Abt. 1, Die Urkunden zur Geschichte des deutschen Etschlandes und des Vintschgaus: Bd. 1 Bis zum Jahre 1200*. Innsbruck: Wagner.

Huter, F. (ed.) 1949. *Tiroler Urkundenbuch: Abt. 1, Die Urkunden zur Geschichte des deutschen Etschlandes und des Vintschgaus: Bd. 2 1200-1230*. Innsbruck: Wagner.

Huter, F. (ed.) 1957. *Tiroler Urkundenbuch: Abt. 1, Die Urkunden zur Geschichte des deutschen Etschlandes und des Vintschgaus: Bd. 3 1231-1253*. Innsbruck: Wagner.

Kaltenbacher, B. 2006. Die Finanzverwaltung Herzog Friedrichs IV. von Tirol. Unpublished Magister thesis, University of Innsbruck.

Klos-Buzek, F. (ed.) 1956. Das Urbar der vorderen Grafschaft Görz aus dem Jahre 1299. Vienna et al.: Böhlau.

Köfler, W. (ed.) 1978. *Die Mittelalterlichen Stiftsurbare des Bistums Brixen 3: Die ältesten Urbare des Zisterzienserstiftes Stams von dessen Gründung bis 1336*. Innsbruck: Wagner.

Kogler, F. 1901. Das landesfürstliche Steuerwesen in Tirol bis zum Ausgange des Mittelalters. *Archiv für österreichische Geschichte* 90: 419-712.

Kritsky, G. 2017. Beekeeping from Antiquity Through the Middle Ages. *Annual Review of Entomology* 72: 249-264.

Kühnel, H. 1977. Aufgaben und Probleme der Realienkunde des Spätmittelalters. *Bericht über den dreizehnten österreichischen Historikertag in Klagenfurt veranstaltet vom Verband Österreichischer Geschichtsvereine in der Zeit vom 18. bis 21. Mai 1976*: 52-57.

Kühnel, H. 1977b. Vorwort, in *Das Leben in der Stadt des Spätmittelalters. Internationaler Kongress, Krems an der Donau, 20. bis 23. September 1976*: 5-8. Vienna: Verlag der Österreichischen Akademie der Wissenschaften.

Laurioux, B. 1996. Cuisines médiévales (XIVe et XVe siècles), in J.-L. Flandrin and M. Montanari (eds) *Histoire de l'alimentation*: 459-477. Paris: Fayard.

Laurioux, B. 2005. *Une histoire culinaire du moyen âge*. Paris: Honoré Champion.

Lemmer, M. and G. Hayer (eds) 1983. *Das Kochbuch der Philippine Welser*. Innsbruck: Pinguin.

Lerner, F. 1984. *Blüten, Nektar, Bienenfleiß: Die Geschichte des Honigs*. Munich: Ehrenwirth.

Lippmann, E. O. 1929. *Geschichte des Zuckers seit den ältesten Zeiten bis zum Beginn der Rübenzucker-Fabrikation: Ein Beitrag zur Kulturgeschichte*. Wiesbaden: Sändig [Reprint of 1970].

Makarovič, G. 1989. Verwendung und Bedeutung von Bienenerzeugnissen in Slowenien, in *Der Mensch und die Biene: Die Apikultur Sloweniens in der traditionellen Wirtschaft und Volkskunst*: 149-172. Vienna: Österreichisches Museum für Volkskunde.

Maleczek, W. 2009. Friedrich IV., Herzog von Österreich, in *Lexikon des Mittelalters 4*: 954. Darmstadt: WBG.

Martini, A. 1883. *Manuale di metrologia ossia misure, pesi e monete in uso attualmente e anticamente presso tutti i popoli*. Turin: Loescher.

Mayer, T. 1919/1920. Beiträge zur Geschichte der tirolischen Finanzverwaltung im späten Mittelalter. *Forschungen und Mitteilungen zur Geschichte Tirols und Vorarlbergs* 16/17: 10-168.

Mersiowsky, M. 2000. *Die Anfänge territorialer Rechnungslegung im deutschen Nordwesten: Spätmittelalterliche Rechnungen, Verwaltungspraxis, Hof und Territorium*. Göttingen: Thorbecke.

Mihelič, S. 1989. Geschichte der slowenischen Bienenhaltung, in *Der Mensch und die Biene: Die Apikultur Sloweniens in der traditionellen Wirtschaft und Volkskunst*: 33-52. Vienna: Österreichisches Museum für Volkskunde.

Obermair, H. and V. Stamm (eds) 2011. *Zur Ökonomie einer ländlichen Pfarrgemeinde im Spätmittelalter: Das Rechnungsbuch der Marienpfarrkirche Gries (Bozen) von 1422 bis 1440*. Bozen: Athesia.

Pechhacker, H. 2003. Die Bienenkunde in Österreich. *Denisia* 8: 15-45.

Prentki, A. 1989. Les traités culinaires et mutations du goût a la fin du Moyen Âge, in R. Jansen-Sieben (ed.) *Artes mechanicae en Europe médiévale. Actes du colloque du 15 octobre 1987*: 199-222. Brussels: Archives et Bibliothèques de Belgique.

Riedmann, J. 1984. Die Rechnungsbücher der Tiroler Landesfürsten, in *Landesherrliche Kanzleien im Spätmittelalter: Referate zum VI. Internationalen Kongreß für Diplomatik*: 315-324. München: Arbeo-Gesellschaft.

Riedmann, J. 1985. Mittelalter, in J. Fontana (ed.) *Geschichte des Landes Tirol 1: Von den Anfängen bis 1490*: 267-684. Bozen et al.: Tyrolia.

Rösener, W. 1985. *Bauern im Mittelalter*. München: Beck.

Rösener, W. 2008. *Leben am Hof. Königs- und Fürstenhöfe im Mittelalter*. Ostfildern: Thorbecke.

Rottleuthner, W. 1985. *Alte lokale und nichtmetrische Gewichte und Masse*. Innsbruck: Wagner.

Santifaller, L. (ed.) 1929. *Brixner Urkunden 1: Die Urkunden Der Brixner Hochstifts-Archive 845 – 1295*. Innsbruck: Wagner.

Santifaller, L. (ed.) 1941. *Brixner Urkunden 2: Die Urkunden Der Brixner Hochstiftsarchive 1295 – 1336: Teil 1: Die Urkunden*. Leipzig/Innsbruck: Hirzel/Wagner.

Schneider, W. (ed.) 2003. *Das Urbar des Heilig-Geist-Spitals zu Bozen von 1420*. Innsbruck: Wagner.

Schneller, C. 1898. *Tridentinische Urbare aus dem dreizehnten Jahrhundert. Mit einer Urkunde aus Judicarien von 1244-1247*. Innsbruck: Wagner.

Schubert, E. 2010. *Essen und Trinken im Mittelalter*. Darmstadt: Wissenschaftliche Buchgesellschaft.

Schulz, A. 2011. *Essen und Trinken im Mittelalter (1000 – 1300): Literarische, kunsthistorische und archäologische Quellen*. Berlin et al.: De Gruyter.

Schwitzer, B. (ed.) 1891. *Urbare der Stifte Marienberg und Münster, Peters von Liebenberg-Hohenwart und Hansens von Annenberg, der Pfarrkirchen von Meran und Sarnthein*. Innsbruck: Wagner.

Simonsfeld, H. 1968. *Der Fondaco dei Tedeschi in Venedig und die deutsch-venetianischen Handelsbeziehungen: Bd. 2 Geschichtliches*. Aalen: Scientia Verlag.

Soler, J. 1996. Le raisons de la Bible: règles alimentaires hébraiques, in J.-L. Flandrin and M. Montanari (eds) *Histoire de l'alimentation*: 73-84. Paris: Fayard.

Spieß, K.-H. 2008. *Fürsten und Höfe im Mittelalter*. Darmstadt: Primus.

Stamm, V. and H. Obermair (eds) 2011. *Zur Ökonomie einer ländlichen Pfarrgemeinde im Spätmittelalter: Das Rechnungsbuch der Marienpfarrkirche Gries (Bozen) von 1422 bis 1440*. Bozen: Athesia.

Stolz, O. (ed.) 1955. *Quellen zur Geschichte des Zollwesens und Handelsverkehrs in Tirol und Vorarlberg vom 13. Bis 18. Jahrhundert*. Wiesbaden: Steiner.

Stolz, O. 1942. Verkehrsgeschichte der Brenner und Reschenstraße. *Großdeutscher Verkehr* 36 (11/12): 270-302.

Stolz, O. 1953. *Geschichte des Zollwesens, Verkehrs und Handels in Tirol und Vorarlberg*. Innsbruck: Wagner.

Stolz, O. 1957. *Der geschichtliche Inhalt der Rechnungsbücher der Tiroler Landesfürsten von 1288-1350*. Innsbruck: Wagner.

Stromer, W. 1995. Binationale deutsch-italienische Handelsgesellschaften im Mittelalter, in S. Rachewiltz and J. Riedmann (eds) *Kommunikation und Mobilität im Mittelalter: Begegnungen zwischen dem Süden und der Mitte Europas (11.-14. Jahrhundert)*: 135-158. Sigmaringen: Thorbecke.

Thurnher, E. and M. Zimmermann (eds) 1979. *Die Sterzinger Miszellaneen-Handschrift*. Göppingen: Kümmerle.

Torggler, A. 2018. Der Küchenbetrieb auf Schloss Tirol in meinhardinischer Zeit aus historischer Sicht, in H. Stadler and E. Flatscher (eds) *Schloss Tirol Archäologie. Die archäologischen Befunde und Funde*: 28-39. Bozen: Athesia.

Voltelini, H. (ed.) 1899. *Die Südtiroler Notariats-Imbreviaturen des 13. Jahrhunderts: Teil 1.* Aalen: Scientia-Verlag [Reprint of 1973].

Voltelini, H. and F. Huter (ed.) 1951. *Die Südtiroler Notariats-Imbreviaturen des 13. Jahrhunderts: Teil 2.* Innsbruck: Wagner.

Warnke, C. 2009a. Honig, in *Lexikon des Mittelalters 5*: 117-118. Darmstadt: WBG.

Warnke, C. 2009b. Wachs, in *Lexikon des Mittelalters 8*: 1888-1890. Darmstadt: WBG.

Website of the Forestry Department of the South Tyrolean Provincial Administration (January 2020): http://www.provinz.bz.it/land-forstwirtschaft/wald-holz-almen/wald-in-suedtirol.asp.

Website of the Forestry Organisation Department of the Federal Province of Tyrol (January 2020): https://www.tirol.gv.at/umwelt/wald/datenundfakten/

Weiss Adamson, M. 1996. Kochrezepte im Codex J 5 (no. 125) der Bibliothek des Priesterseminars Brixen: Edition und Kommentar. *Würzburger medizinhistorische Mitteilungen* 14: 291-303.

Wiesflecker, A. 1986. *Die oberösterreichischen Kammerraitbücher zu Innsbruck 1493-1519: Ein Beitrag zur Wirtschafts-, Finanz- und Kulturgeschichte der oberösterreichischen Ländergruppe.* Graz: DBV Verlag.

Wolfsgruber, K. (ed.) 1968. *Die mittelalterlichen Stiftsurbare des Bistums Brixen 1: Die ältesten Urbare des Benediktinerinnenstiftes Sonnenburg im Pustertal.* Vienna: Böhlau.

Zingerle, I. and J. Egger (eds) 1888. *Österreichische Weistümer 5: Die Tirolischen Weisthümer 4: Burggrafenamt und Etschland 1. Hälfte.* Vienna: Braumüller.

Zingerle, I. and J. Egger (eds) 1891. *Österreichische Weistümer 5: Die Tirolischen Weisthümer 4: Burggrafenamt und Etschland 2. Hälfte.* Vienna: Braumüller.

Zingerle, I. and K. T. von Inama-Sternegg (eds) 1875. *Österreichische Weistümer 2: Die Tirolischen Weisthümer 1: Unterinnthal.* Vienna: Braumüller.

Zingerle, I. and K. T. von Inama-Sternegg (eds) 1877. *Österreichische Weistümer 3: Die Tirolischen Weisthümer 2: Oberinnthal.* Vienna: Braumüller.

Zingerle, I. and K. T. von Inama-Sternegg (eds) 1880. *Österreichische Weistümer 4: Die Tirolischen Weisthümer 3: Vinstgau.* Vienna: Braumüller.

Zingerle, O. 1890. *Meinhards II. Urbare der Grafschaft Tirol.* Vienna: Tempsky.

Chapter 14

Early Irish Law on Bee-Keeping, with Particular Reference to *Bechbretha* 'Bee-Judgements'

Fergus Kelly

School of Celtic Studies, Dublin Institute for Advanced Studies (fkelly@celt.dias.ie)

Summary: The Old Irish law-text entitled *Bechbretha* 'bee-judgements' can be reliably dated on linguistic and historical grounds to the seventh century AD. It provides a great deal of information on the practice of beekeeping in Ireland at this period, and describes the legal mechanisms whereby the ownership of bees was fitted into the traditional framework of the law of neighbourhood. It also discusses the legal consequences if a beekeeper's hives are damaged or stolen, or if his bees sting passers-by without provocation. The final section of the text deals mainly with the legal complexities which arise in cases where a stray swarm is found on another person's land: the claims of beekeeper, landowner and finder of the swarm are ingeniously balanced.

Key Words: Old Irish law; beekeeping; Old Irish; property law

By good fortune, the Old Irish (hereafter OIr) law-text *Bechbretha* 'bee-judgements' survives in its entirety in an excellent fourteenth-century manuscript, now held in the Library of Trinity College, Dublin (ed. Charles-Edwards and Kelly 1983, hereafter *BB*). On linguistic grounds, the original text can be dated to the seventh century (*BB* 12-14). In the manuscript, the text is accompanied by explanatory glosses and commentary, ranging in date from about the twelfth to the sixteenth century.

Honey

The existence of a law-text devoted exclusively to bee-keeping testifies to the importance of honey (Old Irish *mil*) in the early Irish economy. Cane-sugar was not imported into Western Europe until about the twelfth century – except perhaps in very small quantities – and remained for many centuries a luxury of the rich (*BB* 42). Honey was therefore the only sweetener generally available, and occupied an important place in the early Irish diet. A medico-legal text entitled *Bretha Crólige* 'the judgements of blood-lying' describes honey as a relish which enhances a basic diet of bread and milk-products (*Bretha Crólige*, ed. Binchy 1938). This is echoed in an eighth-century saga entitled *Fled Bricrenn* 'The Feast of Bricriu', which refers to wheaten bread baked in honey (Best and Bergin 1929). The health-giving properties of honey were much appreciated by the early Irish. In fact, so much importance was attached to it for this reason that *Bechbretha* contains the proviso that an invalid was legally entitled to receive a ration of honey from his bee-keeping neighbours (*BB* 52 §6). But the lawyers were also aware that in certain medical circumstances, the consumption of honey was not advisable on account of its slight acidity. Thus, *Bretha Crólige* warns against the use of honey in the case of ailments involving the stomach, particularly when there is diarrhoea (*Bretha Crólige* 20 §25 (ed. Binchy)).

Mead and Bragget

We know from the early Irish sagas that mead (OIr *mid*) was the most prestigious alcoholic beverage. Its consumption was mainly associated with the nobility: important people are represented as gathering together for eating, drinking and music in the legendary banqueting hall at the royal site of Tara known as *Tech Midchuarda* 'The House of the Mead-circuit' (Kelly 1997: 356-8). Another alcoholic drink mentioned in our sources is *brocóit* 'bragget', made from malt and honey, which seems to have been intermediate in strength between mead and beer. This beverage may have been introduced from Wales, as the tenth-century Irish scholar Cormac mac Cuilennáin identifies the word as a borrowing from Medieval Welsh *bracaut* (*bragod*).[1] In a comic section in the ninth-century *Triads of Ireland* (ed. Meyer 1906), one type of dispenser of drink is described as *bolcshrónach brocóite*, which might be translated as 'a swollen-nosed bragget-swigger': the implication is that he is in the habit of helping himself generously to his employer's liquor (*The Triads of Ireland* 28 §231). In later Irish folklore, there is frequent mention of a drink called *leann beach* 'bees' beer', made from honeycombs. Sometimes it is associated with the Vikings, but Irish written sources indicate that a honey-beer was popular

---

[1] *Sanas Cormac* 12 §124 (ed. Stokes 1912). Cormac's etymology is generally accepted by modern scholarship: see Bachellery and Lambert 1981, B-94.

in Ireland long before any incursions from Scandinavia. As well as mead and bragget, there are records of the consumption of an unfermented honey and water mixture called *mellit*, probably to be identified as 'hydromel' (Kelly 1997: 113).

Beeswax

Surprisingly, *Bechbretha* makes no mention of beeswax (OIr *céir*, a borrowing from Latin *cera*), a valuable product utilised in the manufacture of candles and writing tablets. By contrast, there are many references to beeswax in Medieval Welsh law-texts. According to the thirteenth-century *Llyfr Blegywryd* an escaped swarm that is found after it has settled is worth four pence. If the finder shows it to the owner of the land, he receives four pence for his trouble, as well as a dinner. Alternatively, he may be given all the wax (*cwyr*) from the colony (*Llyfr Blegywryd* 55.19-23 (eds. Williams and Enoch Powell 1942)).[2]

Arrival of Honey-Bees in Ireland

The early history of the honeybee (*Apis mellifera*) in Ireland and Britain is obscure, and expert opinion is divided as to whether this species was introduced by humans or made its own way from the European continent when there was still a land-connection after the retreat of the Ice Cap. This question was considered by the late Dr Eva Crane OBE, the great expert on the history of bee-keeping, and author of *The Archaeology of Bee-keeping* and *The World History of Bee-keeping and Honey-Hunting*. In a personal communication, she gave her opinion that the honeybee was likely to be native to Ireland (and also to Britain) (*BB* 39, note 1). A different view was held by the eighth-century author of an Irish poem entitled *The Martyrology of Óengus*: he held that a swarm was brought to Ireland from Britain by a saint named Mo Domnóc, who flourished in the seventh century (*Félire Óengusso Céli Dé* 60 §13 (ed. Whitley Stokes 1905)). If true, this would imply that honeybees were quite a recent introduction at the time of the composition of *Bechbretha*. The great legal expert Daniel A. Binchy supported this view, and attributed the extraordinary detail of this text to the fact that the lawyers were dealing with an innovation which they had to fit into the existing legal framework. He also pointed out that the Irish monastic church was responsible for many innovations in farming and gardening practice. In our edition, Thomas Charles-Edwards and I came to the conclusion that – while the possession of bee-hives may well have been a common feature of monastic husbandry – honeybees were present in Ireland at least as early as the arrival of Celtic-speaking colonists

around 500 BC. This conclusion is based mainly on linguistic evidence.

Linguistic Evidence

The Celtic languages have native terms for 'bee', 'honey' and 'mead'. It might be argued that these words originally applied to the wild bumblebee (genus *Bombus*), but this does not adequately explain the linguistic situation. Thus, the OIr term for 'mead' (*mid*, ModIr and Scots Gaelic *miodh*) is cognate with Welsh *medd*, Cornish *meth* and Breton *mez* of the same meaning. This can hardly apply to the produce of the bumblebee, whose meagre stores of honey could never have been adequate for the manufacture of this prestigious drink.[3] An even more telling argument indicating the long-term presence of honeybees in Britain and Ireland is the agreement of words in the Celtic languages for the second swarm to emerge annually from the bee-hive. Thus, an OIr gloss on *Bechbretha* refers to this swarm as *tarbshaithe*, the literal meaning of which is 'bull swarm'. The exact cognate is to be found in MedWelsh *tarwhaidd* and MidBreton *tarvhet*.[4] No other bee has the habit of emitting a number of swarms from the hive each year, so this suggests that acquaintance with honeybees goes back to the time of the arrival of Celtic-speaking people in these islands.

Knowledge of Bee-Reproduction

The complex life-cycle of the honeybee has long been a subject of debate and scientific research. In his *Historia Animalium* the Greek philosopher Aristotle remarks that 'some people call the leaders [of the bees] 'mothers' ... others maintain that copulation occurs among these insects, and that the drones are male and the [worker] bees female' (Aristotle, *HA* 5.21 (trans. Peck). He also records the belief that honeybees collected their young from certain types of flower. In his *Georgics*, Virgil (4. 295-314) put forward the theory that honeybees issued spontaneously from the corpse of a bullock which had been clubbed to death. In medieval times there was similar variety in the theories propounded with regard to bee-reproduction (*BB* 203).

Early Irish sources contain no explicit reference to the queen bee, but an OIr gloss on *Bechbretha* – dating from about the ninth century – refers to an assembly of bulls (*tóla tarb*) in the context of the departure (*inscuchud*) of

---

[2] The Latin version of this passage is given with translation at *BB* 196 §9d.

[3] According to Irish folklore records, the taking of small quantities of honey from various ground-nesting species of bumblebee was a common pastime among children, particularly at hay-making time. These nests were called by local names such as *talmhóg* (from *talamh* 'land') or *cuasnóg* (from *cuas* 'hollow, cavity').
[4] The consensus is that the Breton language originated with an influx of British-speaking immigrants from South-West Britain in the sixth and seventh centuries CE: see Jackson 1967: 1–2 §1.

a swarm of bees (*BB* 64 §25). The context is obscure, but one possibility is that *tarb* 'bull' refers here to the drone. In his article 'Bees in Indo-European languages', D. E. Le Sage (1974) remarks that 'biological knowledge has brought us to dissociate the male bee from that aspect of it which first made it conspicuous to our forebears: its loud and low-pitched buzz'. Consequently, the drone may have been called a *tarb* on account of the bull-like noise with which it is associated. It is possible that naming the second swarm as *tarbshaithe* 'bull-swarm' refers to the greater number of drones present in this swarm (*BB* 117).

In contrast, the Medieval Welsh law-texts on honeybees make no reference to drones but contain terminology suggesting that the true biological function of the queen-bee had been recognised. Texts of North Welsh origin assign the same value to the *modrydaf* as to the whole hive (*henlleu* lit. 'old colony'). The word *modrydaf* derives from **modyr* 'mother' + *bydaf* 'colony of bees', apparently referring to the queen of the hive (Phillips, 1974:119-20).[5] Without her, the whole hive is valueless.[6] This interpretation is supported by a gloss which defines the *modrydaf* as *gwrach* 'old woman', thereby recognising the femininity of the queen bee.[7]

Swarming

In both early Irish and Welsh law-texts we find that considerable attention is paid to the legal problems which may ensue as a result of the honeybees' practice of sending out swarms. The author of *Bechbretha* took the same view as modern bee-keepers, and regarded two strong swarms as the maximum which is sustainable in a normal year. Thus, he counts the first and second swarms as being of value, but dismisses a third swarm as being *meraige*, lit. 'the fool' (*BB* 62 §§19-21). The ownership of swarms is discussed in considerable detail in this text, and it must be admitted that some of the regulations relating to swarming would have been difficult to adhere to in real life. Nonetheless, it seems likely that in practice there was general recognition in the community of the entitlement of the bee-keeper and the owner of the land where the bees have settled to a share in the produce. *Bechbretha* deals with the situation of the bee-keeper who manages to track a swarm from his hive that has established itself in a tree-cavity on the land of a high-ranking neighbour (§37). He is entitled to one third of the produce for three years, but the bees ultimately become the property of the landowner. On the other hand, if the bees have settled on open land or on the branches of a tree, the bee-keeper who tracks the swarm is entitled to two thirds of the produce, and retains ownership of the swarm: the land-owner only gets one third (§41). The same applies if the pursuing bee-keeper manages to capture a swarm with the help of a spread cloth (*brat scarthae*). In the case where a man follows a swarm that is not his and finds the place where it settles, the finder gets one third of the produce while the land-owner and the original owner of the bees each get a third (§43).

Bechbretha also deals with the case where there are many hives in the same place owned by different bee-keepers, and none of them are in a position to swear regarding ownership of swarms which have issued from the hives and have settled on the land of a neighbour. The solution is for the produce of the swarms to be divided in two, with one half going to all the bee-keepers, and the other half to the neighbouring land-owner (§45). If someone finds a stray swarm of unknown origin on land near human habitation, he is entitled to one quarter of the produce for a year: the rest goes to the landowner (§46). If a swarm is found on farmland in a more remote location, the finder gets one half, with the rest going to the landowner (§47). If a swarm is found in forest or inaccessible country, it goes to the finder in its entirety, apart from a share for the church and for the head of the finder's kin-group (§49).

Both Early Irish and Medieval Welsh lawyers devote a good deal of attention to the legal consequences of damaging another person's tree when trying to capture a swarm. *Bechbretha* distinguishes the case of tracked bees which have settled in a tree-cavity from those which have settled on the branches. In the latter case, the bees are only in residence temporarily while scout bees search for a suitable home, and therefore the swarm can be removed without damage to the tree (*BB* 76 §41). In the former case, however, the bees have established a permanent colony in the tree, and can only be removed by cutting the tree-trunk (*BB* 72 §§36-37). In relation to this scenario, a glossator observes 'it is not easy to cut around it (the nest)' (*ní hurusa a imdibe*). A person who cuts into the trunk of a tree on another's land is liable to pay the heavy fines for tree-damage as described in the OIr law-text *Bretha Comaithchesa* 'judgments of neighbourhood'.[8] Medieval Welsh law likewise lays down severe penalties for tree-damage. Thus, Latin Redaction E states that the person who cuts into an oak pays twenty-four pence for making an opening in the tree, and a further twenty-four pence for the honey and the bees (*BB* 197 §1). If he cuts the tree to the ground, an even heavier fine of ten shillings must be paid (*BB* 198 §2).

---

[5] *Bydaf* is most likely a cognate of OIr *betham* 'swarm of bees': see Bachellery and Lambert 1981: B-45 s.v. *bethamain*.

[6] Confusingly, in law-texts of South Welsh origin the term *modrydaf* is applied to the whole colony. See discussion at *BB* 202-03.

[7] See discussion on *gwrach* in note to p. 55.15 in *Llyfr Blegywryd* (ed. Williams and Powell 1967).

---

[8] For the details, see Kelly 1997: 385-9. A new edition of *Bretha Comaithchesa* is being prepared by Thomas Charles-Edwards for the Early Irish Law Series.

Bee-Hives

Throughout the world, a great variety of housing has been provided by humans for honeybees (Crane 1983). The term used for 'bee-hive' in the seventh-century text of *Bechbretha* is *lestar*. It is likely that the author is thinking primarily of a hollowed-out log adapted as a dwelling for honeybees, as the word *lestar* 'vessel' is elsewhere generally applied to wooden containers (Kelly 1997: 110). Another law-text refers to a *bech-dín* 'bee-shelter', which may refer to a number of hives arranged under some sort of roof to keep off wind and weather (*Corpus Iuris Hibernici* II. 384.20 (ed. Binchy 1978)). Later, it seems that bee-hives were more often constructed of woven willow-rods rather than being carved from wood. Thus, legal scholars from the twelfth to sixteenth centuries refer to the bee-hive as *ceis*, a term usually applied to a woven basket (*BB* 44). Another widely used term for bee-hive is *cliabh*, likewise applied to various woven objects, such as a basket, panier, cradle, or coracle (Quin et al. 1983. s.v. *clíab*).

Before the Anglo-Norman Invasion of Ireland in 1169, there is no reference in the written sources to the use of straw in the manufacture of containers and other domestic articles, and it seems that early Irish harvesting methods were unfavourable to the production of straw which could be employed for such purposes (Kelly 1997: 111, 240). Subsequently, however, the linguistic evidence indicates that hives were typically made of oaten straw. The regular word for 'hive' in Modern Irish is *coirceóg* (*corcóg*), which is likely to be a derivative of *coirce* (*corca*) 'oats' (Quin et al.1984, s.v. *corcóg*). A nineteenth-century triad reads: *Trí beaga is fearr: beag na cuirceóige, beag na caerach, agus beag na mná* 'three small things which are best, a small bee-hive, a small sheep, and a small woman'.[9] Perhaps the author of this triad regarded a small hive as most suitable for encouraging swarming and the production of new colonies.

A bee-keeper might keep his hives in various locations. *Bechbretha* refers to bee-hives which are kept in a garden (*lubgort*) or courtyard (*les*). Because these hives are located in proximity to the house, the fine for stealing them is heavy, and fixed at the same rate as household goods (*BB* 84 §§50-51). A lesser fine is payable if the hives are kept in nearby fields, defined as the green or infield (*faithche*). In such cases, the thief is fined at the same rate as would be payable if had stolen large farm animals such as milch cows or oxen (*BB* 86 §§52). However, if the hives are kept in a more remote location outside the green, the culprit is only liable to pay at the same rate as lesser farm animals such as calves or lambs (*BB* 86 §§53).

Bee-Trespass

A unique feature of *Bechbretha* is the attempt which is made to define and regulate 'bee-trespass'. To my knowledge, no other legal code in the world – past or present – has risen to this difficult challenge. Nonetheless, the logic of the early Irish lawyers is undeniable. Honeybees are domestic animals like cows, sheep, hens, geese, etc. whose owners must pay for grazing-trespass on a neighbour's land. Consequently, it makes sense that a bee-keeper should pay for the nectar which his bees have foraged from flowers growing in neighbouring farms. This might seem to be an instance of extreme legal pedantry. However, as the bee-keeper is obliged – after a three-year period of grace – to give a swarm of bees to his nearest neighbours in turn, the effect is to spread the practice of bee-keeping throughout the community. Ultimately, all the landowners in the area can expect to own bee-hives, and the trespasses of their respective bees will cancel out. *Bechbretha* does not deal with the problem of identifying the bees responsible for a particular instance of 'grazing trespass'. A bee-keeper might deny that his bees have been the culprits, thereby obliging the plaintiff to obtain proof of their guilt. A difficult passage in another law-text deals with this conundrum, and describes a procedure whereby the plaintiff – accompanied by witnesses – pursues the offending bees to their hive.[10] One possible interpretation of this text suggests that the plaintiff sprinkles flour onto the trespassing bees to identify them (*BB* 190).[11]

Bee-Stings

The general consensus among bee-keepers is that the honeybee of Europe is a fairly mild-tempered insect and will not attack unless provoked. Nonetheless, it will respond fiercely to any person or animal regarded as posing a threat. *Bechbretha* stresses that the bee-keeper is not liable for any injury sustained by somebody who is robbing or moving a bee-hive (*BB* 66 §27). But if a bee stings a passer-by who is doing no harm or illegality, the bee-keeper is liable. The penalty for this offence seems rather light: the victim is entitled to 'his sufficiency of honey', presumably entailing a generous offering from the honey-combs of the offending hive. However, he must swear an oath that he did not kill the bee which stung him. If he has killed the bee, no penalty is due: the legal view is that the death of the bee compensates for the sting it has inflicted (*BB* 66 §29). A ninth-century glossator on *Bechbretha* makes the point that a bee dies if it leaves its sting in human skin: he stresses that from a legal perspective this is not to be regarded as a case

---

[9] This triad is quoted in Maynooth University Library MS Murphy 48 (3. E. 4) p. 355.

[10] This passage is edited and translated in Appendix 6 of *BB* 189-91.

[11] The case is made here that *blaithe* of this law-text may be genitive singular of an unattested OIr **bláth* 'flour'; cf. Welsh *blawdd*, Breton *bleud* 'flour'.

of the killing of the bee by its victim (*BB* 66 §29, gloss e (B)).

Bechbretha devotes a good deal of attention to cases where a honeybee stings a person in the eye, thereby causing blindness. The text cites the case of Congal, an Ulster king who was stung by a bee causing him to lose the sight of one eye: thereafter he was called Congal Cáech, i.e. 'the one-eyed'. According to early Irish custom, a king who suffered from a physical disability became ineligible to continue in the kingship. Consequently, Congal had to abdicate from the kingship of Tara (*BB* 68 §32.).[12] Bee-keepers have expressed doubt as to the historical accuracy of this account of Congal's injury. In reply to a query, Dr Crane pointed out that the eye-closing reflex in the human is very rapid, and does not allow much opportunity for bee-venom to penetrate to the cornea to inflict serious damage. However, she sent references to two articles in Russian medical journals describing just such an injury, so the account in *Bechbretha* may reflect an actual incident.[13]

Death may result from a sting in cases where the victim has become sensitised to bee-venom. According to Medieval Welsh law (Latin Redaction E), if bees kill a man the entire hive is itself destroyed, and its store of honey is given in reparation. This seems a fairly light penalty. However, if the owner of the offending hive fails to destroy the bees, and allows them to continue to produce honey, he has dishonoured the victim, and must pay the full honour-price or wergeld (*galanas*) of the dead man to his kin (*BB* 198 §§6-7). The OIr text of *Bechbretha* does not deal with the legal implications of death caused by a bee-sting. A MidIr glossator remarks that the 'book' – i.e. *Bechbretha* – refers to the payment of one hive in recompense for blinding, but does not refer to two hives in recompense for killing (*indisid lebar in cis ina chaechad, agus ni hindisenn da chis isin marbad*) (*BB* 68 §30d). The implication here is that in later Irish legal tradition, two hives were regarded as appropriate recompense for a death caused by bees.

Bee-Plagues and Predators

The economic importance of honeybees in early Irish society is indicated by the fact that the occurrence of bee-plague (*bechdíbad*) in AD 951 and again in 993 was held to be of enough significance to be recorded in the Annals of Ulster (*The Annals of Ulster* 396 s.a. 950 (recte 951) §6, 424 s.a. 992 (recte 993) §7 (eds. Seán Mac Airt and Gearóid Mac Niocaill 1983). No details of these outbreaks are supplied but comparison can be

made with the infamous epidemic of 1909-1917 known as 'Isle of Wight disease'. It has been calculated that approximately 90 percent of the indigenous 'Black bees' of Britain and Ireland perished. Italian, Caucasian, Cyprian and other varieties were subsequently imported, so the honeybees of these islands are now of very mixed ancestry. In recent times considerable success has been achieved in reviving fairly pure breeds of Black bee.[14]

In Eastern and Northern Europe, the Brown Bear (*Ursus arctos*) poses a serious threat to bee-hives in some locations, and measures must be put in place to protect them. The archaeological evidence indicates that the Brown Bear inhabited Ireland before the last Ice Age, but did not return after the retreat of the Ice Cap about 10,000 BC. Consequently, it is most unlikely that there were any bears in Ireland when Celtic-speakers arrived during the first millennium BC. Nonetheless – presumably through contact with the Continent or perhaps northern Britain – a knowledge of this animal's existence is apparent in a number of early Irish texts. The Indo-European word for 'bear' is preserved in the common personal name *Art*, cognate with Welsh *arth*, Latin *ursus*, etc. (Vendryes 1959: A-91). Another Irish term for this animal found in an OIr text is *milchobur*, the literal meaning of which is 'honey-lover' (Watkins, 1962: 114-16). The author of this text may never have seen a bear, but knew that it was attracted to honey.

Among native Irish fauna, the only mammal which can pose a serious threat to bees is the badger (*Meles meles*), in Irish *broc(c)*. Badgers are known to attack the nests of honeybees and bumblebees, consuming honey, larvae and adult bees (Neil, 1948: 62-3). According to MidIr legal commentary dating from about the twelfth century, hens may be responsible for three offences within the farm-yard (*lis*). These offences are the destruction of the madder and onion crop (*lot roidh ocus cainninne*) and the soft swallowing of bees (*máethshlucud bech*). In response to a query, Dr Eva Crane expressed surprise at the inclusion of hens as a threat to honeybees (Kelly 1997: 181). She stated that she had come across no instances of damage inflicted by hens on bee-hives or larvae within the hives. She noted that hens are opportunistic feeders, and might well eat dying bees from the ground in front of a hive.

Honeybees and the Church

I conclude with a brief account of the significance of the honeybee in Church tradition. It is likely that most monasteries had their own bee-hives, and we find records of individual monks and nuns who enjoyed a special relationship with the monastic bees. Saint

---

[12] The evidence of the Annals indicates, however, that he retained the local kingship of a North-East Ulster people called the Cruithni: see discussion in Notes at *BB* 123-31.
[13] A text from the eleventh or twelfth century gives a different account of Congal's injury, stating that he was stung as a baby (*Fled Dúin na nGéd* 10.314-18 (ed. Lehmann 1964)).

---

[14] Two hives of Irish Black bees, owned by Seán Byrne, thrive in my orchard in Rathduffmore, Co. Wicklow.

Boecius of Monasterboice is even reputed to have understood their language (*Vita Sanctorum Hiberniae* 97 §xxxi (ed. Plummer 1919)). As we have seen, beeswax was used in the manufacture of candles, an important element in monastic life, both practical and ceremonial. Beeswax features in a delightful Early Modern Irish tale found in the fifteenth-century manuscript *Liber Flavus Fergusiorum*: it tells of a miraculous event which befell an unnamed priest (Gwynn 1905: 82-3). He went one day to tend a sick man, and had the Sacred Host with him. On the way, he encountered a swarm of honeybees, which he gathered up and brought away with him. He laid the Host on the ground and forgot about it. Later he remembered what he had done, and went looking for the Host in anxiety and contrition, but could not find it however much he searched. He subsequently went to Confession, and to atone for his carelessness was required to spend a full year in penitence. At the end of the year, an angel came to him and told him where the Host was to be found. The mystery of the missing Host was then revealed. The bees which the priest had taken with him had returned to where he had left the Host and brought it to their hive. There they lovingly paid it reverence, and constructed from beeswax a fair chapel and an altar and a chalice, as well as models of two priests above the Host. Many people came to see this wonder, and believed in it.

The name of Saint Gobnaid (OIr Gobnait) has a special significance for Irish bee-keepers, as she is the recognised patron of their craft. She is thought to have been born in the sixth century on the island of Inis Oírr (Inisheer, Co. Galway) (Ó Riain 2011). She founded churches in various parts of the country, but is best known for her foundation at Baile Mhúirne (Ballyvourney, Co. Cork). She is reputed have been a bee-keeper, and on one occasion was able to effectively repel aggressors with the assistance of a swarm of her bees. The local veneration of Gobnaid continues to this day, with rounds of the church – in Irish *turas* – being carried out on most Sundays, especially at Whitsuntide. A thirteenth-century wooden image of the saint is kept locally and thought to have miraculous powers. A modern statue at Baile Bhúirne by the sculptor Seamus Murphy features Saint Gobnaid, appropriately surmounted on a straw hive with honeybees depicted on the base.

Bibliography

Bachellery, E. and P.-Y. Lambert (eds) 1981. *Lexique étymologique de l'irlandais ancien: lettre B*. Dublin: Dublin Institute for Advanced Studies / Paris: Centre National de la Recherche Scientifique Paris.

Binchy, D. A. (ed.) 1938. *Bretha Crólige*. *Ériu* 12: 1–77.

Binchy, D. A. (ed.) 1978. *Corpus Iuris Hibernici*. Vols. I–VI. Dublin: Dublin Institute for Advanced Studies.

Charles-Edwards, T. and F. Kelly (eds) 1983. *Bechbretha: An Old Law-tract on Beekeeping*. Early Irish Law Series I. Dublin: Dublin Institute for Advanced Studies. 2008 repr. with additional Appendix.

Crane, E. 1983. *The Archaeology of Beekeeping*. London: Duckworth.

Crane, E. 1999. *The World History of Beekeeping and Honey Hunting*. London: Duckworth.

Fairclough, H. R. (trans.) 1916. *Virgil: Eclogues. Georgics. Aeneid: Books 1-6*. The Loeb Classical Library 63. Cambridge, Mass.: Harvard University Press.

Gwynn, E. J. (ed.) 1905. The Priest and the Bees. *Ériu* 2: 82-3.

Jackson, K. H. 1967. *A Historical Phonology of Breton*. Dublin: Dublin Institute for Advanced Studies.

Kelly, F. 1997. *Early Irish Farming: A Study based mainly on the Law-texts of the seventh and eighth centuries AD*. Early Irish Law Series IV. Dublin: Dublin Institute for Advanced Studies.

Lehmann, R. (ed.) 1964. *Fled Dúin na nGéd*. Mediaeval and Modern Irish Series. Vol. XXI. Dublin: Dublin Institute for Advanced Studies.

Le Sage, D. E. 1974. Bees in Indo-European Languages. *Bee World* 55: 15-26, 46-52.

Mac Airt, S. and G. Mac Niocaill (eds) 1983. *The Annals of Ulster (to A.D. 1131)*. Dublin: Dublin Institute for Advanced Studies.

Meyer, K. (ed.). 1906. *The Triads of Ireland*. Royal Irish Academy. Todd Lecture Series XIII. Dublin: Hodges, Figgis, & Co., Ltd.

Neil, E. 1948. *The Badger*. New Naturalist Series. Collins: London.

Ó Riain, P. 2011. *A Dictionary of Irish Saints*. Dublin: Four Courts Press.

Peck, A. L. (ed.) 1989. *Aristotle: History of Animals*. Books IV-VI. Loeb Classical Library. Harvard University Press. Cambridge. Mass.

Phillips, E. 1974. 'Modrydaf.' *The Bulletin of the Board of Celtic Studies* 25: 119-20.

Plummer, C. (ed.) 1910. *Vita Sanctorum Hiberniae.* Vol.1. Oxford: Clarendon.

Quin, E. G. et al. 1983. *A Dictionary of the Irish Language, Based Mainly on Old and Middle Irish Materials.* Compact edition. Dublin: Royal Irish Academy.

Stokes, W. (ed.) 1905. *Félire Óengusso Céli Dé: The Martyrology of Oengus the Culdee.* Henry Bradshaw Society, vol. 29; repr. 1984. Dublin: Dublin Institute for Advanced Studies.

Stokes, W. (ed.) 1912. *Sanas Cormac: An Old-Irish Glossary compiled by Cormac Ó Cuilennáin,* in O. J. Bergin, R. I. Best, K. Meyer, J. G. O'Keefe (eds) *Anecdota from Irish Manuscripts.* Dublin: Hodges, Figgis, & Co., Ltd.

Vendryes, J. 1959. *Lexique étymologique de l'irlandais ancien: lettre A.* Dublin: Dublin Institute for Advanced Studies / Paris: Centre National de la Recherche Scientifique Paris.

Watkins, C. 1962. Irish *milchobur. Ériu* 19: 114-16.

Williams, S. J. and J. E. Powell (eds) 1967. *Llyfr Blegywryd.* Cardiff: University of Wales Press.

Chapter 15

Arqueología de la apicultura en la Asturias preindustrial

Juaco López Álvarez

Museo del Pueblo de Asturias (juacolopez@gmail.com)

Resumen: Asturias es una región de la España cantábrica en la que los campesinos practicaban dos modelos de apicultura muy diferentes. En una parte del territorio la cría de abejas estaba más desarrollada con colmenares localizados en el monte, gran número de colmenas y la extracción de una parte de los panales ahumando a las abejas. En la otra, las colmenas que tenían unos pocos campesinos estaban junto a la casa y para sacarles la miel y cera se mataba la colonia de abejas. Se estudian estos dos sistemas, y se describen los tipos de colmenas y colmenares, las características y creencias de los apicultores, las prácticas apícolas, la búsqueda y depredación de colmenas silvestres, las vasijas para almacenar la miel, etc. Estos dos modelos de apicultura están hoy completamente extinguidos en Asturias.

Palabras Clave: Apicultura tradicional asturiana; colmenares; colmenas; Asturias

Summary: Asturias is a region of Cantabrian Spain where rural populations used to engage in strikingly different methods of beekeeping. In one part of the territory, beekeeping was highly developed featuring apiaries located on mountains, a large number of hives, and the extraction of a part of the combs through smoking the bees. In another part, some rural populations maintained hives close to their homes to get honey and wax and would subsequently kill off the colony. The following chapter treats these two systems and describes the types of hives and apiaries, the characteristics and beliefs of the beekeepers, their beekeeping practices, the search for and depredation of wild hives, and the vessels to store the honey. These two methods of beekeeping are today completely extinct in Asturias.

Keywords: Traditional Asturian apiculture; apiaries; hives; Asturias

En 1979 comenzó nuestro interés por el estudio de la apicultura en el occidente de Asturias, una región de la España cantábrica. En aquel año todavía existía una importante cría de abejas basada en las viejas colmenas fijas y en unas pocas operaciones que recordaban bastante a las que recomendaba el agrónomo Columela en el siglo I. Fuimos ampliando el área de estudio a toda Asturias y descubriendo diferencias importantes en el manejo de las abejas entre el occidente y el resto de la región. En un espacio geográfico pequeño se desarrollaban los dos modelos de apicultura que existen en Europa según el modo de extraer el producto de las colmenas: matando la colonia de abejas para sacar toda la miel y cera, o ahumando la colmena para extraer solo una pequeña parte de estos productos, dejándoles a las abejas lo suficiente para pasar el invierno. La convivencia de estas dos prácticas tan diferentes me permitió compararlas y analizar las ventajas de una sobre otra. El resultado de aquel trabajo de campo y de la consulta de numerosas fuentes escritas fue el asunto de nuestra tesis doctoral sobre las abejas, la cera y la miel en la sociedad tradicional asturiana, que presentamos en la Universidad de Oviedo en diciembre de 1990; se publicó en 1994 (López Álvarez 1994).

En los años en que se inició el estudio, la apicultura de colmenas de cuadros móviles, difundida en Asturias a partir de los años veinte del pasado siglo por varios sacerdotes, en especial por Carlos Flórez, que desarrolló esta labor desde los servicios agropecuarios de la Diputación Provincial, estaba muy poco extendida. Hoy, en cambio, treinta años después, está apicultura es la única que se practica y la otra, la antigua, ha desaparecido completamente. En consecuencia, todo lo que vamos a describir en este artículo sobre la apicultura en la España cantábrica pertenece a una etapa extinguida y es pura arqueología. Y la mayor parte de los informantes entrevistados en aquellos años han fallecido.

Por otro lado, a partir de aquel estudio se ha ido formando desde 1992 en el Museo del Pueblo de Asturias, fundado en Gijón en 1968, una colección de objetos relacionados con la apicultura tradicional: colmenas de diferentes tipos, cortadores de panales, ahumadores, vasijas para almacenar miel, prensas de cera (Figura 1) y poco más, pues esta apicultura no empleaba un repertorio muy amplio de herramientas o utensilios. Algunas de estas piezas son las que ilustran este artículo.

Figura 1. *Fustes* o prensa manual de cera. Madera de castaño. Col. Museo del Pueblo de Asturias.

Dos apiculturas diferentes: matar las colmenas y ahumar las colmenas

La cría de abejas se llevo a cabo en Asturias de dos modos muy diferentes que están documentados desde mediados del siglo XVIII y que aparecen en dos espacios geográficos separados por el río Narcea. El localizado al oeste vamos a denominarlo "zona occidental" y al otro, "zona centro-oriental". Estas diferencias apícolas se manifiestan en el paisaje rural, el número de apicultores y colmenas, y el distinto manejo que reciben las abejas en una y otra zona.

El paisaje rural asturiano estaba ordenado en unos espacios que aseguraban la subsistencia de los campesinos, cuya base económica eran el cultivo de la tierra y la cría de ganado. Estos espacios productivos se distribuían por el territorio, y eran los siguientes: la huerta, las tierras dedicadas al cultivo de cereales, los prados de siega, las pumaradas y las viñas, y el monte de uso común entre los vecinos. El último espacio era el más extenso y en él se incluían el bosque, el monte bajo y el pasto de altura. Su importancia era vital, pues

de él dependía en gran medida la subsistencia de los campesinos.

En la zona occidental, uno de los aprovechamientos del monte era la apicultura, que se manifestaba en el paisaje con dos construcciones específicas de esta actividad: los *talameiros* y, sobre todo, los *cortinos* (Figura 2). Nada de esto ocurría en el centro-oriente de Asturias, donde los colmenares estaban siempre en el interior de los pueblos o junto a las casas, y no tenían ninguna entidad en el paisaje rural.

El número de campesinos que tenían colmenas es otro aspecto a tener en cuenta para observar las diferencias entre estas dos zonas. En los pueblos centro-orientales había un número reducidísimo de apicultores, a menudo tan sólo eran dos o tres vecinos los que se dedicaban a las abejas. Esta actividad estaba asociada a unas pocas personas o a unas casas determinadas "que desde antiguo" tenían colmenas. Los miembros de estas casas, al casarse fuera, solían llevar una o dos colmenas para continuar la actividad apícola que conocían desde niños.

Figura 2. *Cortín* en el concejo de Allande. Fotografía de Marcelino Lozano.

En cambio, en la zona occidental el número de vecinos que tenían colmenas era grande, pudiendo alcanzar en algunos lugares a más de la mitad del vecindario e incluso a la mayoría de las casas del pueblo. De todos modos, también en esta zona era frecuente asociar la actividad apícola a unas personas o casas determinadas, no por el hecho de poseer colmenas, sino por el gran número que poseían.

La preponderancia del campesino-apicultor en el occidente, que constatamos en el trabajo de campo, se documenta también en el catastro del marqués de la Ensenada realizado en 1752. Las diferencias más notables se daban entre los concejos de la zona centro-oriental y los del interior montañoso del occidente de Asturias: en Casu tenían colmenas en propiedad o las llevaban en aparcería el 16% de los vecinos; en Mieres el número era menor, 12%, y en Somiedo sólo eran el 11%. Mientras que en los concejo de Ibias, Illano, Salime y Allande, todos localizados en el interior del occidente de Asturias, poseían colmenas el 40%, 45%, 58% y 60% de las familias, respectivamente.

En cuanto al número de colmenas que tenían por término medio los apicultores, los resultados de la documentación consultada y del trabajo de campo vuelven a mostrarnos las diferencias entre las dos zonas apícolas. En 1990, los apicultores de la zona occidental tenían por termino medio entre diez y veinte colmenas pobladas de abejas, y se consideraba un buen apicultor al que tenía cincuenta, no siendo raros los que mantenían unas cien colmenas. En el centro-oriente, los campesinos tenían en general entre cinco y diez, alcanzando los apicultores más avezados las veinte o treinta colmenas.

El catastro del marqués de la Ensenada es otra vez la fuente de información más antigua que tenemos para conocer el número de colmenas que poseían los apicultores asturianos a mediados del siglo XVIII. A partir de esta fuente se distinguen tres áreas geográficas:

a) Zona centro-oriental, en la que la mitad o más de la mitad de los apicultores tenían una o dos colmenas. Los que superan la cifra de nueve o diez son una minoría, y existen lugares donde ningún vecino alcanza estas últimas cifras.

b) Zona de la marina occidental, donde en torno al 40% de los apicultores tienen una o dos colmenas, siendo los poseedores de nueve a veinte colmenas el 10%.

c) Zona de la montaña occidental, en la cual el número de apicultores que poseen una o dos colmenas es del 20% y el porcentaje de campesinos con un número superior a siete colmenas oscila entre el 5% y el 10%. Destacan en esta zona los concejos localizados más al interior (Ibias, Allande, Illano, Pezós, Grandas de Salime), en los que un 20% o 30% de los apicultores tienen entre once y veinte colmenas, y alrededor del 10% sobrepasan la treintena.

La causa fundamental de estas diferencias en la apicultura asturiana hay que buscarla en la técnica de manejo que se emplean en una y otra zona, a la que habría que sumar unas causas naturales, como la abundancia de flora melifica en el occidente montañoso, especialmente de castaño y brezo, y razones históricas, al ser la zona occidental la que proveía de cera a los grandes monasterios de Asturias.

En el occidente se aplicaban unas técnicas que servían para mantener y aumentar el colmenar: se recogían los enjambres en mayo y junio, y también se sabía provocar enjambres artificiales cuando era preciso; se quitaba en marzo la cera vieja de la parte inferior de las colmenas, que favorece la propagación de la polilla e impide el desarrollo normal de la cría; se alimentaban las colonias si era necesario y, por último, se extraía una parte del producto cosechado por las abejas, dejándoles lo suficiente para que pudieran pasar el invierno.

Por el contrario, en la zona centro-oriental las operaciones se reducían a dos tareas básicas: la captura de los enjambres en primavera y la recogida completa de la miel y la cera, para lo cual se escogían las colmenas más pesadas y se mataban todas las abejas.

La existencia de estas dos técnicas de extracción, o sea, la de sacar una parte del producto recolectado por las abejas y la de extraerlo todo, aniquilando la colmena, es, desde luego, uno de los hechos que más claramente separa la apicultura del occidente de la que se practicaba en el centro y oriente de Asturias.

La convivencia de estas dos prácticas esta documentada a mediados del siglo XVIII en el mencionado catastro del marqués de la Ensenada, en el que se indica en muchos lugares el modo que se emplea para extraer el producto de las colmenas "según el estilo del país". En unos casos, localizados mayoritariamente en el centro y oriente de Asturias, se dice que "que matan las abejas al tiempo de esquilmarlas" y en otros, que se hallan casi en su totalidad en el occidente, declaran que "sacan el producto de las colmenas ahumándolas sin matarlas".

El trabajo de campo y la escasa documentación escrita sobre el tema muestran que en el centro-oriente no existía el arte de criar abejas, que es como se define la apicultura, sino más bien un aprovechamiento situado a un nivel intermedio entre la recolección y la cría. Este estadio lo definimos como una "depredación controlada", para distinguirlo de la simple depredación de colmenas silvestres, práctica que tuvo en la zona centro-oriental una gran importancia para complementar la escasa producción de miel.

Los campesinos del centro-oriente asturiano tenían las colmenas cerca de casa, para guardarlas del pillaje, pero las atendían muy poco, casi nada, y su cría se dejaba al capricho de la naturaleza. En primavera los campesinos recogían todos los enjambres posibles para mantener el colmenar, rara vez para aumentarlo, debido al sistema de extracción utilizado y a la alta mortandad de colmenas que acaecía durante el invierno o a comienzos de la primavera. En otoño se sacaba el producto de las colmenas, aniquilando a las abejas cuando todavía podían seguir produciendo más miel y cera. Su objetivo era conseguir unos litros de miel para consumirla en casa y para regalarla cuando existía algún compromiso con un vecino, un pariente o alguien ajeno a la comunidad (abogado, médico). La miel no era un comestible cotidiano, sino una sustancia medicinal que se reservaba para combatir ciertas enfermedades de la familia, en especial de las vías respiratorias, y del ganado. Al no poder los campesinos suministrar este producto a las villas y ciudades próximas, era habitual en estas poblaciones la presencia de mieleros de La Alcarria que vendían la miel casa a casa. En cuanto a la cera, la compraban en su mayor parte cereros del norte de la provincia León y suponía una ganancia muy pequeña debido a su escasa producción.

Por el contrario, la eficacia de las labores apícolas practicadas en el occidente de Asturias residía precisamente en la multiplicación y no en la aniquilación de las abejas. Esta característica, que marca la diferencia entre la agricultura y la ganadería, por un lado, y la caza y la recolección, por otro, es una de las razones que explica el mayor desarrollo apícola de esta zona de Asturias.

En el occidente la cría de abejas sí era un objetivo económico de los campesinos. La miel era un comestible habitual, que se consumía en gran cantidad en algunos momentos del año y que se vendía. La cera, por su parte, era un valor de cambio importante: se vendía por dinero, se trocaba por especies y se empleaba para pagar las rentas de la tierra. La abundante producción de cera permitió la formación de una industria, cuya instalación más destacada eran los lagares de cera donde ésta se cocía, exprimía y limpiaba, con la ayuda de un lagar de viga con husillo y pesa. Los comerciantes más activos en este género fueron los mencionados cereros, que procedentes del valle de Ancares (León) compraban la cera a domicilio y alquilaban los lagares para sacar a la provincia de León el producto limpio.

En los concejos occidentales, la naturaleza también influía mucho en el desarrollo de las colmenas: un año malo podía ocasionar la muerte de muchas, pero esa mortandad se compensaba con el empleo de unas prácticas que favorecían la cría y multiplicación de las abejas.

La recolección de miel en colmenas silvestres

Otra práctica relacionada con la apicultura era la recolección de miel de colmenas silvestres, que en la zona central y oriental de Asturias estaba muy generalizada, obteniéndose en algunos casos más miel de esta manera que la que se conseguía con las colmenas domésticas. Había especialistas que se dedicaban a esta tarea en solitario, pero por lo común era una actividad en la que participaban dos o más vecinos que al final se repartían el botín. La localización de una colmena silvestre requería unos métodos de búsqueda que a veces llevaban tiempo, a pesar de que los cazadores conocían bien el medio en el que se movían y las costumbres de las abejas. La búsqueda precisaba de unos conocimientos y unas técnicas de caza específicas. La primera operación consistía en encontrar una fuente donde las abejas acudan a recoger agua. Una vez localizada, los cazadores vigilaban a las abejas, fijándose en el camino que tomaban y siguiéndolas hasta el árbol o peña en donde tenían el nido. Las abejas, según los campesinos, en cuanto cogen el agua van para la colmena. Una vez localizada la dirección de las abejas, los cazadores utilizaban técnicas diferentes para seguirlas: unos, corrían detrás de ellas hasta que las perdían de vista, esperando a que pasaran más en esa dirección; otros, atraían a las abejas colocando miel en su ruta y las seguían de vuelta a la colmena; otros, capturaban varias abejas en un recipiente y cada vez que perdían su rastro soltaban una, y otros, se colocaban en la dirección de las abejas y se gritaban entre ellos cuando pasaba una hasta dar con la colmena silvestre.

Las operaciones de busca se realizaban en el mes de agosto, "porque pasado agosto las abejas beben menos y no son tan fáciles de encontrar" (Cuña, concejo de Teberga).

Las colmenas silvestres estaban por lo común en un árbol hueco o en una covacha de una peña. Los sitios preferidos de las abejas eran, según los cazadores, los troncos de castaños viejos y las peñas orientadas al mediodía y con "pasto" abundante. Las colmenas situadas en el hueco de un tronco o en una rama se depredaban a menudo talando el árbol; una vez tumbado y con la colmena al alcance de la mano se sacaba toda la miel y cera, asfixiando a las abejas y abriendo el hueco con un hacha.

La operación cambiaba bastante cuando la colmena estaba en una peña, en este caso los cazadores iban por la mañana, muy temprano, y tenían que descolgarse desde la cima atados con unas cuerdas que sostenían sus compañeros. Los que bajaban llevaban un caldero amarrado a la cuerda; una vez delante de la cueva asfixiaban a las abejas quemando hierba o azufre en la entrada y a continuación sacaban con una mano todos los panales que podían alcanzar con el brazo.

Los apicultores

En Asturias, a la persona que se le conoce con el nombre de *abeyeiro*, *abeyeru* o apicultor, o se le reconozca como tal, no significa que viva de las abejas, ni tampoco que la mayor parte de sus ingresos económicos provenga de este ramo. Este calificativo se aplica sólo a los campesinos que tienen muchas colmenas y, sobre todo, a aquellos que "entienden bien a las abejas", que año tras año mantienen el colmenar poblado, y que poseen una gran afición por este "ganado".

Normalmente, la cría de abejas está asociada a unas pocas casas en cada pueblo, en la que se hereda la práctica de generación en generación. Es muy raro que un campesino comience a tratar con abejas sin que haya conocido la experiencia de sus antepasados. Más corriente es que los herederos de la casa de un apicultor "no entiendan de abejas", tengan pánico a las picaduras o simplemente no les interese su cría, y así, a la muerte del apicultor, las abejas empiezan a ir a menos hasta que acaban desapareciendo.

El miembro de la familia que atiende las colmenas varía en función de la importancia que tenga su explotación y del espacio donde se lleve a cabo el trabajo, aunque por regla general suele ser una labor masculina. En el occidente, donde las abejas tienen cierta importancia en la economía doméstica y los colmenares están diseminados por el monte, es una actividad exclusiva de los hombres, que atiende directamente el cabeza de familia. Por el contrario, en el centro y oriente de Asturias son frecuentes las mujeres que se ocupan del cuidado de las colmenas en sus casas. En esta zona el aprovechamiento de las colmenas es una actividad secundaria, encaminada sobre todo a conseguir una sustancia medicinal: la miel, y además los colmenares están junto a la casa o en una huerta próxima, es decir, en espacios de trabajo exclusivos de la mujer.

Las personas tenidas por *abeyeiros* no son vecinos normales en la comunidad rural. Sus habilidades para criar y mantener las abejas son conocidas en su parroquia y a veces también en las circundantes. Pero además tienen fama de ser personas "curiosas" para llevar a cabo ciertos trabajos (calzar los arados, cambiar el eje al carro, etc.), algunos de gran responsabilidad en la economía rural, como sucede, por ejemplo, con el injerto de árboles frutales. Son a menudo miembros destacados de la comunidad, que sirven para casi todo y que practican diferentes trabajos: carpintería, cantería o herrería.

Los *abeyeiros* son personas aficionadas y apasionadas de las abejas, muy atraídas por el comportamiento y la biología de estos insectos, tan diferentes al resto de los animales domésticos. Su modo de ser contrasta con la animadversión que les causan a muchos de sus vecinos, que tienen pánico a las picaduras y opiniones muy contrarias a las abejas. Una creencia muy generalizada entre estas últimas personas es la de que las abejas no conocen al amo, es decir que no respetan la autoridad del jefe de la casa, y por tanto rompen la jerarquía que sostiene el mantenimiento y el buen orden de la casería. Su opinión la resumen en una sola frase: "son el peor ganado que existe". Por el contrario, los *abeyeiros* creen que las abejas conocen a los que tratan con ellas, aunque no saben muy bien por qué. Alguno supone que la causa es la continua presencia en el colmenar, y por eso se considera que las abejas de colmenares situados en el monte, lejos de casa y a las que se visita muy poco, son más agresivas (López Álvarez 2005).

Otra creencia que lleva aparejada la animadversión hacia las abejas es la idea de que son poco rentables para la casa y que su provecho es pequeño en comparación al trabajo que dan; esto se refleja en la existencia de refranes, como "En abeyas y en oveyas non metas lo que tengas". En cambio, los *abeyeiros* consideran que las abejas en el sentido económico es el mejor ganado que existe, porque "no cuestan nada y están todo el día trabajando para el amo" (Purón, concejo de Llanes).

Tipos de colmenas

Las colmenas tradicionales son todas verticales y fijas. Reciben los nombres siguientes: *cubu* (del latín *cubus*), *caxiellu* o *caxellu* (del latín *capsa*) y, sobre todo, *truébanu* y *trobo* (del latín *tubus*), y existen tres tipos según su forma y el material con que están fabricadas: colmena de tronco hueco, colmena de tablas y colmena de corcho. Las dos primeras son las más generalizadas, especialmente las que se hacen con un tronco hueco, mientras que las de corcho solo se emplean en unos pocos concejos del extremo occidental (Allande, Grandas de Salime, Pezós, los Ozcos, Boal, Illano) donde hay alcornoques. (Figura 3)

Las fabrican los mismos apicultores durante los meses de invierno (de noviembre a marzo), aprovechando el

Figura 3. Colmenas de tronco hueco y de tablas abandonadas en un *cortín* o colmenar. Fotografía de Marcelino Lozano.

menor trabajo en el campo y porque es la época más propicia para la tala de árboles. Se emplean las maderas de castaño, roble y cerezo. Los apicultores prefieren las dos primeras porque son maderas "muy calientes" y la de cerezo debido a que su "olor es muy del gusto de las abeyas y dan más miel"; de todas maneras, esta última madera es una de las que los campesinos consideran que pone a las abejas agresivas.

La altura de las colmenas oscila entre 45 y 55 cm, y su diámetro entre 35 y 45 cm; en general, se valoran más las pequeñas que las grandes. «Los trobos pequeños tendrán menos [miel] pero aguantan mejor el tiempo y producen más enjambres, porque la casa grande lleva mucho más tiempo llenarla [de abejas]" (Grandas de Salime).

Para hacer las colmenas de tronco se escogen árboles que tengan el corazón podre; una vez talados y troceados se vacía su interior de dos maneras diferentes: una, con la ayuda de una gubia grande que se golpea con un mazo de madera, y otra, hendiendo un trozo de tronco por la mitad, longitudinalmente, con unas cuñas de hierro, y labrando con un hacha y una azuela cada parte, que finalmente vuelven a unirse con dos pares de clavijas de madera que atraviesan interiormente ambas partes, o con dos herraduras viejas que se clavan por el exterior. Normalmente se conserva la corteza del árbol para dar más abrigo a las abejas, aunque algunos, muy pocos, las descortezan para evitar la propagación de insectos. Las colmenas de tronco son las más frecuentes y en muchos lugares las únicas que se utilizan.

Las colmenas de caja se hacen con cuatro tablas que se clavan con clavijas de madera o puntas de hierro. Donde se cosecha corcho no es raro encontrar ejemplares fabricados con dos tablas de madera y dos planchas de corcho. Muchos apicultores no consideran recomendables este tipo de colmenas, pues las tablas con el tiempo se "abarquillan, rajan y desclavan, quedando a menudo las abejas a la intemperie".

Por último, están las colmenas de corcho. Los campesinos distinguen dos clases de alcornoques o sufreiras: la "sufreira blancal", que produce un corcho mejor, y hace colmenas macizas y calientes, y la "sufreira moural", que da un corcho poroso, más propenso a abrirse y tener resquicios. Para las colmenas se prefiere el corcho de los alcornoques situados en laderas solanas. Las épocas de extraer el corcho son los meses de mayo y junio, y de agosto y septiembre, que es cuando el árbol tiene más savia y el corcho se desprende con mayor facilidad. Para evitar la polilla, la extracción debe realizarse con la luna en menguante. Para fabricar una colmena el corcho debe tener dos o tres centímetros de grueso, lo cual solo se consigue con cortezas que tengan en torno a siete años de crecimiento. Antes de poblar de abejas una colmena de corcho es necesario que seque durante

un año, "porque el corcho verde es para las abejas como una casa recién pintada para las personas" (Grandas de Salime).

Las colmenas de corcho son las más apreciadas por los apicultores de los lugares donde existen alcornoques, que las consideran el abrigo más cálido para las abejas. Los campesinos que no tienen alcornoques o los de pueblos próximos en los que no existe esta especie, adquieren el corcho bien comprando a sus dueños la cosecha de este producto y cortándola ellos mismos, bien adquiriendo las colmenas confeccionadas. Hasta finales del siglo XX era habitual ver en las ferias y los mercados del occidente de Asturias vendedores de colmenas y tapas de corcho.

Para fabricar las colmenas de corcho se intentaba sacar la corteza enteriza con solo un corte longitudinal, que se cosía solapando los dos extremos y atravesándolos con pequeñas clavijas hechas con madera de brezo (Erica arborea). En los últimos años se unen con un alambre.

Las colmenas, aparte de las paredes de madera o corcho, se componen de otros elementos. En su interior se colocan las trencas, conocidas en Asturias como fustes (del latín fustis, palo) o xueces, que son unos palos atravesados que sirven para sostener los panales. En el occidente, las colmenas tienen dos pares de fustes cruzados que se colocan a unos quince centímetros de la tapa y de la base, respectivamente; sin embargo, en el resto de Asturias lo normal es que las colmenas tengan dos fustes cruzados y situados a media altura o tan sólo uno, como sucede en los concejos de Lena, Quirós y Morcín, donde se le llama el xuez. Incluso, en muchos pueblos del centro y oriente de Asturias, cuando la colmena es muy estrecha no se pone dentro ningún palo cruzado. El menor número de estos palos, así como su ausencia, es una consecuencia de la práctica de aniquilar la colonia de abejas para obtener la miel y cera, ya que su presencia dificulta la extracción de los panales.

Estos fustes se fabrican con maderas duras y resistentes, tales como brezo (Erica arborea), tejo, acebo o aladierna, y también con avellano. Las maderas de tejo y acebo se considera que, como la de cerezo, ponen a las abejas agresivas, y existen apicultores que hacen los fustes con estas maderas para tener a las abejas en ese estado y así defenderse mejor en el monte.

En la parte inferior de la colmena se abre una muesca para la piquera y en la superior se coloca una tapa, conocida en Asturias como calduya, yérgola o témpanu, que se fabrica con corteza de árbol, madera y corcho. Las cortezas se utilizan especialmente en el occidente de la región y se extraen de abedules, olmos y castaños. Las de abedul son las más habituales; se cortan con una navaja en los meses de julio y agosto, cuando el árbol tiene más savia, se dejan secar durante un tiempo y se colocan en las colmenas con la parte de afuera hacia arriba, "para que no retengan el agua encima del truébano", porque la corteza nunca llega a perder la forma curva.

Las tapas de madera son exclusivas del centro y oriente de Asturias; se hacen con dos piezas y en muchos casos se sujetan por medio de dos clavijas pequeñas que van encajadas en un par de orejeras que tienen las colmenas.

La tapa de corcho es la más valorada por los campesinos del occidente; su grosor es menor que el usado para las colmenas y basta con que tenga tres o cuatro años de crecimiento. Las planchas de corcho se cortan en cuadrados de cuarenta centímetros cada lado y se apilan con un peso encima para aplanarlas. Estas tapas las vendían muchos campesinos de lugares productores de corcho en los mercados y ferias de Berducedo, Navelgas, Cangas del Narcea, etc., a los que acudían a comprarlas los apicultores de los concejos de Tinéu, Cangas del Narcea, Valdés y Allande. También se vendían en comercios rurales de estos mismos concejos. En los últimos años, como los campesinos han dejado de acudir a los mercados con el corcho, son los apicultores los que se trasladan a las zonas de producción para comprar allí las tapas.

Encima de la tapa se coloca una losa de piedra (pizarra, caliza) o varias tejas, con el objeto de proteger la colmena de la lluvia y darle estabilidad con su peso.

Por último, la colmena se coloca sobre una solera de madera o, sobre todo, de piedra, con el fin de asentarla bien y aislarla de la humedad de la tierra, y a continuación se tapan con boñiga de vaca el borde de la tapa y todas las rendijas para que la luz no entre en su interior. Está muy arraigada la creencia de que "las abejas no pueden trabajar con luz y aire". El miedo a que entre la luz en la colmena es la causa de que las de una pieza, hechas de un tronco hueco, sean más estimadas que las fabricadas con un tronco partido o con cuatro tablas.

Colmenares

Las colmenas se colocan en dos lugares muy diferentes del espacio rural: uno, junto a la casa de habitación o en el interior del pueblo, y otro, en el monte y alejadas de los núcleos de población. La primera situación es la más frecuente en Asturias, y las colmenas se colocan en los hórreos o paneras (graneros elevados del suelo); alineadas junto a los muros de las casas o las huertas; en las fachadas de las casas, etc. En todos los casos, las colmenas están orientadas al mediodía y protegidas del viento. También se valora mucho que estén bajo techo, por eso se colocan debajo de los aleros de las casas y los hórreos.

La ubicación de colmenares en el monte es exclusiva del occidente de Asturias, en concreto del territorio situado al oeste de los concejos de Valdés, Tinéu y Cangas del Narcea. Las colmenas se instalan en peñas o lugares de difícil acceso y, sobre todo, en *talameiros* y *cortinos*, construcciones que las protegen de los robos y de los animales, en especial del oso. En general, los apicultores prefieren esta última localización, porque las abejas, aunque están lejos de la vigilancia de sus dueños, tienen más "pasto" o alimento y dan más cantidad de miel y de mejor calidad. En 1752, el 75% de los vecinos del concejo de Illano declara en el catastro del marqués de la Ensenada que tienen sus colmenas fuera del pueblo.

La causa de la existencia de estos colmenares en el monte hay que buscarla en las condiciones naturales tan favorables con que cuenta la apicultura en el occidente de la región, donde los suelos ácidos ofrecen una riqueza de flora melífera muy superior al resto de Asturias.

El *talameiru* es un tipo de colmenar que dejó de utilizarse hace mucho tiempo y del que sólo se conservan ruinas. Consiste en una pequeña torre construida con mampostería de planta cuadrada de dos metros por cada lado y una altura que no sobrepasa los tres metros; en su parte superior se armaba un suelo con unos largos tablones que sobresalían hacia afuera y en los que se asentaban las colmenas. En él entraban un máximo de diez o quince colmenas, muy pocas si lo comparamos con un *cortín*, donde cabían entre treinta y sesenta. Aparece en un área geográfica pequeña repartida por los concejos de Allande, Cangas del Narcea e Ibias, y convivía con los *cortinos*, que eran los colmenares más abundantes y que paulatinamente fueron sustituyendo a los *talameiros*.

Los *cortinos*, también conocidos como *cortines* o *cortíos*, aparecen en todos los concejos situados al oeste de una línea imaginaria formada por la parroquia de Muñás (concejo de Valdés), el concejo de Tinéu y el río Narcea hasta su nacimiento en el concejo de Cangas del Narcea. No existen *cortinos* en ninguna parroquia de la marina del occidente de Asturias. Este mismo tipo de colmenar se extiende por el occidente de la provincia de León y Zamora (Pérez Castro 1994: 26-29; Díaz Otero y Naves Cienfuegos 2010: 23), y por el sur y oriente de Lugo, donde recibe el nombre de *albariza*.

El *cortín* es un cercado de planta redonda, como manifiesta la etimología del término, cuyo diámetro oscila entre los ocho y catorce metros. Esta formado por un muro de piedra en seco de entre 2,5 y 3,75 metros de altura y un espesor de setenta centímetros o un metro, que se levanta siguiendo la pendiente del terreno con el objeto de no dar sombra en su interior y facilitar el movimiento de las abejas. Este muro va rematado

Figura 4. *Cortín* con su pequeña puerta de acceso. Fotografía de Marcelino Lozano.

Figura 5. Escala para entrar en un *cortín* en L'Artosa (Cangas del Narcea). Fotografía de Juaco López.

por arriba con unas losas de pizarra o unos tablones de madera que sobresalen hacia afuera unos cuarenta centímetros y reciben los nombres de *barda* o *veira*. La función de este alero es poner un impedimento a los osos cuando intentan entrar en el colmenar. La mayoría de los *cortinos* posee una puerta de pequeño tamaño (100 x 60-70 cm) con cerradura, que está situada a mitad del muro. En algún caso no tienen vano de entrada y a ellos se accede con una escala. (Figuras 4 y 5)

Los *cortinos* se localizan en las pendientes de las montañas, preferentemente a media ladera para evitar la niebla y facilitar a las abejas el aprovechamiento de la flora en toda la ladera. Lo más común es encontrarlos en los valles pequeños y laterales de los cursos de los ríos, y siempre protegidos de los vientos procedentes del oeste, el norte y el nordeste: "Norte y nordés, y cortar al revés, líbrenos Dios de todos tres" (Fonteta, Allande). En cuanto a la altitud, rara vez están por encima de los 800 m.

Se construyen en laderas solanas con el fin de obtener la mayor insolación posible y normalmente se levantan en una zona de peñas o gleras, por dos razones: por un lado, aprovechar la aridez de las rocas y piedras,

ya que en Asturias la humedad es la peor enemiga de las abejas, y por otro, evitar el transporte de la piedra, que es, junto al barro, el único material empleado en su fábrica y que encarecería mucho su construcción debido a los malos caminos y a lo limitado de los medios de transporte en Asturias. Además, cuando se construyen junto a peñas o crestones es frecuente pegar los *cortinos* a la roca madre y utilizar ésta como cimiento o muro natural. (Figura 6) En su interior el suelo mantiene la inclinación de la ladera para impedir la formación de charcos y, con frecuencia, se construyen unos escalones grandes en los que se asientan las colmenas en hileras, siempre sobre su correspondiente solera de piedra, y en orden descendente, es decir las de atrás siempre están colocadas más altas que las de delante, para facilitar el movimiento de entrada y salida de las abejas.

La principal función de los *cortinos* y *talameiros* era proteger las colmenas de los osos y en menor medida de otros animales (tejones, zorros), entre los que también estaban los rebaños de ganado menudo, que hasta hace pocos años pastaban en el monte bajo. Los campesinos también colocaban en los alrededores de estos colmenares otros ingenios disuasorios para los osos, así como trampas específicas para cazarlos,

Figura 6. *Cortín* en el concejo de Allande. Fotografía de Marcelino Lozano.

conocidas como *pesugos*, que se construían a ocho o diez metros de aquellos colmenares (Lozano Sol 2020: 79-99). Ahora bien, esta estrecha convivencia no estaba sólo en función de proteger a las abejas, sino también en el interés económico que resultaba de la caza de osos, que se veían atraídos por la presencia de colmenas, y cuyos despojos (piel, grasa, carne) se vendían a buen precio (López Álvarez 1994: 70).

La edificación de un *cortín* la afrontaban los campesinos, a menudo, uniéndose dos casas o familias para compartir su coste y beneficio. La ejecución de la obra la realizaban directamente los campesinos, aunque si sus posibilidades económicas se lo permitían se encargaba el trabajo a cuadrillas de canteros de la provincia de Pontevedra (Galicia), que desde el siglo XVIII recorrían la mitad occidental de Asturias, ofreciendo sus servicios por los pueblos. Un testimonio de esto lo dejó escrito Rosendo María López Castrillón (2018: 45), vecino de Riodecoba (Allande), a mediados del siglo XIX: "El cortín del Beyal Longo principiólo Juan el Menor año de 1745. Trabajaron en él más de 20 días gallegos canteros de Pontevedra, lo delinearon a compás con un cordel y una estaca en medio, hicieron mitad de sus paredes".

Para instalar un *cortín* los campesinos buscaban un lugar apropiado, con las características mencionadas (a media ladera, solano, etc.) y donde abundase "el pasto", o sea la flora melífera, y hubiese cerca una corriente de agua de escaso caudal, pero que fuese permanente. Antes de construirlo tenían que asegurarse que el lugar fuese adecuado, pues el trabajo y el gasto eran grandes y su economía no les permitía esfuerzos vanos. Los lugares apropiados no eran muchos, y eso limitaba su construcción: "Todos los vecinos no tenían cortín, porque tiene que ser un sitio apropiao pa que ellas [las abejas] trabajen y esos sitios no se dan. Con el tiempo se estudian los sitios. Aquí van bien, aquí van mal. Los antiguos colocaban colmenas de prueba" (Dou, concejo de Ibias). Pero, además, para edificar un colmenar nuevo era muy importante que no existieran otros en las proximidades, por dos razones que los *abeyeiros* repiten a menudo: en primer lugar, porque si se ponen muy cerca varios *cortinos* las abejas no tienen todas flora bastante y deben ir a buscarla muy lejos con lo que trabajan más, pero producen menos, y también aumentan las posibilidades de que se pierdan en sus salidas; y en segundo lugar, las abejas de *cortinos* muy próximos se pelean y no trabajan, y así pueden morir las colmenas del *cortín* viejo y las del nuevo.

Estas dos circunstancias, unidas al hecho de que la gran mayoría de los *cortinos* se levantaba en los montes de propiedad o uso común de los pueblos, fueron la causa de numerosos pleitos entre los campesinos del occidente asturiano. Según todos los informantes consultados los pueblos mantenían la ley consuetudinaria de que

donde hubiese un *cortín* no se podía edificar otro en las proximidades.

Hasta mediados del siglo XX el número de *cortinos* que poseía una casería dependía de sus recursos económicos. En los siglos XVIII y XIX muchos campesinos del occidente asturiano tenían en propiedad o aforado un *cortín*, o compartían con otros vecinos su beneficio. En el catastro del marqués de la Ensenada (1752) del partido de Berducedo (Allande) aparecen un 60% de vecinos que poseen colmenas, de los cuales el 48% declaran que tienen un colmenar en el monte. Las diferencias de propiedad abarcan desde campesinos que sólo poseen una cuarta parte de un *cortín* hasta algún rentista que es dueño de más de diez. Ahora bien, lo más frecuente no es ninguno de estos extremos, así el 35% poseen un *cortín*, el 23% dos y el 20% tienen la mitad de uno.

Las casas de los campesinos propietarios o pequeños rentistas solían tener varios colmenares, como sucede en la casa de la Fuente en Riodecoba (Allande) que tenía en 1826 sesenta colmenas repartidas en ocho lugares diferentes (López Castrillón 2018: 88). De todas maneras, no debemos pensar que en todos esos lugares la casa tenía un "cortín de pared alta y fuerte", sino que algunos de ellos eran "puestos de abejas", es decir peñas poco accesibles en las que se colocaban las colmenas.

Por último, las casas más ricas de la zona occidental, que poseían varias caserías trabajadas por renteros y eran dueñas de montes enteros, tenían un elevado número de *cortinos* con colmenas cuidadas por criados o dadas en aparcería. Un ejemplo de estas casas era la de Uría en el pueblo de Collada (Allande), que en 1768 era propietaria de nueve *cortinos* y la mitad de otros cuatro, y de unas 190 colmenas pobladas de abejas (López Álvarez 1994: 75-76).

La captura de enjambres

Las dos operaciones más importantes y casi las únicas de la apicultura tradicional eran la captura de los enjambres y la extracción de la miel y la cera. La primera permitía aumentar el número de colmenas, aunque, a menudo, solamente hacía posible el mantenimiento del colmenar con un número estable, debido a la elevada mortandad de colmenas que se registraba en Asturias. No es extraño conocer a *abeyeiros* que han pasado en poco tiempo de tener varias decenas de colmenas a poseer media docena e incluso a quedar sin ninguna. Las causas de tanta mortalidad eran varias, por un lado estaban las enfermedades para las cuales se desconocía cualquier remedio curativo, los años malos con inviernos muy duros y primaveras muy lluviosas, las continuas rapiñas del oso, etc., y por otro, estaba el modo de extraer la miel y la cera que se empleaba en la mayor parte de Asturias, que conllevaba la muerte de toda la colonia de abejas.

Esta situación requería una constante renovación de las colmenas, por lo que el apicultor ponía todo su esfuerzo en procurar que no se le escapasen los enjambre, los cuales dejaban de pertenecer al dueño de la colmena en el momento que se alejaban del colmenar y no eran perseguidos. Una buena parte del prestigio de los *abeyeiros* procedía de su capacidad para capturar los enjambres de su propio colmenar y los que perdidos pasaban por las proximidades de su pueblo; de esta habilidad dependía en gran medida el número de colmenas y su fama como "entendido" en abejas.

En toda Asturias los enjambres más apreciados eran los tempranos, los que salen con las primeras floraciones de primavera: "los ensames del tiempo de la uz moural [*Erica australis*] son muy buenos, porque tienen muito pa comer y en agosto ya puedes dar una cortadura buena". Existen varios refranes, muy extendidos, que aluden a este hecho: "El ensame de mayo vale un caballo" y "La ensambla de mayo vale un caballo, la de xunio un burro, la de xulio un mulo y la de agosto non vale nada" (Caldones, concejo de Gijón).

La captura de enjambres era una labor fatigosa, sobre todo para los apicultores que tenían los colmenares en el monte. En la época en que se preveía que las colmenas iban a enjambrar era necesario ir casi todos los días al colmenar y apostarse allí a la espera. El enjambre sale por la mañana, cuando comienza a calentar el sol, pero si hace viento, si llueve, si hay niebla o el tiempo amenaza tormenta, se sabe de antemano que ese día no saldrá. En cambio, cuando la colmena está barbada o cubierta de abejas en su mitad inferior y éstas están alborotadas, es señal de que la colonia está a punto de enjambrar. Para ese momento, el apicultor tendrá preparado un caldero de agua, una escoba, una sábana blanca, un objeto metálico para producir ruido (un cencerro, una sartén, etc.), y una colmena limpia y untada con alguna sustancia que atrae a las abejas: leche, miel, grasa de cerdo (concejo de Amieva) y *abeyera* o *yerba d'abeyeira* (*Melisa officinalis*), que era el producto más utilizado y la planta que todos los apicultores tenían en su huerta.

En cuanto el enjambre sale volando de la colmena, el *abeyeiro* debe lograr por todos los medios que éste se pose en el suelo, que no huya, para ello le tira tierra y agua y llama su atención con el ruido de una esquila o sartén; asimismo, intenta atraerlo llamándole con frases cariñosas: "¡Pousa, pousa!" (Cangas del Narcea), "¡Pousa galana, pousa galana!" (Allande), "¡Apousa, abeya maestra, apousa!", "¡Posar galanas, posar galanas, venir aquí, a la casa nueva!" (Amieva), "¡Aquí, abeyines, aquí!" (Aller), "¡Queridas, aquí, queridas!" (Lena), "¡Fichas, fichas[1], venir, venir!" (Somiedo) y otras más. Una vez que el enjambre está en el suelo, se le coloca cerca o encima una colmena vacía y preparada de la manera

que dijimos; para meterlo dentro se golpea la colmena con una piedra, "porque el ruido pone en movimiento a las abejas", y se dice con voz pausada: "¡Arriba, arriba galana! ¡Casa nova, casa nova!" (Allande) o frases parecidas. Algunos empujan a las abejas con una rama de helecho o con la mano hasta que logran introducirlas en la colmena vacía.

En los concejos de Ibias, Allande y en algunas parroquias de Cangas del Narcea se utiliza para capturar los enjambres que huyen lejos del colmenar o que se posan en la rama de un árbol, a donde hay que subir a cogerlos, un *queisieyu* o *coxedor*, que es una especie de pequeña colmena (altura: 35-40 cm y diámetro: 30-33 cm) hecha con corteza de *capudre* (serbal, *Sorbus aucuparia*), tilo o abedul y también con corcho; para usarlo es necesario tapar una de sus bocas con un trapo.

En varios concejos del occidente de Asturias (Allande, Boal, Illano, Grandas de Salime, Pezós, los Ozcos) se utiliza un sistema para dividir las colonias de abejas y provocar "enjambres artificiales", que es lo que se denomina "capar la colmena". Los apicultores con varios colmenares y muchas colmenas necesitan recurrir a este sistema para no perder enjambres y no tener que estar "tan sujetos al cortín". Para "capar" una colmena tiene que estar próxima a enjambrar. La operación es sencilla, hay que destapar la colmena por arriba y colocar encima otra vacía y cubierta con un trapo oscuro. Para forzar a las abejas a abandonar la colmena madre se las ahúma por abajo y se golpea con los nudillos. Cuando el *abeyeiro* calcula que ya han pasado bastantes abejas a la colmena nueva, la retira y a continuación tiene que cerciorarse que el *capón* o enjambre partido lleva una reina. Para comprobar la presencia de la reina existen dos métodos: el más generalizado consiste en colocar la colmena que lleva el *capón* encima de una losa de pizarra o un paño negro; si transcurrido un rato se observan sobre este fondo oscuro unos puntos blancos es que la reina está dentro. Para la mayoría de los *abeyeiros* estos puntos son "las cagadas de la maestra", para otros más informados son huevos. Esta manera de provocar los enjambres es una práctica que está documentada en la zona occidental de Asturias desde mediados del siglo XIX. El mismo sistema se empleaba en la apicultura tradicional de otras provincias españolas, y lo recomienda Alonso de Herrera en su *Agricultura General*, publicada en 1513 (1981: 330-333). A partir de este autor lo recogen todos los tratados agrícolas y apícolas publicados en España hasta el siglo XIX.

Otras prácticas apícolas

En el área centro-oriental de Asturias las únicas operaciones que practicaban los apicultores eran la captura de enjambres y la extracción de miel y cera. Sin embargo, en la zona occidental a estas prácticas se

---

[1] *Ficha*, asturiano, 'hija'.

unían otras que favorecían el aumento del colmenar, como los enjambres artificiales, ya mencionados; el *escuciñar* (escarzar) o quitar las ceras viejas y el "cebar" o alimentar a las abejas.

La operación de *escuciñar* o *esconciar* consiste en arrancar y sacar con la mano los panales negros y viejos que van quedando en la parte inferior de la colmena. Según los *abeyeiros* las abejas no dominan la cera vieja y en ella no producen miel, ni cría; quitándola aumenta el espacio disponible en el interior de la colmena y se evita la polilla. Los panales con celdillas viejas y estrechas limitan las puestas de la reina y crían unas abejas más pequeñas. Es opinión general entre los *abeyeiros* que de las colmenas que se *escuciñan* salen en primavera enjambres tempranos y fuertes. La operación se realizaba en los meses de febrero y marzo. El trabajo lo realizaban habitualmente los cereros procedentes de los Ancares (provincia de León), que se ofrecían por las casas y se quedaban con la cera que sacaban; esta cera seca, conocida como de *murrina* o *murría*, no era valorada por los campesinos que no la podían aprovechar en sus casas, sin embargo, los cereros le sacaban toda su sustancia cociéndola y prensándola en los lagares de cera.

El otro cuidado que recibían las colmenas en el occidente era el de "cebarlas". Cuando se veían las colonias flojas o débiles a finales del invierno y durante la primavera, que es la época de cría, se les echaba comida delante de la piquera o por la parte superior de la colmena. Los alimentos más frecuentes eran los siguientes: harina de maíz, castañas cocidas y migadas, caldo de cocer castañas y vino blanco; también se daba azúcar, a veces disuelta en vino, y en menor medida miel. En los pueblos del valle alto del río Ibias era habitual echar en el interior de la colmena, desde arriba, bocanadas de vino, orujo o coñac, para fortalecer a las abejas durante el invierno.

La extracción de miel y cera

En Asturias se empleaban dos modos muy diferentes de sacar la miel y la cera, que conviven al menos desde mediados del siglo XVIII. En uno se exterminaba la colonia de abejas para extraer todos los panales y en el otro se extraía una porción de ellos por la parte superior de la colmena, ahumando a las abejas para que no atacasen al apicultor. El primer procedimiento era el más extendido y abarcaba, como ya hemos mencionado, todo el territorio situado al este del río Narcea.

1. Matar la colmena

La extracción de la miel y cera se realizaba en el área centro-oriental de Asturias durante el otoño, desde finales de septiembre ("Per San Miguel, esmélgase el miel") a noviembre. "Decían que la última luna de octubre era la mejor para escolmear; era cuando estaba la miel hecha, después ya se concentraban y recogíen adentro todos les abeyes, porque viene el frío, y a la miel la perjudiquen. La última luna de octubre es la mejor miel" (La Vita, concejo de Parres). Se esperaba hasta esas fechas porque es en otoño cuando las colmenas tienen almacenada la mayor cantidad de miel, pues las abejas, según comienzan a escasear las flores y a reducirse las horas de luz solar, van ocupando con miel toda la colmena a costa de la cámara de cría.

Las colmenas que se mataban eran siempre las que más pesaban; el campesino iba al colmenar y allí sopesaba todas las colmenas hasta seleccionar las mejores. Este principio poco tiene que ver con una explotación racional, pues, por lógica, las colmenas más pesadas son siempre las que tienen mayor número de abejas y más cantidad de miel, es decir son las que tienen más probabilidades de pasar el invierno y las que producirán más enjambres en primavera.

El número de colmenas que se exterminaba estaba en función de las que integraban el colmenar, normalmente se mataban la mitad si se tenían menos de diez colmenas y un tercio si el número era mayor.

La manera más generalizada de matar las abejas era la asfixia con humo o ahogándolas en agua. Para asfixiarlas se quemaba en la parte inferior de la colmena azufre, hierba verde o mecha de dinamita. En el caso de ahogarlas, se metía la colmena en un saco y se sumergía en un pozo, en el río o en una fuente.

De las colmenas se sacaban todos los panales, y se recortaban las ceras viejas y sucias para que no dieran mal sabor a la miel. Una colmena proporcionaba entre ocho y diez kilos de miel y un poco más de un kilo de cera en rama.

A todos los informantes se les preguntó las razones por las cuales mataban las colmenas y si no conocían otro procedimiento. Una parte de ellos respondió que en su pueblo no se conocía otro método para sacar el producto de las abejas, "nunca nadie se puso a pensar otro sistema mejor, siempre todo el mundo las mató". Otros muchos justificaron plenamente la aniquilación de las abejas, pues, según ellos, si se corta una parte de los panales, las colmenas tienen muchas posibilidades de morir en invierno y, además, es muy probable que no enjambren al año siguiente.

2. Cortar la colmena

En los concejos del occidente de Asturias se cortaban los extremos superiores de los panales, y se ahumaban las abejas para aturdirlas y que no atacaran. Esta operación, que se denomina *esmielgar*, requería muchos más cuidados que la de matar, pues había que asegurar la supervivencia de la colmena.

El número de extracciones dependía de la bondad del año, así como de la altitud y de los recursos naturales del lugar. En los lugares más favorecidos se *esmielgaba* dos veces al año, en julio y septiembre, y en el resto sólo se cortaba una vez, en la mayoría de los casos entre finales de agosto y mediados de septiembre. La costumbre podía alterarse si un año la primavera había sido muy buena para la cría de las abejas, en cuyo caso podía adelantarse la fecha del primer corte a junio. Antes de este mes no se *esmielgaba* nunca. La experiencia había demostrado que era perjudicial para las colmenas debido a que a mitad de la primavera las colonias están criando y las abejas necesitan toda la miel para alimentar a las crías; si en ese momento se les quita se corre el peligro de que la colonia se debilite y muera. El refrán "el que esmelga en el mes de mayo, esmelga pa todo el año" (Samartín del Valledor, Allande) advierte de este peligro. Tampoco era conveniente *esmielgar* pasado el 15 de septiembre, porque no se les dejaba a las abejas tiempo suficiente para recuperarse.

Las opiniones sobre el día más adecuado para ir a cortar son diversas. En general se considera que el mejor momento es coincidiendo con el menguante de la luna, "porque en el menguante sufren menos las abejas y la miel se conserva mejor, no fermenta". Las horas más adecuadas son por la mañana temprano o a última hora de la tarde, cuando calienta poco el sol y las abejas están recogidas. Sin embargo, otros apicultores creen que es mejor al mediodía, "porque con el ganado fuera es más fácil". En cualquier caso, si hay niebla o truenos no puede *esmielgarse*, pues la abejas están "gafas" y "además pueden ponerse locas". Tampoco puede hacerse con el día nublado, ni durante la canícula, porque corren el peligro de coger la "tiña" o "gangrenarse".

La operación de extraer una parte de los panales es muy sencilla: se levantan la losa y la tapa que cubren la colmena, y se ahúma por arriba para que las abejas atonten y bajen a la parte inferior; cuando se observa que no queda ninguna por arriba se cortan los panales con una *esmielgadora* hasta encima de los *fustes* o trencas superiores, sin llegar a tocarlos. En Fonteta (Allande) el modo correcto de cortar los panales era pasando la *esmielgadora* de derecha a izquierda. La cantidad de panales que se saca depende del estado de las colmenas, si están fuertes se corta más, si están flojas se saca menos. Para conocer si las colmenas están cargadas de miel se golpean las tapas con los nudillos: si suena macizo, las colmenas están llenas y pueden cortarse sin problema, si suena a vacío la cosa cambia y a lo mejor no se castran ese año. En general, de cada colmena se extraen dos o tres kilos de miel y medio kilo de cera

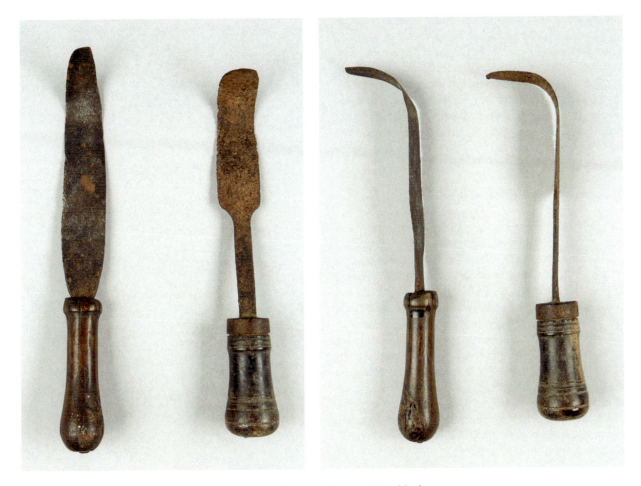

Figuras 7 y 8. *Esmielgadoras*. Col. Museo del Pueblo de Asturias.

en rama. En 1797, Antonio Blanco, vecino de Brañes (Oviedo), recibió un premio de la Sociedad Económica de Amigos del País de Asturias "por haber sacado de 18 colmenas dos arrobas [23 kg] menos dos libras [0,460 kg] de miel", cantidad que corresponde a una media de 1,226 kg por colmena (López Álvarez 1994: 86).

Los *abeyeiros* utilizaban un ahumador y una *cortadeira* o *esmielgadora*. En los últimos años, la mayoría emplea ahumadores industriales de fuelle (*soplón*) en los que se quema boñiga de vaca o trapos de lana, pero hace unas décadas se usaban unos sistemas más rudimentarios: una teja con boñiga encendida; un mechón o *fumeiro* formado por trapos viejos enrollados, que era el modo más frecuente; un *escacio* (liquen epífito) que se arrancaba de los árboles y "daba un humo muy agradable" (concejo de Cangas del Narcea), y una corteza de roble. Estos procedimientos, exigían esfuerzo y atención, pues los apicultores o sus acompañantes tenían que estar soplando casi permanentemente para avivar el fuego y dirigir el humo hacia la colmena.

Las *esmielgadoras* tienen un mango y una hoja de doble filo y curva. Su longitud varía mucho, existen herramientas de 30 cm de largo máximo y otras de 50 cm. El mango casi siempre es de madera. (Figuras 7 y 8)

Las precauciones que tomaban los *abeyeiros* para evitar las picaduras cuando cortaban las colmenas eran pocas, e incluso algunos no tomaban ninguna; por lo común, no llevaban careta, ni guantes. Era normal que se cubrieran la cabeza con una chaqueta y metieran las perneras de los pantalones por entre los calcetines, con el fin de que las abejas no se enredasen en el pelo, ni se introdujeran por el cuello o las piernas. Para no provocar el ataque de las abejas se consideraba muy importante no ponerse nervioso y no hacer movimientos bruscos.

Recipientes para almacenar la miel

La miel se separaba de los panales estrujando estos con las manos. Una vez separada quedaba lista para consumir. Algunos apicultores la dejaban reposar una noche y al día siguiente le quitaban con una cuchara la cima, formada por cera y borra. Pero lo común era comenzar a comer la miel el mismo día que se extraía. Los apicultores que recogían muchas miel destinaban una pequeña parte a la venta, pero era sobre todo un producto que se consumía en casa y que se regalaba. En el occidente de Asturias se empleaba como medicina y alimento cotidiano en algunas épocas del año, y en el resto de la región rara vez pasaba de ser un remedio medicinal con funciones preventivas y, en especial, curativas. "La gente por aquí tenía mucha fe en la miel" (Cabañaquinta, concejo de Aller).

La miel se guardaba en recipientes de madera y cerámica. Los primeros eran barricas de duelas y, sobre

Figura 9. *Canao* para miel, Cangas del Narcea. Madera de castaño. 30,5 x 40,5 cm. Col. Museo del Pueblo de Asturias.

Figura 10. *Canao* para miel, Cangas del Narcea. Madera de castaño. 36 x 21 cm. Col. Museo del Pueblo de Asturias.

Figura 11. Olla para miel, Allande. Madera de castaño torneada. 40,3 x 28,5 cm. Col. Museo del Pueblo de Asturias.

Figura 12. *Trobo* de la miel, Illano. Madera de castaño. 83 x 50 cm. Col. Museo del Pueblo de Asturias.

todo, vasijas de madera hechas en una pieza de madera, bien en el torno, bien vaciadas con una gubia. Estas últimas podían ser de tamaño pequeño (unos 30-36 cm de altura) o muy grandes (hasta 71 cm de altura y 48 cm de dm), como era el caso de los denominados *trobos da mel* o *truébanos del miel*, que son troncos huecos, parecidos a las colmenas que llevan su mismo nombre, con fondo y tapa de madera; la tapa va encajada entre dos orejeras que llevan un orificio por donde se pasa un travesaño de madera que cierra con fuerza el recipiente. Estos grandes envases para la miel solo se empleaban en algunas zonas del occidente interior de Asturias donde se cosechaba mucha cantidad. Todos estos recipientes se fabricaban con madera de castaño y se colocaban en el zaguán de la casa, o en el hórreo o la panera, que son graneros elevados bien ventilados e inaccesibles a los roedores. (Figuras 9, 10, 11 y 12)

En cuanto a los recipientes de cerámica, los más utilizados para guardar la miel eran la olla, el puchero, la tarreña, la *xarrina* y el *pucherín,* que se fabricaban en Asturias, en los alfares de Miranda (Avilés) y Faro (Oviedo). También se empleaban mucho ollas procedentes de Jiménez de Jamuz (León), donde se hacía una cerámica vidriada de mayor calidad que la asturiana. En los concejos más occidentales de Asturias se usaban también recipientes de alfares gallegos y en los del suroeste se generalizarán algunas de

aquellas formas cerámicas a partir del asentamiento en el siglo XVIII de alfareros de Miranda en el pueblo de Llamas del Mouro (Cangas del Narcea). La miel dentro de estas vasijas se guardaba dentro de casa, en un arca cerrada o en una alacena, y en el hórreo o la panera. La boca siempre se tapaba, normalmente con un trapo bien atado al borde, para que no entraran ratones e insectos, en especial hormigas, y para que no fermentara la miel.

Bibliografía

Alonso de Herrera, G. 1981. *Agricultura General que trata de la labranza del campo y sus particularidades, crianza de animales y propiedades de las plantas (1513)*, edición crítica de Eloy Terrón. Madrid: Servicio de publicaciones del Ministerio de Agricultura y Pesca.

Díaz Otero, E. y J. Naves Cienfuegos. 2010. Los colmenares tradicionales del noroeste de España. *AÇAFA On Line* 3: 1-37.

López Álvarez, X. 1994. *Las abejas, la miel y la cera en la sociedad tradicional asturiana*. Oviedo: Real Instituto de Estudios Asturianos.

López Álvarez, J. 2005. ¿Animal doméstico o animal salvaje? La relación del hombre y las abejas, en J. L. Mingote Calderón (ed.) *Animalario. Visiones humanas sobre mundos animales*: 81-99. Madrid: Ministerio de Cultura.

López Castrillón, R. M.ª. 2018. *Las nueve vidas de la casa de La Fuente de Riodecoba. Libro de memoria de una casa campesina de Asturias (1550-1864)*, edición y estudio preliminar de Juaco López Álvarez. Gijón: Museo del Pueblo de Asturias.

Lozano Sol, M. 2020. *Las abejas, con el viento en contra. Hábitat e historia en Allande*. Oviedo: Ediciones Trabe.

Pérez Castro, F. 1994. *Los colmenares antiguos en la provincia de León*. León: Caja España.

Figura 13. Olla de cerámica del alfar de Llamas del Mouro (Cangas del Narcea). 35 x 32 cm. Col. Museo del Pueblo de Asturias.

Figura 14. Olla con tapa de cerámica del alfar de Llamas del Mouro (Cangas del Narcea). 29,5 x 27 cm. Col. Museo del Pueblo de Asturias.

Chapter 16

Approches de l'Archéologie: L'apiculture insolite du nord de l'Espagne

Robert Chevet

Apistoria: Société d'études et de recherches sur l'apiculture traditionnelle (chevet.lisleferme@free.fr)

Résumé: Le chapitre concerne les pratiques apicoles traditionnelles et la construction traditionnelle de ruches et de ruchers dans le nord et en particulier le nord-est de l'Espagne.

Mots clés: Murs à abeille; ruchers; niches; logettes; alvéoles; ruches verticales; ruches horizontales; arnal; Espagne du Nord

Summary: The chapter concerns traditional beekeeping practices and hive and apiary construction in northern and particularly northeastern Spain.

Keywords: bee enclosures (muros apiários); apiaries; bee boles; cubicles; vertical and horizontal hives; *arnas* and *arnales*; Northern Spain

Introduction

La pratique de l'apiculture horizontale est très généralement méconnue dans les régions situées au nord de la Méditerranée bien qu'elle y soit assez répandue. Les cartes établies par Brinkman en 1938 ne signalent que quelques cas très localisés, au nord-est de l'Espagne, et sur une large couronne au nord des Alpes allant de la vallée d'Engadine au Tyrol. En revanche le texte de l'ouvrage mentionne de nombreuses localités ibériques dans ses relevés de ruches régionales (Brinkmann 1938: 49, 97, 125, et 172). Cette absence quasi générale de notoriété est telle que dans l'inventaire des implantations apicoles de la péninsule Ibérique soigneusement établi par frère Adam en 1966, « A la recherche des meilleures races d'abeilles », l'auteur affirme que les ruches primitives « sont toujours placées debout, jamais couchées »(Kehrle et Florence 1980: 80-1). Cette affirmation est largement démentie par les faits.

En réalité, le nord de la péninsule, en particulier porte les traces d'une pratique active de l'apiculture horizontale. Cette pratique est riche en variétés subtiles qui résultent d'une longue antériorité sans qu'il soit réellement possible de discerner actuellement qui a influencé qui et d'où sont venues les novations successives.

Eva Crane qui a traité avec beaucoup de notoriété des multiples facettes de l'apiculture traditionnelle ignorait l'existence de cet îlot de particularités que représente la haute vallée de l'Èbre et les contreforts sud des Pyrénées. Son ouvrage de référence (Crane 1999), consacré à l'histoire de l'apiculture a été publié trop tôt, (en 1999) pour faire mention du vaste ensemble de ruches horizontales que nous décrivons ici. Elle a eu néanmoins l'occasion de la visiter et d'en constater l'importance.

Pour les besoins de notre recherche nous avons sélectionné une large zone au nord de l'Espagne, immédiatement au sud de la frontière pyrénéenne avec la France. Cette zone a été arbitrairement marquée de six sections dans lesquelles les abeilles sont le plus souvent traitées horizontalement. Comme on peut le voir sur la figure n°1, la zone ainsi délimitée s'inscrit globalement dans la vallée de l'Ebre, depuis sa source jusqu'aux derniers massifs montagneux précédant son estuaire. Si l'on considère que le Levant espagnol et la vallée de l'Èbre sont les premiers points touchés par les navigateurs des premier et deuxième millénaires avant notre ère, l'hypothèse de départ serait de considérer cette sélection géographique comme le berceau de la civilisation apportée du Moyen-Orient par les premiers navigateurs de cette région lointaine. Les vagues successives auraient progressivement refoulé vers les montagnes du Nord de la péninsule les pratiques les plus élémentaires de l'apiculture horizontale.

Elles s'étendent d'Est en Ouest, de la vallée du rio Sègre à celle de la Valdavia. Elles sont clairement distinctes à l'Est comme à l'ouest des territoires présentant des ruches anciennes verticales, vanneries en Catalogne et ruches de liège (*cortchos*) dans les Asturies. On notera seulement la particularité du massif des Monegros, d'années. Il est toutefois possible de dresser l'inventaire de ce qui en est encore observable et d'en tirer des hypothèses raisonnables.

Il convient d'abord de définir avec précision la base de cette culture très particulière, celle de cette ruche faite de roseaux tressés qui est appelée localement *l'Arna*. Ce mot, qui n'est pas dérivé du latin, semble provenir d'un héritage très ancien, probablement pré-ibère qui donne le mot *arnal* pour le bâtiment des abeilles et *arnero* pour l'apiculteur.

Ces mots devenus désuets étaient employés autrefois dans toute la province d'Aragon mais ont maintenant perdu de leur usage courant.

La *ruche arna* est un outil domestique fait de vannerie, généralement en osier, tressé autour de neuf arêtes, soit en roseaux, soit en baguettes de bois (genévrier ou chêne vert) d'un

Figure 1. Eva Crane à l'intérieur du rucher rupestre de Barfaluy (Aragon), avant la pose des grilles de protection.

riche en vestiges apicoles récents mais totalement vouée à l'exploitation verticale des ruches.

Malgré la proximité de la frontière avec la France on ne trouve pratiquement jamais de débordement au nord de cette limite naturelle forte que sont les Pyrénées, à l'exception d'une possible construction très détériorée dans les parages du Somport.

Cette région du levant Espagnol et de la vallée de l'Èbre a fait l'objet de mes recherches en apiculture traditionnelle pendant une quarantaine d'années. Peu connue, peu ou pas décrite, elle comporte une grande variété de modèles ayant tous pour point commun, la particularité de traiter les ruches horizontalement. Depuis le début de mes observations, c'est-à-dire vers 1970, où la plupart des ruchers anciens étaient encore en usage ou récemment abandonnés, l'utilisation des ruches à cadres a largement supplanté les exploitations traditionnelles qui étaient encore exploitées par des personnes âgées. Ce constat a justifié l'urgence d'en décrire les pratiques.

L'évolution de la ruche horizontale

Il parait difficile d'écrire un historique précis de l'évolution de la ruche horizontale dans les régions du nord et du nord-est de l'Espagne dans lesquelles elles étaient encore très présentes il y a une cinquantaine

diamètre d'environ 15mm. Le cordon de tressage est en osier, éventuellement remplacé par de longues tiges souples de la vigne vierge. Pour les ruches dont la fabrication est la plus soignée, le dernier tour de garniture est arrêté par une petite pince en osier rigide.

En général le tube ainsi constitué est légèrement tronconique. La partie la plus large, celle qui est placée à l'intérieur du bâtiment ou de l'abri, mesure entre 25 et 28 cm de diamètre, selon les modèles. La partie extérieure, mesure entre 22 et 25 cm.

L'étanchéité du récipient est confortée par un remplissage de tiges de roseau glissé entre les mailles. Pour pouvoir être affecté efficacement au logement des abeilles, la totalité de la surface du récipient, intérieure comme extérieure est enduite d'un pourget spécial qui assure à la fois une étanchéité parfaite et, au profit des abeilles un environnement olfactif qu'elles apprécient. Dans son ouvrage sur les soins réservés aux abeilles, Severino Pallaruelo Campo décrit plaisamment la confection très méticuleuse de ce *pourget* :

«Quand les bœufs et les vaches mangent de l'herbe sèche, en hiver, il faut les suivre pour voir où elles vont bouser. Si les animaux mangent des mauvaises herbes leurs bouses seront inutiles et il ne faut pas les ramasser. Il faut aller chercher la bouse des troupeaux là où les bêtes mangent de la bonne

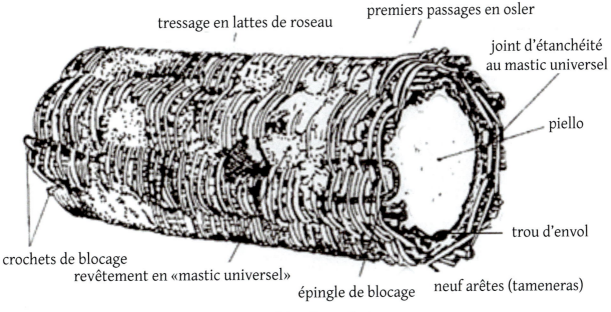

tressage en lattes de roseau

premiers passages en osler

joint d'étanchéité
au mastic universel

piello

trou d'envol

crochets de blocage

revêtement en «mastic universel»

épingle de blocage

neuf arêtes (tameneras)

Figure 2. Schéma d'une ruche *Arna*.

herbe et ramasser les bouses de deux jours. Il faut en préparer une quantité suffisante pour recouvrir la vannerie de la ruche. On commence par mélanger la bouse avec de l'argile bien fine et de l'eau. Ce mélange doit être versé dans une fosse creusée dans le sol, longue d'un mètre avec une profondeur de 30 à 40 cm. On remue bien le mélange dans la fosse avec une houe en brassant d'un côté à l'autre : on le frappe, on le malaxe, on le sarcle, le sépare, l'agite et le bouscule, on le triture, le sillonne et l'aplatit jusqu'à ce qu'il ait l'air tantôt d'un rat, tantôt d'une chose informe douce et souple. L'essentiel est que tout ce travail soit effectué à la lune descendante. En effet c'est à cette condition que le produit durcira parfaitement et que les insectes et parasites ne détérioreront pas l'enduit de la ruche. Quand l'enduit est bien préparé, on plonge le squelette de la ruche dans la fosse jusqu'à ce qu'elle soit recouverte de la douce et onctueuse soupe de boñiga (de bouse). Quand l'opération est terminée la ruche est posée debout et la touche de perfectionnement est obtenue en badigeonnant l'intérieur d'une nouvelle couche de l'enduit avec une branche d'hysope, utilisée comme un long pinceau» (Pallaruelo 1996).

La ruche ainsi préparée reçoit encore quelques accessoires, des barres intérieures pour faciliter le départ des rayons de la réserve, des couvercles avant et arrière qui sont le plus souvent des disques de pierre, les *piellos*, petites lauzes soigneusement martelées pour épouser la forme intérieure des extrémités de la ruche en n'omettant pas de prévoir une petite encoche dans le disque qui servira d'entrée pour les abeilles.

Ainsi fignolée, la ruche est prête pour être mise en position, seule ou avec deux ou trois autres modèles identiques. La, ou les, ruches sont déposées dans un petit abri naturel, un abri sous roche ou petit couloir protégé du vent du nord, propice au passage des abeilles et connu du seul apiculteur.

Ce type d'exploitation des ruches dans les sites rupestres a pratiquement disparu à la fin du XXème siècle. Il était propre aux bergers qui peuplaient les régions montagneuses du nord de l'Aragon et de la Catalogne. Le pastoralisme de ces régions a été complètement modifié depuis une trentaine d'années et l'exploitation des ruches isolées a complètement disparu avec la dernière génération des bergers traditionnels dont le cheptel ne dépassait que rarement cinquante têtes de mouton et une dizaine de ruches.

Pour illustrer cette évolution des modes de vie des habitants de ces villages perdus dans la montagne, je peux citer le cas de Saturnino Cabero qui vivait dans le village de Betorz en exploitant avec son frère José la quarantaine de brebis qu'ils conduisaient dans la montagne environnante. Ils vendaient les amandes de leurs vergers et consommaient le miel de leurs ruches en conservant la cire pour les besoins domestiques.

Saturnino Cabero me racontait qu'il allait deux ou trois fois par an, à pieds avec son âne, vendre ses amandes et un peu de la cire qu'il récoltait très artisanalement, au marché, soit celui de Barbastro soit de Boltaña. « *Combien de temps te fallait-il ? – Siete horas !* (sept heures pour aller, autant pour revenir)».

Ce type d'économie rurale est maintenant abandonné. Les moutons vont toujours dans la montagne mais par troupeaux de 2000 têtes au minimum. Les abeilles butinent toujours le thym et le romarin mais elles sont deux rangées de ruches avec une porte latérale qui permettait un accès facile à la partie arrière des ruches. La capacité moyenne de ces ruchers bâtis était d'une douzaine de ruches.

Les dispositions de ces ruchers ont le plus souvent une orientation au sud ou parfois au sud-est. La largeur de la construction est de l'ordre de 2,50m, parfois 2,80m. La longueur des ruches qui détermine l'épaisseur du mur de façade est comprise entre 70 et 80 cm ce qui laisse à l'intérieur du bâtiment un espace largement suffisant pour pratiquer la récolte. Celle-ci s'effectue en général au moment de Noël ou aux premières neiges et la quantité de miel recueillie dans chaque ruche se situe entre 4 et 7 kg. On remarque la fragilité relative de ces constructions imputables à leur conception. Il s'agissait d'un bâtiment monté sur seulement trois murs, puisque le côté *avant* est totalement dépourvu de maçonnerie et que la toiture de lauzes constitue une charge d'autant plus lourde que les pierres de bordure sont les plus larges et les plus épaisses et qu'elles sont placées sur la partie la plus vulnérable, là où il n'y a pas de mur.

Figure 3. Rucher rupestre dans un abri sous roche. Lieu-dit Canitecho (Sierra de Guara ; Nord Aragon).

exploitées par des apiculteurs bien équipés de ruches à cadre, bardées de hausses, qui produisent le miel à haut rendement. On remarquera que l'usage des plaques de cire gaufrée a pratiquement éliminé la production de cire dans les ruchers modernes. Les quelques ruches restées perchées dans des endroits escarpés ont fini par être démantelées par les animaux sauvages ou balancées dans le vide par les excursionnistes de plus en plus nombreux, peu soucieux de ces espèces de paniers en vannerie qui ne servent plus à rien.

Nous avons vécu depuis une cinquantaine d'année un changement radical de culture dans les zones de haute montagne. La culture apicole qui s'y était attachée, qui était la marque d'une société pastorale vivante de façon quasi autarcique a, elle aussi, disparu.

En descendant vers le sud, c'est-à-dire en se rapprochant de la vallée de l'Èbre la pratique de l'apiculture s'est également modifiée. C'était une apiculture plutôt rurale, orientée vers la consommation domestique mais aussi vers la commercialisation des excédents. Du rucher rupestre éparpillé dans la nature on était passé au petit rucher familial situé à proximité des habitations. C'était une construction en pierres taillées, munie d'un toit à une seule pente couvert de lauzes débordantes. Pas de mur de façade, mais deux fois deux barres de bois supportant

Très rapidement cette production mise dans des conditions modernes de marché a incité les apiculteurs à accroître le volume des bâtiments dédiés aux abeilles. Les locaux ont été doublés, parfois triplés voire davantage.

Les cas les plus spectaculaires sont les ruchers de Roscales de la Peña dans la vallée de la Valdavia, le rucher d'Agreda, dans la région du Moncayo, celui, exceptionnel, du rucher de la zone industrielle du Cinca, dans la banlieue de Barbastro.

Le rucher de Roscalés comporte huit compartiments et se déploie sur une longueur de 25 mètres en abritant près de 160 ruches. Chaque compartiment est aménagé de façon différente en particulier en ce qui concerne le diamètre et la longueur des ruches qui y sont alignées. Il dispose à l'intérieur, d'une chaudière pour le traitement de l'extraction et son toit à double pente abrite un large espace qui facilite confortablement la circulation et l'entretien de l'ensemble.

Le rucher d'Agreda, long de 26 mètres, contient, sur trois rangées, 162 ruches coniques avec des diamètres de base anormalement élevés. Équipé d'un couloir intérieur de circulation très confortable il était encore en état de fonctionnement et parfaitement entretenu à mon dernier passage en 2012.

Figure 4. Rucher monumental de la Zone Industrielle du Cinca (banlieue de Barbastro).

Le rucher de la banlieue de Barbastro, long de 24 mètres ce rucher pouvait contenir environ 300 ruches d'une longueur du diamètre standard mais d'une longueur un peu supérieure à la normale, 130 cm.

Les deux premiers ruchers cités étaient encore en service en 2012, le dernier qui est considéré comme ayant appartenu à l'Évêché de Barbastro a vraisemblablement été abandonné au moment de la guerre civile (1936). Il semble que l'objectif principal des exploitants ait été la production de la cire pour satisfaire à la grande consommation de cierges utilisés dans les cérémonies religieuses.

Une dérive dans l'usage de ces ruchers domestiques a été l'utilisation de ruches en planches, à section carrée ou rectangulaire, allégeant et simplifiant le temps de fabrication.

La plupart des ruchers correspondant à cette étape de la conception des ruchers, montre un mélange de ruches cylindriques en vannerie, de quelques troncs d'arbre creusés et de ruches carrées en planches. Elles marquent le terme de la technique ancestrale fondée sur l'usage de la ruche cylindrique en vannerie. La pratique ultérieure, si elle a conservé le principe de base d'une ruche plutôt horizontale a dérivé progressivement vers des conceptions qui n'ont cessé de s'éloigner du modèle

initial qui était léger et transportable. Elles ont adopté un système orienté vers la réalisation d'un bâtiment spécifique. Le premier degré de cette évolution a consisté à combler, d'un remplissage d'argile et de galets, les interstices entre des ruches cylindriques ; un enduit à la chaux égalisait la surface continue de la paroi extérieure de la construction en ne laissant apparaitre que les trous de vol.

Progressivement, en descendant vers le sud, des ruches en planches de section carrée ou rectangulaire viennent s'ajouter aux ruches-tronc pendant que le volume de remplissage en mortier d'adobe vient à se réduire. Finalement on ne trouve plus d'**étagères** sommaires supportant des cylindres individuels mais un bloc de maçonnerie continu percé d'un alignement de trous et recouvert d'un revêtement de mortier marqué par l'alignement des trous d'envol. Le terme d'*arna* disparait presque complètement du vocabulaire local courant. L'apparence de la structure intérieure du rucher devient en tout point analogue à celle des ruchers *Armarios* de la haute vallée de l'Èbre une succession de cases rectangulaires.

Un autre témoin de cette évolution des structures est donné par l'observation des ajouts de modules de ces constructions. En effet, dans leur quasi-totalité les bâtiments-ruchers ont été agrandis et se composent

Figure 5. Façade de rucher à ruches cylindriques en troncs d'arbre creusés montrant la méthode de remplissage des intervalles (Environs de Bargota Navarre).

duquel commencent à apparaître des structures porteuses de cadres mobiles.

L'héritage de la structure circulaire de la ruche initiale a disparu pour faire place à un logement. C'est le cas très caractéristique du rucher de Mendavia. Il comporte trois panneaux, tous porteurs extérieurement de séries de piqueras alignées à des hauteurs différentes selon les panneaux. Premier constat, chaque panneau présente une organisation externe particulière, il y a donc eu une construction évolutive dans le temps.

Deuxième constat, l'organisation intérieure ne correspond pas du tout à l'apparence externe : le rucher est entièrement cloisonné en grands casiers rectangulaires qui impliquent l'obturation d'une rangée de piqueras sur deux. La leçon de cette réalisation originale est bien la pérennité des lieux d'exploitation et l'adaptation aux *idées modernes*, forme très originale de conversion du patrimoine.

D'une manière générale les ruchers sont situés à l'écart des villages, à l'orée de zones boisées, en bordure de champs cultivés ou dans de vastes clairières au milieu de bois de chênes où la bruyère tapisse le sol.

de deux parties ou davantage, chacune témoignant des usages de l'époque où elles ont été réalisées.

Le rucher le plus caractéristique à cet égard est celui de Roscales, cité plus haut. Il se développe sur huit compartiments différents, chacun montrant une technique améliorée tendant vers une meilleure utilisation de la surface de la façade.

Cette pratique à peu près générale de l'adjonction progressive de compartiments agrandissant le rucher, se justifie apparemment par un besoin accru des produits de la ruche, miel et cire. On notera que les adjonctions ont toujours eu pour effet d'augmenter le nombre de ruches pour une surface de façade constante.

La génération suivante de ces ruchers comporte un grand nombre de variantes qui ont en commun l'utilisation de volumes à section carrée ou rectangulaire, donnant sur leur face arrière l'apparence d'un empilement de caisses en bois ou en maçonnerie légère parallélépipédique à l'intérieur

Cette proximité des espaces boisés favorise considérablement la récupération des colonies à l'essaimage de printemps. De fait on ne voit pas de traces de ruches pièges ou autres artifices destinés à retenir les essaims. La capture de ces essaims que l'on trouve accrochés aux branches basses des chênes se fait à l'aide de paniers tressés soigneusement enduits de cire à l'intérieur.

Dans la plupart des villages la population qui a beaucoup déserté le pays pour s'installer dans les grandes villes, a souvent abandonné les pratiques apicoles. À une époque relativement récente il y avait à peu près un rucher par famille et dans certains endroits comme dans le village de Colmenares, les familles avaient chacune plutôt deux ruchers, voire trois. Ces installations comportaient entre 30 et 60 ruches. Malgré cette désertion importante de l'activité traditionnelle,

Figure 6a et 6b. Rucher de Mendavia et détail.

de nombreux ruchers traditionnels étaient encore en activité en mai 2012. Ils sont entretenus et exploités par des personnes âgées et sont souvent accompagnés de ruches à cadre très actives.

La production en miel de ces ruchers domestiques pourvoyait largement à l'usage familial et avant l'installation de l'électricité, la récolte de cire satisfaisait aux besoins domestiques en éclairage. On constate toutefois que la production locale de cire était excédentaire et qu'il existait un marché de ce produit. L'établissement de nombreux monastères et d'églises organisatrices de grandes cérémonies, a été un grand consommateur de luminaires. Le cas du grand rucher de Roscales dans lequel une chaudière a été installée, laisse entrevoir l'amorce d'un traitement artisanal sur place. Il existait des réseaux de collecte de la cire comme cela se pratiquait dans les Mérindades. La production excédentaire était acheminée vers Santander ou Burgos et vers la ceinture d'abbayes implantées au nord de la province de Palencia. L'installation à Maranchon (Guadalajara) d'un centre de blanchiment de la cire est une confirmation de cette situation (Arias 2005; Lemeunier 2006). Ce petit village, bien placé sur les liaisons charretières entre Madrid et Saragosse s'était spécialisé aux XIX et XXème siècle dans la collecte,

le traitement de la cire vierge, son blanchiment puis ensuite sa diffusion dans tout le nord de l'Espagne.

Grande antiquité de la pratique apicole horizontale

L'archéologie a mis en évidence l'existence d'un trafic commercial entre le Moyen Orient ancien et le Levant Espagnol devenant régulier à partir, au moins, du premier millénaire. On ne peut pas étudier la situation quasi contemporaine de l'apiculture traditionnelle dans la partie nord de l'Espagne qui nous intéresse sans la replacer dans le temps et sans prendre en compte l'impact des influences extérieures.

La région de la côte méditerranéenne d'Espagne, généralement désignée comme *province du Levant*, a connu, dès le milieu du second millénaire avant notre ère, des échanges culturels réguliers avec l'autre extrémité de la Méditerranée.

On peut considérer que le Levant espagnol, complété par la large ouverture voisine du bassin de l'Èbre a constitué le principal point d'entrée de la civilisation du Moyen Orient dans la péninsule Ibérique. Les traces laissées par cet apport culturel nous obligent à nous placer bien avant les grands mouvements de

Figure 7. Intérieur d'une ruche à tubes incorporés
dans la maçonnerie, abandonnée.

population, provoqués par la conquête et l'occupation romaine, puis par l'irruption de l'Islam, et finalement par la reconquête chrétienne. Ces échanges qui se sont poursuivis et développés au cours des siècles, portaient dès leur origine sur des poteries, des objets d'art et aussi des produits de consommation. On sait que le miel et la cire ont été parmi les premiers produits d'échange.

La période préromaine dans la région de Valence a été bien étudiée par les chercheurs de l'Université de Valencia qui y ont consacré des travaux importants. Les archéologues de cette Université ont mis en évidence l'existence d'un trafic commercial entre les ports du Moyen Orient, Tyr et Sidon, puis de Carthage avec la côte de ce Levant Ibérique. Ce trafic est devenu régulier à partir au moins du premier millénaire avant notre ère. Le cas particulier de l'ancienne place forte d'El Puntal dels Llops peut servir d'exemple (Bonet et Mata 1997; cf. Crane 1999: 182). Il est significatif de cet apport de culture. Vers l'an 250 avant J.C., ce village situé actuellement dans la province de Valence, au sommet d'une colline, comptait dix-sept habitations. Sur un terrain long de seulement 45 mètres, les archéologues y ont relevé 17 compartiments correspondant à autant de logements. Sur cette seule surface, ils ont identifié 99 ruches en terre cuite, réparties dans 11 de ces logements présumés. Les ruches, cylindriques, étaient utilisées couchées et devaient être obturées avec un matériau qui n'a pas résisté au temps, mais on a trouvé des traces de hausses dans certains et à l'intérieur de toutes les ruches des stries témoignant du soin, au moment de la fabrication, de faciliter la bonne tenue des rayons à l'intérieur. Ces stries en particulier, ne laissent aucun doute sur la fonction de ces récipients.

L'évaluation du nombre de ruches par familles est de 4 à 6 ruches en terre cuite, sans préjuger de la présence d'autres ruches en matériaux périssables. Ils ont également évoqué l'hypothèse d'un placement de ces ruches situées à l'étage des habitations, plus particulièrement en terrasse. Par rapprochement avec des ruches du même type, trouvées dans des villages de l'Aurès (Afrique du Nord) et à Minorque, les responsables de ces fouilles, Consuelo Mata et Helena Bonet estiment qu'il s'agit d'exploitations domestiques. Elles considèrent que la production de ces ruches, aussi bien le miel que la cire, devait intervenir de façon significative dans l'économie du village. Ces indications et ces chiffres, ainsi que la prise en compte vraisemblable de ruchers extérieurs au village, indiquent des quantités de ruches considérables au regard de la modicité de l'agglomération. On remarque que le lieu-dit est actuellement nommé "el Colmenar", c'est-à-dire, le rucher. Constat qui en dit long sur la permanence des coutumes locales.

L'évocation de cette production suggère une possibilité de commerce non seulement du miel, mais aussi de la cire. Il est important de souligner qu'à cette époque, la cire pouvait être utilisée pour l'éclairage mais aussi pour la fermeture des amphores et autres récipients alimentaires.

Elle était également recherchée pour d'autres usages comme la fabrication artisanale d'objets de bronze traités à la cire perdue. Le cas de Puntal dels Llops n'est pas isolé. Des fouilles menées sur d'autres sites de la même époque ont aussi mis à jour des ruches en céramique tout à fait similaires. L'un de ces sites, toujours juchés au sommet d'une colline, est proche de Puntal dels Llops l'autre, plus au sud, se trouve à proximité de la ville d'Albacete en bordure du fleuve Jucar.

Aucune information géographique ne nous est parvenue sur l'affectation de ces villages stratégiques pendant et après la conquête romaine. Toutefois, on notera que l'écrivain latin Columelle était originaire d'Andalousie et que son traité d'agriculture fait une place à l'apiculture horizontale qu'il est un des seuls à avoir décrit à cette époque.

Nous sommes peu informés sur la localisation de l'activité apicole pendant les six siècles d'occupation musulmane qui ont suivi la longue période romaine.

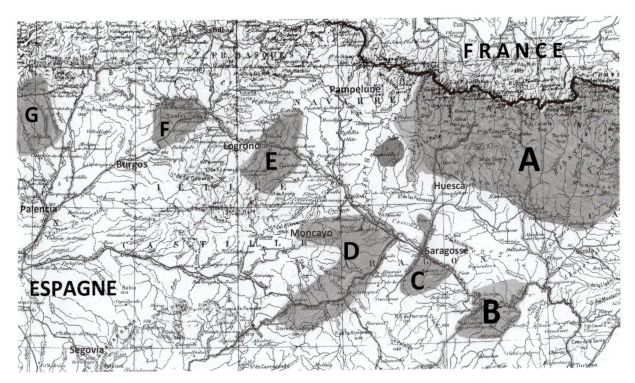

Figure 8. Carte de la vallée de l'Èbre indiquant l'emplacement des régions décrites.

Cependant l'importance de l'usage du miel dans la société musulmane est telle qu'il faut penser que cette activité a perduré pendant les siècles de l'occupation islamique de la péninsule Ibérique et même qu'il y a tenu une grande place. Tout d'abord les références au miel sont nombreuses non seulement dans le Coran mais dans les textes complémentaires, les *hadiz*, qui traitent de médecine, de coutumes, de choses de la vie de tous les jours. La sourate 47 révèle que parmi les bienfaits réservés aux *Pieux*: « *Il s'y trouvera ... des ruisseaux de miel clarifié* » (*Qur'an* 16 :68-9 (trad. Blachère)). Peu de détails concernant l'organisation territoriale de l'apiculture sont parvenus jusqu'à nous. Il semble que, au moment de l'invasion arabe, la transmission des savoirs et des installations relatifs à l'apiculture se soit plutôt bien négociée à la faveur de ce changement de maîtres. Ainsi Juan Pedro Monferrer évoque le cas d'un personnage important de la région de Murcie, un certain Teodomir qui a conclu avec le conquérant Arabe, Abd-dal-A1ziz, un accord écrit stipulant la cession entre autres valeurs de *dos qist de mie* (Monferrer 1991).

En revanche, la transition entre pouvoir islamique et chrétien s'est faite dans la violence. Les artisans de la Reconquête, au nom de la vraie foi, ont systématiquement détruit tous les documents rédigés en langue arabe, et il y en avait beaucoup. Les siècles d'occupation de la péninsule ibérique par des princes islamiques ont été extrêmement féconds. Les cours de Cordoue et de Tolède, entre autres, étaient très brillantes dans les sciences et les arts. La plus grande partie de la littérature publiée pendant cette période est à jamais perdue.

En revanche les techniques et usages de l'apiculture ont été bien conservés. Dès la période de la reconquête, de nombreux textes nous informent sur l'exploitation des ruchers. Des traités publiés en Espagne aux XVIe et XVIIe siècles apportent de précieuses informations. Ce sont entre autres: *Obra de agricultura compilada de diversos autores* de Gabriel Alonso de Herrera, publié à Tolede en 1513, *Tractado breve de la cultivacion y cura de las colmenas* par Luis Mendez de Torres, publié en 1586 à Alcala de Henares et surtout le traité de Jame Gil, *Perfecto y curiosa declaracion de los provechos grandes quedan las colmenas bien administradas* publié en 1621 à Magallon. Ce dernier présente l'intérêt d'avoir été conçu avec l'expérience de l'apiculture au cœur de la zone qui nous concerne. Les indications qu'il nous donne confirment la coexistence de l'exploitation des ruches horizontales et verticales dans un même lieu et l'utilisation de ruchers tout à fait identiques à ceux qui étaient encore en usage il y a une vingtaine d'années seulement. Cette information nous conforte dans la conviction que les traditions apicoles sont particulièrement tenaces. Leur maintien ou leur modification ne se fait pas en fonction des pouvoirs dominants parce qu'ils sont généralement le fait des gens de petite condition et peuvent ainsi perdurer pendant de nombreux siècles.

Sept zones caractérisées par une exploitation de mode horizontal

L'évolution des usages des apiculteurs ne s'est pas faite de manière homogène dans l'ensemble de la zone. Elle varie considérablement en fonction des usages locaux, des évolutions subies dans le temps et vraisemblablement de l'influence des apiculteurs qui en ont été à l'origine. Il en résulte une répartition des ruchers selon les lieux. Pour mieux les appréhender nous avons réparti la région nord Espagne en sept zones.

A) Jacetanie, Serrablo, Sobrarbe et Ribagorza

Ce sont quatre très anciennes provinces Ibères longtemps résistantes à la colonisation romaine. On peut penser que la tradition de l'apiculture y est particulièrement antique. En effet elle est à la mesure de l'âge des civilisations pré-Ibéres qui ont peuplé ces montagnes, quand les hommes peignaient sur les parois des scènes de chasse très rudimentaires. Par vagues successives ces montagnes ont accueilli des populations venues chercher refuge à l'arrivée de conquérants ou de commerçants agressifs remontant la vallée de l'Èbre.

Environ 150 sites ont été inventoriés dans cette zone. Ils ne représentent qu'une partie de la grande richesse d'exploitation locale. L'exploitation des ruches y est d'autant plus rudimentaire qu'on se rapproche de la chaine pyrénéenne, que le relief est plus marqué, l'approche plus difficile

On trouve en effet, dans les régions montagneuses du nord, des abris naturels sommairement aménagés où les ruches étaient posées à même le sol, calées par de grosses pierres et gardées à l'abri du soleil par une petite couverture de branchages, généralement de bruyère.

Le plus remarquable de ces abris est celui de Barfaluy sur la commune de Lecina, qui recèle sur ses parois intérieures des peintures datées de plusieurs milliers d'années. Assez difficile d'accès, ce rucher comportait néanmoins deux rangées de ruches, bien alignées sur des barres de bois encastrées dans une construction rudimentaire.

En descendant les vallées on constate que les ruchers sont de petites cabanes couvertes d'un toit de lauzes avec une ou deux rangées de ruches en roseaux tressés posées sur des barres de bois. Ces petites constructions reproduisent le schéma des abris naturels en donnant plus d'aisance à l'apiculteur. Une porte latérale donne accès à l'arrière des ruches dans un étroit compartiment sombre. Les pierres de bordure de la toiture y descendent très bas ne laissant qu'un mince espace avec la première rangée de ruches. Ces ruches occupaient toute la surface de façade qui était ouverte. L'accès dans un espace d'exploitation,

Figure 9. Ruche posée sur le sol dans un abri sous roche en Sierra de Guara (Nord Aragon).

Figure 10. L'abri de Barfaluy en 1987, avant la pose des grilles de protection. Sierra de Guara, (Haut-Aragon).

situé à l'arrière des ruches, se faisait par une porte latérale. Cette disposition assurait à l'intérieur une certaine obscurité favorable à l'exploitation des ruches. Selon les lieux la récolte se faisait soit à l'intérieur de la ruche, sans déplacer la ruche, en ôtant l'opercule arrière pour prélever les gâteaux de cire, soit en sortant la ruche pour la poser sur un chevalet qui permettait en plus de la récolte, de procéder à quelques réparations de fortune sur les dégradations éventuelles de la « croute » d'enduit. Cette dernière façon de procéder devait se pratiquer au lever du jour, avant que les abeilles de la colonie n'aient commencé à sortir et n'attaquent les apiculteurs.

Plus on descend vers le sud, plus ces constructions deviennent importantes, comportent deux, voire trois compartiments et permettent d'abriter 25 à 30 ruches. Cette progression du volume des ruches traitées culmine dans le rucher de la zone industrielle de la Valle Del Cinca dans la banlieue de Barbastro. Il s'agit d'une construction haute d'environ 5 mètres qui comprend neuf compartiments adossés à un talus escarpé. Chaque compartiment comporte six niveaux, eux-mêmes capables de supporter six ruches. Au total le bâtiment mesure 24 mètres de long et pouvait abriter plus de 300 ruches.

B) Vallées des rios Martin et rio Guadalupe (Caspe, Mequinenza)

Ce secteur est caractérisé par la présence de ruchers à façade fermée à l'intérieur desquels les ruches sont

horizontales et déplaçables. Ces ruchers sont placés dans des abris naturels sous de grandes dalles rocheuses. La totalité de la façade a été murée complètement en ne laissant qu'une porte d'accès et de petits orifices alignés dans la maçonnerie. Ce sont les *piqueras* qui servent au passage des abeilles. À l'intérieur de 'abri, les ruches, qui étaient à l'origine des cylindres de vannerie tressés, sont placées dans des logements délimités par des murets de pierre et des barres de bois transversales. Cette pratique a l'avantage d'offrir aux abeilles une bonne protection contre les variations climatiques et en outre de donner toutes facilités à l'apiculteur pour les opérations de contrôle ou de récolte grâce à la mobilité de chaque ruche.

Placé dans la partie de la plaine de l'Èbre qui borde la grande retenue de Mequinenza, ce secteur est difficile d'accès et jouit de la réputation d'une zone sauvage et riche en flore et faune originales.

Le mode de construction des ruchers n'est pas totalement original dans le nord de l'Espagne. On en trouve également à Uncastillo (secteur de Cinco-Villas) et à Villaroya (secteur Rioja-Navarre). Ce sont cependant des cas assez rares. Hors d'Espagne, c'est le mode d'utilisation traditionnelle sur l'île de Malte.

Dans la province de Saragosse, la sierra de Valdurrios en particulier, bénéficie d'une longue réputation de région mellifère. On y a toujours pratiqué l'apiculture et aujourd'hui encore les apiculteurs de la région viennent

Figure 11. Le grand rucher de la Zone Industrielle de Valle del Cinca près de Barbastro.

y porter leurs ruches en transhumance. La municipalité de Caspe perçoit chaque année un droit de séjour pour ces ruches en transit. En 1988, un apiculteur s'était acquitté de cette taxe pour un total de 800 ruches.

La rareté des vestiges dans ce secteur peut s'expliquer par son isolement et par les difficultés de circulation pour y accéder. La mise en eau de la vallée en 1964 a accentué ces obstacles. Abandonnés depuis au moins 50 ans, la plupart des ruchers traditionnels sont en ruine ou ont disparu.

C) Zuera- Castellar. Las Muelas de l'Èbre moyen

Comme ceux de la zone précédente, ces ruchers ont une façade fermée mais ils ont été construits au fond d'un grand enclos. Le bâtiment consacré aux abeilles est adossé à une surélévation du mur arrière orienté au sud, tenant lieu de générateur d'air chaud qui facilite l'envol des abeilles au départ de la ruche. Les murs de l'enclos qui englobe le rucher ont en moyenne une hauteur supérieure à deux mètres (six à sept pieds).

Ces enclos peuvent être limités en largeur à la dimension du bâtiment ou bien s'étendre plus largement pour former un grand enclos. Selon les lieux ces ruchers peuvent se limiter à un seul compartiment et se développer en limitant l'enclos au prolongement des murs du bâtiment. Ils peuvent aussi, comme dans

les environs de Zuera constituer de grands enclos dans lesquels la partie bâtie se situe au fond de l'espace inscrit dans la clôture. L'espace intérieur est souvent aménagé d'une succession de niveaux pouvant servir soit au dépôt de ruches pièges, soit être planté de plantes mellifères ou d'arbustes aptes à capter les essaims au moment de l'essaimage.

D) Vallée du Rio Jalon et Moncayo

Les villages des vallées des rios Jalon et Jiloca ont adopté le type de rucher *armario* qui constitue le groupe dominant de la partie de la province d'Aragon située au sud du cours de l'Èbre Felix Rivas précise que la pratique de l'apiculture traditionnelle avec des ruchers de type *armario* occupe une vaste zone plus ou moins ovale située dans la partie occidentale de l'Aragon depuis les limites de la Rioja et de la Navarre jusqu'au voisinage des provinces de Castille et de Valence où elle se poursuit (Chevet et Rivas 2005).

Les ruches sont ici des placards de section rectangulaire, inclus dans une façade fermée trouée de *piqueras* pour l'entrée des abeilles. Ces trous de vol se signalent extérieurement par une petite plate-forme comme pour les ruches placards (*alacenas*) que l'on trouve en Galice ou les ruches de type *hornos* utilisée dans une grande partie de la moyenne vallée de l'Èbre L'originalité des *armarios* par rapport aux *alacenas* tient

Figure 12-13. Deux ruchers avec enclos: Bulbuente (près de Tarazona) pour celui de gauche, et au nord de Zuera, pour celui de droite.

souvent *abejares*, alors que les ruchers sont toujours ce que nous avons défini comme *armario*.

Dans toute la zone, l'apiculture ancienne utilisait de manière complémentaire aux ruchers bâtis consacrés aux ruches *armarios*, des ruches verticales qui étaient essentiellement mobiles et destinées au renouvellement des essaims.

On remarque que les ruchers peuvent, selon les villages être placés au cœur même des agglomérations, comme c'est le cas par exemple pour Bubierca, ils peuvent au contraire se situer très loin des maisons d'habitation. Dans ce cas les bâtiments-ruchers disposent d'un espace intérieur assez vaste pour permettre le traitement de la récolte.

Les informations que nous avons pu recueillir indiquent que le même type de rucher est répandu dans toute les vallées affluentes du rio Jalon au moins jusqu'à Medinacelli et du rio Jiloca au-delà de la ville d'Albarracin.

Le massif et les environs du Moncayo constituent un cas particulier de cette zone. Les ruchers présentés ici se situent dans un périmètre de 30 km autour du massif du Moncayo. Dans une dizaine de communes on trouve des ruchers dont le modèle dominant est le rucher à façade fermée contenant des ruches horizontales fixes, englobées dans la maçonnerie.

Près de la moitié de ces ruchers que nous y avons visités est située sur le territoire de la commune de Borja dont une dizaine dans le seul vallon de Los Arbolitas où se trouvent les jardins de la plupart des habitants fortunés de Borja.

Cet endroit nous est connu grâce à l'excellente connaissance de la région d'Alfredo Sanz Villalba. Sa collaboration et ses innombrables relations nous ont été infiniment précieuses. Elles nous ont permis d'enregistrer la remarquable conjonction de l'existence et de l'identification historique de l'un de ces ruchers.

La plupart des familles aisées de Borja y possèdent une petite propriété comportant quelques arbres fruitiers, une bergerie et un rucher. La famille Aznar **détient l'une de ces parcelles au lieu-dit La Ma**taza. Sur cette petite propriété se trouve un rucher à deux

à la structure interne du rucher : les compartiments sont accolés les uns aux autres, séparés par une mince cloison de mortier ou d'adobe, aussi bien latéralement que verticalement. L'aspect intérieur des *armarios* est celui d'un ensemble compact de casiers rectangulaires fermés par des tapes amovibles.

Les *armarios* diffèrent aussi des ruchers de type *horno* par la dimension des ruches et leur disposition. Les ruches *horno* sont des tubes souvent tronconiques placés horizontalement, englobés dans la maçonnerie du mur de façade. Les ruches des *armarios* sont des casiers quadrangulaires. Cependant les termes utilisés sur place pour désigner les ruchers de toute cette zone sont, selon les lieux *hornos* ou *armarios* mais le plus

Figures. 14 et 15. Le rucher de La Mataza daté 1691.

compartiments, en assez bon état pour la partie gauche. Sur les murs de cette partie gauche des inscriptions portent la date 1691. Sur la partie de droite les lettres ne sont plus lisibles.

L'actuel propriétaire nous a présenté des documents de famille donnant pendant les quatre derniers siècles les comptes de leurs propriétés. Dans le registre concernant l'époque inscrite sur le mur du rucher, le texte

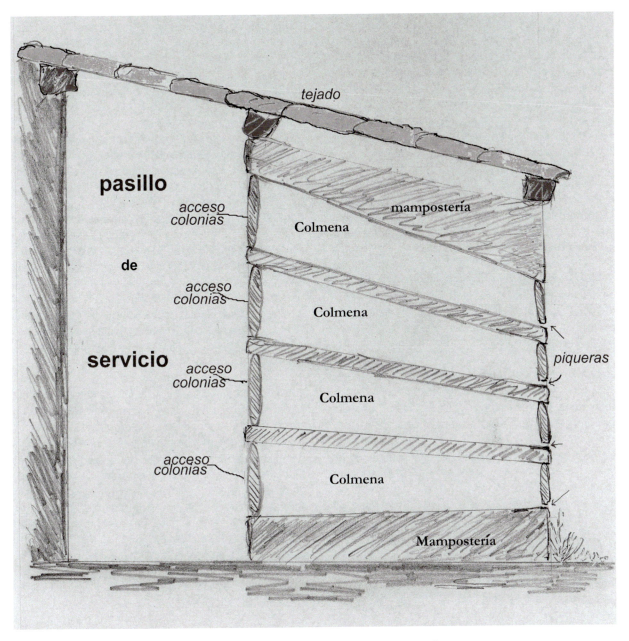

Figure 16. Schéma de rucher du type en usage dans la région du Moncayo.

mentionne plusieurs dépenses relatives à l'entretien du rucher ou de ses accès et quelques recettes de vente de miel ou de cire.

Les anciennes **écritures aussi bien que les inscriptions sur le terrain concordent parfaitement. Elles matérialisent la pérennité de l'usage de l'apiculture dans cette famille.**

L'appellation locale de ces ruchers est *hornos* ou *hornillos* ou parfois *arnales*.

Les ruchers du Moncayo ont la particularité d'être très anciens. La construction de certains d'entre eux est datée de manière incontestable du début du XVIIe siècle et témoigne d'une tradition plus ancienne.

L'abondance des ruchers *hornos* **décrits ne doit pas faire préjuger de l'absence de ruches verticales dans les traditions locales.**

L'usage combiné des deux types de ruches, horizontales et verticales, est attesté dans la région par l'ouvrage de Gil de Magallon, publié en 1621 ainsi que par le témoignage de Morales, le fonctionnaire municipal de Borja chargé jusqu'en 1970 du soin des ruchers privés.

Les emplacements de ruches verticales n'ont pas été conservés, en revanche les ruchers *hornos,* constitués d'un massif de maçonnerie tubulaire ont bien résisté au temps et demeurent visibles des années après l'abandon des locaux et la destruction de leur toiture.

On constatera que la plupart des ruchers décrits sont composés d'une petite construction comportant un bloc de deux ou trois rangées de tubes insérés dans la maçonnerie et à l'arrière des tubes un espace couvert pouvant être hermétiquement clos. Les ruchers sont souvent entourés d'un petit enclos à murs bas et quelquefois de terrasses qui sont plantées d'arbustes. Ces terrasses pouvaient être destinées à recevoir les abeilles au moment de l'essaimage.

Les tubes qui reçoivent les colonies d'abeilles sont légèrement tronconiques ; la partie la plus étroite est à l'avant du bâtiment ; elle est fermée par un bouchon en pierre ou en mortier qui laisse entrer les abeilles par un petit orifice situé à la partie inférieure.

Cet orifice nommé *piquera* est généralement agrémenté d'une margelle extérieure destinée à faciliter l'envol des abeilles. À l'arrière des tubes, un opercule amovible permet la visite de l'intérieur de la ruche et en particulier la récolte.

Les dimensions des tubes sont à peu près les mêmes dans toute la zone décrite : longueur moyenne 110 cm (maximum 120) ; diamètre extérieur 24 cm, diamètre intérieur 33 cm pouvant atteindre 40 cm.

En 2005, sur 45 ruchers visités, trois seulement étaient encore en exploitation. Cependant cet abandon est récent car Morales, le gardien municipal des ruches, était encore en fonction il y a moins de trente ans (Chevet 2005).

E) l'Èbre Moyen

Ce secteur s'étend sur une large bande de territoire de part et d'autre de la vallée de l'Èbre un peu en aval de Logroño. Il concerne les provinces de Navarre et de la Roya et reste très proche de celles de Soria et de Saragosse en Aragon.

Il est limité au nord par les prolongements des Pyrénées en Navarre, à l'ouest par les vallées profondes des rio Lesa et Iregua (sierra de Cameros), à l'est par le massif du Moncayo.

Ce secteur se caractérise par une dominante de vestiges de ruchers bâtis à façade fermée, composés de batteries de tubes horizontaux englobés dans des massifs de maçonnerie qui sont connus localement sous le nom de ruchers *Hornos*.

La tradition populaire et quelques rares vestiges indiquent que la tradition locale faisait également un usage important des ruches verticales en vannerie principalement utilisées sur des banquettes naturelles variables selon les saisons. On les nommait *peones* ou *piones*.

Le principe de fonctionnement des ruchers horizontaux de ce secteur est toujours le même : la façade du rucher offre l'aspect d'un mur droit percé de trous d'entrée soulignés d'une planche d'envol. Les trous donnent aux abeilles un accès immédiat à l'intérieur de la ruche.

Les ruches longues de 95 à 110 cm, font partie intégrale du bâtiment et sont parfaitement inamovibles. À l'intérieur du bâtiment, leurs parties arrière offrent l'aspect d'un mur continu jalonné des disques des opercules. Entre ce mur et l'arrière du bâtiment, ou la paroi rocheuse dans le cas où le rucher est appuyé à un talus. Un couloir de circulation permet à l'apiculteur de surveiller le contenu des ruches. La largeur de ce couloir dépasse rarement un mètre, l'apiculteur n'a pas besoin de davantage d'espace pour retirer les brèches lors de la récolte. En revanche, quand la porte de l'espace intérieur est fermée, l'obscurité y est totale si le constructeur n'a pas prévu une petite fenêtre dans l'un des murs latéraux. On se trouve exactement dans le même schéma d'exploitation que dans les secteurs du Moncayo ou de Zuera.

Toutefois la conception des ruchers de ce secteur n'est pas homogène comme dans le secteur du Moncayo : on trouve bien des tubes obtenus par moulage à l'intérieur d'une construction massive en mortier de chaux, mais il y a également des ruchers obtenus par l'empilement de troncs d'arbre creusé enrobés dans un remplissage d'argile et cailloux ou dans d'autres cas des rayonnages serrés de ruches quadrangulaires séparés par de minces cloisons de mortier à la chaux. En outre, certains ruchers ont été construits en aménageant un abri sous-roche ou du moins l'avancée d'un rocher. La construction se limite alors à la construction d'une façade et d'un accès à l'intérieur.

Ces variantes et la présence dans un même site de méthodes différentes clairement différenciées dans leur date d'origine permettent de conclure à un certain archaïsme comme si la construction d'empilements avait précédé la méthode d'enrobage des tubes dans un massif de mortier, cette méthode ayant elle-même été remplacée par l'utilisation de petites cloisons minces, plus légères.

Les empilements de tubes dans des gangues de mortier forment une structure cellulaire résistante à l'érosion ce qui donne une grande longévité à leurs vestiges. La comparaison de clichés pris à plus d'un demi-siècle d'intervalle montre qu'aucune modification des tubes existants n'est observable. Certains vestiges ne comportant que des groupes de tubes peuvent être plusieurs fois centenaires.

En revanche la nature essentiellement fragile des ruches en vannerie de type *vasos* qui étaient utilisées en transhumance n'a laissé pratiquement aucune trace ce qui rend difficile l'évaluation de l'importance des usages horizontaux et verticaux. Les ruches verticales devaient jouer un rôle important dans la capture des essaims et le renouvellement des colonies à l'intérieur des ruchers horizontaux.

L'existence d'enclos autour de ces ruchers est assez rare et on ne rencontre que peu ou pas de cabanes d'apiculteur à proximité des ruchers. Ceci implique la pratique d'un transport au village des brèches extraites des ruches et d'un traitement de la récolte au domicile de l'apiculteur.

En outre, la plupart des ruchers sont importants et ne correspondent pas à une consommation domestique. Il existait donc dans cette zone un commerce du miel et surtout de la cire. Celle-ci était utilisée à l'éclairage, à la fabrication d'exvotos et dans certaines zones à l'achat par l'armée pour l'imperméabilisation des vêtements de pluie (les cirés).

F) Montes Obarenes

Au nord de la province de Burgos, la *comarca* de Las Merindades est une enclave à l'orographie complexe qui s'enfonce à l'intérieur de la chaîne cantabrique. Ce secteur reçoit naturellement une forte influence méditerranéenne en raison de sa proximité avec les sources de l'Èbre et la présence continue du jeune fleuve à travers la région. Toutefois, il reçoit aussi une forte influence des traditions atlantiques du fait de son voisinage avec les bandes côtières de la région Cantabrique.

Concernant l'apiculture, il s'agit dans les zones frontières du nord de la province, essentiellement de l'usage quasi exclusif du *dujo*. Cette ruche en tronc d'arbre creusé est exploitée verticalement et disposée sur des petits terrassements, couverts et solidement protégés des vents froids du nord et des basses températures des hivers rigoureux.

Il y a plus d'une centaine de ces petits ruchers domestiques dans le nord des Merindades.

Mise à part cette influence des zones du littoral Cantabre, la grande originalité de l'apiculture traditionnelle des Merindades est l'usage d'une ruche horizontale encastrée dans les murs des maisons d'habitation, des bâtiments agricoles, voire parfois de petites maisonnettes spécialement dédiées à leur traitement. Ces ruches sont très généralement des *dujos*, couchés dans la maçonnerie du mur, l'extrémité intérieure débordant d'environ 40 cm. Il s'agit donc de l'adaptation d'un modèle de ruche local, le *dujo*, à un usage observé au sud, dans le bassin inférieur de l'Èbre sans qu'il soit possible de dire qui a influencé qui.

Figure 17. Ruches horizontales en troncs d'arbres, encastrées dans le mur-pignon d'un rucher.

Le caractère archaïque de cette tradition semble la seule évidence avec le fait invariable qu'il s'agit encore d'un avatar de la ruche horizontale élémentaire. Les ruches, des troncs creusés ou parfois des planches grossières vaguement équarries, sont disposées horizontalement en restant séparées assez largement les unes des autres

Cette tradition diffère en cela, à la fois des ruchers de la région de la Valdavia, assez proche au sud-est et de celle des ruchers de Navarre. Dans ces deux cas, les ruches en matière végétale ont été le plus souvent couchées les unes contre les autres de manière jointive, la maçonnerie constituant seulement un remplissage.

Comme dans ces deux zones, assez proches en tradition, les faces externes des ruches affleurent la paroi des bâtiments. Ces ruchers apparemment dédiés exclusivement aux usages familiaux, ne dépassent que très rarement la dizaine de ruches. D'autres sites présentent la même configuration de ruches horizontales encastrées, sont connus dans des provinces voisines, toujours dans une zone au sud de la crête de la Cordillère Cantabrique.

G) La Valdavia

La zone étudiée se situe grosso modo entre la chaîne Cantabrique au nord et les grandes plaines à blé qui se développent à la latitude de Carrion jusqu'à Palencia. Ce pays d'une altitude moyenne oscillant autour de 1000 **mètres**, est marqué par un climat rigoureux, sujet à des variations extrêmes, grands froids et neige en hiver,

fortes chaleurs en été. Le nord de la zone est dominé par les hautes murailles des Picos de Europa qui culminent à près de 3000m. Pendant la moitié de l'année, les vents du nord-est dominants, soufflent un air glacé sur les vastes étendues de collines boisées et de vallées emblavées.

L'eau coule en abondance dans tout ce pays et les rivières qui amènent vers le sud les eaux de la Cordillère marquent le paysage de vallées parallèles ; ce sont les rios, Carrion, Valdavia, Boedo, Pisuerga qui sont tous tributaires du Duero et vont aboutir dans l'Atlantique.

L'apiculture y est traditionnellement développée depuis l'antiquité et ce pays qui a connu une forte colonisation romaine, comme en témoignent les multiples toponymes en *villa* et le site proto-romain important de La Olmeda, a adopté quasi exclusivement les ruches horizontales qui sont nommées *hornillos*. Seuls les villages proches de la frontière de Cantabria ont pratiqué un usage mixte des ruches verticales en troncs d'arbres, les *dujos*, et des mêmes ruches en tronc d'arbre creusé, englobées horizontalement au cœur du mortier d'argile d'un mur de façade.

De manière fréquente, les portes d'accès à l'espace intérieur sont placées dans le mur latéral, à droite de l'édifice, c'est à dire côté Est. Elles sont ainsi abritées des pluies qui viennent majoritairement du sud-ouest.

Comme dans d'autres régions à climat rigoureux, les ruchers sont des bâtiments solidement bâtis, bien calfeutrés. Très peu sont appuyés sur des rochers ou

Figure 18. Grand rucher de Roscales.

Figures 19 and 20. A gauche ruche « type arna » en Anatolie. A droite même type de ruche au Maroc.

Les murs de façade sont entièrement consacrés aux ruches. Du nord au sud de la zone, on constate une petite évolution de la nature de ces ruches. Dans la partie proche de la province de Cantabrie, en Montaña Palentina, les ruches sont des troncs d'arbre couchés et insérés dans la maçonnerie qui occupe une grosse partie du volume. Elles se rapprochent ainsi des ruches les plus archaïques du type *Horno* que l'on trouve dans la moyenne vallée de l'Èbre

De la place de la ruche cylindrique horizontale dans le pourtour de la mer Méditerranée.

Si la ruche de type *arna* que nous avons suivie à travers le nord de l'Espagne est bien le prototype de la ruche traditionnelle de ce pays, telle qu'elle a pu évoluer depuis près de trois millénaires, il est intéressant de se demander pourquoi on la retrouve parfaitement identique dans des régions **à l'extrémité de la Méditerranée** aussi éloignées que la Cappadoce, en Turquie, la région de Meknés au Maroc, dans certaines îles Grecques ou dans les montagnes du Liban On est donc ramené à l'hypothèse d'une transmission ancestrale du savoir-faire à travers la Méditerranée, mais il n'est pas possible dans l'état actuel de nos connaissances d'en déterminer ni les modalités ni l'ancienneté.

des talus. Ils sont généralement placés sur une pente boisée et sont couverts d'un toit de tuiles romaines, à double pente. Les murs les plus anciens sont faits de briques d'adobe, la pierre étant plutôt rare malgré la proximité de la montagne. La couverture est doublée d'un calfeutrage de roseaux tressés qui retiennent une épaisse couche d'argile. Elle est souvent prolongée au-dessus de la façade par un petit auvent soutenu par deux murets en avancée sur les murs latéraux. Les charpentes sont solides. Faites de grosses poutres de chênes, elles ont généralement bien résisté au temps et très peu apparaissent détériorées. Les seuls dommages constatés affectent les murs pignons, souvent tombés et reconstruits avec des briques modernes.

Bibliographie

Arias, A. 2005. La presse à cire de Maranchon: Transfert et restauration. *Cahier d'Apistoria* 4A.

Blachère, R. (trad.) 1957. *Le Coran (al- Qor'ân)*. Paris: Librairie Orientale et Américaine.

Bonet Rosado, H. and C. Mata Parreño. 1997. The archaeology of beekeeping in pre-Roman Iberia. *Journal of Mediterranean Archaeology* 10.1: 33-47.

Brinkmann, W. 1938. *Bienenstuck und Bienenstand in des romanischen Landern*. Hamburg: Hansischer Gildenverlag.

Chevet, R. 2005. *Apicultura tradicional en los alrededores de Borja. Cuadernos de Estudios Borjanos* 48: 271-298.

Chevet R. et F. Rivas. 2005. *Apuntes sobre apicultura tradicional en Aragón*. Zaragoza: Diputación provincial de Aragón.

Crane, E. 1999. *The World History of Beekeeping and Honey Hunting*. London: Duckworth.

Kehrle, A. et P. Florence. 1980. *À la recherche des meilleures races d'abeilles*. Paris: le Courrier du livre.

Lemeunier, G. 2006. *Le blanchiment de la cire*. Apistoria 5B.

Monferrer Sala, J. P. 1991. *La miel en la España musulmana (Al Andalus)*. Vida Apicola 46/47: 64-68; 24-28.

Pallaruelo Campo, S. 1996. Arnales. *Revista del centro de estudios de Sobrarbe* 2: 37-46.

Chapter 17

Historical Beekeeping in Northern Portugal: Between Traditional Practices and Innovation in Movable Frame Hives[1]

Teresa Soeiro

CITCEM (teresasoeiro@sapo.pt)

Abstract: Beekeeping is a small-scale, widely spread activity in northern Portugal since medieval times, more prevalent in rural regions than in urban areas. Apicultural products were used for local consumption and trade, festive occasions, gastronomy, homemade medicines, and religious votive lighting. They were also planned and harvested for inter-regional and international trade routes. In the modern era, the availability of and preference for sugar supplanted honey's privileged place. Portugal pioneered this change of habits. We describe how, on the one hand, methods of traditional apiculture, centred on fixed-comb often cork hives, tended to prevail over newer practices rooted in technological advances or scientific knowledge, developed in Spain and other European countries, which only reached Portugal at the turn of the 19th century. Another century passed until movable frame beehives were adopted based on imported technology. We give an overview of the efforts of private individuals, businesses and governmental organizations to implement new technologies.

Keywords: North of Portugal; honey vs. sugar; methods of traditional beekeeping; scientific observation of bees; implementing movable frame beehive innovation

Introduction

In Northern Portugal, as throughout the country, beekeeping is an omnipresent activity, albeit always more rural than urban. Its products primarily served local consumption and exchange, holiday cuisine, home medicine, or the illumination of the divine; but these products have also been important foci of major interregional and international trade. Honey struggled to compete with sugar as a sweetener, so too pure wax with its late substitutes. Beekeeping has survived the passage of centuries with fluctuations in intensity and spatial distribution, has had moments of enthusiastic technical progress (imported) as well as moments of desperation when observed in the *longue durée*. In the current chapter, we investigate the times, methods, and agents of this apicultural knowledge base in Portugal which was often in dialogue but equally as often in confrontation with the idiosyncrasies of local and traditional practices.

Between Honey and Sugar

'Before sugar became so plentiful as it has been since the Europeans have got possession of the West India islands, honey was much more valuable than it is at present, being then the chief ingredient in general use for sweetening every article of food.' With these words Thomas Wildman began his treatise on beekeeping

(Wildman 1768: XIII). Réaumur, the French scholar, had said so three decades earlier, certainly referring to his own country: 'quoique le miel dont elles [= the bees] font chaque année de grande recoltes, ait beaucoup perdu de l'estime où il étoit dans des temps òu le sucre, aujourd'hui si commun, étoit à peine connu, ce miel nous est cependant encore très-utile; et il a des usages par rapport auxquels le sucre ne pourroit lui être substitué, comme il le lui a été pour les confitures. Mais la consommation que nous faisons de la cire, et qui va journellement en augmentant, ne nous permettroit de penser aux abeilles qu' avec beaucoup de reconoissance,' (Réaumur 1740: 211).

Nowadays, it may seem common place, but we need to keep in mind the history of the dissemination of sugar if we want to address the role played by honey in the modern age. The acquisition of this sweet food was one of the aims of beekeeping; the second objective, the supply of natural wax. Portugal, in the tradition of the Mediterranean, already cultivated sugarcane since the Middle Ages in some parts of the Coimbra region and, above all, in the Algarve, having also come into contact with this product on what is now the coast of Marroco. In the first decades of the fifteenth century, when Portugal became a pioneer of Atlantic expansion, it colonized the islands where the plant was taken, and gave rise to a production directed toward international markets. This trade attained remarkable success beginning from the second half of the fifteenth century. The first hydraulic device for processing sugarcane

---

[1] Translated from Portuguese by David Wallace-Hare.

authorized on the island of Madeira dates to 1452, by 1455 as many as 1600 arrobas[2] of this new product would be produced and in 1490 a staggering 80,000 arrobas, an amount that, for several more decades, would continue to grow (Godinho 2008: 307ff.; Godinho 1983, 4: 75-77; Albuquerque 1994: 15; Vieira 2004).

Before these shipments, sugar - rose sugar, Alexandrian sugar, Bugian (mod. Béjaïa) sugar - was consumed in Portugal, imported at least as early as the time of D. Dinis (1278-1282), but at extremely high prices. It was an apothecary drug within the price range of only the largest purses and was a distinction of the royal storehouse (Arnaut 1986: 46). Honey would also reign, being an accessible sweetener, complemented by frequent consumption of fruits, some of which could be dried, such as figs, which concentrated their sweetness, optimized conservation throughout the year and made them suitable for the export trade, much of it directed to Northern Europe (Magalhães 1970 and 1988).

The *alfelo* was a delicacy that seems to have addicted children and adults even in medieval times. The Bluteau dictionary defines the *alfelo* as 'a mass of white sugar made in the form of little stake,' (Bluteau 1712, 1: 243).[3] There also existed a *alfeloa* of sugar cane honey (*melaço*), called *magana* or 'yellow' (*amarela*). It was produced with honey at first instead of sugar, as the following comment made in the assembly of representatives (*Cortes*) on behalf of the people in a document from 1490. The document complains of the ills caused by the *Alfeloeiros*, or confectioners, 'who come from Castile to this Kingdom to find in them harvests (of honey) and shelter, those who are in every way odious, who bring with them arch villainy injurious to all: the first being to make honey more expensive than if they did not seek it and, wherever they go, they make the children cry and ask their father and mother for money to buy the aforementioned alfelo,'(Arnaut 1986: 106-107).[4] The older ones would steal money to buy it or gamble for the alfelo at dice, a practice that was obviously forbidden. In the Manueline Ordinations (*Ordenações Manuelinas*), a printed compilation of the laws of the kingdom (1st edition 1514), the activity of the Alfeloeiro was forbidden to men and reserved for women, whether carried out at home or for sale on the public road (*Ord. Man.* 1984, 5: 302).

The nobles of the kingdom were no less zealous for sweetness. The infante D. Henrique would celebrate Christmas in 1414 with his brothers, offering him 'all the viands of sugar and preserves to be found in the kingdom,' (Godinho 2008: 308).[5]

Among the elites, in the transition of these centuries (XV / XVI), the choice of honey vs. sugar decidedly fell in favour of the latter, as evidenced in the *Livro de Cozinha da Infanta D. Maria* (*Cookbook of the Infanta D. Maria*), the first compendium of Portuguese recipes we possess. D. Maria of Portugal (1538-1577), granddaughter of King D. Manuel I, would have brought this repository of cooking know-how into the service of her family, when, in 1565, she married Alexandre Farnésio, Duke of Parma (Manuppella 1986). There are sixty-four food recipes and three home remedies for the teeth, throat, and burns. Honey is used only in home remedies for the teeth, in a recipe for a dry cake with spices, the *fartes* or *fartéis* (a type of sweet cake), and in the preserve *flor-de-laranja*, in conjunction with sugar. Sugar already appears in dozens of *manjares de carne* (dishes featuring meat), *de ovos* (dishes featuring eggs) and *de leite* (dishes featuring milk), and in preserved fruits.

At the dawn of a new century, Duarte Nunes de Leão, a jurist and scholar, close to the king, characterized with great insight the situation of the production of various types of honey in Portugal, a country with excellent edaphoclimatic conditions but where this traditional substance had been neglected in favour of the consumption of sugar, which arrived in increasing quantities, coming primarily from the island of S. Tomé and Brazil, whence more than 1,000,000 arrobas came by 1620.

To honey (and wax) the author dedicates the twenty-seventh chapter of his *Descripção do Reino de Portugal* (*Description of the Kingdom of Portugal*), one of the first and most popular chorographs in the country, published in 1610 (Leão 1610: 42v). He begins by telling us: 'There is as much honey in this kingdom as is enough for it [i.e.domestic consumption] and to share with its neighbors [i.e. to export]; there is honey in many places so white and so hard that after it solidifies it looks like very white sugar and cannot be removed from containers without first breaking them; and thus it acquires the shape of the vessel it was in.'[6] Further on, he praises the honey of the Lisbon region for the subtle aroma that it takes on from the roses and orange flowers, and that of the Ourique countryside, which appears threadlike and as if made of clarified sugar.

---

[2] A customary unit of weight = 32 pounds (14,688kg).
[3] uma maça de açucar branco feita a modo de paosinho.
[4] 'os quais se vêm de Castela a estes Reinos por acharem em eles colheita, e abrigo: os quais são em toda a parte mui odiosos, e trazem consigo mui ruins manhas mui empecíveis a todos: a primeira é fazerem o mel caro mais do que seria se os i não houvesse e por onde quer fazem causar os meninos chorar, e pedir a seus pais e mães dinheiro para comprarem a dita alféloa.'

[5] 'todalas viandas d'açucar e conservas que se puderem achar no regno.'
[6] 'De mel ha tanta copia neste reino quanta basta para elle e para repartir com seus visinhos, de que em muitas partes o ha tam branco e tam duro que depois que se congela fica parecendo açucar muito alvo e se nam se pode tirar das vasilhas sem primeiro se quebrarem: e assi fica na figura de que era o vaso em que estava.'

The standard was already, in fact, sugar: the more it resembled it, forming cakes or appearing white and limpid, the higher the possibility that honey from the south of Portugal would have to be revalued.

On the other hand, honey from the rest of the country maintained its old rusticity: 'what they call the bush honey, which is rather thicker and whiter, there is an endless supply throughout all parts of the kingdom, wherever there are weeds and heaths, places in which there are many beehive enclosures (*silhas*) and hives. From such locations is obtained some very tasty honey whenever the bush has myrtle, rosemary, or lavender,' (Leão 1610: 42v).[7] To obtain this wealth and commercialize it, it would only be necessary to *assentar os cortiços* that is, to set up the cork hives. But this method of beekeeping had ceased to be practiced with the same zeal as in the past, as there were no buyers. The Portuguese with some pretension would have stopped consuming honey, the predilection for sugar bears this out. New tastes and social mores, which a century and a half after the beginning of the regular shipments of white gold from Madeira had spread even among social groups with smaller resources, especially in urban areas. Thus 'the marmalades that the older generation made with honey, in very honourable houses, now no tradesman at all wants to eat unless it's made from sugar and with a bit of ambergris or musk, because even in *this* aspect of life there is now ambition and matters of esteem,' (Leão 1610: 42v).[8]

This apparent omnipresence of sugar, desired by many and efficacious for those who could get it, entirely characterizes the first cookbook written and printed in Portugal in 1680, which was successively improved and reprinted until the mid-1800s. We refer to the *Arte de cozinha* (*The Art of Cooking*), by Domingos Rodrigues, head cook in the royal kitchen. It ended up amassing about three hundred different recipes (for soups, meat, fish, vegetables, pastries, sweets, etc.). It must be highlighted that in these recipes honey is barely present, mentioned only in three vegetable dishes - eggplants and carrots *de tigelada* (pudding served in a bowl) and *de potagem* (soup)- in which honey balances the acidic character of the vinegar sauce to obtain a bittersweet taste (Rodrigues 1987).

Unstoppably on the increase, the abundance and availability of sugar would provide the Portuguese with a very peculiar relationship with this product,

supplanting honey as a sweetener and in cooking. The situation would be different in a great number of European countries depending on the size and timing of their dominance of colonial sugar territories. Some populations only widely adopted sugar in the nineteenth and twentieth centuries, after growth in its supply resulting from its extraction from beets, a crop that was available to them (Braudel 1992: 191-193).

That said, in Portugal, honey and sugar were not interchangeable in their own unique context, since in Portuguese culture the former retained a high status as a gift of nature and the food of the gods, which in addition to offering the senses a more complicated sweetness also had distinctive nutritional, prophylactic, and curative virtues. The hive still produced pure wax, the only one accepted in the Catholic Church, and the bees themselves represented an exemplary society whose lessons all people should learn, although the interpretation of those lessons varied with time and according to the ideological leanings of the interpreters (Collomb 1981: 33-37).

Concerning Traditional Apiculture in the Modern Age

There is little detailed information able to be gathered concerning beekeeping in Portugal in the modern age. At the time of D. Manuel I, in the *Ordenações Manuelinas* this activity is mentioned but twice concerning beehives, both to condemn bad practices, laws which would be renewed in the *Ordenações Filipinas* (1st edition, 1603). In Title 42 of book 4 there are prohibitions against contracts of *arrendamento* or leasing in cattle raising, contracts made for income or gain, that is, animals consigned by the owner, for a certain amount of time, to the care of third parties, from whom they receive a certain amount, regardless whether earned or not, sometimes with the right to return the asset or equivalent (*Ord. Man.* 1984, 4: 101; *Ord. Fil.* 1985: 880). Title 97 of the 5th book of the *Ord. Man.* condemns the acquisition of hives with the sole purpose of killing the bees to enjoy the wax. It makes no mention of honey, perhaps because the first product was the target of a more intense and widespread trade. The penalty was scourging or exile and a fine of four times the value of the dead hives (*Ord. Man.* 1984, 5: 101; *Ord. Fil.* 1985: 1225).

The renewal of privilege charters (*forais*), also undertaken during the reign of D. Manuel I, is another rather poor source when it comes to notices of payments in kind of honey or wax in the area north of the Douro. Common to the various municipalities (*concelhos*) of this region is the toll amount to be received for the sale or purchase of external products, the tabulation of which was expressed explicitly in several *concelhos*, namely at Guimarães and Miranda, in the provinces of Entre-

---

[7] 'Do que chamam de mato, que he mais grosso e menos alvo, ha infinda copia per todas partes do reino, onde ha mato, he charnecas, em que ha muitas cilhas de colmeas, de que ha algum mui saboroso quando o mato he de murta, alecrim, ou rosmaninho.'
[8] 'as marmaladas que os antigos fazião de mel, em casas mui honradas, não quer agora qualquer macanico comelas, se não de açucar, e tocadas de ambar e almiscre: que tambem nisto ha agora ambição, e pontos de honrra.'

Douro-e-Minho and Trás-os-Montes, respectively. The heading *azeite e mel semelhantes* ('products like oil and honey') grouped goods such as wax, honey, oil, tallow, dry cheeses and salted butter (and pitch, resin, dark pitch, soap, and vegetable tar), with nine reais (a monetary unit) to be paid for a larger load (*carga maior*) (by horse or mule = 10 arrobas). In another section, the same nine reais was worth a load of sugar and all the preserves prepared with it or with honey, spices and apothecary drugs, dye plants, etc. By way of comparison, an equal load of grain, wine, or salt was taxed at one real, while the same load of fruit and vegetables was taxed at a half real; for dried fruits three reais were charged (Dias 1969, 5: 10-12 and 1961, 2: 2-3).

As examples collected from the charters of the interamnense region, between the Douro and Minho rivers, we have the case of a man from Capela / Canelas (Penafiel), who lived in the village of Vilarinho, isolated in the mountains, 'payment from an uncultivated mountain area from a man who works with his children: two canadas of honey,'(Dias 1969: 23).[9] This area in the Mozinho chain remained, until the 20th century, very devoted to beekeeping and featured transhumant or migratory beekeepers.

On the other hand, wax was a product included in farmstead privilege charters (*foros de casais*) at Cabração and Esturãos (Ponte de Lima) and Cabreiro (Arcos de Valdevez). At Canedo (Celorico de Basto), there was a farmstead (*casal*) of the Beekeeper (*Abelheiro*) (Dias 1969, 5: 23, 92, 111, 180). On the left bank of the Tâmega, in the province of Trás-os-Montes, each of the four farms (*casais*) in the village of Barrondinho (Ribeira de Pena) delivered a pound of wax. Not far away, at Montalegre and in the lands of Barroso, in addition to the share of the universal tax amount that was assigned, each neighbour paid an additional 108 reais if he had animals, equivalent to or in excess of four head of cattle, four beasts (horses and mules), forty sheep, forty goats *or* forty hives (Dias 1961, 2: 75, 53).

Moving to the east, both in town and near the outskirts of Mirandela and at Vilas Boas, the Crown possessed one third of the tithes of the churches. Among the payments in kind that composed this tithe wax was specified as 'dry wax that they call the escarça or estinho.' In the municipality of Macedo de Cavaleiros in the Tua valley there was once a toponym called julgado de Corticos ('court of cork hives'), now a parish (Dias 1961, 2: 15, 20, 77). In the Lamego area, the heart of the old Douro wine region, Rui Fernandes reports that in 1531-1532, 'there is enough honey for the land where it is collected, since

there is no great quantity of it; and of the *agoa rosada* 80 almudes' (Fernandes 2001: 87).[10]

It was mentioned earlier that in the transition to the 17th century, Duarte Nunes de Lião synthesized beekeeping practices in Portugal. These practices were certainly the result of an empirical knowledge with clear inter-generational transmission in the apparently accessible and very common establishment of cork hives (*corticos*) and their eventual grouping in *silhas* (a set of cork hives in an open space).

Since the Middle Ages, in Beira Baixa and the interior of Alentejo, there has existed a local apicultural system called the *malhada*, a term referring to the private and transferable usage of certain areas of the common land (± 1500ha) for the controlled establishment of a number of hives, which in 1368 featured 400 hives per *malhada*, while in 1890 many *malhadas* exceeded 2000 hives (Silbert 1978: 458).

On the other side of the border, in the newly conquered lands, especially in the mountain areas with little human presence in Castilla-La Mancha, Extremadura and Andalusia, the *majada* would have been the model for intense beekeeping. The *majada* system was the basis of important medieval trade, being subject to powerful brotherhoods, such as the Hermandad Vieja de Colmeneros (the Old Brotherhood of Beekeepers) of Toledo, Talavera and Villa Real (Guzmán-Alvarez 2006: 37ff), or that of the wide territory of Seville, which extended to the Portuguese border, with its *ordenanzas de colmenería* (ordinances of beekeeping), from the time by Alfonso X (Méndez de Torres 2006: 117ff).

On the other hand, apiculture seems to have received much more attention in the neighbouring country, where at least since the 1500s treatises on beekeeping were written, either on beekeeping alone or included as a section on beekeeping within larger agricultural manuals, continuing classical knowledge, transmitted and augmented by Muslim scholars of Hispania (Crane 1999: 214ff). In Portugal, Gabriel Alonso de Herrera's work on agriculture, *Obra de agricultura copilada de diversos auctores* (1513), contained seventeen pages (fl.126v-134v) dedicated to beekeeping, blending what the classical authors had written with traditional practices and others more or less invented. Two centuries later, Theobaldo de Jesus Maria continued this tradition, writing a manual called the *Agricultor Instruido* (The Educated Farmer), published in Lisbon in 1730 according to Inocêncio (Silva 1862, 7: 300). This work was directed at the Portuguese public and features guidelines on beekeeping already completely

---

[9] 'paga do monte maninho que lavra com seus filhos duas canadas de me.' A *canada* is a liquid volume = 1/6 pot (1.4 litres).

[10] 'de mel pode aver quanto abaste para a terra de sua colheita, dado que não hé muita cantidade; e de agoa rozada LXXX almudes.' The *almude* is a liquid of 16.8 litres.

anachronistic or even absurd (Jesus Maria 1790: 128-133), as Francisco de Faria e Aragão ironically highlighted in the preface of his own *Tratado* (Aragão 1800: VII). Aragão's work was pioneering in the country and much more up-to-date due to the theoretical and experimental knowledge he had acquired in Central Europe and the international bibliography he engaged with.

Miguel Agustin, in his *Libro de los secretos de agricultura, casa de campo, y pastoril*, dedicates the sixteenth chapter to beekeeping (1st ed. 1617; Agustin 1722). There have been copies of this work in the National Library of Portugal since the 1625 edition. It served as a paradigmatic reference work for Francisco de Faria e Aragão for pointing out how erroneous some interpretations of the classical authors were about bees, and indeed detrimental in their constant repetition, taking as an example the question of the spontaneous generation of new swarms, which those authors recommended took place following the decomposition of the body of an ox that was sewn up and beaten to death (Aragão 1800: 28).

Contrary to the generalist works cited, it is perhaps the case that the treatises specifically on beekeeping written in Spain, such as that of Luiz de Méndez de Torres (1586), may simply not have been accessible in Portugal, even though this work was later integrated under the aegis of Alonso de Herrera, in a widely-circulated edition (Alonso de Herrera 1790). Méndez de Torres was one of the first researchers to recognize the exclusivity of the queen bee in the laying of eggs to furnish the hive with all its members. The *Perfecta y curiosa declaracion, de los provechos grandes, que dan colmenas bien administradas* (*The Complete and Inquisitive Declaration of the Great Benefits that Well-Managed Hives Yield*), by Jayme Gil (Gil 1621), who calls himself an *labrador experimentado* ('experienced professional'), despite its practical character, would also seemingly fail to gain followers in Portugal. Other beekeeping texts of the time were handwritten, cf. that of Francisco de la Cruz (Padilla Álvarez 2002), and, in this condition, even more inaccessible.

The Cork Hive, the *Silha*, and the Traditional Extraction of Honey and Wax

The most typical housing furnished to the swarm by those interested in controlling these insects - for our first dictionary entry defines the bee as a 'species of big fly [...] industrious artificer of honey, and wax' (especie de mosca grande [...] industriosa artifice do mel, e da cera) (Bluteau 1712, 1: 23) – would have been, in fact, the cork hive (*cortiço*), a term synonymous with beehive, which he explains by saying 'the cork house, in which bees make honey. There are sticks placed crosswise, to support the bees, and the combs,'(Bluteau 1712, 2:

578).[11] The definition remains fully up to date, and so specimens continue to be made and used alongside movable frame hives specifically to promote swarming.

These hives consist of a cylinder of cork (*sobreiro*) bark made from a single piece purposely collected from the tree (or from joined plates), about 50-70 cm high and with a variable diameter. The top cover is a disk of the same material. As bees do not like metal, the joints were made with clamps or pins of hard wood (e.g. heather, rock rose), so-called *saraços* or *bios*. The joining of the plates could also be sewn. A wider slab of stone was placed over it, sheltering it from the rain and giving weight and stability to the whole structure. It was sometimes covered with a conical thatched 'hat' called the *corpelas* (Dias 1948: 101; Lorenzo Fernández 1979: 329). At the top and bottom of the hive wall, flight holes were opened for the passage of bees. Inside the hive, two or three crosses (*trancas* or *juízes*) of thin peeled olive branches supported the construction of the combs. A new cork hive might be completed by filling all the cracks of the built structure with clay or dung (*embarrar*). The whole was situated upon a smooth stone slab, called an *alvada*.

The treatise of Alonso de Herrera, from the beginning of the 1500s, had already assessed the cork hive made from the bark of the cork oak as the most appropriate for the swarm, as it obviates great variations in temperature, although he identified other arrangements for regions where this tree was not present, such as plank hives, like chests, wicker hives and coil hives, which must be covered with mud and dung for greater insulation, or ceramic hives, which are much worse because of their lack of thermal insulation (Alonso de Herrera 1790: 235-236). Méndez de Torres (1586) depicts an apiary with aligned upright cork hives and shows a similar methodology (Crane 1999: 216), which in Spain was not as obvious as for Portuguese beekeepers, since in the different regions of the neighbouring country there were highly varied natural conditions and cultural practices (Crane 1999: 312ff.), and not the near uniform conditions attributed to Portugal, with minor variations in border areas, for example that of Vinhais-Bragança where we hear of the use of hollowed out chestnut tree trunks to house bees, a common occurrence in the Astur-Leonese region (López Álvarez 1994: 59ff.).

The critical analysis of the Portuguese cork hive was to be undertaken by Francisco de Faria e Aragão (Aragão 1800: 95-96), who set out to innovate following the European trends of the second half of the 18th century. Although he praised the thermal insulation which the cork hive provides, the author was particularly

---

[11] 'a casa de cortiça, em que as abelhas fazem o mel. Tem huma cruz de páos atravessados, para sustentar as abelhas, e os favos.'

attentive to three of its vulnerabilities. The first would be the choice to not open flight holes, which led the beekeeper not to set the cork hive on a slab base, but to place it on small stones to ensure space for the bees to pass, thus exposing it to predators that could breach the whole perimeter, making defence difficult. This assertion leaves us with some doubts, not only because preserved cork hives have lateral flight holes, but also because of the fact that older illustrations depict them too, such as the one that accompanies *cantiga* nº 128 of the Codex Rico do Escorial das Cantigas de Santa Maria, from the second half of the 13th century, or the depiction of the apiary included in the 16th-century publication of Méndez de Torres (Crane 1999: 216).

The second limitation came from the fineness and irregularity of some cork that, even if coated, could leave passages for the *tinha* (bee moth; *Aphomia sociella)*. The stone base also turned into a pitfall, as it would become too cold in winter and too scorching in summer, fatal for bees loaded down with pollen, with increased damage if there were a second slab set atop the hive, placed, as mentioned earlier, to stabilize and protect it from the rain, but that too heated by the summer sun would melt the upper combs.

Thirdly, one still had to count on the difficulty of executing the *cresta*, that is, the removal of the honey from the hives, and the consequences of it, since it often caused the depletion of honey reserves for the winter, with no time to replace them, and even the death of many bees that were clinging to the combs, so contrary to sensitivity and respect that cultured society held for these creatures. Notwithstanding, the worst way to obtain the annual harvest was that which the *Ordenações* had already condemned at the beginning of the 16th century: killing the entire colony with too much smoke or by immersing it in water in order to then be able to dismantle the contents, without risk of stings, a procedure that continued in the South of Portugal until the 20th century.

There wasn't a beekeeper (*abelheiro*) who wanted it; indeed it was thought that colonies received certain individuals better because they evaluated them as good people and could communicate with them. This welcoming reception became a very important quality, since the beekeeper's protective material was reduced to some old cloth wrapped around his hands / arms and a handkerchief placed under his hat, to cover his face and neck.

As to the location of the cork hives, they could be isolated individually or gathered in small clusters near the house and on the border fence or walls of farm fields, or grouped in large clusters, forming *silhas*, which for Bluteau were 'many hives, put in order' (*muitas colmeias, póstas por ordem*) (Bluteau 1712, 2: 313), as we see in the

aforementioned engravings, positioned under a natural slab or boulder. *Silha* does not necessarily signify an apiary surrounded by a high protective wall, as the best specimens were erected in the 18th and following centuries (if not before), in areas far from villages and at the mercy of large predators, similar to the *alvarizas* of the interior of Galicia and the *cortines* from western Asturias (*Açafa on line* 3, 2010; Caamaño Suárez 2003: 398ff.; López Álvarez 1994). Contador de Argote, in the detailed work which he dedicated to the archbishopric of Braga, published c. 1732, described one of these *silhas* in a wild area of Gerês, closed by a wall of about 4.60m, slanted outwards to the top in order to make it difficult for animals to scale, namely bears (Argote 1738: 390-392).

Near the end of winter and in the spring, when the hills were covered with honey flowers, at different times, depending on their exposure to the sun and the dominant species, the swarms could enter into transhumance, shifted by their owners to facilitate access to these pastures (Menezes 1900: 769). In poor roadways, the hives were placed upon the back of solipeds (horses, asses, mules), on their own wooden frames. If roads were suitable, then the cork hives could be loaded into ox carts for the journey, with the constant hope that there would be no shocks to alarm the bees and make them attack. On the outskirts of Porto, a bee cart (*carro de abelhas*), with twelve cork hives, constituted a counting unit (Barreiro 1922: 522.); with the preparation of a hive for sale and the handing over of a hive to a new owner with its respective colony as a recognized economic activity (Sequeira 1895: 218).

Of the work carried out at the apiary (*colmeal* or *covão*), the most important and frequently assigned task given to professional beekeepers or *abelheiros* would be the harvesting of honey and wax, made with the *crestadeira* (uncapping or harvesting) knife and the help of the smoker (in Entre-Douro-e-Minho this consisted of oak apples or old burlap) blown in the direction of the colony. This harvest took place at two different times of the year: first at the *cresta* (first harvest of honey), at the feast of Santo António (13 June), which yielded the best honey. The second collection time occurred around the time of the feast of S. Miguel (29 September) (Bluteau 1712, 2: 608; 1712, 3: 322). The term *estinha* was used in the aforementioned 16th-century charter of Mirandela, as well as the term *escarça*, which referred to the removal of old combs from the bottom of the cork hive. To uncap the comb (*crestar*), the cap was cut off and the cut made into the new wax combs, laden with honey.

Notwithstanding, it was not only the harvest calendar that determined the quality of the product, as this would still be strongly influenced by the way in which the combs were removed. Turning to Bluteau again, he

Table 1. Honey (Mel) and Wax (Cera) Production in the North of Portugal in the Mid 19th Century

District	1852		1862		1872		Observations
	Mel (kg)	**Cera** (kg)	**Mel** (kg)	**Cera** (kg)	**Mel** (kg)	**Cera** (kg)	
Aveiro	7975	7373	10,872	7872	9515	8414	national (domestic) consumption
Braga	16,899	4737	12,509	4056	-	-	
Bragança	22,680	29,249	24,196	16,257	24,807	14,662	various municipalities (concelhos) export wax: 27,511kg in 1852, 7410kg in 1862 and 8175kg in 1872
Guarda	9760	5152	5127	2015	(1873) 7552	(1873) 2445	national consumption; in 1852 Almendra exports barely 35 arrobas of honey
Porto	7522	7048	(1861) 6827	(1861) 2432	8581	5709	national consumption
Viana Castelo	-	-	11,867	3538	4949	1797	national consumption, V N Cerveira exports only 6 arrobas of honey in 1852
Vila Real	8798	14,276	10,461	7368	18,575	17424	Montalegre exports a small portion of honey and enough wax in 1862; Montalegre and Valpaços export wax in 1872
Viseu	15,741	8695	15,281	9970	-	-	national consumption in 1852, 6 concelhos did not have hives

Source: PT/AHMOP/DGCAM/RA-3S/012 - Mapas de produção de mel e cera, 1852-1859; PT/AHMOP/MOPCI-DGCI/RA/008 - Mapas de produção de cera e mel 1862-1873; *Relatório da Junta Geral do districto do Porto, 1861.* Ramos and Pita 1997.

points out that Francisco de Faria e Aragão yields little information in this regard, commenting that, 'generally speaking, there are two types of honey: white honey and yellow honey. The first is the best, the most attractive and the most pleasant to the taste. It is extracted without using fire, using mats or sheets, from which it flows into clean vessels, which are placed underneath, and where it solidifies. They call this virgin honey; this is the best to eat. Yellow honey does not drain from the combs automatically, but needs to be squeezed over a fire after being wrapped in linen cloth bags with the wax remaining in the bags. White honey is a chest remedy that produces saliva, facilitates breathing, dissolves thick mucus, and relaxes the stomach. Yellow honey is a disinfectant, laxative, digestive, attenuating [of evils] and curative,' (Bluteau 1716, 5: 401).[12]

Beginning in the 1850s and for just over two decades, we come to possess a series of maps detailing honey and wax production collected by the official services, which indicate quantitative data about municipalities,

gathered by district, material that has already been studied at national levels (Ramos and Pita 1997). We use here only the values related to the districts of northern Portugal, where most of the harvest was consumed in the region or in the country, relegating the export of wax, with some significance, to border areas of Trás-os-Montes, in close contact with the professional chandlers (*cereeiros*) of the interior of Galicia and Castile and Leon (López Álvarez 1994). The major national producers were, at the time, the inland and bordering districts of Castelo Branco, Portalegre, Évora and Beja. Portugal would still export a good deal of wax received from its African colonies, which allowed it to supply the markets of Northern Europe and, later, the United States.

From Scientific Observation of Bees to the Dissemination of Movable Frame Hives

In contrast to what was said above, at the turn of the 19th century, some well-known European agricultural manuals were translated and adapted to the realities of Portugal, a few years after their original editions, with chapters dedicated to beekeeping. A product of this innovating movement, for example, was the printing, in 1801, of a version of the *Manuel pratique du laboureur, suivi d'un traité sur les abeillies*, by Chabouillé (1794), prepared in Brazil by José Ferreira da Silva (Chabouillé and Silva 1801), and that of Rozier's much more famous work, entitled *Cours complet d'agriculture* [...] or *Dictionnaire universel d'agriculture* (1781-1805),

---

[12] 'Gèralmente fallando, ha duas castas de mel. Mel branco, e mel amarello. O primeiro he o melhor, o mais fermoso, e mais agradavel ao gosto. Tira se sem fogo, sobre esteiras, ou em lançoes, dos quaes se destilla em vasos limpos, que ficam por baixo, onde se congella; chamão lhe mel virgem; este he melhor para se tomar por boca. O mel amarello, não se destilla por si mesmo, mas depois de envolto em saccos de pano de linho, esprime-se ao lume, e fica a cera nos saccos. O mel branco he peitoral, provoca a saliva, ajuda a respiração, dissolve a pituita grossa, e relaxa o ventre. O mel amarello he detersivo, laxativo, digestivo, attenuante, resolutivo.'

the first volume of which was printed by Francisco Soares Franco at the Real Imprensa of the University of Coimbra in 1804, *com muitas mudanças principalmente relativas a theoria, e ao clima de Portugal* ('with many changes principally relating to theory and to the climate of Portugal') (Franco 1804). In *Gazeta do Campo* (1805) the short and less updated *Guide complete pour le gouvernement des abeilles pendant toute l'anné* (1774) written by Daniel Wildman was translated. With greater or lesser academic and social recognition, all these texts represented a qualitative advance when compared to the aforementioned *Agricultor instruido* of Teobaldo Jesus Maria (1730), or the *Livro da agricultura ou agricultor instruido* (before 1764), which followed the previous work, which João António Garrido, a native of La Rioja living in Lisbon, was reprinting, corrected in successive editions, until at least 1837 (Garrido 1804).

The first two proposals, based on a better scientific and experimental knowledge of entomology, also followed the humanist paradigm for beekeeping, developed from the mid-1600s (Crane 1999: 405ff.), which advocated respect for a hard-working and useful being for humans, which should be offered a comfortable dwelling and one which facilitated the profitable harvest of honey and wax with minimum disturbance to the colony. We were at the time of wooden box hives (*colmeias de alças*), vertical and horizontal in orientation, complex in design but easier to manage than fixed hives because of their partitions. They could boast the quality of luxury quarters, a shrunken down version of buildings, in short, bee houses, in which observation devices were built for instruction and the enjoyment of the affluent classes (including the young and women) and garden ornaments (Thorley 1744; Bromwich 1783; Nutt 1832).

Into this discourse can be inserted the *Tratado historico, e fysico das abelhas* (*Historical* and *Scientific Treatise on Bees*), by Francisco de Faria e Aragão (1726-1806), published in Lisbon in the year 1800, with a partial pre-edition in the periodical *Palladio portuguez, e clarim de Pallas* (1796). However, although it was presented as the first original study by a Portuguese author, and not a translation, it was the result of an investigation carried out when the author, a native of Ferreira de Aves (Sátão, Viseu), lived in Germany and Austria, after the expulsion of the Companhia de Jesus (Society of Jesus) from Portugal in the time of D. José e de Pombal (1759?).

It is his scientific methodology and this knowledge of the *estrangeirados* that it transmitted, as he also did in other domains, with studies on gnomonics and electricity (Bernardo 1998: 20).[13] In his *Tratado*, almost everything

would seem new to the few national beekeepers capable of reading it, if they did not have access to similar foreign works. Specifically regarding the installation of the swarm, he critically reviews the faults of the traditional cork hive and the injurious 'fixist' practices,[14] contrasting them with modern European proposals, which he adopts, transmitting detailed instructions for the construction and management of a vertical hive of wooden boxes (Aragão 1800: 101-104).

Returning to Portugal in 1783, he settled in his native land, where he set up the small physics lab he brought from Germany and started a botanical garden, having also had built a *casa de madeira e envidraçada* (a wooden house with glass windows) in which to put swarms so he could continue studying its behaviour (Silva 1870, 9: 287), a subject he had already addressed in the aforementioned work, recommending the placement of an observation port on the hive wall (Aragão 1800: 107)).

Before him, Teodoro de Almeida (1722-1804), from the Congregação do Oratório, listened to by King D. João V in public awareness sessions on the natural sciences, mentioned, in volume 5 of the *Recreação filosofica ou diálogo sobre a filosofia natural* (*Philosophical Recreation or a Dialogue on Natural Philosophy*) (1761), the experience of observing bees through a glass-walled hive. As he affirms, despite recognizing that his knowledge rested on book learning: 'For some time I'd been curious about having a cork hive with observation ports, in which I observed many things that countryfolk were ignorant of.'[15] He explains further: 'I made a square box, long and with enough capacity to resemble a cork hive; on the four walls I had four glasses installed: on the outside I protected them with individual wooden doors.'[16] When the doors were closed, the hive was dark and the bees worked at their leisure. They were only opened to observe them for short periods of time (Almeida 1819: 186,188). This scientific practice of observing bees dates back to the middle of the 17th century (Crane 1999: 380).

As a final example, we select a late *inventor*, the Franciscan Brother Manoel da Senhora das Dores Penella, who is certainly an author least afflicted by self-praise and his work the *Obra prima: Nova descuberta, abelhas em habitação de vidro sem se occultarem e memoria do que fazem as abelhas em enxame dentro de sua habitação* (*The Masterpiece: A New Discovery, Bees in a Dwelling of Glass*

---

[13] *Estrangeirados* were Portuguese intellectuals in the 18th and 19th centuries who, often after sojourns abroad in Europe and elsewhere, returned to Portugal and introduced new scientific and philosophical ideas.

[14] The term *fixista* refers to the use of fixed comb hives as opposed to movable frame hives, cf. the opposite, *mobilista*.

[15] 'Eu já tive por alguns tempos a curiosidade de ter hum cortiço de vidraças, no qual observava muitas cousas que ignorão os camponezes.'

[16] 'Fiz huma caixa quadrada, comprida e com bastante capacidade para supprir as vezes de hum cortiço; nas quatro paredes fiz pôr quatro vidros: por fóra delles os defendi com suas portas de páo.'

Without Being Hidden and a Record of What the Bees Do in the Swarm within Their Quarters). Edited in 1823 and offered to Infante D. Miguel, president of the Royal Academy of Sciences, in this study Manoel Penella reports the continual risk to which he was exposed until, in 1815, landing on 'a safe and permanent way of introducing a swarm of bees into a perfectly transparent, white glass sleeve; and another easy and constant means of preserving and propagating, without everything being hidden so far as can be seen through the transparent pure glass, as when the swarm entered it.'[17] We are not able to gauge the public interest in this, but the *habitação de vidro*, with its respective colony, would end up being sent to the Jardim da Bemposta, where it remained exhibited to the public. The masterpiece was thus available, with the discoverer not caring to use it to deepen the investigation, since some of the most debated issues in scientific circles, such as the reproduction of bees, seemed impractical and, for unknown reasons, unappreciated.

A more open and pragmatic mind in this respect characterized Francisco Pereira Rubião, from Trás-os-Montes, a Bachelor of Medicine from the University of Coimbra, a rural proprietor and industrial entrepreneur, who was forced into exile during the Civil War that opposed absolutist and liberal monarchists (Queiroz 2005: 16). In Paris, he became friends with the director of the *Journal des connaissences usuelles et pratiques* and through him came into contact with the Nuttian hive, invented by the Englishman Thomas Nutt, which offered high profitability (110kg of honey and wax / year) and easy handling, without killing any bees, nor needing fumigations and cutting (Nutt 1832). Armed with this information, he provided it to his fellow citizens, advising them on the difficult choice among the range of models available. In the publication, printed in Paris- *Colmea Nuttiana, importada de França*, he justified the option, and instructed how to build and manage the hive and offered to supply it, manufactured by himself.

The impact of these new beekeeping practices on the productivity of apicultural workers in the country is obscure, but it is anticipated to be very limited, more a curiosity and prestige item for the enlightened owner than appealing to the thousands of farmers used to dealing with cork hives. As the Viscount de Carnide, founder of the Royal Association of Portuguese Agriculture and the *Gazeta dos Lavradores*, would already say in the 1880s, technical and scientific advances: 'unfortunately were little attended by beekeepers

of that time and by those who followed,' (Viscount of Carnide 1888).[18] His assertive hope lay in greater repercussions of the recent 'mobilist' beekeeping. In 1884, at the Lisbon Agricultural Exhibition, it appeared as a novelty presented by *Quinta Regional de Sintra*, different equipment (i.e. imported) necessary for this renovation (*Catalogo* 1884: 608-609). However, since the middle of the century, the liberal state implemented statistical collection, the results of which are related to production (honey and wax) in Northern Portugal, and are summarized in Table 1.

The Proposition of the Movable Frame Hive in Portugal

Thus, in Portugal movable frame beekeeping did not result from an internal process of knowledge construction and experimentation, but seems rather abruptly imported, along with its scientific support, the technical means to carry it out and everything else that befit the contemporary beekeeper, wherever they existed. It flourished in the North of Portugal, in the final decades of the 19th century (Menezes 1900: 774) and has as its top proponent the book by Eduardo Sequeira, from 1895, with the title *As abelhas: Tratado de apicultura mobilista (The Bees: A Treatise on Movable Frame Beekeeping)* (Sequeira 1895). However, considering the dedications in the book, we are conveyed to a circle of enthusiastic pioneers in this respect: Alberto Velozo de Araújo, Bento Carqueja and Cristiano Van-Zeller.

The first, in the text of a conference given at the Ateneu Comercial do Porto in 1896, declared himself a mobilist (advocate of movable frame hive beekeeping) five years prior, because 'it is modern and rational beekeeping, the only method of keeping bees that I can and must acknowledge and one that is used in all the civilized nations of the world,' (Ajaújo 1896: 22).[19] Revisiting the situation in the various countries, the question arose for Ajaújo: 'Now do we, gentlemen, have beekeeping? We have to have it. Until recently, there existed among us only cork hive keepers who knew how to steal the laboriously obtained harvest from the bees in a brutal way and often put the finishing touches on their rudeness by a savage act – they would suffocate their own bees!'(Ajaújo 1896: 31-32).[20] This crime was still practiced in the South of Portugal, without the Royal Association, the Institute of Agronomy and all

---

[17] 'Hum firme e constante meio de introduzir hum enxame de abelhas em huma manga de vidro branco, e perfeitamente transparente: e outro facil e constante meio de se conservar, trabalhar, e se propagar, sem que se occulte, quanto se possa observar pelo vidro transparente, e puro, como quando nelle entrou o enxame.'

[18] 'Infelizmente foram pouco attendidos pelos apicultores d'aquelle tempo e pelos que se seguiram.'
[19] 'É a apicultura moderna e racional, a única cultura de abelhas que posso e devo admitir e que se usa em todas as nações civilizadas do mundo.'
[20] 'E nós, senhores, temos apicultura? Havemos de a possuir. Até ha pouco tempo só havia entre nós possuidores de cortiços que sabiam roubar ás abelhas d'um modo brutal a colheita laboriosamente obtida e muitas vezes completavam a rudeza por um acto selvagem - asphyxiavam as abelhas!'

Table 2. Portuguese Hives (1934)

MOVABLE FRAME HIVES	Internal Dimensions (length x width x height) in mm		Number of Frames	Frame Dimensions (length x width x height) in mm	
	body	box		body	box
Prática	450 x 370 x 310	450 x 370 x 160	10	420 x 280	420 x 150
Reversível	380 x 380 x 246	380 x 380 x 245	10	340 x 202	340 x 202
Lusitana	380 x 380 x 313	380 x 380 x 165	10	340 x 275	340 x 128
Mira	370 x 300 x 220	370 x 300 x 220	10	267 x 195	267 x 195
Nunes da Mata	430 x 382 x 225	430 x 382 x 225	12	340 x 195	340 x 195

Source: Graça 1933:44.

the institutions of the capital having endeavoured to prevent it through the introduction of mobilism.

Velozo de Araújo continues with praise for Bento Carqueja, a journalist and professor at the Escola Normal and the University of Porto, who had operated a *revolução pacífica* by installing didactic apiaries. Cristiano Van-Zeller, a business owner and winemaker, would be the first to own movable frame hives in Portugal. Another early adopter was Abílio Torres, who was a doctor at the Vizela hot springs, in the park of which he installed a beehive for public education. But Eduardo Sequeira would always be considered the major proponent for this early stage of widespread movable frame hive adoption in Portugal, for his books, his work in the administration of the beehive at the Municipal Garden and for his promoting efforts that led the Companhia Horticolo-Agricola to set up a warehouse for movable frame hives and beekeeping equipment, which he advertised widely to the public and to those who could sell them directly or ship them by train.

This initial enthusiasm led to the production of versions of this imported equipment, which supposedly modified them to the natural and handling conditions existing in Portugal and reduced costs, but it was also a way around the issue of patents. For this purpose, for example, Eduardo Sequeira made some modifications to the Gariel hive, which started to be manufactured for the Companhia Horticolo-Agricola, Quinta das Virtudes (Porto), marketing it as the Sequeira hive, suitable for small ventures, for placement in gardens and usage by ladies (*senhoras*) (Sequeira 1895: 226). The Castro Portugal, Lusitana and Prática hives derive from the Dadant-Blatt model and also belong to northern companies (in Porto and Famalicão); while the Mira or Alentejo hive and the Nunes da Mata hive became more common in the South. The Reversível hive follows the American Root hive (popularized by the business owner Amos Root), with reduced dimensions (Menezes 1900: 776ff; Graça 1933: 45-46).

Manuel Tavares de Sousa, creator of the Lusitana hive, after a journey to the United States, set up a small semi-artisanal factory at his residence at Rio Mau (Penafiel),

where he replicated various apicultural tools and equipment, an initiative that is at the heart of modern industrial units of this branch in the region (Soeiro 2006-2007: 132ff.). And this is not a unique example.

This progress dynamic was also a source of literature for different audiences, from scientific books to small advertising brochures, clearly indebted to the foundational work of Eduardo Sequeira. However, few people like him had the objectivity to assume that a prompt move to mobilism would not, in the short term, be an obvious reality, as confirmed by the investigation to follow. Hence, he suggested that 'in order for the transition to movable frame hives not to be made – the only rational method and the one which yields a favorable result of more than fifty percent –traditionalist farmers should use at least fixed straw hives, wider than the abominable cork hive,'(Sequeira 1895: 219).[21] These hives were large and capable of being supered with boxes, something unknown in Portugal. Nonetheless the 'abominable' traditional cork hive *could* also be improved, if partitioned into three overlapping bodies, which were added to fit the honey season. According to Sequeira, 'this would already be a great step forward and a good learning experience, which would make the management of movable frame hives much easier, which will soon need to fatally replace fixed hives, as they are doing in all cultured countries of the old and new world,' (Sequeira 1895: 221).[22]

State Action in Favour of Movable Frame Hives

The traditional beekeeping practices, dispersed and grounded in routine, will, in reality, prevail until the middle of the 20th century, although signs of change are obvious. This change was also supported by the

---

[21] 'Para não ser feita de um salto a transição para as colmeias moveis - o unico methodo racional e que dá um resultado favoravel de mais de cincoenta por cento, - devem os lavradores mais rotineiros utilisar ao menos as colmeias fixas de palha, mais amplas que o abominável cortiço.'

[22] 'Isto já seria um grande progresso e uma bôa aprendisagem, que tornaria muito mais facil o manejo das colmeias moveis, que em breve tempo hão-de fatalmente supplantar as colmeias fixas, como o estão fazendo em todos os paizes cultos do velho e novo mundo.'

Table 3. Number of Cork Hives (Cortiços) and Movable Frame Hives (Colmeias), Per District, in the 20th Century

District	1933-1935		1952-1954		1968	
	Cortiços	Colmeias	Cortiços	Colmeias	Cortiços	Colmeias
Aveiro	20,784	1414	21,631	3563	17,872	4678
Beja	34,926	309	19,452	9315	23,137	5037
Braga	14,283	1132	21,785	2344	19,280	2394
Bragança	27,672	187	22,356	6947	15,176	7237
Castelo Branco	41,053	481	29,806	12,965	29,801	9497
Coimbra	33,555	460	23,912	8490	12,829	4567
Évora	23,593	596	23,300	9478	15,593	7542
Faro	24,996	438	20,184	5296	19,287	9964
Guarda	16,712	767	14,196	3662	8431	2597
Leiria	18,578	151	14,291	2116	7507	2619
Lisboa	4319	320	4227	2173	1402	1285
Portalegre	18,801	1138	15,923	6197	17,173	6718
Porto	16,684	1438	20,166	2507	17,954	4051
Santarém	28,171	827	19,756	11,075	11,297	7648
Setúbal	18,374	160	12,897	9132	4261	4248
Viana do Castelo	31,182	1085	19,608	3941	14,227	3676
Vila Real	27,481	436	26,293	5349	13,727	7734
Viseu	31,608	1021	31,935	7374	21,586	5477
TOTAL	432,772	12,360	361,718	111,924	270,540	**96,969**
Variation			(- 16,42%)	(+ 805,53%)	(- 25,18%)	(- 13,36%)
	445,132		473,642	(+ 6,40%)	367,509	(- 22,41%)

Source: *1º Inquérito Apícola Nacional, 1933-1935*; *Inquérito às Explorações Agrícolas do Continente, 1952-1954*, 1, 1953: 270-271; 2, 1954: 438-441; 3, 1955: 190-191; *Inquérito às Explorações Agrícolas do Continente, 1968: explorações de 20 e mais hectares: Continente, 1971: 26-27*; *Inquérito às Explorações Agrícolas do Continente, 1968: explorações de menos de 20 hectares: Continente, 1973: 26-27*.

State which had come to value beekeeping and its main products (Menezes 1900). The State reserved a place for it in agricultural exhibitions, in advertising new technical means available for farming, in academic studies on the national economy, in promoting beekeeping associations, in public demonstrations of the usefulness of this *indústria caseira*, demanding neither in space required nor investment, simply requiring dedication, something well within the reach of the small farmer and could even profitably occupy women, without removing them from home.

Among state interventions concerning beekeeping during the period of government of the Republic (1910-1926), the publication of Law No. 149, passed May 1, 1914, approved the creation of *postos de apicultura* (offices of apiculture), nurseries containing bee forage plants (bee gardens) (*plantas melíferas*), and customs protection and the granting of tax exemptions for beekeeping. These measures resulted from the efforts of Senator José Nunes da Matta, a well-known supporter and proponent of mobilism (Matta 1915).

The support network on the ground came into effect later, in the transition from the Dictatorship to the Estado Novo (decree nº 20.417 of the 20th of October

1931), with the creation, in 1932, of a dedicated service within the Ministry of Agriculture, at the Instituto Superior de Agronomia, based in Tapada da Ajuda (Lisbon). It consisted of the Posto Central de Fomento Apícola (Central Office of Apicultural Promotion) and the Comissão Central de Apicultura (Central Commission on Apiculture), which brought together beekeepers, technicians and representatives of state economic organizations. It had widespread delegations in the country, acting in close proximity, such as the Regional Commissions and new cooperatives / beekeepers' associations, which the government protected, rewarding the first twenty to be formed with a subsidy of 200 movable frame hives.

For a synthesis of the situation on the ground over the course of the 20th century, we have at our disposal the results of state surveys, although they reveal different formulations and response rates, some aimed specifically at the sub-sector of beekeeping and others targeting all agricultural and husbandry activities. The most significant – *1º Inquérito Apícola Nacional*, would come out shortly after the installation of the aforementioned state service, with the work going on from 1933 to 1935. In the following decades, the results were updated through the answers to questions that are part of the series *Inquérito às Explorações Agrícolas do Continente*, from which we selected the one carried out in 1952-1954 and one from 1968, reflecting the beginning of the intense abandonment of the fields, the combined result of migration to the cities in search of employment in industry and the tertiary sector, coupled with the exodus towards European countries with more dynamic economic growth.

The results of 1935 show that, when beekeeping services were being installed, the national average still ranged between 1 movable frame hive to 35 cork hives. That said, this figure conceals very different situations at the micro level of districts and municipalities, requiring further detailed analysis. Indeed, for the former, the ratio varies between values less than 1 movable frame hive to 15 cork hives in Porto, Braga, Aveiro and Lisbon, slightly surpassing it in Portalegre, areas with committed proponents of mobilism, to more than 1-100 in Setúbal, Beja, Leiria and Bragança, and the Northeastern district of Trás-os-Montes, where it would be 1 in 148. In 1905, José Augusto Tavares, abbot of Carviçais (Torre de Moncorvo), noted that in this municipality of Bragança there were still no movable frame hives (Teixeira 1905-1908: 636).

And so, considered synoptically at the national level, the cork hive clearly prevailed, from which, on average, 2.5 kg of honey and 2 kg of wax per year were obtained, whereas with a movable frame hive, at least 10 kg of honey were obtained for the same period. According to Luiz Quartin Graça, who directed the work of the *Inquérito* (Graça 1935a: 15), the value of the swarms added to the potential of the new technical means in the possession of beekeepers if utilized would have amounted to 18,000,000.00$ escudos, while that of annual production would be around 8,000,000.00$, as it was estimated at 1,200,000 kg of honey and 1,000,000 kg of wax, of the most appreciated yellow colour from Trás-os-Montes (Graça 1935b: 14).[23]

The western or European honey bee (*Apis mellifera*) was then represented in Portugal by four groups with different characteristics, commonly designated as follows: black bees (*abelhas pretas*), docile and zealous for work; brown bees (*abelhas pardas*), resistant to winter; Bizkayan bees (*abelhas biscaínhas*), whose name refers to a region in the Basque country, typical of mountainous areas, resistant to the bee moth; mouse bees (*abelhas ratinhas*), smaller and of a glossy black tone, grumpy but hardworking and with a tendency to swarm, behaviors, according to Graça, similar to those attributed to the groups of seasonal workers that migrated from Central Portugal to the fields of the south, who were known as

mice (*ratinhos*); and lastly Italian swarms (*Apis Ligustica*), which were rare (Graça 1935a: 15).

Melliferous flora, once varied and abundant, were affected by the expansion of cereal agriculture, which has been in a dynamic growth phase since the end of the 19th century, especially in the South and Centre of Portugal. But this paucity of melliferous flora was being partially offset by the advance of forestation and the cultivation of ornamental plants, which became alternative food sources for bees.

The two decades that followed the 1º *Inquérito*, decades of strong population growth and State intervention in the economy, still with great weight in the primary sector, would come to be the moment of the most committed public support for the integration of movable frame hives. Questions about beekeeping, now inserted in the *Inquérito às Explorações Agrícolas do Continente*, which ran between 1952 and 1954, obtained very different results, not only quantitative (for example, the ratio dropped to 3.23 cork hives for each movable frame hive and the number of these grew more than 800%), but also in the geographic distribution of preference for each way of installing swarms. In Porto, Braga and Aveiro the number dropped to 8, 9.3 and 6 cork hives per movable frame hive, which had stood out in the initial adoption of mobilism, trail, with the largest deviation from the national average, together with Leiria (6.75) and, the shortest deviation in Viana, Vila Real and Viseu (4.98, 4.91 and 4.3 cork hives to one movable frame hive), that is, almost the entire North of Portugal. By contrast, the districts of the Centre and South advanced with large productions (Castelo Branco, Coimbra, Évora) and others that had the conditions to invest in technical innovation, such as Santarém, Setúbal and Beja. Yet even in these regions, there are a greater number of farms with only cork hives, compared to those using the two systems or only movable frame hives.

A decade later and the ideal of a stable rural Portugal, of pluriactive farmers committed to meeting the food needs of their families and the nation, was going up in smoke. In the subsector under analysis, the loss in production was over 106,000 cork hives and movable frame hives, corresponding to a drop of more than 25% in the first case, not offset by the adoption of new movable frame hives. While movable frame hives were also reduced by more than 13%. The national average increased to 2.8 cork hives for each movable frame hive and from the comparison between districts, it stands out. The degree of abandonment of this activity in areas previously of major production makes this clear, such as in Coimbra, Évora, Santarém and Setúbal; the technical regression in beekeeping at Beja, as opposed to the advance of mobilism, for example, in Vila Real and Porto; and Braga's resistance to transformation

---

[23] The *escudo* was a unit of currency in use prior to the euro; 1 euro = 200,00 escudos.

Table 4. Farmsteads with Cork Hives and/or Movable Frame Hives by District in 1952-1954

District	Farms with Only Cork Hives		Farms with Only Movable Frame Hives		Farms with Both		
	Farms	cortiços	Farms	colmeias	Farms	cortiços	colmeias
Aveiro	3219	17,161	462	1533	526	4470	2030
Beja	817	10,377	238	2655	596	9075	6660
Braga	5103	19,951	314	968	411	1834	1376
Bragança	1587	15,327	222	1642	667	7029	5305
Castelo Branco	3127	21,971	680	5796	837	7835	7169
Coimbra	3710	18,192	570	2682	787	5720	5808
Évora	695	13,174	110	1508	417	10,126	7970
Faro	2114	16,260	284	1928	453	3924	3368
Guarda	1720	11,316	300	1512	392	2880	2150
Leiria	2562	11,579	184	631	421	2712	1485
Lisboa	823	3327	263	1493	142	900	680
Portalegre	927	9380	188	1975	552	6543	4222
Porto	4662	17,894	308	1019	455	2272	1488
Santarém	2537	12,020	622	3916	1 182	7736	7159
Setúbal	422	5766	138	2334	339	7131	6798
Viana do Castelo	3033	15,760	533	2350	448	3848	1591
Vila Real	2117	20,520	199	1825	499	5773	3524
Viseu	4741	23,478	820	2997	1 016	8457	4377

Source: *Inquérito às Explorações Agrícolas do Continente, 1952-1954.*

(8 cork hives for each movable frame hive), with the addition of movable frame hives being insignificant.

A long road, but as Eduardo Sequeira had predicted, inevitable. Portugal would enter fully into mobilism. This does not mean that at the edge of a field, on a wall, even in apiaries of movable frame hives, we will not still today find, in the North of the country, some cork hive of the old order, intact, a millennial tradition.

Sources

Catalogo da Exposição Agricola de Lisboa em 1884. Lisbon: Imprensa Nacional, 1884.
Ordenações Filipinas. Lisbon: Fundação Calouste Gulbenkian, 1985.
Ordenações Manuelinas. Lisbon: Fundação Calouste Gulbenkian, 1984.
PT/AHMOP/DGCAM/RA-3 S/012 - Mapas de produção de mel e cera, 1852-1859.
PT/AHMOP/MOPCI-DGCI/RA/008 - Mapas de produção de cera e mel 1862-1873.
Portugal: Instituto Nacional de Estatística - Inquérito às Explorações Agrícolas do Continente, 1952-1954. Lisbon, 3 volumes, 1953-1955.
Portugal: Instituto Nacional de Estatística - Inquérito às Explorações Agrícolas do Continente, 1968. Lisbon, 10 volumes, 1971-1973.
Relatório da Junta Geral do districto do Porto, 1861. Porto, 1862.

Bibliography

Agustin, M. 1722. *Libro de los Secretos de Agricultura, Casa de Campo, y Pastoril.* Barcelona: En la Imprenta de Juan Piferrer.
Albuquerque, L. (dir.) 1994. *Dicionário de História dos Descobrimentos Portugueses.* Lisbon: Caminho.
Almeida, T. 1819. Recreação Filosofica ou Diálogo sobre a Filosofia Natural, 6ª ed., tomo 5, Lisbon: Na Impressão Regia.
Alonso de Herrera, G. 1513. *Obra de Agricultura.* Alcala de Henares: Arnao Guillen de Brocar.
Alonso de Herrera, G. 1790. *Agricultura General.* Madrid: Por Don Josef de Urrutia.
Aragão, F. F. 1800. *Tratado Historico, e Fyzico das Abelhas.* Lisbon: Na Offic. da Casa Litteraria do Arco do Cego.
Araujo, A. V. 1896. *A Abelha e a sua Utilidade.* Santo Tirso: Typ. do «Jornal de Santo Thyrso».
Argote, H. C. 1738. *De Antiquitatibus Conventus Bracaraugustani.* Lisbon: Typis Sylvianis, Regalis Academiae.
Arnaut, S. D. 1986. *A Arte de Comer em Portugal na Idade Média.* Lisbon: IN-CM.
Barreiro, J. 1922. *Monografia de Paredes.* Porto: Tipografia Mendonça.
Bernardo, L. M. 1998. Francisco de Faria e Aragão e a electricidade no séc. XVIII. *Gazeta de Física* 21: 19-25.
Bluteau, R. 1712-1716. *Vocabulario Portuguez e latino.* [vol. 1 A]. Coimbra: No Collegio das Artes, 1712; [vol. 2 B-C]. Coimbra, No Collegio das Artes, 1712; [vol. 3

D-E]. Coimbra: No Collegio das Artes, 1712; [vol. 5 L-N]. Lisbon: Na Officina de Pascoal da Sylva, 1716.

Braudel, F. 1992. *Civilização Material, Economia e Capitalismo, Séculos XV-XVIII: As Estruturas do Quotidiano.* Porto: Edições Afrontamento.

Bromwich, B. A. 1783. *The Experienced Bee-keeper, Containing an Essay on the Managemente of bees.* London: Printed for Charles Dilly.

Caamaño Suárez, M. 2003. *As Construccións da Arquitectura Popular. Património Etnográfico de Galicia.* Santiago de Compostela: Consello Galego de de Colexios de Aparelladores e Arquitectos Técnicos.

Chabouillé, M. e J. F. Silva. (transl.) 1801. *Manual Pratico do Lavrador: com hum Tratado sobre as Abelhas, por Chabouillé.* Lisbon: Na Typographia Chalcographica e Litteraria do Arco do Cego.

Collomb, G. 1981. La ruche humaine, em J. Cuisinier (ed) *L' Abeille, l'Homme, le Miel et la Cire:* 33-37. Paris: Éditions de la Réunion des Musées Nationaux.

Crane, E. 1999. *The World History of Beekinping and Honey Hunting.* London: Routledge.

Cruz, F. 2002. Tratado breve de la cultivación de las colmenas. *Cuadernos de Etnologia de Guadalajara* 34: 9-25.

Dias, J. 1948. *Vilarinho da Furna: uma Aldeia Comunitária.* Porto: INIC/CEEP.

Dias, L. F. C. 1961. *Forais Manuelinos do Reino de Portugal e do Algarve: Trás-os-Montes.* Beja: Edição do autor.

Dias, L. F. C. 1969. *Forais Manuelinos do Reino de Portugal e do Algarve: Entre Douro e Minho.* Fundão: Edição do autor.

Fernandes, R. 2001. *Descrição do Terreno ao Redor de Lamego Duas Léguas.* Lamego, Beira Douro: Associação de Desenvolvimento do Vale do Douro.

Franco, F. S. 1804. *Diccionario de Agricultura.* Coimbra: Na Real Imprensa da Universidade, tomo 1.

Garrido, J. A. 1804. *Livro de Agricultura.* Lisbon: Na Impressão Regia.

Gil, J. 1621. *Perfecta y Curiosa Declaracion, de los Provechos Grandes, que Dan las Colmenas Bien Administradas y Alabanças de las Abejas.* Zaragoza: Por Pedro Gel.

Godinho, V. M. 2008. *A Expansão Quatrocentista Portuguesa.* Lisbon: Publicações Dom Quixote.

Godinho, V. M. 1983. *Os Descobrimentos e a Economia Mundial,* vol. 4. Lisbon: Editorial Presença.

Graça, L. Q. 1933. *Noções de Apicultura.* Lisbon: Emprêsa Nacional de Publicidade.

Graça, L. Q. 1935a Elementos do Pôsto Central de Fomento Apícola VI - 1º Inquérito Apícola Nacional, 1933-1935. *Boletim de Agricultura* 3ª série, 4: 14-35.

Graça, L. Q. 1935b. *Portugal: Etat Actuel de l'Apiculture: Action du Ministère de l'Agriculture.* Lisbon: Ministère de l'Agriculture.

Guzmán-Álvarez, J. R. 2006. Introducción de la edición original de 1586 del Tratado de cultivación y cura de las colmenas, in L. Méndez de Torres, *Tratado Breve de la Cultivación y Cura de las Colmenas. Ordenanzas de Colmerería de la Ciudad de Sevilla y de su Tierra*: 13-54. Junta de Andalucía.

Jesus Maria, T. 1790. *Agricultor instruido com as prevençoens necessarias para annos futuros, recupilado de graves autores.* Lisbon: Na Officina de Lino da Silva Godinho.

Leão, D. N. 1610. *Descripção do Reino de Portugal.* Lisbon: Impresso com licença por Jorge Rodriguez.

Lima, A. et al. 2010. Muros, entre as abelhas e os ursos. *Açafa on line* 3.

López Álvarez, X. 1994. *Las Abejas, la Miel y la Cera en la Sociedad Tradicional Asturiana.* Oviedo: Real Instituto de Estudios Asturianos.

Lorenzo Fernández, X. 1979. Etnografia: Cultura material. In Otero Pedrayo, R. (eds). *Historia de Galiza,* 2. Madrid: Akal Editor.

Magalhães, J. R. 1970. *Para o Estudo do Algarve Económico Durante o Século XVI.* Lisbon: Edições Cosmos.

Magalhães, J. R. 1988. *O Algarve Económico 1600-1773.* Lisbon: Editorial Estampa.

Manuppella, G. (coment.) 1986. *Livro de Cozinha da Infanta D. Maria.* Lisbon: IN-CM.

Matta, J. N. 1915. *Apicultura Pratica Mobilista.* Lisbon: Livraria Ferin.

Méndez de Torres, L. 2006. *Tratado Breve de la Cultivación y Cura de las Colmenas. Ordenanzas de Colmerería de la Ciudad de Sevilla y de su Tierra.* Junta de Andalucía.

Menezes, A. A. T. 1900. L' apiculture, em B. C. C. Costa e L. Castro (eds) *Le Portugal au Point de Vue Agricole:* 763-782. Lisbon: Imprensa Nacional.

Nutt, T. 1832. *Humanity to Honey-bees, or, Practical Directions for the Management of Honey-bees.* Wisbech, H. and J. Leach, for the author.

Padilla Álvarez, F. 2002. Los conocimientos apícolas del Hermano Francisco de la Cruz. *Cuadernos de Etnologia de Guadalajara* 34: 27-69.

Penella, M. 1823. *Obra Prima. Nova Descuberta, Abelhas em Habitação de Vidro sem se Occultarem e Memoria, do que Fazem as Abelhas em Enxame dentro de sua Habitação.* Lisbon: Em a Nova Impressão da Viuva Neves e Filhos.

Queiroz, J. F. F. 2005. A Companhia de Artefactos de Metais estabelecida no Porto (1837-1852). Para o estudo monográfico de uma fundição pioneira. *Arqueologia Industrial* 4ª série 1: 15-72.

Ramos, C. R. e L. Pita. 1997. A apicultura em Portugal no século XIX. *Vipasca* 6: 55-94.

Reaumur [R.-A.] 1740. *Memoires pour Servir à l' Histoire des Insectes.* Tomo 5. Paris: De l' Imprimerie Royale.

Rubião, F. I. P. 1835. *Colmea Nuttiana, Importada de França.* Paris: na Typographia de Guiraudet.

Rodrigues, D. 1987. *Arte de Cozinha.* Lisbon: IN-CM.

Sequeira, E. 1895. *As abelhas. Tratado de Apicultura Mobilista.* Porto: Typographia Social.

Silbert, A. 1978. *Le Portugal Méditerranéen à la Fin de l' Ancien Régime.* 2ª ed. Lisbon: INIC.

Silva, I. F. 1862-1870. *Diccionário Bibliographico Portuguez,* tomo 7 (1862) e tomo 9 (1870), Lisbon: Na Imprensa Nacional.

Soeiro, T. 2006-2007. Em busca do doce sabor. *Portugalia*, Nova série 27/28:119-158.

Teixeira, T. 1905-1908. Ethnographia transmontana. *Portugalia* 2: 627-638.

Thorley, J. 1744. *Melisselogia. Or, the Female Monarchy.* London: Printed for the Author; and sold by N. Thorley; and by J. Davidson.

Vieira, A. 2004. *Canaviais, Açúcar e Aguardente na Madeira, Séculos XV a XX.* Funchal: Secretaria Regional do Turismo e Cultura/ CEHA.

Visconde de Carnide. 1888. *Estudos Agrícolas.* Lisbon: Imprensa Nacional.

Wildman, D. 1805. Guia completa para a direcção das abelhas em hum anno inteiro. *Gazeta do Campo* 4: 107-137.

Wildman, T. 1768. *A Treatise on the Management of Bees; wherein is Contained the Natural History of those Insects; with the Various Methods of Cultivating them, Both Antient and Modern and the Improved Treatment of Them.* London: Printed for the author, and sold by T. Cadell.